BASSETT'S ENVIRONMENTAL HEALTH PROCEDURES

Environmental health law is a wide-ranging, detailed and complex body of law within the UK. *Bassett's Environmental Health Procedures* is an established and essential reference source which provides an accessible entry into enforcement and administrative procedures for environmental health. The main legal procedures used in the environmental health field are presented as flow charts supported by explanatory text. This new edition comes at a time of extensive legislative change, and is the only book of its kind for environmental health practitioners.

The structure of this eighth edition has been revised for ease of use, with each chapter now addressing a single topic instead of a piece of legislation. It also introduces legal guidance for environmental health practitioners to prepare them for the court prosecutions that are an essential part of their work. The book has been updated throughout to reflect new practices, legislation and statutory guidance, including:

- Primary authorities
- Authorisations for public water supplies
- Infectious disease control
- Port health
- RIDDOR
- Environmental permitting
- Environmental damage
- Imported food
- Empty homes
- Licensing of housing
- Licensing of gambling activities.

Environmental health officers/practitioners and students will find this essential reading. It will also be a valuable reference for all those whose responsibilities demand they keep abreast of current environmental health practices.

Bill Bassett authored the previous seven editions of *Environmental Health Procedures*, the first of which appeared in 1983. His dedication to navigating and clarifying the relevant regulations and laws in this book has aided countless environmental health practitioners in the UK and beyond. In this latest edition, he has passed the baton to Tim Deveaux, but remained on hand to help both the new author and the publisher throughout. The eighth edition continues in the spirit of Bill's classic book, and is the first to be named after him.

Tim Deveaux is an independent environmental health practitioner with 36 years of public sector experience in a wide range of areas in environmental health and sustainability. His experience includes developing and implementing strategies to change behaviour, and protecting the environment and human health, particularly in planning for, and implementing, climate change and sustainability solutions.

BASSETT'S ENVIRONMENTAL HEALTH PROCEDURES

Eighth edition

TIM DEVEAUX

Routledge
Taylor & Francis Group
LONDON AND NEW YORK

First edition published 1983 by Chapman & Hall
Second edition 1987
Third edition 1992
Fourth edition 1995

Fifth edition 1998 by E & FN Spon
Sixth edition 2002

Seventh edition 2007 by Taylor & Francis
This edition published 2014 by Routledge
2 Park Square, Milton Park, Abingdon, Oxon, OX14 4RN

and by Routledge
711 Third Avenue, New York, NY 10017

Routledge is an imprint of the Taylor & Francis Group, an informa business

British Library Cataloguing in Publication Data
A catalogue record for this book is available from the British Library

Library of Congress Cataloging-in-Publication Data
Deveaux, Tim, author.
Bassett's environmental health procedures/Tim Deveaux. – Eighth edition.
p.; cm.
Environmental health procedures
Preceded by: Environmental health procedures/W.H. Bassett. 7th ed. 2007.
Includes bibliographical references and index.
I. Bassett, W. H. Environmental health procedures. Preceded by (work): II. Title.
III. Title: Environmental health procedures.
[DNLM: 1. Environmental Health–legislation & jurisprudence–Great Britain. 2. Public Health–legislation & jurisprudence–Great Britain. WA 33 FA1]
KD3359
344.4104–dc23
2013020856

ISBN: 978-0-415-63971-2 (hbk)
ISBN: 978-1-315-87113-4 (ebk)

Typeset in Times
by Wearset Ltd, Boldon, Tyne and Wear

MIX
Paper from responsible sources
FSC® C013056

Printed and bound in Great Britain by
TJ International Ltd, Padstow, Cornwall

Contents

List of flow charts x
Subject index of procedures xvii
Preface xxv
Extract from the preface to the first edition xxvii
Acknowledgements xxviii

Chapter 1: Environmental health legal structure and process **1**
Environmental health law: an introduction 1
The framework for enforcement policies 1
Enforcing authorities 8
Enforcement options 9
Alternative actions in enforcement 13
Primary authorities 18
Service of notices by local authorities 24
Professional practice 29
Preparing a case 40
Court procedure 53

Chapter 2: Environmental protection **58**
Air quality
Clean Air Act 1993 58
General procedural provisions – Clean Air Act 1993 58
Height of chimneys serving furnaces 60
Height of chimneys not serving furnaces 62
Control of grit and dust from furnaces 64
Measurement of grit and dust from furnaces 67
Prohibition of dark smoke etc. from chimneys 69
Prohibition of dark smoke etc. from industrial and trade premises 72
Smoke control areas 74
Burning of crop residues 80
Obtaining information about atmospheric pollution 84

Environment Act 1995 87
General procedural provisions – Environment Act 1995 87
Local authority reviews of air quality 89
Air quality management areas 94

Environmental damage
Environmental damage 98

Land 105
General procedural provisions: Environmental Protection Act 1990 105
Contaminated land 106

Noise
Noise: introduction 116

Control of Pollution Act 1974 119
General procedural provisions – Control of Pollution Act 1974 119
Consents for the use of loudspeakers in streets 121
Noise from construction sites 123
Noise abatement zones: designation by local authorities 128
Noise abatement zones: measurement and registration of noise levels and consent to exceed them 130
Noise abatement zones: noise level determinations for new buildings 133
Noise abatement zones: noise reduction notices 135

Noise Act 1996 137
Noise from certain premises at night 137
Seizure of equipment used to make noise unlawfully 142

Clean Neighbourhoods and Environment Act 2005 145
Audible alarms notification areas 145
Anti-Social Behaviour Orders 148

Pollution prevention
Environmental permitting 153
Permitting of activities by Local Authorities 158

Chapter 3: Food safety **176**
Food control and hygiene legislation: legal structure 176
Registration of food business establishments 180
Approval of product-specific establishments 183
Hygiene improvement notices 188
Hygiene prohibition orders 191
Hygiene emergency prohibition notices and orders 194
Remedial action notices and detention notices 196
Detention and seizure of food 199
Imported food 203
Sampling of food for analysis and examination 208
Closure notices on mollusc harvesting areas 212

Chapter 4: Health and safety **216**
General procedural provisions 216
Improvement and prohibition notices 221
Reporting of injuries, diseases and dangerous occurrences 226
Notification of cooling towers and evaporative condensers 234
Seizure of dangerous articles or substances 237

Chapter 5: Housing and health **240**
Structure of the law dealing with housing standards and their enforcement 240
General procedural provisions 240
Housing standards: the housing health and safety rating system (HHSRS) 244
Housing standards for accommodation of the homeless 250

Enforcement options
Enforcement 251
Defective premises 253
Ruinous and dilapidated buildings 255
Dangerous buildings 257
Protection of buildings 259
Empty dwelling management orders 262
Demolition of buildings 267
Demolition orders 270
Clearance areas 274
Hazard awareness notices 278
Improvement notices 280
Notice of emergency remedial action 286
Prohibition orders 289
Emergency prohibition orders 293
Overcrowding notices for non-HMOs 296
Overcrowding notices for HMOs 300

Housing licensing
Selective licensing of residential accommodation 303
Houses in multiple occupation 311
Houses in multiple occupation declarations 313
Licensing of houses in multiple occupation 316
Designation of areas by local authorities for the additional licensing of HMOs 322
Interim management orders 325
Final management orders 332
Temporary exemption notices for HMOs 338
Rent repayment orders 340
Private landlords' accreditation schemes 343

Miscellaneous
Means of escape from certain high buildings 346

Temporary accommodation
Licensing of camping sites 346
Licensing of caravan sites 350

Chapter 6: Public health **355**

Animal welfare
Animal boarding establishments 355
Dangerous wild animals 358
Dog breeding establishments 364
Pet shops 368
Riding establishments 372

Disease control
General procedural provisions: Public Health (Control of Disease) Act 1984 377
Infectious diseases control 378
Key to powers available for the control of disease 378
Proper officer 380
Notification of diseases 380
Local authority powers 387
Power to order health measures in relation to persons, things and premises 394
Disinfection or decontamination of things and premises 401
Cleansing of filthy or verminous premises, articles and persons 403
Removal of persons in need of care 406
Skin piercing and other special treatments 410

Dog control
Dog control orders 414
Seizure of stray dogs 418

Exhumations
Exhumations 422

Removal of unauthorised campers
Removal of unauthorised campers 425

Pest control
Control of rats and mice 429

Port health
Infectious diseases control – aircraft 430
Infectious diseases control – trains 436
Infectious diseases control – ships 437

Sanitation and drainage
General procedural provisions: Public Health Act 1936 and 1961 445
General procedural provisions: Building Act 1984 447
Defective sanitary conveniences 449
Sanitary conveniences: provision/replacement 451
Sanitary conveniences at places of entertainment etc. 453
Conversion of earth closets etc. to water closets 456
Blocked private sewers 458
Stopped-up drains, private sewers, etc. 460
Defective drainage to existing buildings 463
Repair of drains, private sewers, etc. by local authorities 465
Paving of yards and passages 467
Overflowing and leaking cesspools 469

Statutory nuisances
Statutory nuisances 471

Waste
Waste on land 481
Litter 484
Litter abatement orders 487

Litter abatement notices 490
Litter clearing notices 492
Street litter control notices 495
Litter offences 498
Removal of noxious matter 501
Accumulations of rubbish 503
Removal and disposal of abandoned refuse 503
Removal of controlled waste on land 507
Abandoned trolleys 509
Abandoned vehicles 512
Receptacles for commercial and industrial waste 517
Receptacles for household waste 520
Scrap metal dealers 522

Water supply

General procedural provisions: Water Industry Act 1991 526
Control over public water supplies 527
Control over private water supplies 529
Water supply for new houses 534
Authorisations for quality standards for public water supplies 536
Closure or restriction of polluted water supply 536

Chapter 7: Public safety **539**

Licensing Act 2003

Alcohol and entertainment licences: Licensing Act 2003 539
Personal licences 542
Premises licences 545
Club premises certificates 549
Permitted temporary activities 549
Closure of licenced premises 554

Gambling Act 2005

Gambling Act 2005 558
Premises licences 561
Casino licence applications 578
Temporary use of premises for gambling 579
Gaming and gaming machines in alcohol-licensed premises 583
Permits for gaming and gambling machines in clubs and miners' welfare institutes 592
Permits to family entertainment centres for the use of certain lower-stake gaming machines 595
Permits for prize gaming 597
Registering small society lotteries 601

Miscellaneous

Controls over trading on Sundays at large shops 606
Street trading 610
Street collections 617
Sex establishment licences 619
Pleasure boat licensing 626

Flow charts

FC1	Primary authorities – Enforcement action – Sections 22 to 31 Regulatory Enforcement and Sanctions Act 2008	21
FC2	Inspection plans – Section 30 Regulatory Enforcement and Sanctions Act 2008	23
FC3	Height of chimneys serving furnaces – Sections 14 and 15 Clean Air Act 1993	61
FC4	Height of chimneys not serving furnaces – Section 16 Clean Air Act 1993	63
FC5	Control of grit and dust from furnaces – Sections 6 to 9 Clean Air Act 1993	65
FC6	Measurement of grit and dust from furnaces – Sections 10 and 11 Clean Air Act 1993	68
FC7	Prohibition of dark smoke etc. from chimneys – Section 1 Clean Air Act 1993	70
FC8	Prohibition of dark smoke etc. from industrial and trade premises (other than from chimneys) – Section 2 Clean Air Act 1993	73
FC9	Smoke control areas: directions by the Secretary of State – Section 19 Clean Air Act 1993	75
FC10	Smoke control areas: declaration by local authority – Section 19 Clean Air Act 1993	76
FC11	Smoke control areas: adaptations to fireplaces etc. – Section 24 Clean Air Act 1993	78
FC12	Burning of crop residues – Crop Residues (Burning) Regulations 1993	81
FC13	Obtaining of information about atmospheric pollution – Sections 35 to 40 Clean Air Act 1993	85
FC14	Local authority reviews of air quality – Sections 80 to 86 Environmental Protection Act 1990	90
FC15	Air quality management areas (AQMAs) – Sections 83 and 84 Environment Act 1995	95
FC16	Request for action by interested parties – Environmental damage – Regulation 29 Environmental Damage (Prevention and Remediation) Regulations 2009	103
FC17	Environmental damage – Regulations 13 to 23 Environmental Damage (Prevention and Remediation) Regulations 2009	104
FC18	Identification of contaminated land – Part 2A Environmental Protection Act 1990	107

FC19	Remediation of contaminated land – Part 2A Environmental Protection Act 1990	108
FC20	Statutory nuisances – control over noisy parties – Section 79 Environmental Protection Act 1990	117
FC21	Statutory nuisances – certain noise in the street – Section 80 Environmental Protection Act 1990	118
FC22	Consents for the use of loudspeakers in streets – Section 62 Control of Pollution Act 1974 and Section 8 Noise and Statutory Nuisance Act 1993	122
FC23	Noise from construction sites – Sections 60 and 61 Control of Pollution Act 1974	125
FC24	Noise abatement zones: designation by local authorities – Section 63 and Schedule 1 Control of Pollution Act 1974	129
FC25	Noise abatement zones: measurement and registration of noise levels and consent to exceed them – Sections 63 to 67 Control of Pollution Act 1974	131
FC26	Noise abatement zones: noise level determinations for new buildings – Section 67 Control of Pollution Act 1974	134
FC27	Noise abatement zones: noise reduction notices – Section 66 Control of Pollution Act 1974	136
FC28	Noise from certain premises at night – Noise Act 1996	139
FC29	Seizure of equipment used to make unlawful noise – Section 10 Noise Act 1996	143
FC30	Audible alarm notification areas (ANAs) – Sections 69 to 80 Clean Neighbourhoods and Environment Act 2005	146
FC31	Anti-social behaviour orders (ASBOs) – Sections 1 to 4 Crime and Disorder Act 1998	149
FC32	Permitting of scheduled activities by local authorities – Regulation 14 and Schedule 5 Environmental Permitting Regulations 2010	160
FC33	Handling of commercially confidential and national security information by local authorities – Regulations 47 to 55 Environmental Permitting Regulations 2010	161
FC34	Transfer of permits – Regulation 21 and Schedule 5 Environmental Permitting Regulations 2010	164
FC35	Variations to permit conditions – Regulation 20 and Schedule 5 Environmental Permitting Regulations 2010	166
FC36	Application for surrender of an environmental permit (Part A processes) – Regulation 25 Environmental Permitting Regulations 2010	167
FC37	Notification of surrender of an environmental permit (Part B processes) – Regulation 24 Environmental Permitting Regulations 2010	168
FC38	Enforcement and suspension notices – Regulations 36 and 37 Environmental Permitting Regulations 2010	169
FC39	Local authority powers to prevent or remedy pollution – Regulation 57 Environmental Permitting Regulations 2010	171

FC40	Revocation of permits by local authority – Regulations 22 and 23 Environmental Permitting Regulations 2010	172
FC41	Applications to local authority to exclude commercially confidential information from public register – Regulations 48 to 55 Environmental Permitting Regulations 2010	174
FC42	Registration of food business establishments – EC Regulation 852/2004 on the Hygiene of Foodstuffs	182
FC43	Approval of product-specific establishments by food authorities – EC Regulation 882/2004	185
FC44	Hygiene improvement notice (HIN) – Regulation 6 Food Hygiene (England) Regulations 2006	190
FC45	Hygiene prohibition orders (HPOs) – Regulation 7 Food Hygiene (England) Regulations 2006	192
FC46	Hygiene emergency prohibition notices and orders (HEPNs and HEPOs) – Regulation 8 Food Hygiene (England) Regulations 2006	195
FC47	Remedial action notices (RANs) and detention notices (DNs) – Regulation 9 Food Hygiene (England) Regulations 2006	198
FC48	Detention and seizure of food notices – Section 9 Food Safety Act 1990	200
FC49	Imported food – Regulations 27 to 42 Official Food and Feed Control (England) Regulations 2009	205
FC50	Sampling of food for analysis and examination – Food Safety Act 1990 and Food Safety (Sampling and Qualifications) Regulations 1990	210
FC51	Closure notices on mollusc harvesting areas	214
FC52	Improvement notices – Section 21 Health and Safety at Work etc Act 1974	223
FC53	Prohibition notices – Section 22 Health and Safety at Work Act 1974	224
FC54	The reporting of injuries, diseases and dangerous occurrences – RIDDOR 1995	231
FC55	Notification of cooling towers and evaporative condensers – Notification of cooling towers and evaporative condensers Regulations 1992	236
FC56	Seizure of dangerous articles or substances – Section 25 Health and Safety at Work etc. Act 1974	238
FC57	Defective premises – Section 76 Building Act 1984	254
FC58	Ruinous and dilapidated buildings – Section 79 Building Act 1984	256
FC59	Dangerous buildings – Sections 77 and 78 Building Act 1984	258
FC60	Protection of buildings – Sections 29 to 32 Local Government (Miscellaneous Provisions) Act 1982	260
FC61	Interim and final empty property management orders – Sections 132 to 138 Housing Act 2004	265
FC62	Demolition of buildings – Sections 80 to 83 Building Act 1984	268
FC63	Demolition orders (DOs) – Sections 265 to 275 Housing Act 1985	271

FC64	Clearance areas (CAs) – Sections 289 to 298 Housing Act 1985 as amended	275
FC65	Hazard awareness notices (HANs) – Sections 28 and 29 Housing Act 2009	279
FC66	Improvement notices (INs) – Sections 11 to 19 Housing Act 2004	281
FC67	Emergency remedial action (ERA) notices – Sections 40 to 42 and 45 Housing Act 2004	287
FC68	Prohibition orders (POs) – Sections 20 to 27 Housing Act 2004	290
FC69	Emergency prohibition orders (EPOs) – Sections 43 to 45 Housing Act 2004	294
FC70	Overcrowding notices for non-HMOs – Sections 324 to 344 Housing Act 1985	297
FC71	Overcrowding notices (ONs) for houses in multiple occupation	301
FC72	Designation of areas for selective licensing of rented dwellings – Sections 80 to 84 Housing Act 2004	305
FC73	Selective licensing of rented dwellings – temporary exemption notices – Section 86 Housing Act 2004	307
FC74	Selective licensing of rented dwellings – Sections 87 to 93 Housing Act 2004	308
FC75	House in multiple occupation (HMO) declarations – Sections 255 and 256 Housing Act 2004	315
FC76	Licensing of houses in multiple occupation (HMOs) – Sections 55 to 71 Housing Act 2004	317
FC77	Designation of areas for additional licensing of houses in multiple occupation – Sections 55 to 60 Housing Act 2004	323
FC78	Interim management orders (IMOs) – Sections 102 to 112 Housing Act 2004	327
FC79	Final management orders (FMOs) – Sections 113 to 122 Housing Act 2004	333
FC80	Temporary exemption notices (TENs) – Section 62 Housing Act 2004	339
FC81	Rent repayment orders – Sections 96 to 98 Housing Act 2004	341
FC82	Means of escape from certain high buildings – Section 72 Building Act 1984	347
FC83	Licensing of camping sites – Section 269 Public Health Act 1936	349
FC84	Licensing of caravan sites – Caravan Sites and Control of Development Act 1960	351
FC85	Animal boarding establishments – Animal Boarding Establishments Act 1963	356
FC86	Dangerous wild animals – Dangerous Wild Animals Act 1976	361
FC87	Dog breeding establishments – Breeding of Dogs Act 1973 as amended	365
FC88	Pet shops – Pet Animals Act 1951	370
FC89	Riding establishments – Riding Establishments Act 1964 as amended	373

FC90	Notification of diseases – Health Protection (Notification) Regulations 2010	383
FC91	Requirement to keep a child away from a school – Regulation 2 Health Protection (Local Authority Powers) Regulations 2010	388
FC92	Request for cooperation for health protection purpose – Regulation 8 Health Protection (Local Authority Powers) Regulations 2010	391
FC93	Restriction of contact with and access to dead bodies – Regulations 9 and 10 Health Protection (Local Authority Powers) Regulations 2010	393
FC94	Part 2 orders – Public Health (Control of Disease) Act 1984	398
FC95	Disinfection or decontamination of premises and things on request of owner – Regulations 4 to 7 Health Protection (Local Authority Powers) Regulations 2010	402
FC96	Cleansing of filthy or verminous premises – Section 83 Public Health Act 1936	404
FC97	Cleansing of verminous persons and their clothing – Section 85 Public Health Act 1936	405
FC98	Removal of persons in need of care: normal procedure – Section 47 National Assistance Act 1948	408
FC99	Removal of persons in need of care: accelerated procedure – Section 47 National Assistance Act 1948	409
FC100	Skin piercing and other special treatments – Sections 13 to 17 Local Government (Miscellaneous Provisions) Act 1982	411
FC101	Dog control orders (DCOs) – The Clean Neighbourhoods and Environment Act 2005 Sections 55 to 67 and the Dog Control Orders (Procedures) Regulations 2006	415
FC102	Seizure of stray dogs by the public – Section 150 Environmental Protection Act 1990	419
FC103	Seizure of stray dogs by local authorities	420
FC104	Exhumations – Burial Act 1857	423
FC105	Removal of unauthorised campers – Section 77 Criminal Justice and Public Order Act 1994	427
FC106	Control of rats and mice – Sections 2 to 7 Prevention of Damage by Pests Act 1949	431
FC107	International travel – infectious disease control – Public Health (Aircraft) Regulations 1979	433
FC108	Infectious disease control – Public Health (Ships) Regulations 1979 (as amended)	438
FC109	Ship sanitation control certificates – Public Health (Ships) Regulations 1979 (as amended) Regulations 18A to 18D	441
FC110	Defective sanitary conveniences – Section 45 Public Health Act 1936	450
FC111	Sanitary conveniences: provision/replacement – Sections 64 and 65 Building Act 1984	452
FC112	Sanitary conveniences at places of entertainment etc. – Section 20 Local Government (Miscellaneous Provisions) Act 1976	454

FC113	Conversion of earth closets etc. to water closets – Section 66 Building Act 1984	457
FC114	Blocked private sewers – Section 35 Local Government (Miscellaneous Provisions) Act 1976	459
FC115	Stopped-up drains, private sewers, etc. – Section 17 Public Health Act 1961	461
FC116	Defective drainage to existing buildings – Sections 59 and 60 Building Act 1984	464
FC117	Repair of drains, private sewers, etc. by local authorities – Section 17 Public Health Act 1936	466
FC118	Paving of yards and passages – Section 84 Building Act 1984	468
FC119	Overflowing and leaking cesspools – Section 50 Public Health Act 1936	470
FC120	Statutory nuisances: local authority action – Section 80 Environmental Protection Act 1990	474
FC121	Statutory nuisances: action by aggrieved persons – Section 82 Environmental Protection Act 1990	475
FC122	Litter abatement orders (LAOs) – Sections 89 and 91 Environmental Protection Act 1990	489
FC123	Litter abatement notices – Section 89 Environmental Protection Act 1990	491
FC124	Litter clearing notices (LCNs) – Section 92A Environmental Protection Act 1990	493
FC125	Street litter control notices – Sections 93 to 95 Environmental Protection Act 1990	496
FC126	Litter offences – Sections 87 and 88 Environmental Protection Act 1990	500
FC127	Removal of noxious matter – Section 79 Public Health Act 1936	502
FC128	Accumulations of rubbish – Section 34 Public Health Act 1961	504
FC129	Removal and disposal of abandoned refuse – Section 6 Refuse Disposal (Amenity) Act 1978	506
FC130	Removal of controlled waste on land – Section 59 Environmental Protection Act 1990	508
FC131	Abandoned trolleys – Section 99 Environmental Protection Act 1990	510
FC132A	Part 1 Abandoned vehicles: removal – Sections 3 and 4 Refuse Disposal (Amenity) Act 1978	513
FC132B	Part 2 Abandoned vehicles: disposal	514
FC133	Receptacles for commercial and industrial waste – Section 47 Environmental Protection Act 1990	518
FC134	Receptacles for household waste – Section 46 Environmental Protection Act 1990	521
FC135	Scrap metal dealers – Scrap Metal Dealers Act 1964	523
FC136	Private water supply: notice procedure – Regulation 18 Private Water Supplies Regulations 2009	531

FC137	Private water supply: action in the event of failure – Part 3 Private Water Supplies Regulations 2009	533
FC138	Water supply for new houses – Section 25 Building Act 1984	535
FC139	Closure or restriction of polluted water supply – Section 140 Public Health Act 1936	537
FC140	Licensing Act 2003 – Framework	540
FC141	Personal licences – Sections 111 to 135 Licensing Act 2003	543
FC142	Premises licences – Sections 11 to 59 Licensing Act 2003	546
FC143	Club premises certificates (CPC) – Sections 60 to 97 Licensing Act 2003	550
FC144	Permitted temporary activities (PTA) – Sections 98 to 110 Licensing Act 2003	552
FC145	Closure of licensed premises – Sections 40 and 41 Anti-Social Behaviour Act 2003	555
FC146	Premises licences – Gambling Act 2005 and Gambling Act 2005 (Premises Licences and Provisional Statements) Regulations 2007	562
FC147	Review of a premises licence – Gambling Act 2005 and Gambling Act 2005 (Premises Licences) (Review) Regulations 2007	574
FC148	Temporary use of premises for gambling – Gambling Act 2005 and Gambling Act 2005 (Temporary Use Notices) Regulations 2007	580
FC149	Registration of small society lotteries – Schedule 11, Part 5 Gambling Act 2005	602
FC150	Controls over trading on Sundays at large shops – Schedules 1 and 2 Sunday Trading Act 1994	607
FC151	Controls over loading and unloading at large shops on Sundays – Section 2 Sunday Trading Act 1994	609
FC152	Street trading: designation of – Schedule 4 of Local Government (Miscellaneous Provisions) Act 1982 (LG(MP)A)	613
FC153	Street trading: licences – Schedule 4 of Local Government (Miscellaneous Provisions) Act 1982 (LG(MP)A)	614
FC154	Street trading: consents – Schedule 4 of Local Government (Miscellaneous Provisions) Act 1982 (LG(MP)A)	615
FC155	Sex establishment licences – Section 2 and Schedule 3 Local Government (Miscellaneous Provisions) Act 1982	620

Subject index of procedures

Abandoned refuse
Removal – Section 6 Refuse Disposal (Amenity) Act 1978 *FC129*

Abandoned vehicles
Removal – Section 6 Refuse Disposal (Amenity) Act 1978 *FC132*

Accidents
Reporting of Injuries, Diseases and Dangerous Occurrences Regulations 1995 as amended (2012) *FC54*

Accumulations
Abandoned refuse (in open air) – Refuse Disposal (Amenity) Act 1978 *FC129*
Abandoned trolleys – Environmental Protection Act 1990 *FC131*
Abandoned vehicles – Refuse Disposal (Amenity) Act 1978 *FC132*
Accumulation of rubbish (seriously detrimental) – Public Health Act 1961 *FC128*
As statutory nuisances – Environmental Protection Act 1990 *FC120*
Cleansing of filthy or verminous premises – Public Health Act 1936/61 *FC96*
Removal of controlled waste from land – Environmental Protection Act 1990 *FC130*
Removal of noxious matter – Public Health Act 1936 *FC128*

Air pollution and air quality
Air quality management areas – Environment Act 1995 *FC15*
Burning of crop residues – Environmental Protection Act 1990 *FC12*
Control of grit and dust from furnaces – Clean Air Act 1993 *FC5*
Height of chimneys not serving furnaces – Clean Air Act 1993 *FC4*
Height of chimneys serving furnaces – Clean Air Act 1993 *FC3*
Local authority reviews of air quality – Environment Act 1995 *FC14*
Measurement of grit and dust from furnaces – Clean Air Act 1993 *FC6*
Obtaining information about atmospheric pollution – Environmental Protection Act 1990 *FC13*
Prohibition of dark smoke from chimneys – Clean Air Act 1993 *FC7*
Prohibition of dark smoke from industrial and trade premises – Clean Air Act 1993 *FC8*
Smoke control areas – Clean Air Act 1993 *FCs 9–11*

Animal boarding establishments
Licensing – Animal Boarding Establishments Act 1963 *FC85*

Anti-social behaviour (for noise)
Anti-social behaviour orders – Crime and Disorder Act 1998 *FC31*

Audible alarms
Audible alarms notification areas – Clean Neighbourhoods and Environment Act 2005 *FC30*

Camping sites
Licensing – Public Health Act 1936 *FC83*
Removal of unauthorised camping – Criminal Justice and Public Order Act 1994 *FC105*

Caravan sites
Licensing – Caravan Sites and Control of Development Act 1960 *FC84*
Cesspools
Overflowing and leaking – Public Health Act 1936 *FC119*

Chimney heights (see air pollution and air quality)

Clean air (see air pollution and air quality)

Cooling towers (and evaporation condensers)
Notification of Cooling Towers and Evaporative Condensers Regulations 1992 *FC55*

Consents
For the use of loudspeakers on streets – Control of Pollution Act 1974 *FC22*

Contaminated land
Identification – Environmental Protection Act 1990 *FC18*
Remediation – Environmental Protection Act 1990 *FC19*

Crop residues
Burning of – Environmental Protection Act 1990 *FC12*

Dangerous wild animals
Licensing – Dangerous Wild Animals Act 1990 *FC86*

Dark smoke (see air pollution and air quality)

Defective premises
As statutory nuisance – Environmental Protection Act 1990 *FC120*
Dangerous buildings – Building Act 1984 *FC59*
Defective premises – accelerated nuisance provisions – Building Act 1984 *FC57*
Demolition of buildings – Building Act 1984 *FC62*
Means of escape from certain high buildings – Building Act 1984 *FC82*
Protection of buildings – Local Government (Miscellaneous Provisions) Act 1982 *FC60*
Ruinous and dilapidated buildings – Building Act 1984 *FC58*

Demolition
Demolition of buildings – Building Act 1984 *FC62*
Demolition orders – Housing Act 1985 *FC63*

Dilapidated buildings (see defective premises)

Disease control
Cleansing of filthy and verminous premises – Public Health Act 1936/61 *FC96*
Cleansing of verminous persons and their clothing – Public Health Act 1936 *FC97*
Disinfection or decontamination of premises and things on request of owner – Health Protection (Local Authority Powers) Regulations 2010 *FC95*
Notification of diseases – Health Protection (Notification) Regulations 2010 *FC90*
Part 2 orders – Person, thing or premises infected or contaminated

etc. – Health Protection (Part 2A orders) Regulations 2010 *FC94*
Removal of persons in need of care – Section 47 National Assistance Act 1948 *FC98, FC99*
Request for cooperation for health protection purpose – Health Protection (Local Authority Powers) Regulations 2010 *FC92*
Requirement to keep a child away from a school – Health Protection (Local Authority Powers) Regulations 2010 *FC91*
Restriction of contact with and access to dead bodies – Health Protection (Local Authority Powers) Regulations 2010 *FC93*

Dogs
Dog control orders – Clean Neighbourhoods and Environment Act 2005 *FC101*
Licensing of dog breeding establishments – Breeding of Dogs Act 1973 *FC87*
Seizure of strays – Environmental Protection Act 1990 *FC102, FC103*

Drains (see also sewers)
Blocked drains – Public Health Act 1961 *FC115*
Blocked private sewers – Local Government (Miscellaneous Provisions) Act 1976 *FC114*
Conversion of earth closets – Building Act 1984 *FC113*
Defective drainage to existing buildings – Building Act 1984 *FC116*
Paving of yards and passages – Building Act 1984 *FC118*
Repairs by local authorities – Public Health Act 1961 *FC117*
Sanitary conveniences at places of entertainment – Local Government (Miscellaneous Provisions) Act 1976 *FC112*

Dust (see air pollution and air quality)

Emergency prohibition orders
For food safety – Food Hygiene Regulations 2006 *FC46*
For housing – Housing Act 2004 *FC69*

Empty dwelling management orders
Housing Act 2004 *FC61*

Enforcement notices
For food safety – Food Safety Act 1990 *FCs 44–47*
For health and safety – Health and Safety at Work Act 1974 *FC52*
For permitted activities – Pollution Prevention and Control (England and Wales) Regulations 2000 *FC38*

Environmental damage
Environmental Damage (Prevention and Remediation) Regulations 2009 *FC17*
Request for action by interested parties – Environmental Damage (Prevention and Remediation) Regulations 2009 *FC16*

Exhumations
Burial Act 1857 *FC104*

Food
Approval of product-specific establishments by food authorities – European Community Regulation 852–854 *FC43*
Closure notices on mollusc harvesting areas – European Community Regulation 852–854 2004 *FC51*
Detention and seizure of food notices – Food Hygiene Regulations 2006 *FC48*
Detention notices – Food Hygiene Regulations 2006 *FC47*
Hygiene emergency prohibition notices and orders – Food Hygiene Regulations 2006 *FC46*
Hygiene improvement notices –

Food Hygiene Regulations 2006 *FC44*
Hygiene prohibition orders – Food Hygiene Regulations 2006 *FC45*
Imported food – Official Food and Feed Control (England) Regulations 2009 *FC49*
Registration of food business establishments – European Community Regulation 852–854 *FC42*
Remedial action notices – Food Hygiene Regulations 2006 *FC47*
Sampling of food – Food Hygiene Regulations 2006 *FC50*

Food hygiene and control (see food)

Gambling
Premises licences – Gambling Act 2005 and Gambling Act 2005 (Premises Licences and Provisional Statements) Regulations 2007 *FC146*
Registration of small society lotteries – Gambling Act 2005 *FC149*
Review of a premises licence – Gambling Act 2005 and Gambling Act 2005 (Premises Licences) (Review) Regulations 2007 *FC147*
Temporary use of premises for gambling – Gambling Act 2005 and Gambling Act 2005 (Temporary Use Notices) Regulations 2007 *FC148*

Grit (and dust) (see air pollution and air quality)

Health and safety at work
Improvement notices – Health and Safety at Work Act 1974 *FC52*
Notification of cooling towers and evaporative condensers – Health and Safety at Work Act 1974 *FC55*
Prohibition notices – Health and Safety at Work Act 1974 *FC53*
Reporting of deaths, injuries, diseases and dangerous occurrences – Health and Safety at Work Act 1974 *FC54*
Seizure of dangerous articles and substances – Health and Safety at Work Act 1974 *FC56*

Houses
Clearance areas – Housing Act 1985 *FC64*
Demolition orders – Housing Act 1985 *FC63*
Emergency prohibition orders – Housing Act 2004 *FC69*
Emergency remedial action notices – Housing Act 2004 *FC67*
Hazard awareness notices – Housing Act 2004 *FC65*
Improvement notices – Housing Act 2004 *FC66*
Overcrowding notices in non-houses in multiple occupation – Housing Act 1985 *FC70*
Prohibition orders – Housing Act 2004 *FC68*

Houses in multiple occupation (HMOs)
Designation of areas for additional licensing of HMOs – Housing Act 2004 *FC77*
Designation of areas for selective licensing of rented dwellings – Housing Act 2004 *FC72*
Final management orders – Housing Act 2004 *FC79*
HMO declarations – Housing Act 2004 *FC75*
Interim management orders – Housing Act 2004 *FC78*
Licensing of HMOs – Housing Act 2004 *FC76*
Means of escape from certain high buildings – Building Act 1984 *FC82*
Overcrowding notices for HMOs – Housing Act 2004 *FC71*

Rent repayment orders – Housing Act 2004 *FC81*
Selective licensing of rented dwellings – Housing Act 2004 *FC74*
Selective licensing of rented dwellings – temporary exemption notices – Housing Act 2004 *FC73*
Temporary exemption orders – Housing Act 2004 *FC80*

Improvement notices
For food hygiene – Food Hygiene Regulations 2006 *FC44*
For health and safety – Health and Safety at Work Act 1974 *FC52*

Information
About air pollution – Environment Protection Act 1990 *FC13*
About ownership etc. of premises – Local Government (Miscellaneous Provisions) Act 1976 *page 28*

Injuries
Reporting of injuries, diseases, and dangerous occurrences regulations 1995 (as amended) *FC54*

Integrated Pollution Prevention and Control (IPPC) (see pollution)

Licensing
Animal boarding establishments – Animal Boarding Establishments Act 1963 *FC85*
Camping sites – Section 269 Public Health Act 1936 *FC83*
Caravan sites – Caravan Sites and Control of Development Act 1960 *FC84*
Closure of licensed premises – Licensing Act 2003 and Anti-Social Behaviour Act 2003 *FC145*
Club premises certificates – Licensing Act 2003 *FC143*
Context of Licensing Act 2003 – Licensing Act 2003 *FC140*
Dangerous wild animals – Dangerous Wild Animals Act 1976 *FC86*
Dog breeding establishments – Breeding of Dogs Act 1973 *FC87*
Permitted temporary activities – Licensing Act 2003 *FC144*
Personal licences – Licensing Act 2003 *FC141*
Pet shops – Pet Animals Act 1951 *FC88*
Premises licences – Licensing Act 2003 *FC142*
Riding establishments – Riding Establishments Act 1964 *FC89*
Street trading – Local Government (Miscellaneous Provisions) Act 1976 *FCs 152–154*

Litter (also see accumulations)
Abandoned trolleys – Environmental Protection Act 1990 *FC131*
Litter abatement notices – Environmental Protection Act 1990 *FC123*
Litter abatement orders – Environmental Protection Act 1990 *FC122*
Litter clearing notices – Environmental Protection Act 1990 *FC124*
Litter offences – Environmental Protection Act 1990 *FC126*
Street litter control notices – Environmental Protection Act 1990 *FC125*

Loudspeakers
Consent for the use of loudspeakers in streets – Control of Pollution Act 1974 *FC22*

Noise
Audible alarms notification areas – Clean Neighbourhoods and Environment Act 2005 *FC30*

Consents for the use of loudspeakers in streets – Control of Pollution Act 1974 *FC22*
From construction sites – Control of Pollution Act 1974 *FC23*
Noise abatement zones – Control of Pollution Act 1974 *FCs 24–27*
Noise from certain premises at night – Noise Act 1996 as amended by the Anti-Social Behaviour Act 2003 and the Clean Neighbourhoods and Environment Act 2005 *FC28*
Noise nuisances – Environmental Protection Act 1990 *FCs 20, 28, 120*
Noisy parties – Environmental Protection Act 1990 *FC20*
Seizure of equipment – Noise Act 1996 *FC29*

Notification
Of cooling towers and evaporative condensers – Health and Safety at Work Act 1974 *FC55*
Of deaths, injuries, diseases and dangerous occurrences – Reporting of Injuries, Diseases and Dangerous Occurrences Regulations 1995 as amended *FC54*
Of diseases – Public Health (Control of Disease) Act 1984 *FC90*

Nuisances (see statutory nuisances)

Overcrowding
In houses – Housing Act 1985 *FC70*
In houses in multiple occupation – Housing Act 1985 *FC71*

Permits
For scheduled activities – Pollution Prevention and Control Regulations 2000 *FC32*

Persons in need of care and attention
Removal to care – National Assistance Act 1948 *FC98, FC99*

Pest control
Prevention of Damage by Pests Act 1949 *FC106*

Pet shops
Licensing – Pet Animals Act 1951 *FC88*

Pollution
Applications to exclude commercially confidential information from public registers – Pollution Prevention and Control Regulations 2000 *FC41*
Enforcement and suspension notices – Pollution Prevention and Control Regulations 2000 *FC38*
Handling of commercially confidential information – Pollution Prevention and Control Regulations 2000 *FC33*
Local authority powers to prevent or remedy pollution – Pollution Prevention and Control Regulations 2000 *FC39*
Permitting of scheduled activities by local authorities – Pollution Prevention and Control Regulations 2000 *FC32*
Revocation of permits – Pollution Prevention and Control Regulations 2000 *FC40*
Surrender of permits (Part A processes) – Pollution Prevention and Control Regulations 2000 *FC36*
Surrender of permits (Part B processes) – Pollution Prevention and Control Regulations 2000 *FC37*

Transfer of permits – Pollution Prevention and Control Regulations 2000 *FC34*
Variations to permit conditions – Pollution Prevention and Control Regulations 2000 *FC35*

Port health
Infectious disease control in aircraft – Public Health (Aircraft) Regulations 1979 *FC107*
Infectious disease control in ships – Public Health (Ships) Regulations 1979 (as amended) *FC108*
Ship sanitation control certificates – Public Health (Ships) Regulations 1979 (as amended) *FC109*

Prohibition notices
For health and safety – Health and Safety at Work Act 1974 *FC53*

Prohibition orders
For food hygiene – Food Hygiene Regulations 2006 *FC45*
For housing – Housing Act 2004 *FC68*

Rats and mice
Control of – Prevention of Damage by Pests Act 1949 *FC106*

Receptacles for waste
For commercial and industrial waste – Environmental Protection Act 1990 *FC133*
For household waste – Environmental Protection Act 1990 *FC134*

Registration
Food premises – European Community Regulation 852 2004 *FC42*
Scrap metal dealers – Scrap Metal Dealers Act 1964 *FC135*
Skin piercing – Local Government (Miscellaneous Provisions) Act 1976 *FC100*

Riding establishments
Licensing – Riding Establishments Act 1964 *FC89*

Rubbish (see accumulations)

Ruinous buildings (see defective premises)

Sanitary conveniences
At places of entertainment – Local Government (Miscellaneous Provisions) Act 1976 *FC112*
Defective sanitary conveniences – Public Health Act 1936 *FC110*
Provision and replacement – Building Act 1984 *FC111*

Scheduled activities under Integrated Pollution Prevention Control (see pollution)

Scrap metal dealers
Registration – Scrap Metal Dealers Act 1964 *FC135*

Sewers (see also drains)
Blocked private sewers – Local Government (Miscellaneous Provisions) Act 1976 and Public Health Act 1961 *FC114*
Repair of private sewers by local authorities – Public Health Act 1961 *FC117*

Shellfish
Closure of mollusc harvesting areas – European Community Regulation 852/854 *FC51*

Skin piercing
Registration – Local Government (Miscellaneous Provisions) Act 1976 *FC100*

Smoke (see also air pollution)
Smoke control areas – Clean Air Act 1993 *FCs 9–11*
Smoke nuisances – Environmental Protection Act 1990 *FC120*

Statutory nuisances
Action by aggrieved persons – Environmental Protection Act 1990 *FC121*
Action by local authority – Environmental Protection Act 1990 *FC120*
Defective premises – Building Act 1984 *FC57*

Street trading
Licensing, designation and consents – Local Government (Miscellaneous Provisions) Act 1976 *FCs 152–154*

Sunday trading
Controls over loading/unloading at large shops – Sunday Trading Act 1994 *FC151*
Controls over trading at large shops – Sunday Trading Act 1994 *FC150*

Trolleys
Abandoned shopping and luggage trolleys – Environmental Protection Act 1990 *FC131*

Waste (see also accumulations and litter)
Receptacles for commercial and industrial waste – Environmental Protection Act 1990 *FC133*
Receptacles for household waste – Environmental Protection Act 1990 *FC132*
Removal of controlled waste from land – Environmental Protection Act 1990 *FC130*

Water supply
Closure or restriction of polluted supply – Public Health Act 1936 *FC139*
For new houses – Building Act 1984 *FC138*
Private water supply – action in the event of failure – Private Water Supplies Regulations 2009 *FC137*
Private water supply – notice procedure – Private Water Supplies Regulations 2009 *FC136*

Preface

Each edition of this work records the ever-increasing breadth of environmental health law and complexity of its content. This edition continues to reflect these trends and also indicates just how substantial legislative changes in environmental health law have been since the publication of the seventh edition in 2008. These changes have not been peripheral but have, again, affected in a major way the main activities of environmental health departments, as outlined below.

In editing this edition, the opportunity has been taken to restructure the book to reflect the curriculum organisation of an environmental health degree advocated by the Chartered Institute of Environmental Health, with additional parts on legal structure and process and public safety.

The book has been divided into the following parts:

- Legal structure and process
- Environmental protection
- Food safety
- Health and safety at work
- Housing and health
- Public health
- Public safety.

There are new provisions for authorisations for water quality standards for public water supplies, environmental permitting, RIDDOR, infectious disease control through health protection regulations and official feed and food regulations.

Several new procedures have been added to provide a wider coverage of matters administered by environmental health practitioners in local government.

Additional sections have been added for:

- primary authorities – how local authorities can work with businesses for better consistency and improving public health;
- alternative actions in enforcement and achieving remedies, with a section on works in default written by Tim Everett;
- professional practice – how environmental health practitioners may conduct themselves in an enforcement role;
- preparing a prosecution – the bare essentials in preparing a case for court;
- court procedure – what an environmental health practitioner may expect in court;
- environmental damage – encompassing the Environmental Damage (Prevention and Remediation) Regulations 2009, where local authorities have a role;
- imported food – through the Official Feed and Food Regulations 2009;

- private landlord registration schemes and accreditation schemes – working with the private rented housing sector;
- empty homes – using legislation to deal with the problems of empty properties;
- selective licensing of houses in multiple occupation and rent repayment orders to control the problems brought by certain types of housing;
- exhumations – using the Burial Act to ensure safe exhumation where it is necessary;
- cleansing of filthy and verminous premises – additional to the provisions for persons;
- port health controls for aircraft, ships and international trains;
- street collections, sex shop licences and pleasure boat licences cover more licensing controls operated by environmental health practitioners;
- gambling – this section has been added because the environmental health service is a statutory consultee in the processing of gambling-related premises licences.

It should be stressed that this work is an interpretation of the law and is not the law itself. The recent legislation on which the procedures are based involves both a change of policy approach to the legal requirements and a new enforcement regime, and together this appears to be more complicated and less easy to interpret than the legislation replaced. The editor would appreciate any comments on the work and suggestions as to alternative interpretations will be very helpful.

Tim Deveaux
April 2013

Extract from the preface to the first edition (1983)

The idea for this work came after many exasperating years of having to wade through Acts of Parliament or legal encyclopediae in order to find the legal answer to a fairly simple practical problem which faced me as an environmental health officer. The author has also been conscious of a need for a reference text for both practising and pupil environmental health officers, community physicians and others which gives basic information about environmental health procedures in a simple form. It has also been suggested to the author that other people outside the environmental health profession, including lawyers, law centre administrators, surveyors, estate agents, elected local authority members and members of the general public who become involved in environmental health problems, would welcome a procedural analysis of this type.

This book attempts to meet all those needs and provide a means of identifying procedures relating to the many facets of environmental health practice.

Each of the main procedures has been reduced to a diagrammatic form and included with it is text on the basic issues concerned with that procedure. By this means, the steps involved in the different procedures can be quickly followed and individual stages identified. Related procedures have been cross-referenced. At the beginning of each chapter, the general procedural provisions and definitions relating to the procedures contained in that chapter are indicated. Where a procedure differs either procedurally or in definition, this is indicated either in the diagram or in the text.

W.H. Bassett
Exeter
1983

Acknowledgements

I would like to thank the following Environmental Health Practitioners for their advice, assistance and invaluable input into this edition.

Graham Jukes and Tina Garrity for access to legal information; Martin Walker on Port health – ships; Iain Pocknell on Port health – aircraft; Elaine Rudman on licensing; Julie Barrett on law and legal procedure; Mark Elliott on Welsh legislation; Patrick Mackie on Scottish legislation; Andy Statham on enforcement framework; Carol Deveaux and Jenny Morris on food safety; Emma Gall, Vanessa Shiplee, Amy Campbell and Alice Tomasevic for comments on the 7th edition; Paul Christer and Steve Ramshaw on environmental protection; Bob Mayho and Mandy Reed on housing updates; Howard Price on noise, air quality, land and environmental permitting and Janine Knighton on exhumations.

Chapter 1

ENVIRONMENTAL HEALTH LEGAL STRUCTURE AND PROCESS

ENVIRONMENTAL HEALTH LAW: AN INTRODUCTION

The procedures interpreted in this book are based on primary and secondary legislation as it applies in England. In some cases, they will also apply in the rest of the UK but this varies because of the separate legislative powers of the National Assemblies in Northern Ireland, Scotland and Wales. In Northern Ireland and Scotland, the primary legislation as well as the secondary legislation may be different, but the Welsh Assembly has powers only in respect of the making of secondary legislation (e.g. regulations).

In each procedure, the applicability of the procedure to the rest of the UK is indicated in the section headed 'Extent'.

In the case of some procedures relating to food safety, the environmental health service will be directly enforcing EC Regulations without there being implementing regulations in the UK.

THE FRAMEWORK FOR ENFORCEMENT POLICIES

In undertaking the enforcement of environmental health legislation and forming their individual enforcement policies, local authorities need to take account of government policy, government agencies and initiatives taken at national level which seek to guide their enforcement policies. The most significant are outlined below and it will be seen that, in total, they form a very substantial influence over local authority enforcement activity.

The Hampton Review

A review was commissioned by the Chancellor in 2004 to review regulatory inspection and enforcement with a view to reducing the administrative cost of regulation consistent with maintaining effective regulatory outcomes. The final report was made in March 2005 and is the basis of much of enforcement policy today. The principal findings included:

- entrenching the system of risk assessment throughout the regulatory system;
- reducing the need for inspections by up to one third;
- reducing the number of forms regulators send out by up to 25 per cent;
- making more use of advice by applying the principle of risk assessment;
- substantially reducing the need for form filling;
- applying tougher and more consistent penalties where these are deserved;
- reducing the number of regulators that businesses have to deal with;
- entrenching reform by requiring all new policies and regulations to consider enforcement and to use existing structures wherever possible;
- creating a new business-led body at the heart of government to drive implementation of the recommendations and challenge departments on their regulatory performance.

This resulted in the creation of the Better Regulation Commission, Better Regulation Executive and Local Better Regulation Office, but the principles of better regulation have been embodied in a new body created to take better regulation forward (see below).

Better regulation

The Legislative and Regulatory Reform Act 2006 imposes a duty on any person exercising a specified regulatory function to have regard to the five principles of good regulation. The principles provide that regulatory activities should be carried out in a way which is:

- transparent;
- accountable;
- proportionate;
- consistent; and
- targeted only at cases in which action is needed (section 21).

Section 22 of the Act enables a Minister of the Crown to issue a Code of Practice relating to the exercise of regulatory functions (the 'Regulators' Compliance Code'). This section imposes a duty on any person exercising a specified regulatory function to have regard to the Regulators' Compliance Code – April 2008, when determining general policies or principles and in setting standards or giving general guidance about the exercise of other regulatory functions.

The Legislative and Regulatory Reform (Regulatory Functions) Order 2007 specifies the regulatory functions to which these duties apply.

Regulator's compliance code

The principles of the code are as follows:

Risk assessment – Regulators, and the regulatory system as a whole, should use comprehensive risk assessment to concentrate resources on the areas that need them most.
Inspection – No inspection should take place without a reason.

Data requirements – Businesses should not have to give unnecessary information, nor give the same piece of information twice.
Handling suspected breaches – The few businesses that persistently break regulations should be identified quickly, and face proportionate and meaningful sanctions.
Advice – Regulators should provide authoritative, accessible advice easily and cheaply.
Supporting economic progress – Regulators should recognise that a key element of their activity will be to allow, or even encourage, economic progress and only to intervene when there is a clear case for protection.
Accountability – Regulators should be accountable for the efficiency and effectiveness of their activities, while remaining independent in the decisions they take.

Better Regulation Delivery Office

The Better Regulation Delivery Office (BRDO) was created on 1 April 2012 as an independent unit within the Department for Business, Innovation and Skills. They are guided by the Representative Steering Group (RSG). They operate across the United Kingdom and support the Coalition Government's Strategy for Better Regulation.

BRDO promotes a regulatory environment in association with business in the delivery of regulation and developing practical tools for regulators.

This means:

- providing a centre of expertise for primary authorities and extending the scheme to enable more businesses to participate;
- simplifying delivery of regulation for business, regulators and government;
- providing a forum for business engagement to shape regulatory policy and delivery;
- supporting Local Enterprise Partnerships to tackle regulatory delivery issues at the local level;
- providing policy advice to UK and Welsh ministers on regulatory delivery.

The Enforcement Concordat

This agreement between the local authority associations and the government was reached in March 1998 and is supported by the British Retail Consortium. The Concordat is not mandatory but the vast majority of all organisations within the scope have adopted it. It deals with:

- standards of level of service and performance;
- openness in the provision of information and the use of plain language;
- helpfulness in working with those affected by environmental health laws;
- well-publicised procedures for dealing with complaints about enforcement;
- proportionality to ensure that enforcement action is related to risk levels;
- consistency by enforcing in a fair, equitable and consistent manner;
- procedures to guide the enforcement action of individual officers.

Most local authorities have embraced the Enforcement Concordat into their enforcement policies. The principles are embedded in operational procedures in local authorities and guidance has been produced to help them – 'Applying the Regulators' Compliance Code and Enforcement Concordat: Local Better Regulation Office briefing for local authority regulatory services' (March 2008).

Best value

This concept involves continually improving how all functions, including those involving enforcement, are undertaken by an authority. The scheme is given statutory force in Part I of the Local Government Act 1999 and involves processes that ensure functional reviews include challenge, comparison, consultation and competition. Benchmarking, which is the continuous process of measuring services against leaders in the field, allowing the identification of best practices, is central to all aspects of this review process. The principles of best value have, to a large extent, been embodied in local government culture and the framework of performance management.

Comprehensive Performance Assessment (CPA) and Comprehensive Area Assessment (CAA)

These schemes looked at how well services, including environmental health, were delivered and how they impacted on the delivery of services. The scheme is based on audit, and categorised local authorities into excellent, good, fair, weak and poor. The judgements were based on three separate elements:

- annual use of resources assessment;
- annual service assessment;
- a periodic corporate assessment.

Performance management was supported by the use of performance information based around the use of Best Value Performance Indicators (BVPIs), which were rationalised to a suite of indicators, many of which are retained by local authorities to measure their performance internally. CPA and CAA are no longer a priority for the government or local authorities.

Performance management and competence

The Local Better Regulation Office (LBRO) (now the Better Regulation Delivery Office, an independent unit within the Department for Business, Innovation and Skills) produced the Local Authority Regulatory Services Excellence Framework. Local authorities can measure service delivery quality based on the achievement of set targets, identified through consultation with their customers. Many have adopted quality management systems, some obtaining certificated systems through British Standards and ISO.

'An Introductory Guide to Performance Management in Local Authority Trading Standards and Environmental Health Enforcement Services' is a best-practice guide published in 1999. It aims to assist local authorities with best value targeted performance management and includes matters relating to enforcement. The guide identifies the following themes as being central to good performance:

- Protection of the wider agenda (i.e. councils leading and energising local communities)
- Transparency and consistency
- Quality and value
- Delivery and review.

The LBRO also published the 'Common Approach to Competence for Regulators' in 2011, which incorporates the Regulators' Development Needs Analysis (RDNA) and Guide for Regulators Information Point (GRIP) tools, which are available to ensure professional competency in regulatory activities.

The Food Standards Agency

The Agency sets and monitors standards of performance for food authorities in enforcing food safety laws. They may make reports on that performance, including guidance on improving standards, and direct that the food authority should publish, within a specified time, details of actions they will take to comply. They may also require each food authority to provide information and to make its records available for inspection.

In 2000, the Agency published a Framework Agreement on Local Authority Food Law Enforcement, which applies to the whole of the UK. It has been amended several times, the last in April 2010. This has four main elements:

- The standard – which sets out the requirements for the planning, organisation, management and delivery of local authority food law enforcement services.
- Service planning guidance – on matters that include:
  1. Service aims and objectives
  2. Background
  3. Service delivery
  4. Resources
  5. Quality assessment
  6. Review.
- A monitoring scheme – which sets out the arrangements for the Agency to obtain enforcement information from the local authorities. The Agency collects key data on how each local authority is delivering feed and food law enforcement, on an annual basis. This is achieved through the Local Authority Enforcement Monitoring System (LAEMS), which was introduced in April 2008.
- An audit scheme – which provides for a rolling programme of audits by the Agency. Authorities are audited against the Feed and Food Law Enforcement Standard in the Framework Agreement, which sets out the minimum standards of performance expected from local authorities across the full range of their feed and food law enforcement activities. The scheme is implemented on a UK basis, with the Agency in England, Scotland, Wales and Northern Ireland each coordinating their own audit programme.

In addition, statutory Codes of Practice issued under section 40 of the Food Safety Act 1990 contain mandatory provisions as to how and when certain actions are to be taken. These are identified in the appropriate procedures.

Earned autonomy

The Food Standards Agency is exploring the concept of earned autonomy.

The aim of a food business should be to operate to maximise food safety and ensure the highest possible level of consumer protection. Under earned autonomy, each business should be rewarded for operating at higher levels of food safety by being permitted carefully controlled variations in the delivery of official controls. This concept is yet to be fully developed.

Food information regulation

Food information regulation will come into effect on 13 December 2014, with the nutrition labelling provisions coming into effect two years later. These provisions will replace the food labelling regulations and will contain enforcement powers. The dates of implementation are as follows:

- 13 December 2014 – for the general labelling rules and, if provided, a nutrition declaration must use the format set out in the EU FIC (European Union food information for consumers) from this date;
- 1 January 2014 – for the rules on the composition and labelling of minced meat;
- 13 December 2016 – for the rules on the mandatory nutrition declaration needed for most pre-packaged foods.

The responsibility for these provisions is as follows:

- in Scotland and Northern Ireland, the Food Standards Authority;
- in Wales, the Food Standards Authority, except nutritional labelling, which will be dealt with by the Welsh government;
- in England, DEFRA, for general food labelling not related to food safety; the Department of Health for nutritional labelling; and the Food Standards Authority, which leads on food safety aspects of food labelling and liaison with food authorities with regard to food labelling law enforcement.

Food hygiene rating scheme

This scheme is voluntary throughout the UK. However, the Welsh Assembly is considering a mandatory scheme which includes mandatory display of the awarded score. If agreed, this is likely to be operational in 2014.

The Health and Safety Executive (HSE)

The Health and Safety Executive (HSE) have advised local authorities that the following elements are essential for a local authority to discharge its duty as an enforcing authority under the Health and Safety at Work Act 1974:

(a) a clear published statement of enforcement policy and practice;
(b) a system for prioritised planned inspection activity;
(c) a service plan;
(d) capacity to investigate accidents and respond to complaints;

(e) arrangements for benchmarking performance with peer local authorities;
(f) a trained and competent inspectorate;
(g) liaison and cooperation with the Lead Authority Partnership Scheme.

Section 18(4) of the Health and Safety at Work Act 1974 requires local authorities to act in accordance with guidance issued under that section by the HSE. 'Health and Safety in Local Authority Enforced Sectors – HSE Guidance to Local Authorities' was issued to local authorities in October 2002, giving such statutory guidance on the matters listed above.

The latest HSE Statement on Enforcement Policy was published in February 2009, which local authorities are required to reflect in their policies, the main elements of which are:

(a) proportionality – relating risks to enforcement action;
(b) targeting – of inspections to those areas containing the most serious risks;
(c) consistency of approach;
(d) transparency – helping duty holders to understand what is expected of them;
(e) accountability – policies and standards by which local authorities can be judged.

The HSE have produced an enforcement guide (2003), which is useful for other enforcement agencies.

Both the Food Standards Agency and the HSE are working to target inspections of high-risk activities and premises, with dwindling resources.

Guidance produced in May 2011 ('Joint Guidance for Reduced Proactive Inspections') is intended to assist local authorities in determining the use of proactive interventions to achieve both improved health and safety outcomes for each locality and the government's regulatory reform agenda.

A joint statement between the Food Standards Agency and the HSE, produced in 2011, aims to reduce inspections of business and introduce more local authority efficiencies to help concentrate resources. It states that combining interventions (health and safety, and food safety inspections) would only be appropriate where the local authority's health and safety and their food intervention programmes coincide.

The HSE is currently consulting on proposals for a National Local Authority Enforcement Code – Health and Safety at Work for England, Scotland and Wales. The code has been developed in response to the Löfstedt review recommendation for the HSE to be given a stronger role in directing local authority health and safety inspection activity to ensure local authority regulators take a more consistent and proportionate approach to enforcement. It sets out:

- government expectations on a risk-based approach to targeting health and safety regulatory interventions;
- a principle-based framework that recognises the respective roles of business and the regulator in the management of risk;
- the risk-based approach to be followed by local authority regulators that will provide business with a consistency of approach.

Environment Agency

Some local authorities work with the Environment Agency in issues such as fly tipping by shared enforcement under a strategic enforcement policy in a geographical area. Other areas of similar work could be the focus of coordinated enforcement.

Local authority enforcement policies

Working within the directions and advice contained in the framework described above, each enforcing authority should adopt its own enforcement policy to guide departments and individual officers as to the way in which enforcement is to be carried out. This should include how decisions are to be reached regarding which enforcement option to adopt in individual cases.

Such policies will also include provisions for dealing with disputes and complaints arising from enforcement activity.

ENFORCING AUTHORITIES

There is no single, legal definition of an environmental health authority and responsibility for the enforcement of each procedure is identified either in the particular piece of primary legislation or in the regulations made under it. However, there are only minor differences in the definitions of local authorities in the many pieces of legislation. References to licensing authorities, food authorities, etc. usually mean a local authority.

Generally, the lowest tier of local authority for the area, excluding parish councils, enforces the environmental health law dealt with in this book. The authorities charged with the implementation of most of the procedures in the book are:

(a) In England:
   - (i) Unitary authorities
   - (ii) District councils in areas where there is a county council
   - (iii) County borough councils
   - (iv) The Common Council of the City of London
   - (v) The London boroughs

(b) In Wales and Scotland, the unitary councils

(c) In Northern Ireland, the city councils, borough councils and district councils

(d) The council of the Isles of Scilly for:
   - (i) Caravan Sites and Control of Development Act 1960
   - (ii) Health and Safety at Work etc. Act 1974
   - (iii) Local Government (Miscellaneous Provisions) Act 1982
   - (iv) Public Health (Control of Diseases) Act 1984
   - (v) Environmental Protection Act 1990
   - (vi) Food Safety Act 1990
   - (vii) Noise Act 1993
   - (viii) Criminal Justice and Public Order Act 1994

(ix) Sunday Trading Act 1994
(x) Licensing Act 2003
(xi) Environmental Permitting Regulations 2010

(e) In London, the Sub-Treasurer of the Inner Temple or Under-Treasurer of the Middle Temple for:
(i) Control of Pollution Act 1974
(ii) Health and Safety at Work etc. Act 1974
(iii) Public Health (Control of Diseases) Act 1984
(iv) Clean Air Act 1993
(v) Noise Act 1993
(vi) Licensing Act 2003.

Where there is a substantial difference in definition, it is given in the relevant chapter.

A list of the definition sources in legislation can be found in Appendix 1.

ENFORCEMENT OPTIONS

The flow charts in this book indicate only the statutory remedies/provisions prescribed in the legislation that contains the particular procedure. There will also be a number of other remedies/provisions available depending upon the particular circumstances, and the options are given below. In addition, there are a number of statutory remedies that are available generally and these are also included here.

No action

In some procedures, once a local authority is satisfied that a particular set of circumstances exists, it is required by law to take the next step of statutory action, usually the service of a statutory notice – for example, in the case of the existence of a statutory nuisance under section 80(1) of the Environmental Protection Act 1990. These situations are usually identified in the legislation by the use of the word 'shall' in respect of action required of the local authority. In other cases, the action to be taken, or a decision to take no action, is at the discretion of the local authority.

Oral warning

It may be felt that, in the circumstances of a particular contravention, it can be adequately remedied by speaking to the person responsible and asking for action to be taken. Such actions should be recorded.

Informal letters/notices

These may be used to confirm the existence of contraventions and to ask for them to be remedied. Informal notices are in effect a letter given the form of a notice but nevertheless are not part of a statutory procedure and no offences are committed by not complying with them.

Statutory notices

These are notices served under the provisions of a particular procedure. They must be in accordance with the requirements of that procedure, and with the general provisions covering such notices, and are legally enforceable. Non-compliance will lead to the possibility of prosecution and/or the taking of further steps specified in the legislation. They are the most usual form of statutory remedy used in environmental health law.

Formal cautions

These are applicable to the enforcement of all criminal law and may sometimes be appropriate in the enforcement of environmental health legislation. Their purpose is to deal quickly and simply with less serious offences, diverting them from unnecessary court action and reducing the chances of re-offending. The detailed provisions are contained in Home Office circulars 18/1994 and 30/2005. The essential elements are as follows:

1. Cautioning decisions are at the discretion of the enforcing authority.
2. Cautions should be used only where they are likely to be effective and their use is appropriate for the offence.
3. They should not be used for indictable-only offences.
4. Consideration should be given to:
   - the nature and extent of the harm caused by the offence;
   - whether the offence was racially motivated;
   - any involvement of a breach of trust;
   - the existence of a systematic and organised background;
   - the views of the persons offended against.
5. Cautions should be recorded.
6. Multiple cautioning of the same offender should not generally be used.
7. Before a caution is administered:
   - there must be evidence of the offender's guilt;
   - the offender must admit the offence;
   - the offender must understand the significance of the caution and give informed consent to being cautioned;
   - consideration must be given to the public interest.
8. Officers giving the caution, which is of course done in person, should hold a position of seniority.

Whilst continued offending after the receipt of a formal warning is not an offence in itself, the fact will both guide subsequent decisions regarding that individual person, organisation or company and be taken into account by a court in any later proceedings.

Prosecutions

Environmental health law is a branch of criminal law, and prosecution of offenders in the courts is the usual last stage for most of the procedures dealt with here. The decision to prosecute or not is a matter of discretion for the local authority, guided by its own enforcement policy, which will have taken into account statutory and other guidance from the Health and Safety Executive and the Food Standards Agency. The Code for Crown Prosecutors issued by the Director of Public Prosecutions, whilst mainly aimed at police enforcement, provides useful guidance on the process of deciding whether prosecution should be taken.

Cases taken for offences against Housing Acts procedures are normally a matter initially for the Residential Property Tribunal, whereas most other environmental health procedures are dealt with at the initial stages by the magistrates' court. Higher courts will become involved in the event of appeals against decisions made by these courts, although there are situations (e.g. with the abatement of statutory nuisances) where a local authority may proceed directly to the High Court in order to seek a remedy.

Regulatory penalties and sanctions

Part 3 of the Regulatory Enforcement and Sanctions Act 2008 gives regulators sanctioning powers as follows:

- fixed monetary penalties – such fines will be imposed by a regulator in respect of low-level instances of non-compliance;
- discretionary requirements, which include:
  - variable monetary penalties – requiring a person to pay a monetary penalty, the value of which is determined by the regulator;
  - compliance notices – requiring a non-compliant business to undertake certain actions to bring themselves back into compliance; and
  - restoration notices – requiring a person to undertake certain actions to restore the position, as far as possible, to the way it would have been had regulatory non-compliance not occurred.
- stop notices – requiring a person to cease an activity that is causing serious harm or presents a significant risk of causing serious harm and has given rise, or is likely to give rise, to regulatory non-compliance; and
- enforcement undertakings – an agreement offered by a person to a regulator to take specific actions related to what the regulator suspects to be an offence.

Fixed-penalty offences

There has been a move towards this type of action particularly in the Environmental Protection Act 1990 and, more recently, in the Clean Neighbourhood and Environment Act 2005. The use of fixed penalty notices is a discretionary power given to authorised officers in some enforcement procedures as an alternative to prosecution and fine. The local authority may specify the amount of the fine to be collected within statutory bands as detailed in the Environmental Offences (Fixed Penalties) (Miscellaneous Provisions) Regulations 2007. Where the local authority does not take this option, a default amount is specified in the regulations. The receipts from such notices may be used to defray the cost of designated environmental functions and are detailed in the Environmental Offences (Use of Fixed Penalty Receipts) Regulations 2007.

Closure/withdrawal of licence, etc.

It is sometimes the case that, in addition to prosecution to exact a penalty, the local authority may give consideration to the cancellation of the licence/approval it may have issued authorising the activity where the offence has taken place, e.g. approval of meat product establishments and pet shop licences.

THE ENVIRONMENTAL INFORMATION REGULATIONS 2004

These provide a general right of access to environmental information, subject only to withholding in the public interest where that outweighs the public interest in disclosure, and applies to many of the procedures dealt with in this work.

DELEGATION OF AUTHORITY

The Local Government Act 1972 allows a local authority to delegate its powers and responsibilities to committees, sub-committees and officers. In respect of environmental health law enforcement, a clear scheme of delegation to officers is vital if investigation and remedial action is to be swift and effective. Such a scheme will require the formal approval of the individual local authority. It will include matters like the exercise of powers of entry and inspection, the service of statutory notices and the institution of legal proceedings. There is also a need for the scheme to include the designation of officers to whom the Act or regulations concerned give authority to act, e.g. the designation of proper officers, authorised officers and officers who will be inspectors under the Health and Safety at Work etc. Act 1974.

POWERS OF ENTRY

The powers of entry and investigation that relate to each procedure are given either in the section on General Procedural Provisions at the beginning of each chapter or within the text relating to the procedure itself.

THE COMMISSIONER FOR LOCAL ADMINISTRATION (THE OMBUDSMAN)

Each procedural flow chart indicates the stages at which the courts may become involved in dealing with appeals, prosecutions, etc. In addition to this, any local authority action, or lack of it, is subject to investigation by the Local Government Ombudsman. Whilst the Ombudsman cannot challenge the decision itself, e.g. not to serve a statutory notice, he may do so in respect of the way in which the decision was reached. Whilst he does not have the power to change the local authority decision, the Ombudsman can issue a public report criticising the authority and is also able to indicate a level of compensation which he feels ought to be made.

ALTERNATIVE ACTIONS IN ENFORCEMENT

Fixed-penalty tickets

Fixed-penalty tickets are a halfway house between a warning letter and prosecution.

Each local authority needs to decide in their enforcement policy how many fixed-penalty tickets can be issued to the same person before further action should be taken. The discretion of the individual officer can determine the most appropriate course of action in every case, but local authorities should consider the practicalities of enforcing any policy they decide to adopt in this area.

Mortgagees

Mortgagees usually regard any action that attracts a statutory notice as being action that devalues a property. The Housing Act 2004 and the Town and Country Planning Act 1990 actually require that notices are not only served on owners and occupiers but also on the mortgagee, so that they can take such steps as they consider appropriate to protect their investment.

Where legislation demands that notices are served on mortgagees, the local authority must do that, but where the legislation is silent on the point, there is nothing to preclude the local authority from sending the mortgagee a copy of any notice or letter for information. Any mortgagee who considers that their investment is being jeopardised will normally write to the mortgagor, requiring them to take such steps as are necessary to comply with the notice or be regarded as defaulting on the mortgage.

Clearly this only works where a property is subject to mortgage, which can be determined by a simple Land Registry check.

Planning and how environmental health can benefit

The six tests for the efficacy of planning conditions are that they must be:

- necessary – would the planning permission be refused if the condition was not attached? If yes, the condition is valid; if no, it is not;
- necessary to planning;
- relevant to the development to be permitted – its purpose is to ensure that conditions imposed are directly related to controlling the development, and cannot be used to achieve something that is covered by other legislation. You may not, for example, use a planning condition to prevent emission of dark smoke from a premise, since the Environmental Protection Act covers such

emissions and to try to control them by planning condition is therefore unenforceable;
- enforceable;
- precise;
- reasonable in all other aspects.

How can environmental health practitioners (EHPs) get maximum benefit from planning conditions?

EHPs must identify where problems may arise during the development, or once the development is operational, and design a condition that is enforceable and will ameliorate the problems identified. For example, they may consider problems caused by noise or dust during the course of a development, and design a condition to deal with this – it may, for instance, cover hours of working, or suppression techniques for dust or lights on site, or number of vehicle movements permitted where residential streets are to be used. This is lawful since statutory nuisance law will not provide a remedy and, subject to the wording being precise and reasonable, will be enforceable.

1. Ensure consultation with planners at a sufficiently early stage.
2. Consider any environmental health problems that may arise either during the construction or during the operation of the development.
3. Check that there are other powers that will sort out the identified problems.
4. Draft an appropriate and lawful condition.
5. Engage with planners to ensure that the proposed condition and the reasons for it are understood and the condition is incorporated into the planning permission.

Section 106 Town and Country Planning Act 1990

Section 106 allows for a planning obligation or planning gain. Planning gain can only be sought if:

- it is relevant to planning and is directly related to the proposed development;
- it would make the proposed development acceptable in planning terms; and
- planning conditions or other procedures cannot achieve this.

Therefore, a planning obligation should represent a benefit for the land and/or the locality. Developers cannot be asked to solve existing problems, e.g. to provide a new facility, although they may be asked to contribute towards resolving an existing problem if the proposed development would make things worse.

A Section 106 Agreement cannot require the developer to do something that can be required by other legislation, but what it can do is achieve something by way of public health gain.

Works in default

Carrying out works in default (WID) is often a cost-effective option when statutory notices are not complied with and, in some cases, when urgent steps need to be taken. The local authority has these powers under a variety of different legislation, with Environmental Health, Building Control, Planning and Highways being the main service areas for such rechargeable works. In some situations, the costs incurred become an automatic charge on the property. In most cases, the local authority can sue for debt in the civil courts.

The process followed needs to be clear and well documented, as any costs being recovered potentially have to be justified to the courts. Where the local authority decides to delay reclaiming the costs and relies on a charge on the premises, the case needs to be monitored at least yearly to ensure that time limits are not breached and that the debt plus interest is still covered by the value of the property. When there is no charge on the premises, there are still several options to collect the debt. Local authorities are entitled to include in the costs all its own reasonable costs in carrying out the WID, so that potentially some of the normal service budget can be reclaimed.

As with other means of enforcement, how the local authority will use these powers should be included in its relevant policies. Carrying out WID is always discretionary,[1] even where the statutory notice is a mandatory one, so the local authority needs to show it has exercised its discretion properly.

The main Environmental Health powers to do WID are covered by:

- Public Health Acts 1936 (particularly Part XII) and 1961;
- Prevention of Damage by Pests Act 1949;
- Building Act 1984 (various, section 107 for the creation of a charge);
- Environmental Protection Act 1990 (section 81 as amended);
- Housing Act 2004 (schedule 3, part 3).

This list is not exhaustive, and various Local Government Acts as well as local legislation create other opportunities.

Time limits

All civil law cases are subject to the appropriate time limits for bringing the action. Under section 5 of the Limitation Act 1980, all 'simple' debt/contract cases have to be brought within six years of the cause of action. For WID, the courts have said this means six years from the date the local authority completes the relevant works.[2] Where a charge is created, this is covered by section 20 of the Act, which gives 12 years for these additional powers to be used. Again for WID, the courts have confirmed that the 12 years runs from when the demand becomes operative[3] – which is immediately upon service for the Public Health Act and Building Act (notices served on owners), and after 21 days (if no appeal) for the Environmental Protection Act (owners again) and Housing Act. In any event, interest is generally only recoverable once the demand has been served.

Why use WID?

It may be a more effective sanction. It gets the job done – with likely more satisfied tenants/neighbours, etc. A prosecution and conviction for an offence may still not lead to the works being done. It may cost the recipient more and therefore be a more effective deterrent than if they are simply fined.

It may be quicker than prosecuting – in some areas, getting a defended prosecution heard can take many months. It may be easier than prosecuting due to the lower standard of proof and rules of evidence. It may be cheaper than prosecuting, and there is a better chance of getting your costs back.

In most cases where there is an offence as well as WID powers, it is possible to pursue both. The local authority should cover the interplay between these sanctions in its relevant policies.

What to consider when using WID?

- Was the original notice correctly served and signed by the right person? If not, then as there is never an obligation to appeal against an invalid notice, it is unlikely any costs can be recovered if the local authority does the works[4] (so remember to check, including delegated powers for both the notice and the power to authorise WID);
- The likelihood the works will otherwise get done, including the track record of the recipient;
- The scale of the works;
- The ability to specify the right works – if the property has deteriorated since the service of the notice and/or more extensive repairs are now found, it may be necessary to reserve the notice;
- The impact on the owners, occupiers and neighbours – this should now include any implications under the Human Rights Act 1998;
- The contribution to any corporate policies and objectives – this could include dealing with empty and eyesore properties.

It may be sensible to warn those affected that the local authority is minded to take such action and give them a short time to make representations. There is often a formal right of appeal against the demand after the works are done so the local authority needs to be able to show it has acted reasonably. Someone who has such a right of appeal cannot use judicial review to challenge such decisions,[5] but those without such a right (e.g. any tenants) may be able to use this procedure.

Records

It is necessary to correctly record all enforcement decisions where these count as executive decisions, even when taken under delegated powers. Some legislation (e.g. housing) are always executive decisions, others (e.g. nuisance) may be under the local choice arrangements. In practice, it is sensible to record all WID decisions in case of challenge.

The arrangements to carry out the works will also need to be recorded, including details of any tendering or use of approved lists of contractors. These must conform to the local authority's own standing orders on letting contracts.

Records of all the time spent on arranging the works are needed, including letting the contracts, supervising the works and liaising with owners and tenants, etc. The costs can also include preliminary inspections, highways management, security, consulting fees, scaffolding, roof removal works and road closures.[6]

Local authorities have often underestimated the true costs, including overheads and support staff. They are all potentially recoverable. Local authorities often do not use time recording for their Environmental Health staff, so a system for WID can be adopted based on arrangements used for other professional staff, such as lawyers.

Remember – the local authority does not usually have the ability to add costs in after the demand has been served!

Once all the details have been collated and the total charge calculated, the demand should be served on the right person (which may not be the person who received the original notice), and any others who are entitled to receive a copy.

The original statutory notice should have been included on the Local Land Charges register maintained by the local authority. Details of the WID expenses should also be lodged with the Local Land Charges to warn anyone carrying out a search before buying the property. Failure to do this may make the costs irrecoverable against a new owner.

Remember – in the appropriate cases, an automatic charge on the property comes into existence as soon as the demand is served/becomes operative. It is not necessary to register these at the Land Registry. Such a charge binds all estates and interests in the property and takes precedence over any existing mortgage, etc.[7]

Getting the money back where there is no automatic charge on the property

An early decision is needed on pursuing debt through civil courts if the demand is not successful in getting payment or an agreed payment plan.

Action to collect a debt is taken in court (fast track <£25,000, small claims <£10,000). Both of these are likely to be dealt with by the County Court, and small claims can be dealt with without lawyers (and the legal costs cannot be reclaimed). Complex cases and those above £25,000 in value may be dealt with in the High Court.

Once a County Court Judgement is obtained, the debt can be collected in several ways:

- through bailiffs;
- attachment of earnings;
- third party debt order;
- applying for a Charging Order – if the debtor has assets such as property or shares (interim and final orders have to be registered at Land Registry!). This has the same effect as a charge on a property but it is at the discretion of the court and ranks behind existing mortgages, etc.

If all this fails, the local authority can factor the debts (i.e. sell them on to someone else). If the debt has to be written off, this is likely to impact negatively on the service accounts. While EHPs may think that debt collection is not their problem, failure to run the process properly may discourage the local authority from using it in future.

Getting the money back where there is an automatic charge on the premises

Apart from using the normal debt collection process, the local authority can also enforce such a charge by:

- taking possession and selling the property (this is what banks etc. tend to do);
- putting in a receiver and collecting any rents;
- seeking a vesting order, which transfers the property to the local authority without further payment (this is foreclosure, and it wipes out any other equity in the property).

County Court procedure for recovering a debt

The local authority should attempt to recover the outstanding debt using the County Court. If the identity of the debtor is known, action can be commenced against him by summons alleging that the debt is owed to the local authority. The summons, with particulars of the claim attached, is issued by the court and the procedure then effectively runs itself. Should the debtor fail to respond to the summons within the prescribed time, the local authority can gain judgement in default. The judgement can then be enforced by the local authority using the court bailiffs.

The process is not without difficulties. The debtor may seek to defend the action, in which case a hearing of the matter will follow. The debtor may take no part in the process at all, and after judgement is awarded for the local authority and bailiffs engaged, it may become clear that the debtor has no goods against which the debt can be levied. The local authority may end up with nothing at the end of the process; however, the advantage of a County Court judgement over a charge on property is that the judgement is personal, and restrictive on the debtor. He will be prevented from obtaining or extending a mortgage and from obtaining credit, which will often focus the mind of the debtor rather more than a charge on a property.

PRIMARY AUTHORITIES

References

Regulatory Enforcement and Sanctions Act 2008 – sections 22 to 31.
Primary Authority Guidance LBRO, March 2009.

Extent

These provisions apply in England, Wales, Scotland and Northern Ireland.

Scope

Where a person or company carries on an activity in the area of two or more local authorities, and each of those authorities has the same relevant function in relation to that activity, the two parties can agree to set up a primary authority agreement.

Relevant function

The categories of relevant functions (enactments) are:

- age-restricted sales;
- agriculture;
- animal establishments and companion animal welfare;
- consumer credit;
- environmental protection;
- explosives licensing;
- fair trading;
- farm animal health;
- food safety and hygiene;
- food standards;
- general licensing;
- health and safety;
- housing;
- metrology;
- petroleum licensing;
- pollution control;
- product safety;
- road traffic.

(See schedule 3 for the full list of legislation.)

Primary authorities

The Secretary of State may nominate a local authority to be the primary authority for the exercise of the relevant function in relation to the regulated person (section 25).

Nomination of primary authorities

The Secretary of State may only nominate a local authority if the:

- authority and the regulated person have agreed in writing to the nomination; or
- regulated person has requested the Secretary of State to make a nomination for the exercise of the relevant function in relation to the regulated person; and
- Secretary of State considers the authority suitable for nomination.

The Secretary of State may in particular consider as suitable:

(a) the local authority in whose area the regulated person principally carries out the activity in relation to which the relevant function is exercised;
(b) the local authority in whose area the regulated person administers the carrying out of that activity.

Before nominating a local authority, the Secretary of State must consult:

(a) that authority; and
(b) the regulated person.

The Secretary of State must have particular regard to any representations made by a local authority about the resources available to it.

The Secretary of State may at any time revoke a nomination if it is considered that the authority is no longer suitable for nomination, or it is appropriate to do so for any other reason. The Secretary of State must maintain a register of nominations (section 26).

Primary authority functions

The functions of a primary authority are giving advice and guidance to the regulated person in relation to the relevant function and giving advice and guidance to other local authorities with the relevant function as to how they should exercise it in relation to the regulated person.

The primary authority may make arrangements with the regulated person as to how it will discharge its function (section 27).

Enforcement action

An enforcing authority must notify the primary authority before taking any enforcement action against the regulated person concerning the relevant function as soon as it reasonably can (sections 28 and 29).

If the primary authority determines within five working days that the proposed enforcement action is inconsistent with advice or guidance previously given by the primary authority (generally or specifically), it may within that period direct the enforcing authority not to take the enforcement action (section 28(2)).

If the enforcing authority is not directed and continues to propose to take the enforcement action, it must inform the regulated person.

The enforcing authority may not take the proposed enforcement action:

- at any time during the relevant period (five working days);
- at any time after the end of that period, if it is directed by the primary authority.

Where an enactment limits the period within which the enforcing authority may take the proposed enforcement action, any time during which the authority is prohibited under this section or schedule 4 from taking the action is to be disregarded in calculating that period (section 28).

FC1 Primary authorities – Enforcement action – Sections 22 to 31 Regulatory Enforcement and Sanctions Act 2008

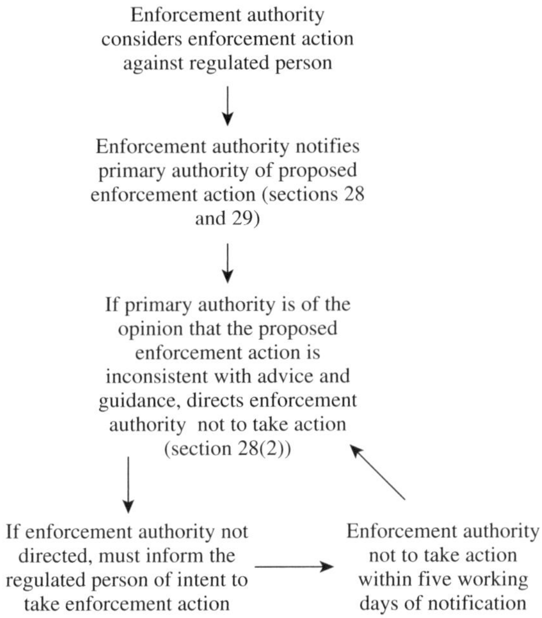

Note

1. If the enactment relating to the enforcement action has a time limit for enforcement, the five working days rule is ignored.

Enforcement action: exclusions

Where the enforcement action is required urgently to avoid a significant risk of serious harm to human health or the environment (including the health of animals or plants) or the financial interests of consumers and their application would be wholly disproportionate, the requirements above do not apply (section 29).

Inspection plans

Where a relevant function consists of or includes a function of inspection, the primary authority may make an inspection plan containing recommendations as to how a local authority with the function of inspection should exercise it in relation to the regulated person.

An inspection plan may set out:

(a) the frequency at which, or circumstances in which, inspections should be carried out;
(b) what an inspection should consist of.

Before making an inspection plan, the primary authority must consult the regulated person.

Where a primary authority has made an inspection plan, it must, if the Secretary of State consents to the plan, bring the plan to the notice of the other local authorities with the function of inspection.

A local authority (including the primary authority) exercising the function of inspection in relation to the regulated person must have regard to a plan.

Before a local authority, other than the primary authority, exercises the function of inspection in relation to the regulated person otherwise than in accordance with a plan, it must notify the primary authority with reasons for doing so. A primary authority may from time to time revise a plan (section 30).

Power to charge

The primary authority may charge the regulated person such fees as it considers to represent the costs reasonably incurred by it in the exercise of its functions in relation to the regulated person (section 31).

Definitions

Enforcement action means:

(a) any action which relates to securing compliance with any restriction, requirement or condition in the event of breach (or putative breach) of a restriction, requirement or condition;
(b) any action taken with a view to or in connection with the imposition of any sanction (criminal or otherwise) in respect of an act or omission;
(c) any action taken with a view to or in connection with the pursuit of any remedy conferred by an enactment in respect of an act or omission.

FC2 Inspection plans – Section 30 Regulatory Enforcement and Sanctions Act 2008

Relevant period means:

(a) the period of five working days beginning with the day after that on which the primary authority is notified; or
(b) such longer period beginning with that day as the Secretary of State may direct (section 28).

SERVICE OF NOTICES BY LOCAL AUTHORITIES

Generally, notices must be in writing, dated and signed by an authorised officer. A facsimile signature can be used. A notice must be served on the person responsible, which is usually defined in the legal provision in each Act or regulation. Usually the notice will specify the contravention of the law or a defect which needs to be remedied. It may also specify the works to be done to rectify the contravention or defect.

Addressing a notice on individuals

The notice is addressed to the individual concerned, using the full name, not initials, and the envelope addressed likewise. Any failure to comply with the notice can be dealt with against the person named on the notice and the person to whom it is addressed can be prosecuted.

Bodies corporate

Notices can be served on and summonses issued against a number of bodies corporate, including limited companies, unincorporated bodies and local authorities. The name on the notice should be that of the company, with its registered office being given as its address. Naming the company ensures that it is the company that is required to comply with the requirements of the notice.

The name and address on the envelope is a different matter. A company secretary is the legal persona of a company upon whom notices can be served. They can be sent the notice in an envelope and will accept delivery on behalf of the company. Envelopes containing notices should therefore be addressed to the company secretary with the company name and address. This ensures that somebody with appropriate authority accepts the notice. In the case of local authorities, it is usually the chief legal officer who is authorised to accept documents on behalf of the authority, and documents addressed to that person by name or by job title are deemed as having been properly served on the authority concerned.

In the cases of unincorporated bodies, clubs and school governing bodies, it is wise to ask on whom service of the notice may be served. Where the body has a formal constitution, this person should be identifiable; where it does not, naming all of the managing committee on the face of the notice and serving each individual ensures that service is properly made.

Some Acts will specify that service shall be served in a particular way and on a particular party, and failing to comply with that laid-down method will mean that effective service will not be achieved.

If service of the document cannot be proved, no prosecution, based on allegation of non-compliance with the requirements of the notice, can proceed.

A notice that has been addressed on its face to the wrong party is a ground of appeal against it. Addressing a notice to the company secretary rather than to the company itself gives rise to that ground of appeal.

Service

This may be effected by one of the following:

- (i) delivery to the person;
- (ii) in the case of service on an officer of the council, leaving it or sending it in a pre-paid letter addressed to him at his office;
- (iii) in the case of any other person, leaving it or sending it in a pre-paid letter addressed to him at his usual or last known residence;
- (iv) for an incorporated company or body, by delivering it to its secretary or clerk at its registered or principal office, or sending it in a pre-paid letter addressed to him at that office;
- (v) in the case of an owner by virtue of the fact he receives the rackrent as agent for another person, by leaving it or sending it in a pre-paid letter addressed to him at his place of business;
- (vi) in the case of the owner or occupier of a premises, where it is not practicable after reasonable enquiries to ascertain his name and address or where the premises is unoccupied, by addressing it to the 'owner' or 'occupier' of the premises (naming them) to which the notice relates and delivering it to some person on that premises, or, if there is no-one to whom the notice can be delivered, by affixing it or a copy of it, to a conspicuous part of the premises (Public Health Act 1936, section 285).

Proof of receipt and/or service of notices should be obtained.

The Public Health Act 1936 (sections 283 to 285) is the basis of service of notice in environmental health law and notices served in any of these ways will be held in later proceedings to have been properly served, but it is possible to serve in some other way and prove service in subsequent proceedings.

The main provision for service of notice is found in the Local Government Act 1972 and, unless otherwise excluded by a particular provision of another Act, is available in addition to methods of service available under other legislation.

1. Subject to subsection (8) below, subsections (2) to (5) below shall have effect in relation to any notice, order or other document required or authorised by or under any enactment to be given to or served on any person by or on behalf of a local authority or by an officer of a local authority.
2. Any such document may be given to or served on the person in question either by delivering it to him, or by leaving it at his proper address, or by sending it by post to him at that address.
3. Any such document may:
   - (a) in the case of a body corporate, be given to or served on the secretary or clerk of that body;

(b) in the case of a partnership, be given to or served on a partner or a person having the control or management of the partnership business.

4. For the purposes of this section and of section 26 of the Interpretation Act 1889 (service of documents by post) in its application to this section, the proper address of any person to or on whom a document is to be given or served shall be his last known address, except that:
   (a) in the case of a body corporate or their secretary or clerk, it shall be the address of the registered or principal office of that body;
   (b) in the case of a partnership or a person having the control or management of the partnership business, it shall be that of the principal officer of the partnership;

   and for the purposes of this subsection the principal office of a company registered outside the United Kingdom or of a partnership carrying on business outside the United Kingdom shall be their principal office within the United Kingdom.
5. If the person to be given or served with any document mentioned in subsection (1) above has specified an address within the United Kingdom other than his proper address within the meaning of subsection (4) above as the one at which he or someone on his behalf will accept documents of the same description as that document, that address shall also be treated for the purposes of this section and section 26 of the Interpretation Act 1889 as his proper address.
6. ...(Repealed)
7. If the name or address of any owner, lessee or occupier of land to or on whom any document mentioned in subsection (1) above is to be given or served cannot after reasonable enquiry be ascertained, the document may be given or served either by leaving it in the hands of a person who is or appears to be resident or employed on the land or by leaving it conspicuously affixed to some building or object on the land.
8. This section shall apply to a document required or authorised by or under any enactment to be given to or served on any person by or on behalf of the chairman of a parish meeting as it applies to a document so required or authorised to be given to or served on any person by or on behalf of a local authority.
9. The foregoing provisions of this section do not apply to a document which is to be given or served in any proceedings in court.
10. Except as aforesaid and subject to any provision of any enactment or instrument excluding the foregoing provisions of this section, the methods of giving or serving documents which are available under those provisions are in addition to the methods which are available under any other enactment or any instrument made under any enactment.
11. In this section 'local authority' includes a joint authority, an economic prosperity board, a combined authority, a joint waste authority, a police and crime commissioner and the Mayor's Office for Policing and Crime. (Section 233 Local Government Act 1972)

The provisions for the service of notice are included in some other Acts and regulations. A list of the sources of these provisions is included in Appendix 2.

AUTHENTICATION OF DOCUMENTS

The following provisions of the Local Government Act 1972 will apply unless a particular piece of legislation has provisions which cover these issues.

1. Any notice, order or other document which a local authority is authorised or required by or under any enactment (including any enactment in this Act) to give, make or issue may be signed on behalf of the authority by the proper officer of the authority.
2. Any document purporting to bear the signature of the proper officer of the authority shall be deemed, until the contrary is proved, to have been duly given, made or issued by the authority of the local authority. In this subsection the word 'signature' includes a facsimile of a signature by whatever process reproduced.
3. Where any enactment or instrument made under an enactment makes, in relation to any document or class of documents, provision with respect to the matters dealt with by one of the two foregoing subsections, that subsection shall not apply in relation to that document or class of documents.
4. In this section 'local authority' includes a joint authority, an economic prosperity board, a combined authority, a joint waste authority, a police and crime commissioner and the Mayor's Office for Policing and Crime. (Section 234 Local Government Act 1972)

AUTHORISATION OF OFFICERS

Local authorities must have a system of authorisations and delegations. Environmental health practitioners have the power to carry out their functions because that power is given to them by the local authority by whom they are employed.

They are authorised by their employing local authority to carry out a number of functions, and whilst they are carrying out that function they have the right to do whatever the statute says they can and no more. This authorisation should be a council minute signed by the leader or chair, naming the authorised person by full name, stating that they are authorised by the council, and stating for what purpose the person is authorised.

If an environmental health practitioner is not authorised, anything done by way of enforcement, whether serving a notice or bringing proceedings is void – it has no lawful standing.

Such actions taken by unauthorised officers open the council and the officer up to claims for trespass, suing for the cost of doing any work done required by the officer and any losses to the business or even misrepresentation.

Practitioners should *always* have a certified copy of the authorisation with them while at work and acting on behalf of the local authority.

The keeping of accurate records of authorisation is vital, not least because the Office of Surveillance Commissioners (OSC) will be expecting to see records of authorisations. Environmental health departments may wish to design their own forms, but as a guide there are excellent Authorisation, Renewal of Authorisation and Cancellation of Authorisation forms on the Home Office website (www.gov.uk/government/organisations/home-office/series/ripa-forms--2), which can be tailored to suit the needs of an environmental health service.

INFORMATION REGARDING OWNERSHIPS ETC.

The Local Government (Miscellaneous Provisions) Act 1976 makes provision for a local authority to require information to be provided about the ownership of property etc. Where a local authority, in performing functions under any legislation, requires information regarding interests in any land, it may serve a notice, specifying the land and the legal provision containing the particular function under which it is acting, requiring the person concerned to declare, within not less than 14 days, the nature of his interest and the names and addresses of all persons who he believes have interests.

The notice may be served on any of the following:

(a) the occupier;
(b) the freeholder, mortgagee or lessee;
(c) the person directly or indirectly receiving the rent;
(d) the person authorised to manage the land or arrange for its letting.

Failure to comply with the notice or the making of false statements carries a maximum penalty of level 5 on the standard scale (section 16).

These powers are given to county councils, district councils, London borough councils, the Common Council, the Council of the Isles of Scilly, police authorities, joint authorities, police and crime commissioners, the Mayor's Office for Policing and Crime, an economic prosperity board, a combined authority, joint waste authorities, the London Fire and Emergency Planning Authority and parish and community councils (section 44(1)).

FURTHER READING

This book is an interpretation of particular legislative procedures that are used in environmental health work. For a study of the wider legislative context in which these procedures are to be exercised, the reader is referred to:

1. *Legal Competence in Environmental Health*, Terence Moran, 1997, E & FN Spon, London.
2. *Clay's Handbook of Environmental Health 20th edition*, Stephen Battersby, Editor, 2011, Spon Press, London.

PROFESSIONAL PRACTICE

Legal process

Legal process is to make sure, before something is done by 'authority', whether a local council or the police, that the views, concerns and objections of individuals and groups who are affected by the decision are heard and the rights of those whose freedoms, whether personal or over property or goods, may be curtailed by the action are protected. EHPs need to understand their responsibilities and duties in order to comply with the requirements of the legal process.

Enforcement policy

Local authorities must have an enforcement policy; however, it must be noted that an enforcement policy guides but does not direct. It is among other issues to be taken into account when considering how to decide on action. A local authority should have regard to its enforcement policy, as a guide promoting consistency and equity of treatment. Decisions about prosecution are for the prosecutor to make. Enforcement policies are drafted and agreed, taking into account the various government and agency policies and guidelines (see pages 1–8).

Local authorities should agree, through its cabinet, a policy to demonstrate that their procedures comply with the right to a fair trial (Article 6 of the European Convention on Human Rights).

The policy should describe the manner in which defendants will be dealt with, and the considerations that will be taken into account when determining disposal of matters.

Examples of matters to be taken into account are

- previous convictions for like offences;
- previous Home Office cautions;
- the severity of the offence alleged;
- previous convictions by other departments of the council which may lead to the view that the individual will not cooperate with the council;
- the degree of criminality involved or likelihood of re-offending.

A policy will ensure that the local authority itself is satisfied that all those against whom it is considering taking action are fairly treated.

When the policy has been approved and adopted, there needs to be a good and compelling reason for deviation from its policies.

The local authority can clearly demonstrate that its reasoning is Article 6 compliant if challenged by a defending solicitor, or by an aggrieved 'victim' who considers that an alternative course to that adopted should have been taken.

This policy is for guidance only, and in every case officers must use discretion. In most cases, the discretion will determine that the course identified in the policy is appropriate, but where it is not, the reasoned justification for deviation should be recorded and can then be cited where allegations of rigid adherence to policy without consideration of other factors are made.

Consider the risk to complainants in the event of them being identified by those they have complained of, and the steps they have taken to prevent this happening. The policy should clearly outline how you will protect identities. A simple example is, in the event of noisy parties, it is not wise to park outside the house of the complainant, go into the house and then immediately go to the house of the party-giver; instead, the policy should advise parking some distance away, and not attending one premises directly from the other.

The policy should consider the wording to be used in letters, indicating that the complainant's name and address should not be revealed and that all letters should be worded in such a way as to prevent identification of complainants – geographical references to the location of complainants should be as oblique as possible or, preferably, no mention should be made of the complainant at all and reliance should be placed on the evidence of environmental health officers.

Competence

Officers who are authorised by the local authority must hold a relevant qualification and be sufficiently competent in the particular field of operation. Local authorities must ensure that the officers operating on their behalf are competent to do whatever is required of them. Government agencies, such as the Food Standards Agency and the Health and Safety Executive, set criteria and guidance for local authorities to guide them in ensuring officers are competent to carry out the function assigned to them. It will be necessary to show that knowledge and skills are held and practised, refresher training has been done and practical skills are up to date, as well as showing how knowledge is maintained.

Inspections leading to potential prosecution

When carrying out an inspection which is likely to lead to a prosecution to establish matters of fact, speak to no-one, but note all of the potential offences. Questions as to identity of persons and possibly ownership of the premises can be asked, depending on the charges being contemplated, but nothing about the offences observed.

The potential defendant can then be asked to attend for an interview under caution where questions about each of the offences observed can be put to him. There is no guarantee of results from this process as defendants may not accept the invitation, or may give no comment during interviews. The alternative is to lay informations citing all of the offences observed and run the risk of defences being laid out in court.

Police and Criminal Evidence Act (PACE): PACE notebooks and contemporaneous notes

The PACE (Police and Criminal Evidence) page-numbered notebook helps to demonstrate that contemporaneous notes were made when the writer says that they were.

The role of the PACE notebook is to tie the inspection sheet to the date of the inspection. The inspecting officer should record in his or her PACE notebook the date and time and address of the inspection and the fact that an inspection sheet was used for the details of the inspection. The inspection sheet can be given a unique code, which should appear both on the sheet and in the PACE notebook, and which will tie the two together. The PACE notebook should record time in and time out; the inspection sheet should show the detail of the inspection and its findings. The content of the PACE notebook can either be exhibited or alternatively listed as unused material in a court case. The defence should know of its existence and they can see it if they want to.

In court, a well-kept PACE notebook shows that the officer has credibility, knows what they are doing and knows why it is important. Well-presented and completed PACE notebooks also protect officers from interrogation about when notes were made. They protect not just the evidence but also the officer giving it.

Good contemporaneous notes allow officers to refresh their memory before giving evidence about what they saw, did and heard, and are a valuable resource when writing witness statements.

Witness statements are disclosed; contemporaneous notes are disclosable. Descriptions recorded in the note should appear in the same way in the statement.

Note: Dating and timing of input of information into a data collector is not usually a problem; data collectors usually have a date and time-recording mechanism which, subject to the usual rules on the admissibility of computer-generated information, will be sufficient to establish when the information was input.

Caution

A person who is suspected of an offence must be cautioned before any questions about it (or further questions, if it is his answers to previous questions that provide grounds for suspicion) are put to him for the purpose of obtaining evidence which may be given to a court in a prosecution.

It is only necessary to caution before asking questions that may give rise to answers of evidential value, whether directly or indirectly. It is not necessary to caution before asking questions for other purposes – for example, to establish identity, or ownership or responsibility for a vehicle.

Wording of the caution

'You do not have to say anything, but it may harm your defence if you do not mention, when questioned, something you later rely on in court. Anything you do say may be given in evidence.'

The Code of Practice on PACE notes that minor deviations from its exact wording do not constitute breaches of the requirements of the Code of Practice, provided the general sense of the caution remains.

It remains at the absolute discretion of the person being questioned as to whether they say anything. What do you do if the defendant clearly does not understand? Do not, under any circumstances, try to explain the caution to the defendant yourself. If he has a solicitor with him, let the solicitor explain it. If the defendant does not have a solicitor, terminate the interview and let him get one.

The purpose of the caution is to protect the defendant. The duty of the judge in court is to ensure that the prosecutor has obtained the evidence with which he intends to convict the defendant in a fair and lawful manner. A court that feels that a lack of understanding on the part of the defendant was exploited, whether intentionally or not, is likely to rule that the evidence has been unfairly obtained and is inadmissible.

Correspondence

E-mails

Unlike a telephone conversation, which is generally not recorded, an e-mail has permanent form; once generated, it is discoverable, whether as part of a case file or under the Freedom of Information Act.

The e-mail may contain information that is defamatory or damaging to the local authority case. It may contain personal comments about the defendant or, more importantly, it may hold information that could be damaging to the substance of the council case.

E-mails should be treated with the same care as a letter or memo. Language should be formal and polite, and the normal rules of professional courtesy should apply. Generally, they should relate to a single issue, so that they can be printed off, or retained in the relevant file. Multi-issue e-mails should be avoided, since they will have to be filed in a number of files, and will introduce unrelated material into those files. E-mails need to be treated as what they are – documents of potential evidential value – and given the respect they merit in consequence.

Without prejudice

A letter written and headed ‘without prejudice’ will contain something which, were it to come into the public domain, could potentially be damaging for the writer, such as an admission against his interest, or a proposal for settlement of a matter. If the letter is accepted, this is the end of the matter; if the letter is not accepted, the fact that it is written ‘without prejudice’ means that the recipient cannot later use the letter in evidence against the writer to demonstrate the offer made, or the weakness of the writer’s position.

The use of ‘without prejudice’ correspondence or discussions means that lawyers can deal with each other on a pragmatic basis, and can discuss matters candidly without damaging their client’s position whilst doing so.

Covert surveillance

In some situations, it may be necessary for local authority enforcement officers to act in a covert manner in the investigation of possible breaches of environmental health legislation, e.g. the illegal sale of food that is unfit for human consumption.

Regulation of Investigatory Powers Act 2000 (RIPA)

The law governing the use of covert surveillance techniques by public authorities is contained in the Regulation of Investigatory Powers Act 2000 (RIPA). When public authorities, such as local authorities, need to use covert surveillance techniques to obtain private information about someone, they do it in a way that is necessary, proportionate and compatible with human rights.

RIPA applies to a wide range of investigations in which private information might be obtained, including terrorism, crime, public safety and emergency services.

RIPA's guidelines and codes apply to actions such as:

- intercepting communications, such as the content of telephone calls, e-mails or letters;
- acquiring communications data – the 'who, when and where' of communications, such as a telephone billing or subscriber details;
- conducting covert surveillance, either in private premises or vehicles (intrusive surveillance) or in public places (directed surveillance);
- the use of covert human intelligence sources, such as informants or undercover officers;
- access to electronic data protected by encryption or passwords.

The Regulation of Investigatory Powers (Directed Surveillance and Covert Human Intelligence Sources) Order 2010 stated that such surveillance carried out by local authorities must be by officers of at least director, head of service, service manager or equivalent status, or by an officer designated as being responsible for the management of an investigation. Guidance to local authorities is given in the Home Office guidance to local authorities in England and Wales on the judicial approval process for RIPA and the crime threshold for directed surveillance.

From 1 November 2012, local authorities are required to obtain judicial approval prior to using covert surveillance techniques. Local authority authorisations and notices under RIPA will only be given effect once an order has been granted by a Justice of the Peace in England and Wales, a sheriff in Scotland and a district judge (magistrates' court) in Northern Ireland.

Additionally, from this date, local authority use of directed surveillance under RIPA will be limited to the investigation of crimes which attract a six month or more custodial sentence, with the exception of offences relating to the underage sale of alcohol and tobacco.

Generally, local authorities are only concerned with carrying out directed surveillance 'for the purpose of preventing and detecting crime or preventing disorder' – for example, making observations in respect of fly tipping or the sale or distribution of counterfeit goods.

When considering an allegation of statutory nuisance, evidence is collected to support the serving of a notice, not to prevent or detect crime, or prevent public disorder. That being the case, it is not appropriate to use the powers in RIPA to authorise directed surveillance since the activity falls outside the terms of the specific grounds available.

It is not until the notice is served and, subject to withstanding any appeal against it, has taken force that RIPA becomes relevant. Once evidence is being gathered, this shows that the terms of the notice are being breached and the evidence is being gathered '*for the purposes of detecting crime*', i.e. the proving that the notice is being breached and therefore an offence is being committed.

If an officer wants to carry out an investigation, he/she must first ask him/herself if what he wants to do constitutes covert directed surveillance. If it does, RIPA applies and he/she will need to have that activity authorised. He/she then needs to consider whether the purpose of the surveillance is to prevent or detect crime, or to prevent disorder. If it is neither of these, then the proposed surveillance cannot be authorised under RIPA and therefore cannot lawfully be carried out. The simplest test is to ask whether you could commence proceedings based on the evidence you would obtain. If you can, RIPA authorisation is required. If you cannot, you cannot legitimise the proposed covert directed surveillance and should not therefore carry it out.

Directed surveillance must be:

(i) covert;
(ii) not intrusive;
(iii) for the purpose of a specific investigation; and
(iv) carried out in such manner as to result in obtaining private information about a person, otherwise than by way of an immediate response to circumstances.

Covert surveillance – RIPA advises that surveillance is covert only if it is carried out in such manner that the subjects are unaware that it is, or may be, taking place (section 26(8)(a)).

So, suppose you tell the subject that surveillance *will* be taking place and how it will be carried out. Let us suppose that when an abatement notice is served, the local authority write a covering letter advising that they will be carrying out monitoring to ensure compliance, and this monitoring will take the form of periodic visits, to be carried out at any time of day or night, by inspectors who will engage in non-intrusive monitoring. The 'target' could not complain that the surveillance was covert, since he not only knows it will be taking place, he has been advised how it will be done and by whom. All he does not know is when. This knowledge is enough to take the surveillance out of the covert category and make it overt, and hence preclude the need for authorisation. It must be noted that this procedure has not been tested in court and may be subject to challenge.

It is essential for authorising officers to ensure that they give proper consideration to the reasons given for seeking authorisation to conduct directed surveillance. The authorising officer must be able to say why they believed what they were told about the complaint, what convinced them and what further inquiries, if any, they required to be made. Did they consider whether there was an alternative to directed surveillance and, if so, did they require that it was explored? If not, why not? It is essential that there is a clear paper trail that demonstrates the thinking of the authorising officer.

There is a higher threshold for directed surveillance with the exception of two trading standards offences. Local authority applications will only be granted for offences that attract a maximum custodial sentence of six months. Magistrates must approve any RIPA-regulated activity before it begins. This applies in England, Wales and Northern Ireland and to communications data in Scotland.

Home Office guidance says EHPs can continue to use covert techniques provided they have considered other less intrusive avenues first. Officers need to present magistrates with documentation and respond to questions by the officer with most knowledge of the case and show that covert surveillance is both proportionate and necessary. It is important to consider the questions that may be put by the magistrates – in other words, have less intrusive measures been used and subsequently failed?

Stake-outs

The Court of Appeal in the case of R v Johnson[8] declared that there are two requirements for stake-outs.

First, the officer in charge of the inquiry must be in a position to give evidence that he visited the premises to be used and ascertained the attitude of the occupiers of the fact that the premises were to be used, and their attitude to the fact that a surveillance operation had been in place would have to be disclosed, which may then lead to the identification of both the premises and the occupants. He/she may also inform the court of any local difficulties in obtaining assistance from members of the public that may have caused the use of the particular premises chosen.

The second requirement is that a more senior officer should be in a position to give evidence that, immediately before the trial, he had visited the premises and ascertained whether the occupiers were still of the same view and, whether they were or not, what their attitude was to the possible disclosure of the use made of the premises and of facts which may lead to the identification of both premises and occupiers.

The lessons of R v Johnson for local authority officers are:

1. Make sure the local authority's enforcement policy takes account of the rules in R v Johnson; and
2. Be pragmatic, and use local authority premises or vehicles for surveillance where possible.

The Protection of Freedom Act 2012 includes provision for making codes of practice on surveillance, regulation of surveillance using CCTV and powers of the Secretary of State to amend powers of entry.

Obstruction

Inspecting officers often encounter being partially obstructed by being delayed just enough to mean that gaining access to a premises is pointless.

Delayed consent, however willingly given, does not constitute real consent, and taking up such an invitation to enter makes later pleading an allegation of obstruction in court difficult.

First approach

When immediate consent to enter is not given, officers could walk away on the grounds that delayed entry is pointless, and prosecute for obstruction, based on the facts of the initial refusal. The defendant did not allow immediate access and, as such, this is an offence.

Second approach

Stage 1 – If officers want to enter a premises and are being frustrated by being given only delayed consent, they could write to the owner and/or occupier telling them that they want to enter, that they have the power to do so and that when they ask to be allowed entry, access is required immediately. Owners etc. are then forewarned of what officers expect and can take legal advice should they want to.

Importantly, officers also have a paper trail that can be produced in evidence, showing the court that they have advised owners/occupiers of the position, and their claims not to know or to understand would therefore be groundless.

Stage 2 is to turn up unannounced and ask to be allowed in. If officers are delayed, even slightly, they should just go away. Do not wait. Then prosecute, alleging obstruction. Then repeat the process until the message gets through. Fines for obstruction can be significant, commonly in the order of £1,000. Being fined once may cause the defendants to consider their position; being fined several times will usually make a real difference.

Warrants

Obstruction of an officer in the course of their duties under legislation is an offence. Furthermore, the same Acts make provision for officers who have been refused entry, or who anticipate that entry will be refused, to make an application to the magistrates' court for a warrant authorising entry.

Obtaining a warrant is a straightforward procedure. An application is made to the magistrates' court; evidence as to why the warrant is required is given on oath, before the magistrates, usually in camera (in private); and subject to the magistrates being satisfied that the application has merit, the warrant will be issued. Where the premises in question is domestic, the property owner must be given 24 hours' notice of the hearing application, in order that they may attend, should they wish. This requirement does not apply to commercial premises. Where an EHP can make out a sound case for the issue of a warrant, it will be granted. To execute a warrant, an officer is required to show it to the owner of the premises. If the property owner still refuses entry in the face of a valid warrant, they are in contempt of court and should be prosecuted.

The requirement that must be satisfied when seeking a warrant is that the officer seeking to execute the warrant must have reasonable belief that an offence is being committed in the premises or on the land to which the warrant relates. The reasonableness of this view will be tested by the magistrate granting the warrant; therefore, it must be sufficiently robust to satisfy that inquisition.

Reasonable belief is that attributed to a reasonable man with reasonable judgement. Where there is dispute as to what is and is not reasonable, the judge in court will be asked to arbitrate, and it is for the judge to decide if, on the facts as known at a particular point in time, a prosecutor had reasonable grounds to believe that an offence had been committed.

Deciding to prosecute

Prosecutors must ask themselves the following:

1. On the evidence, is there a *prima facie* case to answer?

And subject to the answer being yes:

2. Is it in the public interest to prosecute?

The person making the decision to prosecute must be independent of the case, and must demonstrably be so.

Is there a prima facie case?

A lawyer will normally decide this – legal advice as to whether a certain series of facts constitutes an offence is definitive. A lawyer who is a member of an investigation team is not independent from the case, so a different, uninvolved lawyer, chief EHP or whoever is considered to be an appropriate person will have to review the file when it is completed and consider whether to prosecute. The independent case review must demonstrably be completely independent.

The Code for Crown Prosecutors 2004, which contains the Full Code Test, is a more sophisticated test, with guidance, as to whether it is appropriate in a particular case to commence proceedings.

Is it in the public interest to prosecute?

This is for the officer in the case to consider. Sometimes there will be good reasons for prosecuting a case – for instance, the publicity generated by doing so may be disproportionate to the offence itself but may influence businesses. However, it may also have a detrimental effect on the way the council is viewed by the public. Overall, it is the officer who must decide how best and most appropriately to deal with the offence.

Things to take into consideration

Courts will usually look at the overall offending when it imposes sentence. This will include the number and seriousness of offences; however, the court will not look favourably on a large number of informations.

Where the defendant has indicated that he will be pleading guilty, the prosecutor can pick out a couple of specimen offences and then ask if the defendant is prepared to ask the court to take the remaining counts into consideration. Where he is, they are tabulated, and after his plea is accepted, they are passed to the court to be taken into consideration when sentence is considered. The time and resource saved by this practice attracts discount off sentence, but the offences to be taken into consideration (TiCs) remain recorded against the defendant, and can be cited as part of his previous offending record in any future cases.

Where the defendant has pleaded guilty, he will usually accept the TiCs, which will have been put to him on his plea being indicated. It is an option for the prosecutor, even in the face of a not guilty plea to list a number of TiCs and offer them to the defence on the basis that the defendant will accept them in the event of his being convicted against his plea. If he is not convicted, the main offences and the TiCs will fall, but where such an offer is made, the prosecutor will generally be confident that he will succeed at trial, and therefore the risk of major loss is minimal. This provision should be in every local authority's enforcement policy.

'Shall have regard to'

You must have regard to something when the circumstances which cause it to be relevant actually exist. Therefore, if the requirement is that, when considering how a potential prosecution should be disposed of – whether to instigate court action or to offer a formal caution – the officer making the decision may be required by an enforcement policy to take into consideration the character of the defendant, the officer need only consider the character of the defendant when making that decision.

Consideration is subjective – it requires free thought and an independent decision.

What the officer is required to do is to consider what he or she knows about the character of the defendant – good or bad – and to consider how that defendant will respond to whatever method of disposal is preferred.

The officer making the decision should record the reason for the decision being made, showing that he or she has taken the character of the defendant into consideration in its making the requirement.

You must be able to show that you have given it some weight, and that the consideration you gave the issue was true consideration that did impinge on the decision-making process, not just a token consideration because you had to do so.

Failure to take account of, consider or have regard to things that you are required to take account of etc. can cause your decision to be challenged through the mechanism of judicial review.

Any authority that does not consider all of the relevant issues, particularly those it is specifically required to consider, whether they are codes of practice, statutory guidelines, planning policy guidance, etc., can be accused of flawed decision making, have its decision overturned by the High Court and be returned to it for reconsideration.

Where directory words are used, officers should ensure that what is required is actually done, so that those affected by a decision, whilst perhaps not liking the decision itself, cannot challenge the process by which the decision was made.

Prompt action

Time between end of compliance period and check for completion of works

If an officer is trying to impress on the court the importance and urgency that necessitated tackling a statutory nuisance, the officer has also to demonstrate that he/she acted expeditiously throughout. This means that the officer went back to the site as soon as the compliance period ended to check for completion of works.

Anything less risks giving the impression that the matter had ceased to be important – in effect, the service of the notice rather than compliance with it resolved the issue. At the end of the compliance period, the site should be visited to assess compliance. If a matter is important enough to warrant action, it is important enough to follow through.

Time lapse between offence and bringing case to trial

Summary proceedings must be commenced by the laying of an information within six months of the prosecutor having sufficient evidence to make a *prima facie* case, and in either way or indictable matters, they should be laid within 12 months. Failure to comply may lead to a challenge based upon the case being out of time.

Officers must ensure that all proceedings are commenced as expeditiously as possible. Where there is the possibility of delay, for whatever reason, the reason for the delay and the manner in which the prosecutor has attempted to mitigate it should be documented in order that the court may be appraised, if necessary. In such scenarios, the defence will be to have the prosecution dismissed without consideration of the evidence on the basis of prejudice through delay, and breach of Article 6 of the European Convention on Human Rights.

It is highly unlikely that the court will be sympathetic where lack of progress is attributed to lack of time or resources, or pressure of other work. Valid reasons for unavoidable delay will have to be presented.

In some cases, commencement dates are prescribed within the legislation, to take account of the fact that covert illegal action may not be discovered for a significant period, e.g. the Food Safety Act 1990 – commencement must be either three years from the commission of the offence or within one year of its discovery by the prosecutor, whichever is the shorter (section 34).

What constitutes commencement? When is a case considered to be started? The date a case is commenced is the date on which the information making the allegation is laid.[9] For all practical purposes, this is the date on which the informations arrive at the magistrates' court. When they are delivered by hand, that date is clear.

When they are sent by post, certain rules apply to when delivery is assumed to have taken place. Similarly, it is now established that when informations are laid by electronic mail, they are laid on being sent, not on being received.[10]

When is an offence discovered? In some cases, it will be very clear – when an officer finds a breach of the legislation. Further, if an officer decides that they will not commence proceedings in respect of the breach but will serve a notice requiring remediation, it is not until they revisit at the end of the compliance period and discover that the requirements of the notice have not been satisfied that time begins to run on an allegation of non-compliance. In both cases, the date of discovery is clear and can be pinpointed – the date of the initial visit, or the date of the inspection to ascertain compliance.

The date of the discovery of an offence is the date upon which an investigating officer has reasonable ground to believe that an offence has been, or is being, committed.

PREPARING A CASE

Collecting evidence

You must first collect evidence of fact from witnesses, identify the reasons/causes of the breach or incident, then speak to those who may be responsible for the breach or incident, and finally gather interview information.

Evidence is all the material on which either party seeks to rely on to prove their case, i.e. it is what the court sees or hears (e.g. testimony or oral evidence, documents, objects, videos, photographs).

Gather all documents related to the cases so long as you can provide witnesses who can tell the court about their origin and content. Never write on original correspondence – using sticky notes that can be removed is preferable.

Appendix 3 shows a method of ensuring the relevant, admissible and fair evidence is collected for a particular offence. Each row should address an individual element of the offence that needs to be proved (first column). The second column shows the details of the offence element that you have collected. The remaining columns to the right are self-evident in that these are for the sources and references, ensuring that the evidence is well organised and can be used in court for efficient reference.

Disclosable documents

Disclosure is the process of informing the defence of unused material which has been recorded or retained by the authority and not disclosed in evidence capable of undermining the prosecution case or assisting the defendant, e.g. notebooks, draft versions of witness statements if different to the final version, interview records, communication with experts and any material casting doubt on the reliability of the evidence or a witness.

Each party to a case is required to inform the other side of the existence of all disclosable documents that are or have been in their possession or control.

Always keep your notebook tidy and neat as the relevant pages of officers' notebooks are disclosable. If the book is untidy and disorganised, and the court sees it, it can be used to suggest a less than professional attitude.

Internal office documentation must also be disclosed – file notes or telephone messages written on whatever was to hand when the call was received are part of the case file, and are therefore disclosable.

Anything that relates to a prosecution, or goes onto a case file, has the potential to be disclosed; therefore officers should ensure that nothing other than relevant information is included.

The process of disclosure requires the prosecutor to provide to the defence a copy of all relevant material pertaining to a case. The material is divided into:

- material upon which reliance will be placed;
- material which is disclosed but upon which reliance will not be placed;
- other material which exists, is relevant, but will not be disclosed.

Into this third class goes material for which legal professional privilege is claimed. Advice from in-house lawyers needs to be kept separate from the case notes so that it attracts legal professional privilege.

Exhibits

Give the court articles of real evidence, in the form of photographs, plans or maps of the location, and real 'things' that they can look at, weigh in their hands and scale against the rest of the scene. This is likely to influence the court more than witness statements.

Any physical evidence taken should be bagged and tagged, and then stored in a lockable evidence cupboard, to which access is restricted.

A register should be kept indicating:

- when the item of evidence was put in the cupboard;
- by whom it was put in the cupboard;
- if and when the item was taken out of the cupboard, by whom and for what purpose – for example, an item of food removed to be inspected by a company representative.

Photographs/videos

The role of the witness is to give evidence as to what was seen or heard; the court makes decisions of fact. Give short factual statements of what you saw to introduce a photograph or video. It is vital to let the photograph/video speak for itself. The rule is 'simply exhibit the photograph, and exhibit the photograph simply'.

Identification evidence

The leading case for identification is that of R v Turnbull.[11] In this case, the Court of Appeal laid down guidelines that should be applied where identification evidence is disputed.

Condensed, they are:

A Amount of time the suspect was observed by the witness
D Distance between suspect and witness
V Visibility at the time of observation
O Obstructions between witness and suspect
K Known previously (i.e. does the witness know the suspect)?
A Any significant reason for the witness to remember the suspect?
T Time elapsed since witness saw the suspect and description written down
E Error or discrepancies in the witness's description.

This is a list of all those points that a witness identifying a suspect *must* cover when making their identification, whether orally in the witness box or in a witness statement made as part of the case if their identification evidence is to be relied on. If one of these is not covered, the judge is obliged to warn the jury about the risks of relying on the identification evidence, and he may still do this even if all of the points are covered and the issue of identity is critical. It is a misdirection should he fail to do so and grounds for appeal.

Enforcement officers should subject their own identification evidence to this test to be sure that it will stand up in court.

Proving competence and authority

It is necessary to prove that the officer taking the case or who is the main witness is competent to do so. It needs to be shown that the officer holds a relevant qualification and is sufficiently competent in the particular field of the case. It may be necessary to show that knowledge and skills are held and practised, refresher training has been done and practical skills are up to date, as well as showing how knowledge is maintained.

Where newly qualified officers take a case, knowledge is the most up to date and skills, once obtained, can be practised often. Where officers practise within their field of expertise and keep their skills and knowledge up to date, it is very difficult to challenge competency successfully.

It is necessary to show that the officer is duly authorised and was acting within their authority (see page 27).

Witness statements

A witness statement is a document recording the evidence of a person, which is signed by that person to confirm that the contents of the statement are true. A statement should record what the witness saw, heard or felt. In the case of the EHP, this would include all of the exhibits that he collected in the course of the inquiry. You should rely on notes made at the time of the incident when making a witness statement. Lawyers expect to find a very close connection between the notes and the text of the witness statement.

Officers should consider producing a witness statement at timely intervals. Breaking the case up into manageable amounts makes the witness statement likely to be more reliable and therefore less vulnerable to challenge. Make sure in your witness statement you make it clear that you are familiar with the leading case, the current Code of Practice and the relevant guidance, i.e. the key legal background to the case. You should check all facts written in the statement are correct, with the correct date and supported by evidence. Never add or remove things from your statement. Describe behaviour and observable facts in the statement and avoid professional terms and jargon, using everyday language. Ensure any opinion stated is within your expertise and be aware of any inconsistencies.

Witness statements can be made under two pieces of legislation:

- Section 9 of the Criminal Justice Act 1967 states that the contents of a written statement will be admissible, without the witness attending court to give oral evidence.
- Section 20(2)(j) of the Health and Safety at Work Act 1974 gives officers the power to require any person whom the officer has reasonable cause to believe will be able to provide information relevant to his examination or investigation to answer such questions thought fit to ask, and to sign a declaration of the truth of the answers.

Witness statements should be on prescribed forms. They should describe the sequence of events in a logical order, usually chronological. They should be certified by the witness that the evidence is true. They should be short and the information in it should be relevant to the offence.

Witness statements from lay people

It is important to impress on witnesses that they should only say what they know to be true, and thereafter to leave it to the court to determine effect.

The first step is to make sure that witnesses know that they should not discuss the matter with each other. Quite clearly you cannot stop this happening, All you can do is make sure you advise them, in simple and understandable terms, not to talk to anyone about their evidence. Stress the importance of it being their evidence and remaining uncontaminated. Make sure they understand your advice. The second step is to impress on a witness who is about to provide a witness statement the need to ensure that they can give their own evidence of what they are about to write by highlighting that their evidence will not be accepted per se, but will be challenged in cross-examination. If the witnesses come under pressure in the witness box and concede that they have chatted with other witnesses about their evidence, on their head be it. They will also have to concede that you told them not to, either in cross-examination or under re-examination, so at the very least you can demonstrate you did try to prevent evidence contamination.

Can you write the statement for the witness? You can write down the statement, at the dictation of the witness, but you must only record what you are told. You may not add bits, and you should not sanitise the use of English to make the statement more readable. A defence lawyer can usually tell whether a witness is using his own words or words are being used for him, and any suspicion that a statement has been doctored or sanitised will lead to the witness who made it being called and cross-examined on it.

The first thing you can do is talk the witness through their evidence. Get them to go through what they saw, heard, etc. in chronological order. You cannot impress on them what is important and what is not because to do so is to affect their judgement, but at least they will have had an opportunity to rehearse what they will say before committing it to paper. If the witness does not remember something vital, you can push very gently – e.g. 'Did you not say to me … that may be worth recording.' You cannot remind them; only they can remember.

Where a witness writes their own statement, the rule is that you can guide the pen but you may not push it. It is perfectly acceptable to suggest to the witness that he writes his statement in chronological order because it makes sense to record the story as it unfolded. Given that you will have talked it through as suggested above, this will ensure that the witness has a clear picture in his mind, recalls as much as possible and records it in admissible form. When the witness has finished writing, you can read over the statement and point out that he may have forgotten to mention something he has told you orally, giving him the chance to include it, provided the exchange to which reference is made actually took place.

How much assistance can you give to a witness who is about to appear and give oral evidence for the authority? Coaching is not allowed. Officers must be very vigilant to ensure that nothing that they do or say can be construed as witness coaching.

Sub judice

Information that is relevant to a case that has not yet come before the court is *sub judice*, literally '*in the course of trial*', and must not be discussed in the public domain. It must be preserved intact, to be considered in court by a jury or lay bench who are hearing its contents for the first time, and can form a view as to its validity in context of the hearing and alongside all other relevant evidence.

The reason for the rule of *sub judice* is simple – it allows arbitrators to consider evidence on the basis of what they hear and see in court. Where information has been in the public domain, it is virtually impossible for an individual not to form a view about it themselves and for their own judgement not to be coloured by what they have heard from others.

Expert witnesses

Experts, and indeed environmental health practitioners, who are considering offering themselves as expert witnesses must be aware of the rules governing the conduct of expert witnesses and must abide by them. Failure to do so may damage the credibility of the expert witness in the eyes of the court, their client or others who may have considered engaging him. Expert witnesses hold a very privileged position in litigation, and with this comes responsibilities.

An expert witness may give the court evidence of his or her opinion, unlike a witness to fact, who is restricted to purely factual evidence. In order to give opinion evidence, the expert must know what each witness has said, and in order to facilitate him having this knowledge he is allowed to sit in the well of the court and listen to the evidence of the witnesses as to fact. He may then comment on that evidence when he gives evidence by, for example, indicating which scenario presented in the evidence is more likely, in his view, to be the case.

The expert witness has more freedom in court than the lay witness.

It is incumbent both on those who act as expert witnesses and on those who chose to offer expert witnesses in support of their cases to ensure that the restrictions of the role are understood.

Where an expert is asked a question or considers a proposition that takes him outside the core of his field of expertise, he should advise the court that he cannot give expert evidence on the point.

What is an expert witness?

An individual must be an expert in a particular and relevant field. Clear evidence of expertise in the form of qualifications, papers published, etc. must be provided to satisfy the court of expertise. The court would look for evidence that the expert was so regarded by his peers. The court will also consider whether her expertise is up to date – an individual who has not updated their knowledge will not be regarded as an expert. Equally, the court will want to know how much experience an expert has.

However, it is in the discretion of the court as to whether an expert witness is required.

The purpose of an expert witness is to help the court by providing an expert opinion based on the facts as presented by the lay witnesses, and if the court feels that it (or the jury) is capable of making a decision without such assistance, it will refuse to allow an expert witness to be called. If it does agree, the court also has to be persuaded that the person tendered as an expert witness is in fact someone who can be regarded as an expert.

Selecting an expert witness

It is important to strictly define that area in which the assistance of expert evidence is required.

It is necessary to be clear about the exact areas of evidence upon which expert opinion is required and to ensure that the expert chosen has sufficient expertise to consider all of them.

It should be remembered that a particular expert's field of expertise may not necessarily be as wide as the area upon which expertise is required, and in some cases more than one expert may be required.

Ensure that the court will accept the expert chosen:

- Is his expertise is up to date?
- Is his expertise mainly theoretical or does he have field experience?
- Is he mainstream and does he speak as the majority in this field or not?

Consider whether you can understand your expert:

- Is what your expert has to say comprehensible to the court and the jury?
- Can your expert speak so that lay jurors or magistrates understand him and will be persuaded by what he says?

Take the advice of your expert:

- Does he agree with your evidence and your methods?
- Are there more tests he would recommend?
- Does he suggest another approach?
- Does his opinion conflict with your evidence?
- Does your expert know what is expected of him?

Expert rules

An expert is there to assist the court and his overriding duty is to the court, not to the party calling him. He must stay within his area of expertise and must inform the court if he speaks on matters outside that, when his evidence will not be treated as expert evidence.

It is incumbent on him to make it clear to the court when he speaks outside his area of expertise, since that is within his knowledge more than anyone else's.

The expert is required to refer to what facts and assumptions he has used and made when coming to his conclusions and also to state what material facts do not support his conclusions.

Where the expert has not been able to research the issue fully because, for example, insufficient data is available or because the theories relied upon may not be widely accepted, he should note that his view is provisional.

If, after exchange of statements, the expert changes his mind on a material matter, he should tell the other side and the court as soon as possible.

Photographs, plans, surveys and other documents relied on by the experts in coming to their stated views must be exchanged at the same time as the reports, so that each side can come to a fully advised view of the other's position.

There is an exception where the law says the relevant officer can act on an opinion. EHPs are lay witnesses and can only give evidence of fact unless the law says otherwise.

Importantly, the EHP will give opinion evidence on only one point: whether, in his or her opinion and based upon all of the factual evidence he or she gives as a lay witness, something is prejudicial to health or is a nuisance, for example.

However, EHPs can be regarded as expert witnesses, but only if they have specialised in a particular field and have reached a very high level in that field.

Professional witnesses

In cases where it is perceived that there may be some difficulty persuading lay persons to give evidence, local authorities can consider the use of professional witnesses.

Professional witnesses, usually from an investigation agency, undertake monitoring, as well as ensure that the evidence gathered will be admissible and available. In such cases, the cost of engaging the witnesses can legitimately be claimed as investigation costs when the matter goes to court.

Provided the witnesses are properly briefed as to what the investigation requires, and can be demonstrated to have no interest in the outcome of it, they are a more certain source of evidence than lay persons.

Interviews under caution (IUCs)

Before carrying out interviews, you should prepare well. Prepare your questions in advance. Ensure you have the ability to record the questions and answers in a notebook, interview record form or tape recorder.

The right to legal advice

An interview under caution is evidence and is admissible for the truth of its content in court.

When a person attends an interview voluntarily, he must be informed about his continuing right to consult privately with a solicitor and that free independent legal advice is available.

It is essential he is properly and appropriately advised by someone who not only understands the law but understands the way in which it works in the context of an interview under caution.

Why carry out an interview under caution?

IUCs are governed by the requirements of PACE Code of Practice B.

A prosecutor carries out an IUC when he has *prima facie* evidence to suggest that an offence has been committed and that the interviewee is either the sole accused or has played a part in a multi-handed offence.

Interviews under cautions are not necessary if you already have enough evidence to establish a *prima facie* case against the proposed defendant. Conducted properly, an IUC should fill gaps in the inquiry by establishing the role of the accused, providing the interviewer with details not known at the time of the offence or, in some cases, allowing the interviewee to clear himself and eliminate himself from the inquiry. An IUC should not be done until the obvious ways of obtaining evidence have been exhausted.

Ending an interview under caution

The interview must cease when 'the officer in charge of the investigation … reasonably believes there is sufficient evidence to provide a realistic prospect of conviction for that offence'.

Interviewing officers conducting IUCs need to carefully consider every answer that they get and think about the cumulative effect of the answer. If that answer gives them enough evidence to believe that they have a reasonable prospect of a successful prosecution, they must stop the interview. There is no discretion to do otherwise and the case will not be assisted if they do.

At the end of the interview, invite the interviewee to sign the notes or the seal on the tape. Record the time the interview started and ended, any breaks taken and all persons present.

Evidence of previous condition and failure to comply

To introduce the previous nature of the premises in a prosecution is to open a witness statement with a sentence such as 'XYZ premises is rated as Risk

Category 1 because of previous issues with …'. Risk assessment-based inspection is standard practice and there should be no problem in introducing these facts.

You can also include a potted inspection history to the statement as an exhibit. For example, 'This visit was the third within a period of x months. The inspection history of the premises is attached as Exhibit AB1', but no further detailed comment on the inspection history should be made.

Claiming costs

'Where … any person convicted of an offence before the Crown Court, the Court may make such order as to the costs to be paid by the accused to the prosecutor as it considers just and reasonable' (Prosecution of Offenders Act 1985 s.18(1)c).

Provided a local authority can show that its claim for costs is just and reasonable, there is no *prima facie* reason for the costs not being awarded. Costs start to run when the environmental health practitioner enters the premises on a routine or investigatory visit. It is important when collating a costs application to accompany a prosecution that everything that can justifiably be claimed as costs of the investigation is included.

A complete schedule of costs should include:

- costs of the time spent by the environmental health practitioner in the investigation;
- administrative staff costs in typing notices, making copies and transcribing interview notes;
- costs of photo processing and analysts' cost;
- costs incurred in service of documents by, for example, personal service or recorded delivery post;
- any conferences or meetings with the council's legal team should be costed, with the costs of both environmental health practitioner and lawyer being included in the claim;
- telephone calls relating to the matter, whether internally, with other officers of the council or with the defendant or his legal representative;
- expenses that are incurred in attending on witnesses and taking statements etc.
- costs of environmental health practitioner, lawyers and witnesses in attending at the hearing.

Try, so far as you can, to estimate what the defendant may claim and cover it off.

For example, if he has had food hygiene training, advise your lawyer. He can then refer to the training and the knowledge that the defendant should have in consequence of it in his address to the court if there is a plea of guilty tendered or in cross-examination if the matter goes to trial.

Include the time spent advising and assisting the defendant. This has a dual purpose: it undermines any claim that he did not know and could not be expected to know what was required, and further rebuts any claim that the council has been unhelpful or uncooperative.

If you can show that he has made substantial profits from his activity, advise the prosecutor to give details of your calculation to the court – it will have to be either accepted or denied and if denied will have to be explained.

Give your lawyer as much background information as you can and make sure potential 'areas of dispute' are addressed either in the facts of the case or during cross-examination.

If the defence, having been sent a copy of the prosecution's application for costs, do not tell the prosecutor or the court that they propose to object to some or all of them before the verdict is delivered, they may not raise objection to any part of the costs application.

Local authorities are required by rules deriving from the *Associated Octel* judgement to serve their schedule of costs on the defendant at the earliest opportunity, in order that the defence may decide what parts, if any, they wish to challenge as unjust or unreasonable. (R v Associated Octel[12])

Victim personal statements

Victim personal statements (VPS) give victims an opportunity to say how the crime has affected them, information which the judge or magistrates can take into account when imposing sentence. Details of their use can be found in the Consolidated Criminal Practice Direction – 8 July 2002.

They can be of considerable assistance to the court when considering sentence in a case where there has been a workplace death or serious injury in outlining the human effects of what appears to be a 'mere' breach of a technical regulation or requirement.

There are a number of rules regarding collation of a VPS:

1. The victim must be told about the scheme and offered the opportunity to make a VPS before the final disposal of the case.
2. The victim cannot be compelled to give a VPS; victims must wish to give such a statement. VPSs are often intensely personal statements and not every victim wishes to go through the process of quantifying the effect of the incident on their lives and their thoughts for the future at what is a very traumatic time for them.
3. Should the victim wish to make a VPS, it must be served in the proper form, being either in the form of a witness statement (section 9 Criminal Justice Act 1967) or contained in an expert's report – for example, when the witness has been receiving counselling or psychiatric help consequent to the incident.

The VPS, once made, is served on the defendant or his solicitor and on the court prior to sentence. The court may then take account of the effects of the incident on the victim when considering sentence.

Local authorities proposing to take a VPS should consider seeking professional assistance either from the police or from trained counsellors to ensure that the statement is taken properly and is a full reflection of the victim's position, but also to ensure that the process is no more traumatic for the victim or the person taking the statement than it needs to be.

Evidence suggests, however, that victims who do produce a VPS feel that their views are taken into account, and that the sentence reflects this. Where a VPS would be appropriate, local authorities should assist the court and the victim by obtaining and using them.

VPS can also apply to environmental crime – for example, on costs of clear-up, effect on the local community, short- and long-term effects on the environment and those affected by the damage. The Environment Agency can give guidance on costs and timescales of recovery. Also considered should be local tourist boards (in cases of oil spills onto holiday beaches or areas of outstanding natural beauty, the wider tourist industry in an area may suffer), chambers of trade and local interest groups.

The prosecutor should speak on behalf of the environment as a victim and should put before the court details of all of the damage caused by environmental crime and all of the impacts, short and long term, on the environment and to all of those who are affected by damage to the environment.

Sentencing

It will help the court greatly, and shorten hearings, if a case summary can be supplied by the prosecutor at the time of the issue of proceedings. This document should set out the facts on which the prosecution rely and contain aggravating and mitigating features. These are known as Friskies Schedules and have their origin in a prosecution brought by the Health and Safety Executive.[13] In this case, the Court of Appeal made strong recommendations about presentation of health and safety cases, which should assist the judge to understand the aggravating and mitigating features of the case and to pass an appropriately tailored sentence.

The rules are:

1. The prosecutor should list, in writing, the facts of the case and the aggravating features that he says are present.
2. This list should be served on the defendant and the court.
3. In the event of the defendant pleading guilty, the defence should submit a similar document, but in their case outline those mitigating features they say are present.
4. If the prosecution and the defence agree on the aggravating and mitigating features, they should commit this to writing and put the agreement to the court. Where there is no agreement, the judge will hear argument on the disputed points, form a view and advise the parties of his view, which would be relevant in any appeal against the sentence handed down.

The Court of Appeal has also declared what constitutes aggravating and mitigating factors in health and safety cases, in the case of Howe and Son.[14]

The court stepped outside the facts of the case to list other features that might be relevant to sentence, but noted that, depending on the circumstances of the case, it would be for the court to decide whether they were aggravating or mitigating, and to what extent. These included degree of risk and extent of danger (i.e. whether death or injury was foreseeable), gravity and extent of breach, whether it was an isolated incident or continued over a long period, and the defendant's resources and the effect of a fine on its business, but it was noted that this list is not exhaustive.

Whilst both the cited cases were brought under health and safety legislation, it will assist the court to take a similar approach irrespective of the legislation under which proceedings are brought. Applying the same principle by providing information that will assist the magistrates or judge when considering sentence should ensure that the sentence is tailored to reflect the degree of criminality involved in the offence, which is in the interests of prosecution, the defendant and the general public's belief in the criminal justice system.

Laying informations

A prosecutor who wants the court to issue a summons must serve an information in writing on the court officer or, unless other legislation prohibits this, present an information orally to the court, with a written record of the allegation that it contains (Criminal Procedure Rules 2011 7.2).

An information is a document which states in clear and unambiguous language the offence with which the accused is charged. There should be only one offence in the information. You should accurately describe the name and address of the accused and the specific offence with which the accused is charged in ordinary language, avoiding technical terms. It should include the appropriate section of the Act or regulation contravened. The information should also include the date of offence where possible. If the date is not accurately known, then the phrases 'on or about' or 'between date 1 and date 2' can be used. Finally, refer to the section where the penalty for the offence is contained.

Briefing solicitors or Counsel

Counsel is required to keep a distance from the client – he or she can only be contacted via the instructing solicitor. Solicitors also need to be briefed before they come to court. It is therefore important that the brief contains all of the necessary information, and copies of all of the relevant documents, as well as clear and unequivocal advice as to what the client wants.

Solicitors or Counsel may have no knowledge at all of the case or the facts surrounding it and have to be told everything, particularly those things that, as professionals, EHPs consider to be taken as read, or common knowledge, because for someone outside the field of environmental health they are not.

If you want Counsel to attend at court and represent the local authority, it is necessary to be clear and to say so. It is also necessary to underline what else you want – for example, that an application for costs (to include Counsel's reasonable fees) should be made, that a Confiscation Order should be sought, etc. and to advise what would be acceptable by way of plea bargain, should it be offered.

Where Counsel is asked to advise, it is essential to be clear about what advice you actually want.

It is important to include some practical information which solicitors or Counsel would be keen to have, such as the fact that the lay witness is reluctant and very nervous, or that the defendant has a violent temper. This allows them to be ready for whatever may happen, and to plan accordingly.

Proving the case

In a court case, all that is necessary is to know what you have to prove and to show, by way of evidence, that you can do so.

It is very important to present all the relevant facts to the court. The courts want a presentation of all of the relevant facts – for example, in the case of environmental damage, the court should be made properly aware of environmental damage and understand how long the environment would take to recover from a particular incident, or the true cost of remediation. A plea of guilty should not prevent all of the relevant facts coming before the court, with sufficient detail, so that a bench of lay magistrates can understand the complexity of the damage caused by the incident and the scale and cost of remediation.

The information laid before the court has to be comprehensive and understandable, and, for example, in the case of environmental damage, it should take account of the following features, which demonstrate aggravating damage:

- adverse effects on human health, flora and fauna;
- noxious and persistent damage by pollutants;
- the need for extensive clean-up operations; and
- the effect of the incident on lawful activities.

Previous convictions of a like nature are also aggravating features.

Any financial gain to the defendant as a result of the incident should be demonstrated as it should be reflected in the fine, such that polluters cannot make enough from their activities to pay any fine and still be in pocket as the result of their activities.

The court wants:

- projections of clean-up costs and times, particularly since the cost of clean-up all too often falls to publicly funded bodies;
- evidence of the polluting incident, e.g. colour photographs or video footage can have a significant effect on those who do not have regular contact with such incidents;
- estimates of the financial benefit to the defendant (these are very valuable).

Prosecution file

The following list is an example of what could be contained in a prosecution file:

- file front sheet to include:
  - defendant's name and address, date of birth, gender, occupation, nationality
  - defendant's previous convictions
  - alleged offences with details of offence and relevant legislation
  - date of alleged offence

    - date of laying information
    - officer's authorisation details
    - case summary
    - evidence – witness statements, photographs and physical evidence
    - witness reliability
    - whether evidence is admissible, substantial and reliable
    - whether there is a reasonable prospect of conviction
    - whether it is in the public interest to prosecute;
- initial witness assessment (whether witnesses are vulnerable and require support);
- witness non-availability;
- witness statements;
- exhibit list, including documents;
- record of interviews;
- unused material likely to undermine the case;
- schedule of prosecution costs;
- officer's contemporaneous notes.

If there is a not guilty plea, add:

- schedule of sensitive material;
- schedule of non-sensitive unused material;
- disclosure report;
- update on witness non-availability.

COURT PROCEDURE

The magistrates' court is the usual court in which local authorities take prosecutions. There are three lay magistrates sitting in court and they are assisted by the clerk of the court. The clerk of the court sits in front of the magistrates and gives them legal and procedural advice. The clerk also deals with general administrative work.

When in court, before the session starts, you should stand up when the magistrates or judge enter the court. The magistrates will normally bow and you should bow immediately afterwards. Sit down after the magistrates sit down.

The clerk will read the charge to the defendant and the defendant will plead guilty or not guilty.

Where the defendant pleads guilty, the prosecuting lawyer will read a statement of facts relating to the case. This includes submissions of the seriousness of the case, notice of the offence, description of the circumstances giving aggravating and mitigating circumstances, costs and compensation claims. The defendant's solicitor will then plead mitigation. The magistrates will then normally impose a sentence but they may ask for more information before passing sentence.

Where a not guilty plea is made, the prosecution will present its case, calling witnesses, and the defendant's lawyer will cross-examine those witnesses. Then the defendant's lawyer presents this defence, including witnesses. The prosecution lawyer will cross-examine the defendant's witness.

An EHP who has given evidence and sits in court may find the evidence of the defendant incredulous, or the comments of the defence solicitor in mitigation farcical, and may feel frustrated in the extreme that he or she cannot counter it. The temptation to roll your eyes or shake your head in disbelief is considerable, as it may be seen as the only way in which the evidence being advanced can be challenged. The temptation must be resisted because behaviour in court that may influence the magistrates or jury can form the ground for appeal. Be impassive to the antics of the defence.

Giving evidence

The prosecuting lawyer will ask the main witness, probably the EHP taking the case, questions about themselves and the details of the offence or offences based on the witness statement. When answering the lawyer, face the bench or the jury to ensure they understand what you are saying as they will be making the decision on the case. In the Crown Court, make eye contact with one or two of the jury. If they think the witness is looking at them, they concentrate. Crisp, sharp evidence engages juries.

If you are asked questions by the magistrates, address them collectively as 'your worships' and individually as 'Sir' or 'Ma'am'. Crown Court judges should be addressed as 'My Lord' or 'My Lady'. Appear serious, caring, flexible, well-informed, fair and reasonable in giving evidence.

When giving oral evidence, it is absolutely critical that the witness says enough to establish that those things they are required to prove are indeed demonstrated, and that this is done sufficiently robustly to rebut any challenge that may be put in cross-examination.

When giving evidence in the magistrates' court, watch the clerk. If he or she is writing, they consider the point sufficiently important to record it. If they are not, they do not, so stop. Slow down if the clerk seems to be writing too fast and may be missing some important points. The same applies in the Crown Court and in other forms of administrative tribunal – if the judge or the inspector is writing, the point is important; carry on, slowly. If they are not, stop.

Watch your own lawyer. If they are nodding encouragingly, keep going. If they look as if they want to speak, let them. The rule is to keep it short, keep it simple and keep it relevant.

When it comes to language in court, the rule is no acronyms, no jargon and no unexplained use of technical terminology.

Juries like clear, unfussy information in understandable chunks and not embellished with unnecessary details. The witness needs to get the message across as one who speaks their language.

Opinion

Opinion is a judgement or belief not founded on certainty or proof, or an evaluation or judgement given by an expert. Opinion is the view you hold – the judgement you form, which need not be certain or proved to be true. If an officer can show he is properly informed and has considered his view, his opinion is valid, and he is entitled to act on it. Some legislation, such as the Health and Safety at

Work Act 1974, allows officers to act on opinion. See expert witnesses and opinion (pages 44–46).

Role of the lawyer

The role of the local authority lawyer is to take the local authority's instructions and act so as to give effect to those instructions.

The lawyer must always advise the local authority of the likely outcome of the proposed action, and the likely outcome of any alternatives that he may propose, but at the end of the day it is always for the client (local authority/EHP) to decide which action should be taken.

The instructing department must be free to exercise their professional judgement as to the disposal of matters. The role of the legal department is to follow their clients' instructions. The role of the instructing department is to take legal advice, consider it and then make a decision, taking account of all of the relevant considerations.

NOTES

1. Elliott v Brighton [1980] 79 LGR 506.
2. Swansea City Council v Glass [1992] 3 WLR 123.
3. Kensington & Chelsea RLBC v Khan [2002] EWCA civ 279.
4. Graddage v Haringey LBC [1975] 1 WLR 241, and Sterling Homes (Midlands) Ltd v Birmingham CC [1996] Env. LR 121.
5. R v Mansfield DC ex parte Ashfield Nominees Ltd [1999] E.H.L.R. 290.
6. Swindon BC v Forefront Estates Ltd [2012] EWHC 231.
7. Paddington Borough Council v Finucane [1928] Ch567, Bristol CC v Virgin [1928] 2 K.B. 622 and Westminster CC v Haymarket Publishing Ltd 1[981] 1 W.L.R. 677.
8. R v Johnson [1988] 1 WLR 1377.
9. Beardsley v Giddings (1904) 1KBD847.
10. R v Pontypridd Juvenile Court (unrep).
11. 1976 63 Cr Ap R 132 [1977] QB 224(CA).
12. R v Associated Octel Co Ltd. (Costs) TLR Nov 15 1996 646.
13. R v Friskies Petcare Ltd (2000) 2 CAR(S) 401.
14. R v Howe and Son (Engineering) Ltd [1999] 2 All ER 249.

APPENDIX 1

Local authority definitions

The definition of a local authority is very similar in all Acts of Parliament and regulations. The references for their definition are below.

- Animal Boarding Establishments Act – section 5(2)
- Breeding of Dogs Act 1973 – section 5(2)
- Building Act 1984 – sections 91 and 126
- Caravan Sites and Control of Development Act 1960 – section 29(1)
- Clean Air Act 1993 – sections 55 and 64
- Control of Pollution Act 1974 – section 73(1)
- Criminal Justice and Public Order Act 1994 – section 77(6)

- Dangerous Wild Animals Act 1976 – section 7(4)
- Environment Act 1995 – sections 91, 82 and 105
- Environmental Damage (Prevention and Remediation) Regulations 2009 – regulation 2
- Environmental Permitting (England and Wales) Regulations 2010 – regulation 6
- Environmental Protection Act 1990 – section 78A(9) – Contaminated land
- Environmental Protection Act 1990 – section 4 and section 30 (waste authorities)
- Food Safety Act 1990 – section 5, and Food Hygiene (England) Regulations 2006 – regulation 2
- Health and Safety at Work etc. Act 1974 – section 53
- Housing Act 2004 – section 261
- Licensing Act 2003 – section 3
- Local Government (Miscellaneous Provisions) Act 1982 – sections 13(11) 20(1) and 44(1) 29(4)
- National Assistance Act 1948 – sections 33 and 64
- Noise Act 1993 – section 11(1)
- Notification of condensers etc. – regulation 2
- Pet Animals Act – section 7(3)
- Prevention of Damage by Pests Act 1949 – section 1
- Public Health Act 1936 – sections 1(2) and 91
- Public Health (Control of Diseases) Act 1984 – section 1
- Refuse Disposal (Amenity) Act 1978 – section 11
- Riding Establishments Act – section 6(4)
- Sunday Trading Act 1994 – section 8
- Water Industry Act 1991 – section 219

APPENDIX 2

Service of notices references

- Building Act 1984 – sections 92 and 94
- Control of Pollution Act 1974 – section 105
- Environment Act 1995 – sections 123 and 124
- Environmental Permitting (England and Wales) Regulations 2010 – regulation 10
- Environmental Protection Act 1990 – section 160
- Food Safety Act 1990 – sections 49 and 50
- Health and Safety at Work etc. Act 1974 – section 46
- Housing Act 2004 – section 246
- Licensing Act 2003 – section 113
- Local Government Act 1972 – sections 233 and 234
- Prevention of Damage by Pests Act 1949 – section 10
- Public Health Act 1936 – sections 283 to 285
- Public Health (Control of Diseases) Act 1984 – section 60
- Water Industry Act 1991 – section 216

APPENDIX 3

Evidence form – Section <XX> Act <Year>

Ref.	*Offence element (point to prove)*	*Details of the offence element Case law/definitions/ interpretation*	*Evidence of the offence element*			
			Name of witness giving evidence of the element	*Relevant exhibit no. (if any), witness initial and number*	*Page number and paragraph number of witness statement*	*Page number of interview under caution*
1	Name					
2	Address					
3	Date and time of offences					
4						
5						
6						
7						
8						
9						
10						

You must ensure that there is sufficient evidence to prove that each element of the offence has been committed.
The evidence must be referred to in a witness statement for it to be admissible in court.
The exhibit(s) must be attached to witness statements for it to be admissible in court.
Taken together, the witness statements must cover every element of the offence – so some witnesses may prove some elements and others may prove other elements or reinforce the same elements where necessary.

Chapter 2

ENVIRONMENTAL PROTECTION

CLEAN AIR ACT 1993
GENERAL PROCEDURAL PROVISIONS

Extent

These procedures are applicable in England, Wales and Scotland, with the substitution of authorities and bodies appropriate to the structure of government in that country, but not to Northern Ireland (section 68).

Power of entry

An authorised officer of the local authority may at any reasonable time enter any land or vessel in order to:

(a) perform functions conferred by the Act;
(b) determine if these functions should be performed; or
(c) carry out inspections, measurements and tests and to take away samples.

This power does not apply to entry into private dwellings except in relation to works of adaptation of fireplaces in a Smoke Control Area (SCA) where a notice had been served under section 24(1) and seven days' notice of entry has been given. Where:

(a) admission has been refused; or
(b) refusal is apprehended; or
(c) the land or vessel is unoccupied; or
(d) the occupier is temporarily absent; or
(e) the case is one of emergency; or
(f) application for admission would defeat the object of entry

the local authority, after giving (except in emergency) seven days' notice to any occupier, may make application to a Justice of the Peace by way of sworn information in writing. The Justice of the Peace may by warrant authorise entry, if necessary by force (sections 56 and 57).

The penalty for obstruction of an authorised officer is a fine not exceeding level 3 (section 57(6)).

Crown premises

Although the procedures in this chapter cannot be applied to Crown premises, the local authority is required to bring the following situations to the attention of the responsible minister where the premises or vessel concerned is Crown property:

(a) dark smoke or grit and dust outside a SCA;
(b) any smoke within a SCA;
(c) any smoke nuisance;
(d) any dark smoke from a vessel.

The minister must inquire into the circumstances and take all practicable means of action (section 46).

Relationship to the Environmental Protection Act 1990 and the Pollution Prevention and Control Act 1999

The procedures in this chapter cannot be applied to a prescribed process under the Environmental Protection Act 1990 and activities regulated under the Pollution Prevention and Control Act 1999 (section 41 and 41A Clean Air Act 1993).

Definitions

Black smoke means smoke which, if compared in the appropriate manner with a chart of the type known … as the Ringelmann chart, would appear to be as dark as or darker than shade 4 on the chart (Dark Smoke (Permitted Periods) Regulations 1958 Reg. 2(2)).

Chimney includes structures and openings of any kind from or through which smoke, grit or fumes may be emitted and, in particular, includes flues, and references to the chimney of a building includes references to a chimney which serves the whole or part of a building but is structurally separate from the building (section 64(1)).

Dark smoke means smoke which, if compared in the appropriate manner with a chart of the type known … as the Ringelmann chart, would appear to be as dark as or darker than shade 2 on the chart (section 3(1)).

Fireplace includes any furnace, grate or stove, whether open or closed (section 64(1)).

Fumes means any airborne solid matter smaller than dust (section 64(1)).

Practicable means reasonably practicable having regard, among other things, to local conditions and circumstances, to the financial implications and to the current state of technical knowledge and 'practicable means' includes the provision and maintenance of plant and its proper use (section 64(1)).

Prejudicial to health means injurious, or likely to cause injury, to health (Public Health Act 1936 section 343(1)).

Private dwelling, except as far as the context otherwise requires, means any building or part of a building used or intended to be used as such, and a building or part of a building shall not be deemed for the purposes of this Act to be used or intended to be used otherwise than as a private dwelling by reason that a person who resides or is to reside in it is, or is to be required or permitted to reside in it, in consequence of his employment or holding of an office (section 64(4)).

Smoke includes soot, ash, grit and gritty particles emitted in smoke (section 64(1)). Exempted appliance and authorised fuels lists are included in the following web links: http://smokecontrol.defra.gov.uk/appliances.php and http://smokecontrol.defra.gov.uk/fuels.php

DEFRA aims to commence a formal consultation process of the Clean Air Act in summer 2013, followed by a second public consultation containing detailed proposals in 2014.

HEIGHT OF CHIMNEYS SERVING FURNACES

References

Clean Air Act 1993, sections 14 and 15.
Clean Air (Height of Chimneys) (Exemption) Regulations 1969.
Clean Air (Units of Measurement) Regulations 1992.
HMIP Technical Guidance Note D1 (01).

Extent

These provisions apply in England, Scotland and Wales.

Scope

This procedure applies to chimneys attached to furnaces used to burn:

(a) pulverised fuel; or
(b) any other solid matter at a rate of 45.4 kilograms/h or more; or
(c) any liquid or gaseous matter at a rate of 366.4 kilowatts/h or more (section 14(1) and (2)).

Approval of chimney height

In relation to chimneys from these furnaces, the occupier of a furnace of a building or the person having possession of any fixed boiler or industrial plant outside of a building, must not use that furnace unless the height of its chimney has been approved by the local authority and any conditions of approval are complied with (sections 14(2)–(5)).

In relation to furnaces served by a chimney which was constructed before 1 April 1969, this requirement only applies where the combustion space of the furnace has been increased since that date or the furnace served by that chimney has been replaced by a larger one (schedule 5, paragraph 7).

FC3 Height of chimneys serving furnaces – Sections 14 and 15 Clean Air Act 1993

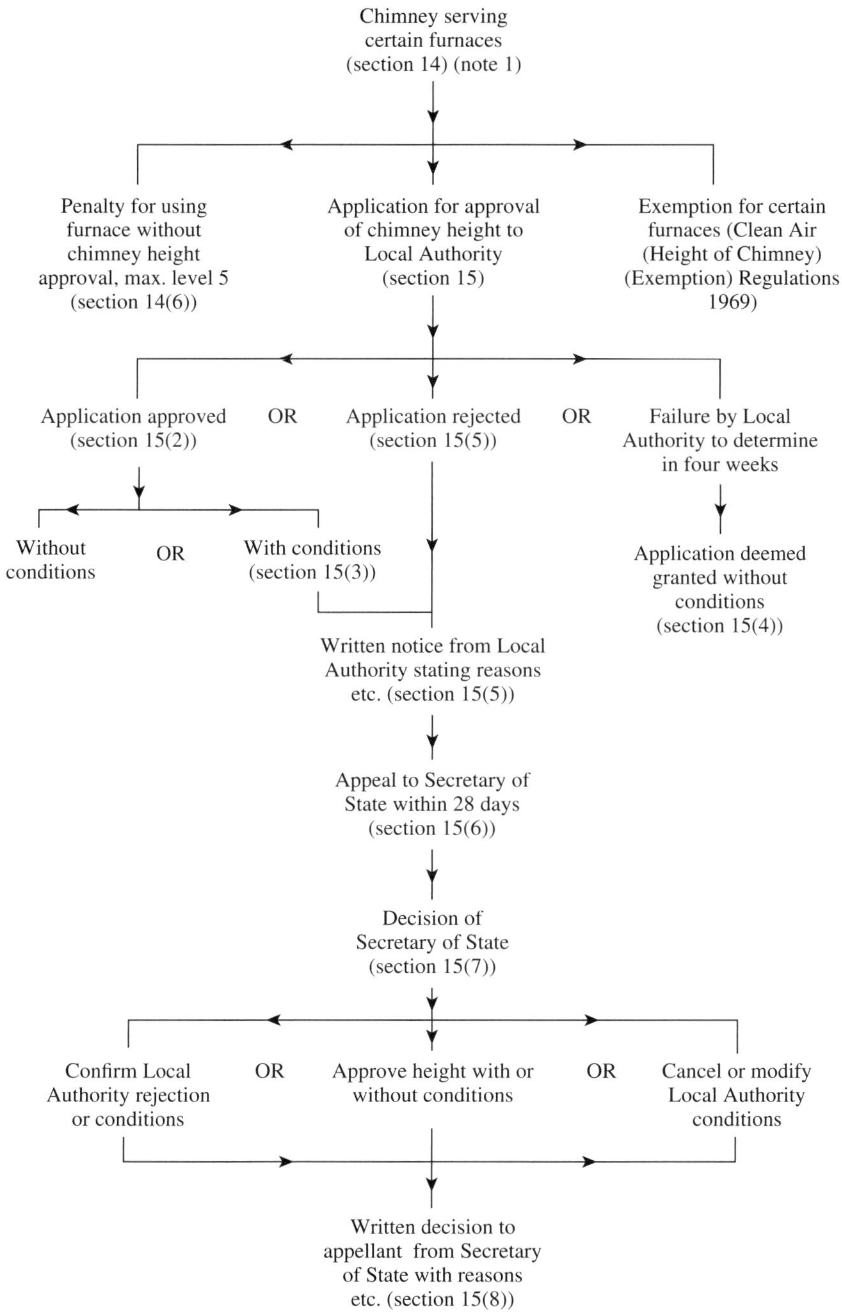

Note

1. For height of chimneys not serving furnaces, see FC4.

Exemptions

The Clean Air (Height of Chimneys) (Exemption) Regulations 1969 specify the following uses of boilers or industrial plant where chimney height approval is not required:

(a) temporarily replacing any other boiler or plant which is under inspection, maintenance or repair, or is being rebuilt or replaced;
(b) providing a temporary source of heat or power during building operations or for investigation and research;
(c) providing products of combustion to heat other plant to an operating temperature;
(d) providing heat or power for mobile agricultural plant (regulation 3).

Criteria for decision

The local authority is required to reject the application unless it is satisfied that the height of the chimney will be sufficient to prevent, as far as practicable, the smoke, grit, dust, gases or fumes from becoming prejudicial to health or a nuisance having regard to:

(a) the purpose of the chimney;
(b) the position and description of nearby buildings;
(c) the levels of neighbouring ground;
(d) any other matters requiring consideration in the circumstances (section 15(2)).

Refusals and conditions

If the application is not approved or conditions are attached to approval, the local authority is required to give written notification of its reasons to the applicant within four weeks of the application. Conditions may relate to the rate and/or quality of the emissions from the chimney. Where the application is rejected, the local authority is also required to state the lowest height which it is prepared to accept conditionally, unconditionally or both (section 15(4) and (5)).

Similar requirements are placed upon the Secretary of State in relation to decisions following appeals (section 15(7) and (8)).

HEIGHT OF CHIMNEYS NOT SERVING FURNACES

References

Clean Air Act 1993, section 16.
HMIP Technical Guidance Note D1 (01).

Extent

These provisions apply to England, Scotland and Wales.

FC4 Height of chimneys not serving furnaces – Section 16 Clean Air Act 1993

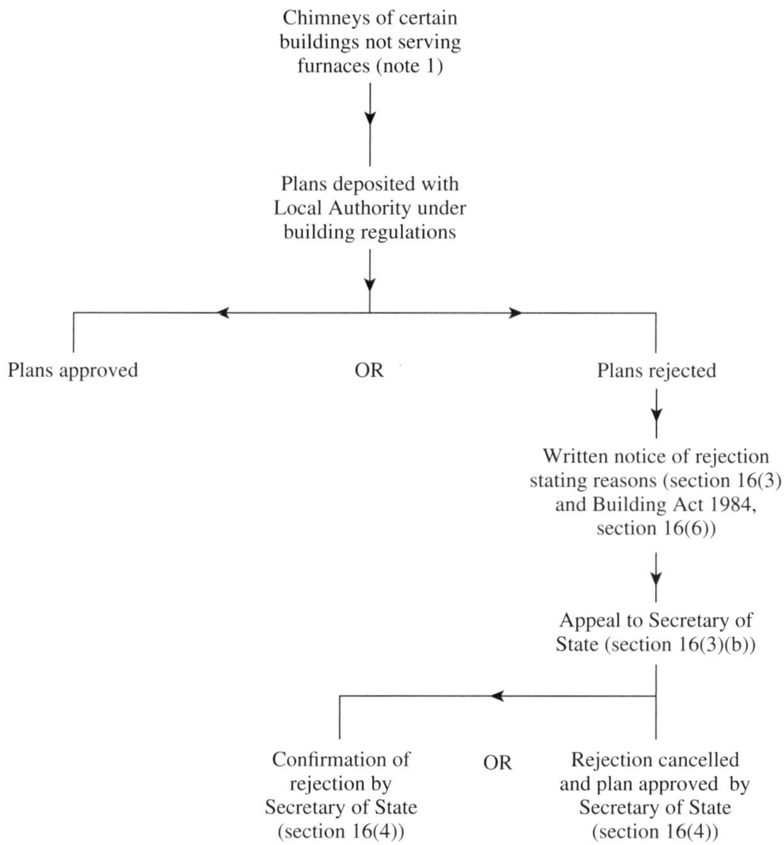

Note

1. For height of chimneys serving furnaces, see FC3.

Scope

This procedure applies where plans deposited under building regulations for buildings other than:

(a) residences; or
(b) shops; or
(c) offices

show the proposed construction of a chimney, other than one serving a furnace for carrying smoke, grit, dust or gases from a building (section 16(1)).

Where plans are not deposited because the 'building notice' procedure is being used, the application of this chimney height approval procedure may be dealt with as a linked power.

Criteria for decision

The local authority is required to reject the plans unless it is satisfied that the height of the chimney will be sufficient to prevent, as far as practicable, the smoke, grit, dust or gases from becoming prejudicial to health or a nuisance having regard to:

(a) the purpose of the chimney;
(b) the position and description of buildings nearby;
(c) the levels of neighbouring ground;
(d) any other matters requiring consideration in the circumstances (section 16(2)).

CONTROL OF GRIT AND DUST FROM FURNACES

References

Clean Air Act 1993, sections 5 to 9.
Clean Air (Arrestment Plant) (Exemption) Regulations 1969.
Clean Air (Emission of Grit and Dust from Furnaces) Regulations 1971.
Clean Air (Units of Measurement) Regulations 1992.

Extent

These provisions apply to England, Scotland and Wales.

Requirement to fit arrestors

Furnaces, except those designed solely or mainly for domestic purposes and with a maximum heating capacity of less than 16.12 kilowatts, used in buildings to burn:

(a) pulverised fuel; or
(b) any other solid matter at a rate of 45.4 kilograms/h or more; or
(c) any liquid or gaseous matter at a rate of 366.4 kilowatts/h or more

FC5 Control of grit and dust from furnaces – Sections 6 to 9 Clean Air Act 1993

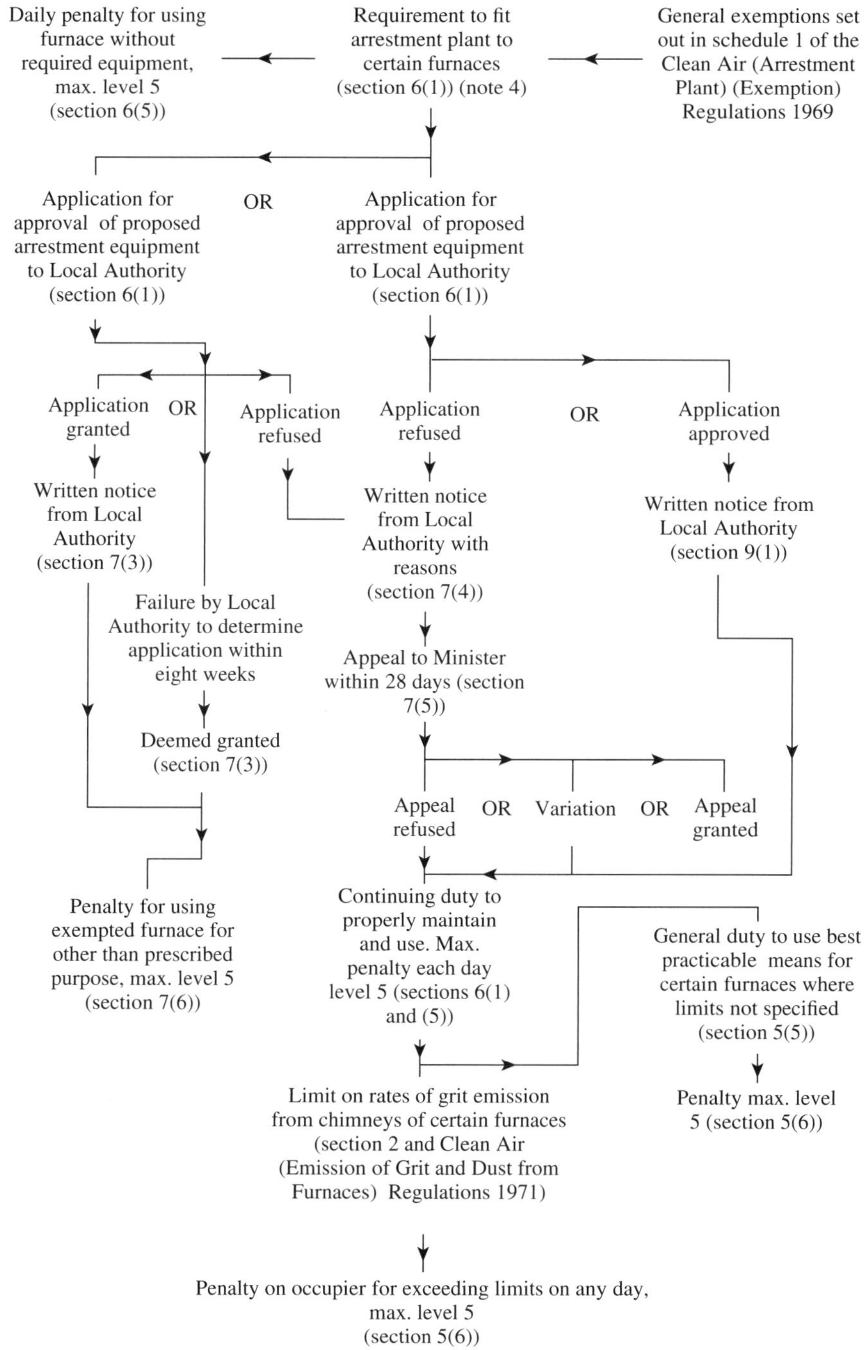

Notes

1. For grit, dust etc. emitted in smoke constituting statutory nuisance, see FC121.
2. For grit, dust etc. not emitted in smoke causing statutory nuisance, see FC121.
3. For measurement etc. of grit and dust, see FC6.
4. Local authority's powers to obtain information about furnaces and fuel under section 12 apply to this procedure.

are to be provided with plant for arresting grit and dust which has been approved by the local authority and be properly maintained and used (section 6(1)).

Domestic furnaces burning pulverised fuel or solid fuel at a rate of 1.02 tonnes/hour or more are also required to be provided with grit arrestment plant approved by the local authority (section 8(1)).

Exemptions

Exemption from the requirement to fit arrestment plant may be by either:

(a) falling into the categories of exempted furnace set out in schedule 1 Clean Air (Arrestment Plant) (Exemption) Regulations 1969, which also set out the information to be provided in the application for exemption; or
(b) specific application for exemption to the local authority for furnaces falling outside the regulations in (a) above, in which case the local authority will need to be satisfied that the furnace can be operated without being prejudicial to health or a nuisance without arrestment plant (section 7).

Approval by local authority to proposed equipment

In situations where arrestment plant is required for either domestic or non-domestic furnaces, the plant must be approved by the local authority or installed in accordance with plans and specifications submitted to and approved by the local authority (sections 6(1) and 8(1)).

Emission of grit and dust

The occupier of any building containing a furnace, other than one designed solely or mainly for domestic purposes, with a maximum heating capacity less than 16.12 kilowatts/h which burns solid, liquid or gaseous matter must take all practicable means for minimising the emission of grit and dust from the chimney; otherwise he is committing an offence (section 5(5)).

This general requirement does not, however, apply to furnaces where maximum emission limits have been specified by the Clean Air (Emission of Grit and Dust from Furnaces) Regulations 1971.

Although there is no similar requirement for those domestic furnaces which are required to have arrestment plant, that plant must be properly maintained and used (section 8(1)).

Defences

In proceedings for an offence under the regulations, it is a defence to prove that the best practicable means have been used to minimise the emission (section 5(4)).

Furnaces outside buildings

These provisions also apply to furnaces attached to a building or fixed to or installed on any land and in these cases the person responsible for compliance is the person having possession of the boiler or plant (section 13).

MEASUREMENT OF GRIT AND DUST FROM FURNACES

References

Clean Air Act 1993, sections 10 to 12.
Clean Air (Measurement of Grit and Dust from Furnaces) Regulations 1971.
Clean Air (Units of Measurement) Regulations 1992.

Extent

These provisions apply to England, Wales and Scotland.

Information

In connection with the provisions relating to grit arrestment plant and the measurement of grit and dust, local authorities are empowered to require any occupier to give them reasonable information about any furnaces used and the fuels burned following written notice giving not less than 14 days to do so (section 12(1)).

Measurement by occupiers

These provisions apply only to furnaces used to burn:

(a) pulverised fuel; or
(b) any other solid matter at a rate of 45.4 kilograms/h or more; or
(c) any liquid or gaseous matter at a rate of 366.4 kilowatts/h or more.

The regulations detailing measurement etc. are applied to an individual furnace by written notice from the local authority to the occupier (or, in relation to outside furnaces, the person having possession) (sections 10(1) and 13(2)).

Requirement for local authority to undertake measurement

This applies only to furnaces burning:

(a) solid matter, other than pulverised fuel, at a rate less than 1.02 tonnes/h; or
(b) liquid or gaseous matter at a rate less than 8.21 megawatts/h.

In these cases, the occupier by written notice may request the local authority to undertake the making and recording of measurements (section 11(2)).

The occupier remains responsible, however, for any necessary adaptations to the chimney to allow measurements to be taken.

FC6 Measurement of grit and dust from furnaces – Sections 10 and 11 Clean Air Act 1993

Local authority notice on occupier directing in relation to certain furnaces that the provisions of Clean Air (Measurement of Grit and Dust from Furnaces) Regulations 1971 apply (section 10(1)) (note 2)

→ Subsequent duty to comply with the regulations. Penalty max. level 5 and a cumulative penalty (note 4) (section 10 (3))

Local authority may subsequently revoke its direction but renew it later by further notice (section 10(6))

Local authority notice requiring necessary adaptations to chimney and provision of apparatus for measurement and recording giving at least six weeks for compliance (paragraph 1, schedule to regulations)

For specified smaller furnaces, occupier may serve written notice on local authority requiring them to make and record measurements. This may subsequently be withdrawn (section 11(2))

Local authority make and record measurements (section 11(2))

AND

Occupier must make any necessary adaptations to chimney after min. six weeks' written notice from local authority (section 10(2)(b))

Local authority give at least 28 days' written notice to occupier to undertake measurements (paragraph 2(1), schedule to regulations)

Occupier to give at least 48 hours' written notice of start of measurements (paragraph 2(2), schedule to regulations). Local authority may be present during measurement (section 10(4))

Occupier to keep written record of measurements (paragraph 2(3), schedule to regulations)

AND

Copy of measurements to local authority within 14 days (paragraph 2(3), schedule to regulations)

Notes

1. For general powers of local authority regarding information about atmosphere pollution, see FC13.
2. There is no appeal against the local authority notice applying the provisions of the regulations.
3. These provisions also apply to outdoor furnaces, section 13.
4. In relation to offences which are a repetition or continuation of an earlier offence, there is a penalty of £50 max. daily.

Measurements etc.

The Clean Air (Measurement of Grit and Dust from Furnaces) Regulations 1971 lay down the detailed requirements. Although power is given to the Secretary of State in the Act to include the measurement of fumes, the regulations apply only to grit and dust.

The notice from the local authority requiring measurements may require them to be taken from time to time or at staged intervals, but the local authority cannot normally require measurements from a chimney in excess of once each three months unless the local authority thinks that this is necessary to obtain the true level of emission (paragraph 2(4) schedule to regulations).

The records to be kept of measurements must show:

(a) the date;
(b) the number of furnaces discharging into the chimney;
(c) results of measurements in pounds per hour grit and dust emitted and the percentage of grit in the solids emitted (paragraph 2(3) schedule to the 1971 regulations).

PROHIBITION OF DARK SMOKE ETC. FROM CHIMNEYS

References

Clean Air Act 1993, section 1.
Dark Smoke (Permitted Periods) Regulations 1958.
Dark Smoke (Permitted Periods) (Vessels) Regulations 1958.

Extent

These provisions apply in England, Scotland and Wales.

Scope

It is an offence to emit dark smoke from the chimney of any building (including houses) on any one day and this includes chimneys serving the furnaces of boilers or industrial plant which are either attached to buildings or for the time being fixed to or installed on any land (section 1). Emissions lasting not longer than the following are, however, to be left out of account:

(a) Dark Smoke (Permitted Periods) Regulations 1958 – applies to emissions from chimneys other than on vessels:

No. of furnaces	*Aggregate in any eight-hour period*	
	Without soot blowing	*With soot blowing*
1	10 min.	14 min.
2	18 min.	25 min.
3	24 min.	34 min.
4 or more	29 min.	41 min.

FC7 Prohibition of dark smoke etc. from chimneys – Section 1 Clean Air Act 1993

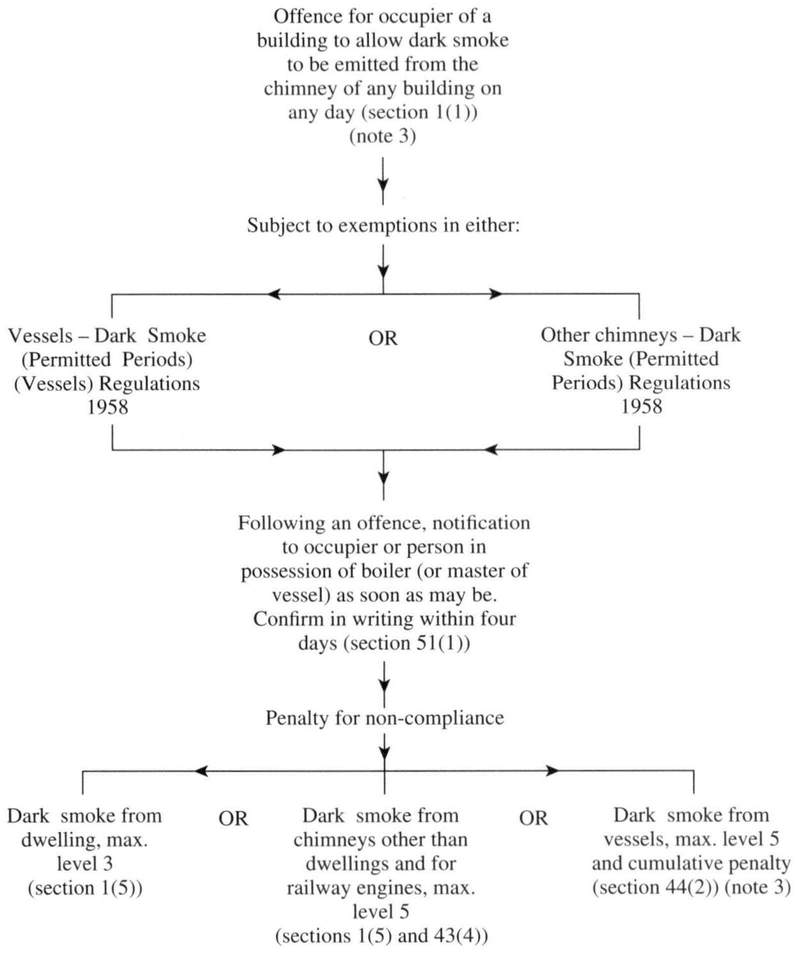

Notes

1. For dark smoke from industrial or trade premises (other than from chimneys), see section 2 of the Clean Air Act 1993 (FC8).
2. This procedure also applies to vessels and railway engines (Clean Air Act 1993, sections 43 and 44).
3. In relation to offences which are a repetition or continuation of an earlier offence, there is a penalty of max. £50 daily (section 50).

but in any event emissions must not exceed four minutes' continuous dark smoke (other than by soot blowing) or two minutes' black smoke in aggregate in any 30-minute period.

(b) Dark Smoke (Permitted Periods) (Vessels) Regulations 1958. These regulations apply only to vessels and details of the exemptions for the emission of dark smoke are set out in the schedule and vary depending upon the class of boiler. The exemptions do not permit the continuous emission of dark smoke for in excess of four minutes (ten minutes for certain natural-draught, oil-fired boilers) or the emission of black smoke for more than three minutes in any 30-minute period.

Defences

In any proceedings under section 1, it will be a defence to prove that the contravention was solely due to:

(a) lighting up from cold and all practicable steps were taken to prevent or minimise the emission; or
(b) failure etc. of the furnace that could not reasonably have been foreseen or prevented; or
(c) suitable fuel being unavailable and all practicable steps taken to prevent or minimise the emission; or
(d) any combination of (a)–(c) above; or
(e) there was a failure by the local authority to comply with the notification provisions (below) (sections 1(4) and 51(3)).

Person responsible

This is:

(a) for chimneys of buildings – the occupier of the building (section 1(1));
(b) for chimneys not attached to buildings – the person having possession of the boiler or plant (section 1(2));
(c) for vessels – the owner and master or other person in charge of the vessel (section 44(2));
(d) for railway engines – the owner (section 43(2)).

Notification of offences

Upon becoming aware that an offence under section 1 is being or has been committed, an authorised officer of the local authority must:

(a) notify the person responsible as soon as may be; and
(b) confirm the offence in writing before the end of the four days following the day on which the offence was committed (section 51).

PROHIBITION OF DARK SMOKE ETC. FROM INDUSTRIAL AND TRADE PREMISES (OTHER THAN FROM CHIMNEYS, E.G. TRADE BONFIRES)

References

Clean Air Act 1993 sections 2 and 33.
Clean Air (Emission of Dark Smoke) (Exemption) Regulations 1969.

Scope

These provisions apply only to industrial and trade premises where the smoke is emitted other than from a chimney, e.g. bonfires etc. It is an offence for dark smoke to be omitted from these premises but this does not apply to dark smoke from a chimney covered by section 1 and to exempted material prescribed in regulations by the Secretary of State (sections 2(1) and (2)).

Their main use is to control industrial bonfires.

Industrial or trade premises are defined as meaning premises used for any industrial or trade purposes or premises not so used on which matter is burnt in connection with any industrial or trade process (section 2(6)).

Where any material is burned on industrial and trade premises and this is likely to produce dark smoke, there is an assumption that dark smoke has been emitted unless the occupier or person who caused or permitted the emission can show that no dark smoke was emitted (section 2(3)).

Exemptions

These are specified in the Clean Air (Emission of Dark Smoke) (Exemption) Regulations 1969 and include:

- demolition waste (ABC);
- waste explosive (AC);
- matter burnt in connection with fire research and training (C);
- tar, pitch, asphalt and the like burnt in the laying of surfaces (C);
- carcases of diseased animals or poultry (AC);
- contaminated containers associated with veterinary or agricultural purposes (ABC).

Certain conditions must occur for these exemptions to apply. The conditions are listed below and the letters in brackets above must apply for the exemption to be valid.

- Condition A: that there is no other reasonably safe and practicable method of disposing of the matter.
- Condition B: that the burning is carried out in such a manner as to minimise the emission of dark smoke.
- Condition C: that the burning is carried out under the direct and continuous supervision of the occupier of the premises concerned or a person authorised to act on his behalf.

FC8 Prohibition of dark smoke etc. from industrial and trade premises (other than from chimneys) – Section 2 Clean Air Act 1993

Offence for occupier of industrial or trade premises to emit dark smoke on any day (section 2) (note 1)

↓

Exemptions provided by Clean Air (Emission of Dark Smoke) (Exemption) Regulations 1969

↓

Notification of offence to occupier by local authority confirmed in writing within four days of offence (section 51(1))

↓

Penalty max. £20,000 (section 2(5)) (note 3)

Notes

1. For procedure relating to dark smoke emitted from chimneys, see FC7.
2. For special offences relating to cable burning, see section 33.
3. The maximum penalty was increased to this amount by schedule 22, paragraph 195 of the Environment Act 1995 and came into effect on 1 April 1996.

Defences

It is a defence in proceedings to prove that:

(a) the contravention was inadvertent and that all practicable steps had been taken to prevent or minimise the emission of dark smoke; or
(b) the notification procedure had not been followed (section 2(4)).

Notification of offences

The notes on page 73 are also applicable here (section 51).

There is a specific offence for cable burning. Section 33 of the Clean Air Act 1993 makes it an offence liable to a fine of level 5 unless it is subject to Part I of the Environmental Protection Act 1990 or section 2 of the Pollution Prevention and Control Act 1999.

SMOKE CONTROL AREAS

Although these provisions remain in force, those relating to the making of new smoke control areas (SCAs) are rarely used since the problem of smoke pollution from domestic chimneys has been largely overcome by earlier use of orders and the changing nature of domestic fuels. However, there is still the need to enforce existing orders and these procedures indicate how this may be done.

The current approach to the reduction of air pollution is through air quality management and the making of air quality management areas (see page 94).

References

Clean Air Act 1993, sections 18 to 29.
Smoke Control Areas (Exempted Fireplaces) Order (Various).
Smoke Control Areas (Authorised Fuels) Regulations (Various).

Extent

These provisions apply to England, Scotland and Wales.

Directions by Secretary of State

Where the Secretary of State (or in Scotland the Scottish Environmental Protection Agency) considers it expedient to abate pollution by smoke and the local authority concerned has not exercised its powers to make SCAs sufficiently or at all, he may, after consulting the local authority, direct it to submit a programme to him. Following approval of its proposals by the Secretary of State or, in its default, following the determination of a programme by the Secretary of State, the local authority is under a statutory obligation to comply (Clean Air Act 1993 section 19).

FC9 Smoke control areas: directions by the Secretary of State – Section 19 Clean Air Act 1993

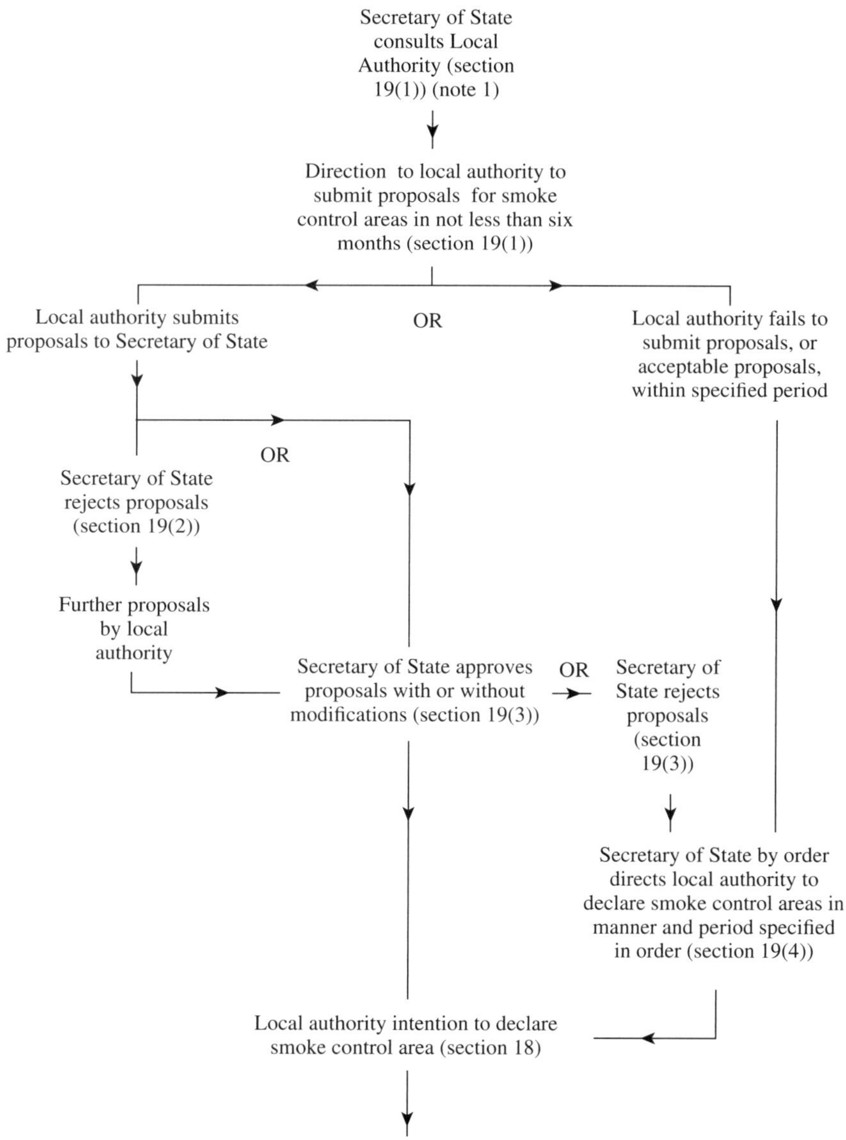

(Continued on next page)

Note

1. In Scotland, the powers of the Secretary of State are exercised by the Scottish Environmental Protection Agency and the procedure chart needs to be read accordingly.

FC10 Smoke control areas: declaration by local authority – Section 19 Clean Air Act 1993

(Continued from previous page)

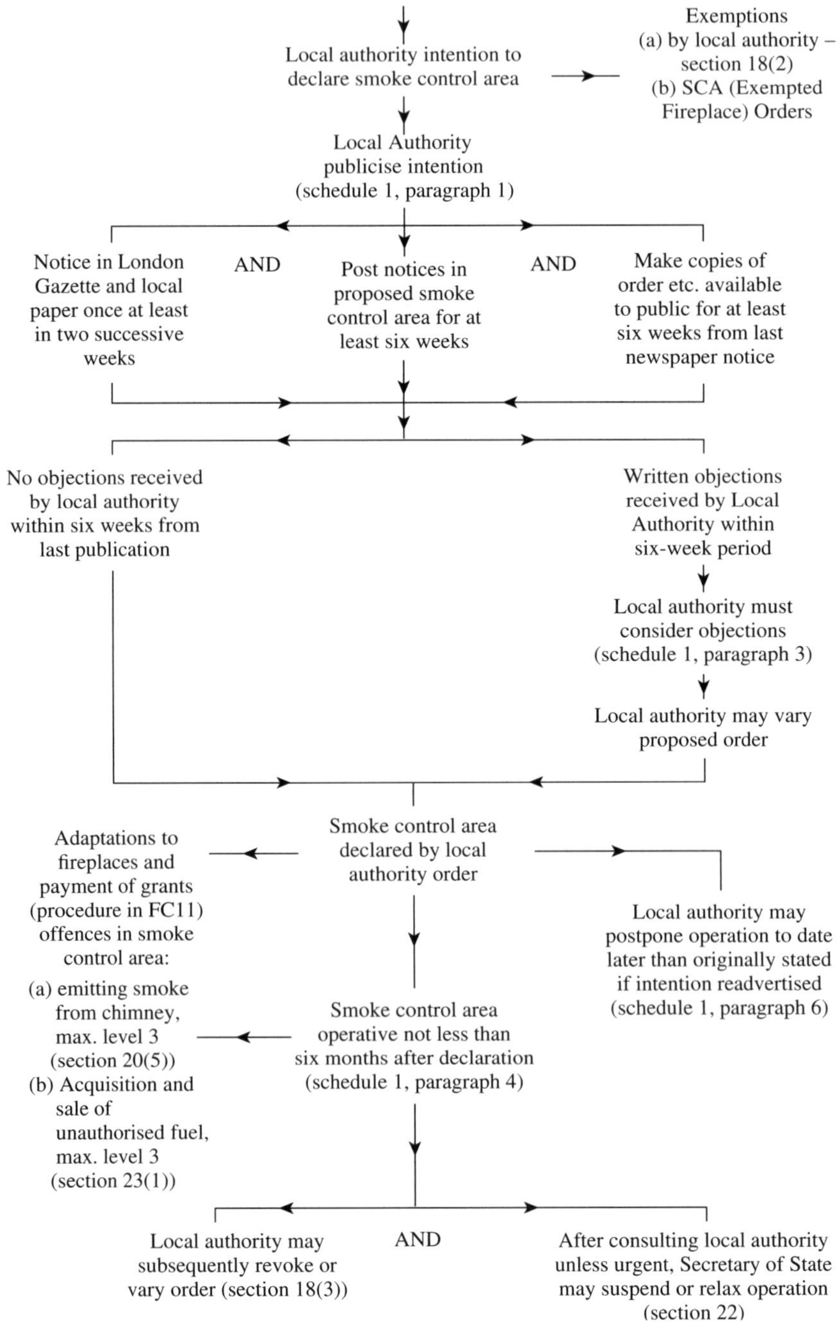

These powers have been used to secure compliance with the EC Air Quality Directive on sulphur dioxide and suspended particulates 80/779/EEC.

Smoke control areas

These may relate to the whole or part of a local authority area (section 18(1)) and in them the emission of smoke from the chimney of a building, or a chimney serving a fixed boiler or industrial plant is generally prohibited (sections 20(1) and (2)).

Exemptions

Exemptions from the general requirement not to emit smoke can come about in the following ways:

(a) The burning of an authorised fuel, i.e. one designated by one of the SCA (Authorised Fuels) Regulations, is a defence against proceedings (section 20(4) and (6)).
(b) By the use of one of a class of fireplaces exempted by a SCA (Exempted Fireplaces) Order following the Secretary of State being satisfied that the appliance is capable of burning other than authorised fuels without producing any or a substantial quantity of smoke (section 21).
(c) Exemption granted by the local authority in respect of a specified or class of building or specified or class of fireplaces, upon such conditions as the local authority may specify (section 18(2)).
(d) Exemptions granted by the local authority for the purpose of investigation and research (section 45).

Offences

Within a SCA, the following are offences:

(a) emitting smoke from the chimney of a building etc. (section 20(1) and (2)) subject to the exemptions listed above;
(b) acquiring solid fuel, other than authorised fuel for use in a building, fixed boiler or industrial plant in a SCA, unless intended for use in an exempted building or fireplace (section 23(1)(a) and (b));
(c) selling by retail solid fuel, other than authorised fuel, for delivery to a building or for any fixed boiler or industrial plant in a premises in a SCA. In this case the person cannot be convicted if he had reasonable grounds for believing:
    (i) that the building was exempted; or
    (ii) that the fireplace or plant for which the fuel was intended was exempted (section 23(1)(c) and (5)).

Upon becoming aware that an offence under section 20 relating to emissions of smoke in SCAs has been committed, an authorised officer of the local authority must:

(a) notify the person responsible as soon as may be; and
(b) confirm the offence in writing before the end of the four days following the day on which the offence was committed (section 51).

FC11 Smoke control areas: adaptations to fireplaces etc. – Section 24 Clean Air Act 1993

Private dwellings in smoke control area

↓

Local authority notice on owner or occupier requiring adaptation of fireplace (Clean Air Act 1993, section 24(1))

↓

Appeal to magistrates' court within 21 days (Public Health Act 1936, section 290(3))

↓

Failure to execute works within time specified

↓

Penalty max. level 4 and £2 daily (Public Health Act 1936, section 290(6))

AND

Local authority may undertake works in default (Public Health Act 1936, section 290(6))

↓

Recovery of three-tenths of expenses (section 24(3) and Public Health Act 1936, sections 291 and 293)

Note

1. The provisions of Part 12 Public Health Act 1936 relating to appeals against, and the enforcement of, notices requiring the execution of works apply to notices in this procedure (section 24(2)) (pages 445–447).

Adaptation of fireplaces in private dwellings

The notice procedure may be used where adaptations are required to avoid contraventions of section 20 and these may be served on the owner or occupier of the private dwelling either between the date of making the order and its operation, or after the date of operation. The period allowed for compliance must be reasonable and not less than the appeal period, i.e. 21 days. The provisions of Part 12 of the Public Health Act 1936 apply to the service and enforcement of these notices, but costs of carrying out work in default can be recovered only to an amount up to three-tenths of the expenditure by the local authority (section 24(3)).

Grants

The local authority is required to pay grants to owners or occupiers of private dwellings to assist with the adaptation of fireplaces where this is required to avoid contraventions of section 20, provided that the works are carried out to the local authority's satisfaction before the order becomes operative, unless the local authority has served a notice as above (section 25).

The normal grant is seven-tenths of necessary expenditure, but the local authority may pay more at its discretion. Payments to occupiers other than owners for moveable cooking or heating appliances are made at seven-twentieths initially and the remainder after two years, provided that the appliance is still present (schedule 2, paragraph 1). The local authority has a discretion to make up to 100 per cent grants for adaptations in churches, chapels or buildings used by charities (section 26).

Exchequer contributions towards the local authority's costs ceased with orders coming into operation from 31 March 1996 other than in exceptional cases where, for example, the local authority may breach 80/779/EEC and funding is not available from the local authority's own resources. Otherwise local authority funding requirements arising from its SCA programme will be recognised by the Secretary of State for England by the use of the Supplementary Credit Approval System and in Wales as part of the annual Basic Credit Approval for that local authority (DoE Circular 9/93).

Meaning of 'adaptations'

In relation to the payment of grants and to notices, 'adaptations' include:

(a) adapting or converting a fireplace; or
(b) replacing any fireplace by another fireplace or by some other means of heating or cooking; or
(c) altering any chimney which serves any fireplace; or
(d) providing gas ignition, electric ignition or any other special means of ignition; or
(e) carrying out any operation incidental to any of the operations (a)–(d) above, and includes works both inside and outside that dwelling.

The works must be reasonably necessary in order to make what is, in all circumstances, suitable provision for heating and cooking without contravention of section 20 (section 27).

Designated appliances

Classes of heating appliance designated by a local authority for its area, or by the minister either generally or in the area of defined local authorities, which would impose undue strain on the fuel resources available, do not qualify for local authority grant (schedule 2, paragraph 2).

BURNING OF CROP RESIDUES

References

Environmental Protection Act 1990, section 152.
The Crop Residues (Burning) Regulations 1993.

Extent

The procedure applies in England, Wales and Scotland but not in Northern Ireland (section 164).

Scope

There is a prohibition on the burning of the following crop residues:

(a) cereal straw;
(b) cereal stubble;
(c) residues of:
  (i) oil-seed rape;
  (ii) field beans harvested dry; and
  (iii) peas harvested dry (regulation 4 and schedule 1).

Exemptions

The burning of the crop residues specified above is allowed, however, for the purposes of:

(a) education and research;
(b) disease control or the elimination of plant pests where a notice has been served under article 22 of the Plant Health (Great Britain) Order 1993;
(c) the disposal of straw stack remains or broken bales (regulation 4).

Following a review of the exemptions from these regulations in 1996, the government announced that the non-inclusion of linseed residues would continue. Farmers are, however, urged to follow the safety rules on allowable burning.

Restriction on allowable burning

Where any of the specified crop residues are to be burned, this must be carried out in strict compliance with the following restrictions and requirements:

FC12 Burning of crop residues – Crop Residues (Burning) Regulations 1993

Offence to burn certain crop residues on agricultural land (The Crop Residues (Burning) Regulations 1993 – (regulation 4))

→ Penalty for unlawful burning, max. level 5 (regulation 6)

↓

Exemptions specified for:
(a) education and research;
(b) disease/pest control; and
(c) disposal of straw stack remains or broken bales
(regulation 4)

↓

Notification to be given to:
(a) Environmental Health Department of district council;
(b) occupiers of adjacent premises; and
(c) air traffic control of any aerodrome with a perimeter fence within 800 m of the area to be burned
at least one hour but not more than 24 hours before the commencement of the burning (schedule 2, paragraph 6)

AND

Burning to be carried out in accordance with specified restrictions and requirements (text) (regulation 5 and schedule 2)

↓

Offence for non-compliance, max. level 5 (regulation 6)

1. No crop residue may be burned:
   (a) during the period between one hour before sunset and the following sunrise; or
   (b) on any Saturday, Sunday or bank holiday.
2. No crop residue may be burned if the area to be burned extends, in the case of cereal straw or cereal stubble, to more than ten hectares, and in any other case to more than 20 hectares.
3. No crop residue may be burned unless:
   (a) the area to be burned is surrounded by a fire-break, which borders on that area and which, in the case of cereal straw or cereal stubble, shall be at least ten metres wide and in any other case at least five metres wide;
   (b) any building, structure or other thing mentioned in paragraph 4(b) or (c) below which lies within the area to be burned is surrounded by a fire-break of the relevant width referred to in sub-paragraph (a) above which borders on any crop residues which are to be burned; and
   (c) in the case of any land ('intervening land') between a fire-break to be established in accordance with sub-paragraph (a) or (b) above and any other land or any building, structure or other thing mentioned in paragraph 4(b), (c) or (d) below which lies within the relevant distance there mentioned of the area to be burned or, as the case may be, of any crop residues within that area:
      (i) the intervening land is cleared of all crop residues; or
      (ii) all crop residues on the intervening land are incorporated into the

   soil before burning takes place.
4. No crop residue may be burned if the area to be burned:
   (a) is less than 150 metres from any other area in which crop residues are being burned;
   (b) in the case of cereal straw or cereal stubble, is less than 15 metres, and in any other case less than five metres from:
      (i) the trunk of any tree (including any tree in coppice or scrubland);
      (ii) any hedgerow;
      (iii) any fence not the property of the occupier of the land upon which the burning is carried out;
      (iv) any pole which is or may be used to carry telegraph or telephone wires;
      (v) any electricity pole, pylon or substation;
   (c) in the case of cereal straw or cereal stubble, is less than 50 metres, and in any other case less than 15 metres from:
      (i) any residential building;
      (ii) any structure having a thatched roof;
      (iii) any building, structure, fixed plant or machinery which could be set alight or damaged by heat from the fire;
      (iv) any scheduled monument which could be set alight by the fire;
      (v) any stack of hay or straw;
      (vi) any accumulation of combustible material other than crop residues removed in the making of a fire-break;
      (vii) any mature standing crop;

(viii) any woodland or land managed as a nature reserve;
(ix) any building or structure containing livestock;
(x) any oil or gas installation on or above the surface of the ground; or
(d) is less than 100 metres from:
(i) any motorway;
(ii) any dual carriageway;
(iii) any A-road;
(iv) any railway line.

5. No crop residue may be burned unless all persons concerned in the burning operation are familiar with the provisions of the regulations and, except where an emergency arising during the operation renders it impracticable, each area to be burned is supervised by at least two responsible adults, one of whom having experience of burning crop residues shall be in general control of the operation.
6. No crop residue may be burned unless there is available at the area being burned:
   (a) not less than 1,000 litres of water in one or more mobile containers, together with means of dispensing the water for fire-fighting purposes in a spray or jet at a rate of 100 litres per minute; and
   (b) not fewer than five implements suitable for use in fire-beating.
7. No crop residue may be burned unless every vehicle used in connection with the burning is equipped with a suitable and serviceable fire extinguisher.
8. No crop residue may be burned unless reasonable precautions have been taken to ensure that the fire will not cross a fire-break.
9. Ashes of burnt cereal straw or cereal stubble shall not, without reasonable excuse, be allowed to remain on the soil for longer than 24 hours after the time of commencement of the burning, but shall be incorporated into the soil:
   (a) within that period; or
   (b) in a case where to do so would be likely, having regard to wind conditions, to cause nuisance, as soon as conditions allow (schedule 2).

Notification of intended burning

Before any allowable burning takes place and so far as is reasonably practicable, notice must be given to:

(a) the Environmental Health Department of the District Council where the burning is to take place;
(b) occupiers of any adjacent premises; and
(c) the air traffic control of any aerodrome with a perimeter fence within 800 metres of the area to be burned.

Such notice is to be given at least one hour or not more than 24 hours before the commencement of the burning (schedule 2, paragraph 6).

Power of entry

No such powers are provided in respect of this procedure.

Definitions

Crop residue means straw or stubble or any other crop residue remaining on the land after harvesting of the crop grown thereon.
Fire-break means an area of ground of which the surface consists wholly or mainly of substances other than combustible material (regulation 2(1)).

OBTAINING INFORMATION ABOUT ATMOSPHERIC POLLUTION

References

Clean Air Act 1993, sections 35 to 40 (as amended by the Pollution Prevention and Control Act 1999).
Control of Atmospheric Pollution (Research and Publicity) Regulations 1977.
Control of Atmospheric Pollution (Appeals) Regulations 1977.
Control of Atmospheric Pollution (Exempted Premises) Regulations 1977.

Extent

These provisions apply to England, Wales and Scotland. Section 30 applies to Northern Ireland.

General provision for obtaining information etc.

A local authority may obtain information about pollutants and other substances in the air from any premises, other than private dwellings or caravans, in any of the following ways:

(a) issuing notices on occupiers under section 36; or
(b) measuring and recording itself; or
(c) entering into an arrangement for occupiers to measure and record information on behalf of the local authority (section 35(1)).

Whilst local authorities are able to obtain information under section 36 about processes subject to part 1 of the Environmental Protection Act 1990 and activities subject to regulation under section 2 of the Pollution Prevention and Control Act 1999, they are prevented by section 35(3) from undertaking any investigations into any emissions from those processes and activities.

The local authority may also undertake, or contribute towards, investigation and research relevant to air pollution and publish information (section 34(1)).

Duty to consult

In exercising any of its powers to obtain information, the local authority must consult from time to time, at least twice in each financial year, with trade and business and their representatives and any other persons conversant with air pollution problems or having an interest (section 35(4)–(6)).

FC13 Obtaining of information about atmospheric pollution – Sections 35 to 40 Clean Air Act 1993

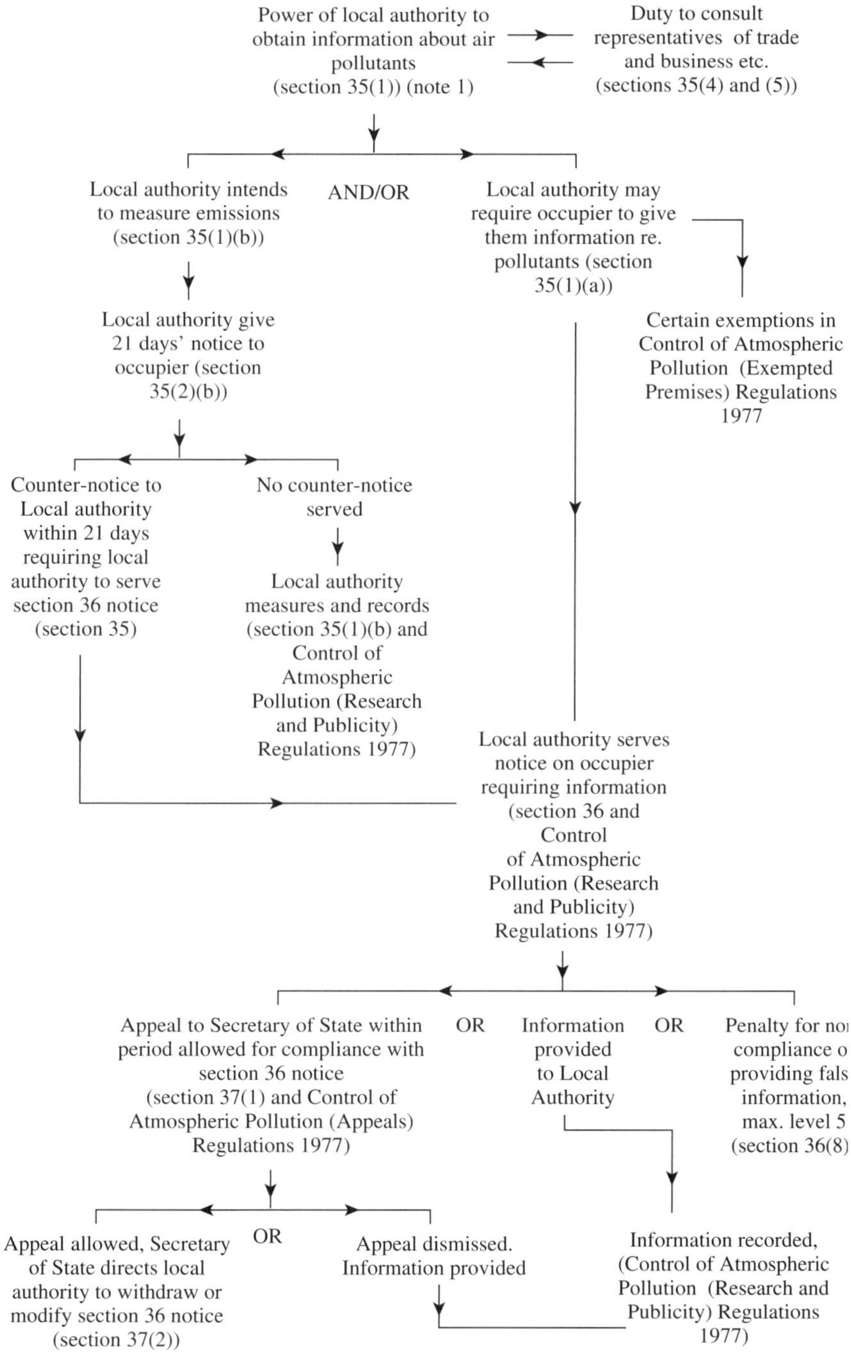

Note

1. For powers to require information about grit and dust, see Clean Air Act 1993, section 12.

Measurement by local authorities

Before exercising their powers to measure pollution, the local authorities must, unless proceeding by agreement, serve 21 days' notice on the occupier, stating:

(a) the kind of emissions in question; and
(b) the steps to be taken to measure and record.

There is no appeal against this notice but a counter-notice may be served within the 21 days requesting the local authority to serve notice on them under section 36, requiring the occupier to provide information about air emissions (section 35(2)).

The local authority must register the information recorded (Control of Atmospheric Pollution (Research and Publicity) Regulations 1977 regulation 6). This register must be open to public inspection free of charge (section 38(5)).

Section 36 notices

These may be served on occupiers, other than occupiers of exempted or private dwellings, and require the provision of any information relating to the emission of the following pollutants and other substances from any chimney, flue or other outlet to the atmosphere:

(a) SO_2 or particulate matter derived from any combustion process where the material being heated does not contribute to the emission;
(b) any gas or particulate matter derived from any combustion process where the material being heated contributes to the emission;
(c) any gas or particulate matter derived from any non-combustion process or similar industrial activity (Control of Atmospheric Pollution (Research and Publicity) Regulations 1977, regulation 3).

There are also more detailed requirements as to the type of information of these kinds which may be sought in the notice.

The period allowed for compliance must be at least six weeks. No one notice can cover information for a period exceeding one year and periodic information may be requested at not greater than quarterly intervals (section 36(4) and (5)).

Certain Crown premises are exempted from these provisions and these are listed in the Control of Atmospheric Pollution (Exempted Premises) Regulations 1977, but otherwise the procedure applies to Crown property, except that there is no power of entry (section 36(6)).

Appeals against section 36 notices

Appeal is to the Secretary of State by the recipient of the notice or other person having an interest on the following grounds:

(a) the information would prejudice to an unreasonable degree a private interest by disclosing a trade secret;
(b) the information would be contrary to the public interest;
(c) the information is not immediately available and cannot be made so without incurring undue expenditure (section 37(1)).

The procedure for dealing with appeals is laid down in the Control of Atmospheric Pollution (Appeals) Regulations 1977.

Recording of information

Information obtained by the local authority following section 36 notices must be kept in a register and adequately indexed (Control of Atmospheric Pollution (Research and Publicity) Regulations 1977, regulation 6). This register must be open to public inspection free of charge (section 38(5)).

Definition

Emissions of substances – references to the emission of substances into the atmosphere are to be construed as applying to substances in a gaseous or liquid or solid state, or any combination of those states (section 40(a)).

ENVIRONMENT ACT 1995
GENERAL PROCEDURAL PROVISIONS

Extent

Part 4 of the Act applies in England, Scotland and Wales but not in Northern Ireland (section 125(7)).

Powers of entry

An enforcing authority may authorise suitable persons to exercise any of the following powers:

(a) to enter at reasonable times (or in emergency at any time) any premises which he has reason to believe it is necessary for him to enter (seven days' notice required for residential premises);
(b) to take with him other persons and equipment (seven days' notice to take heavy equipment);
(c) to make examination and investigations;
(d) to direct premises to be left undisturbed for the purpose of examination or investigation;
(e) to take measurements, photographs and tape recordings for the purpose of examination and investigation;
(f) to take samples of any substances or of the air, water or land in and around premises;
(g) to subject to process or test any article or substance which appears to have caused or be likely to cause pollution of the environment or harm to human health;
(h) to take possession of substances etc., as in (g) above to examine it, ensure that it is not tampered with or to use as evidence;

(i) to require persons to give information;
(j) to require the production of records;
(k) to require persons to afford facilities and assistance;
(l) any other power necessary to carry out the purposes of entry (section 108).

The purpose for which the authorised officer may demand entry (and exercise the above powers) is to determine whether any provision of the pollution control enactment is being, or has been, complied with, to exercise a pollution control function of the authority or to determine whether and how such a function should be exercised.

In relation to local enforcing authorities, the pollution control powers to which these provisions relate are:

(a) Part 1 (Air Pollution Control by Local Authorities) and Part 2A (Contaminated Land) of the Environmental Protection Act 1990;
(b) Regulations made under Part 4 (Air Quality) of the Environment Act 1995, e.g. the Air Quality (England) Regulations; or
(c) Regulations made under the European Communities Act 1972 so far as they relate to pollution, e.g. the Air Quality Limit Values Regulations;
(d) Regulations made under section 2 of the Pollution Prevention and Control Act 1999, i.e. the Pollution Prevention and Control Regulations.

Where entry has been refused or is apprehended, the authorised officer may apply to a Justice of the Peace for a warrant. In exercising a warrant the authorised officer must secure any unoccupied premises and there are also provisions for the payment of compensation in certain cases (sections 108 and 110 and schedule 18).

Enforcing authorities

These are:

(a) the Secretary of State;
(b) the Environment Agency and the Scottish Environment Protection Agency;
(c) a waste collection authority;
(d) a local enforcing authority.

A local enforcing authority means:

(a) a local enforcing authority within the meaning of Part 1 of the Environmental Protection Act 1990 (integrated pollution control and air pollution control by local authorities);
(b) a local authority within the meaning of Part 2A of that Act (contaminated land);
(c) a local authority for the purposes of Part 4 of the Environment Act 1995 (air quality);
(d) a local authority under regulations made under section 2 of the Pollution Prevention and Control Act 1999.

In relation to the procedures in this chapter, the local authorities referred to above are:

(a) unitary authorities in England, Scotland and Wales, including London borough councils; and
(b) district councils that are not unitary councils (sections 82(1), 91 and 108).

LOCAL AUTHORITY REVIEWS OF AIR QUALITY

References

Environment Act 1995, sections 80 to 91 and schedule 11.
The Air Quality (England) Regulations 2000 (as amended).
The Air Quality (Scotland) Regulations 2000 (as amended).
The Air Quality (Wales) Regulations 2000 (as amended).
DEFRA Guidance Notes:

(a) LAQM. PG(09) Policy guidance, DEFRA;
(b) LAQM. TG(09) Technical guidance, DEFRA.

Extent

These provisions apply to England, Wales and Scotland.

Scope

This procedure deals with the statutory requirement on local authorities to review the quality, and the likely future quality, of air within the authority's area (section 82(1)).

National Air Quality Strategy (NAQS)

The Secretary of State is required to prepare and publish a strategy containing policies with respect to the assessment and management of the quality of air.

Such a strategy takes account of UK obligations under the EC treaties and of international agreements and includes statements of:

(a) air quality standards – concentrations of pollutants in the atmosphere which can broadly be taken to achieve a certain level of environmental quality;
(b) air quality objectives – these provide a framework for determining the extent to which policies should aim to improve air quality and indicate the progress which can be made towards air quality standards; and
(c) measures to be taken by local authorities and others to achieve the objectives (section 80(1)–(5)).

The first strategy was published in March 1997 with revisions in 2000 and 2003 (CM 4548). The latest version was published in July 2007. It sets out certain

FC14 Local authority reviews of air quality – Sections 80 to 86 Environmental Protection Act 1990

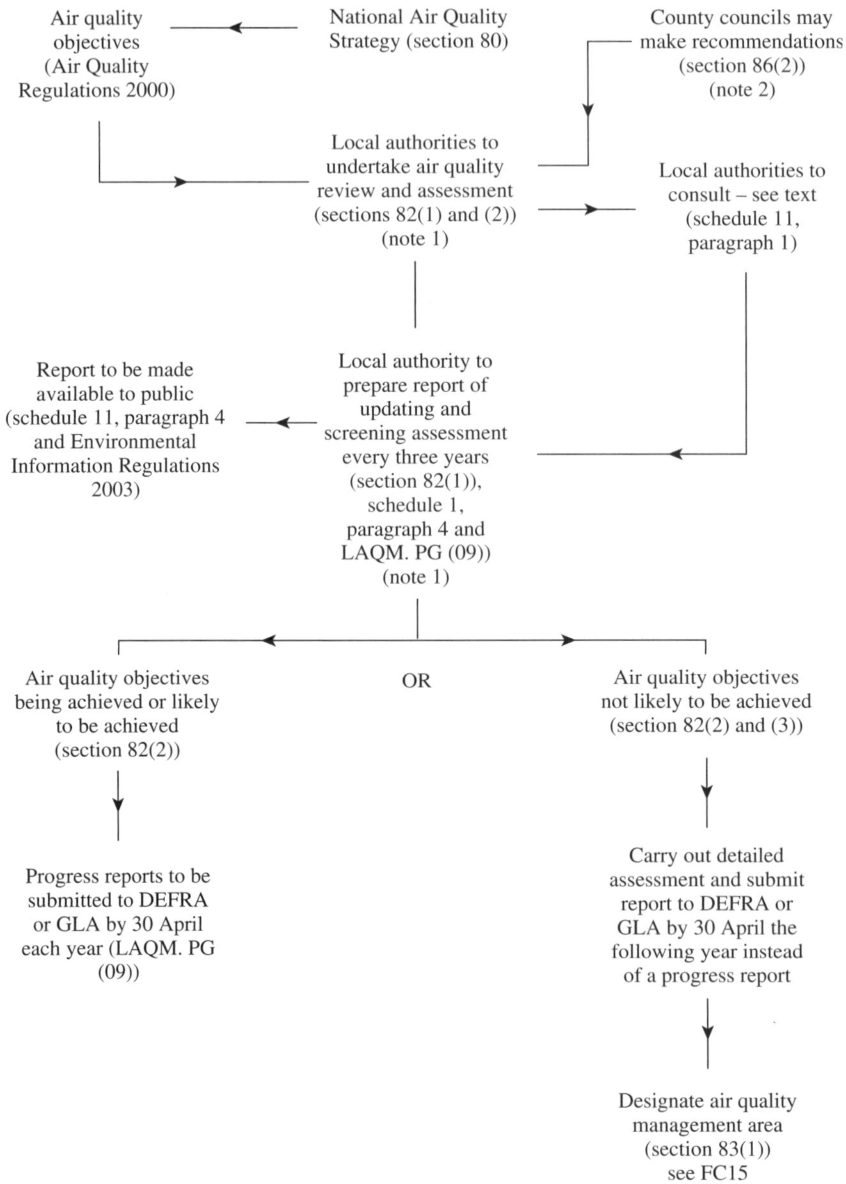

Notes

1. The Secretary of State or Scottish Environment Protection Agency has reserve powers in the event of local authority default (section 85).
2. The county council (where one exists) should be involved throughout the air quality management process.

standards and specific objectives for key air pollutants to be achieved in the long term (up to 2050).

The air quality objectives are set out in the Air Quality (England) Regulations 2000, as amended by the Air Quality (England) (Amendment) Regulations 2002, which provide the statutory basis for the air quality objectives under local air quality management in England (see Table 1 in LAQM. PG (09)). Local authorities should note that they also have a duty to continue to work towards meeting the air quality objectives beyond the deadlines set in the regulations.

Note that there are separate air quality regulations for Wales, Scotland and Northern Ireland.

London Air Quality Strategy

Working within the NAQS, the Greater London Authority (GLA) is required to produce an Air Quality Strategy for London, which will explain how, working with London borough councils and other bodies, the national objectives will be achieved. The first strategy was issued by the GLA in September 2002, entitled 'Cleaning London's Air – the Mayor's Air Quality Strategy'. The latest update was in December 2010 and sets targets for 2015 and 2020.

The Mayor's Air Quality Strategy does not replace the duties of the London borough councils, who will have to take account of it in undertaking their air quality management functions (Greater London Authority Act 1999).

Local air quality strategies

There is no statutory duty on local authorities to prepare a local strategy but they are urged to do so. Guidance on this is given in LAQM. PG(09) at chapter 2.

Local authority reviews and assessments

Each of the following local authorities (as enforcing authorities) is required to review periodically air quality in its area:

(a) Unitary authorities (in England, Wales and Scotland, including London borough councils);
(b) District councils which are not unitary authorities (but see 'functions' of county councils on page 93) (sections 82(1) and 91).

LAQM. PG(09) is guidance issued by the Secretary of State under section 88(1) of the Environment Act 1995, and local authorities must have regard to it when carrying out their local air quality management duties. The government and the Mayor of London recommend that this policy guidance is made available to all local authority departments. The chapters covering transport and planning are relevant to those working in various local government departments, such as environmental health, land use planning, economic development and transport planning.

Every local authority shall review the air quality within its area, both at the present time and the likely future air quality (section 82). Local authorities should designate an air quality management area where air quality objectives are not being achieved, or are not likely to be achieved, within the relevant period, as set out in the Air Quality (England) Regulations 2000 (section 83). Once an area has been designated, the local authority should carry out an assessment and then develop an action plan for the air quality management area (section 84).

All local authorities are expected to undertake a review and assessment every three years. Authorities should follow a two-step approach when carrying out review and assessment.

At Step 1, all authorities are required to undertake an updating and screening assessment. Where an authority identifies a risk that an air quality objective will be exceeded at a relevant location, the local authority is required to proceed to Step 2 – a detailed assessment.

To ensure continuity in the local air quality management process and fill the gaps between the three-yearly requirement to carry out a review and assessment, local authorities are required to prepare review and assessment progress reports in the years when they are not carrying out an updating and screening assessment. Where a detailed assessment is being undertaken, a short progress report should be provided within the detailed assessment for those areas not covered by the detailed assessment (LAQM (PG) (09) paragraph 1.7).

The most recent round of assessments etc. started in April 2012 and the next round starts in April 2015.

If DEFRA or the Mayor of London (as applicable) does not accept the conclusion contained within a local authority's report, the authority will be invited to provide written comments justifying their decision within a specified deadline set out in the appraisal letter. This will involve a response within a short deadline in keeping with the need to complete the process as quickly as possible.

Local authorities are able to amend or revoke an existing air quality management area order at any time (section 83 (2)). In order to make a significant amendment or revoke an air quality management area, the local authority is required to submit a detailed assessment report clearly outlining the evidence for changes in the likelihood of exceedence of the objectives occurring and demonstrating the cause of these changes, such as changes to the source of the pollution and/or better monitoring/modelling information. DEFRA and the Mayor of London also expect the authority to consult all the relevant statutory consultees, local stakeholders, businesses and members of the public. Those local authorities should submit their further reports for appraisal showing the monitoring results and other evidence to justify their decision to take action.

Where it is accepted by DEFRA and/or the Mayor of London that the revocation or amendment is justified, local authorities will be expected to take the relevant action within four months following receipt of comments from DEFRA or the Mayor of London on detailed assessments or further assessments.

Where an air quality management area is revoked, local authorities should consider drawing up a local air quality strategy to ensure air quality issues maintain a high profile locally and to respond to any public expectations (chapter 2 LAQM PG 09).

Consultation

In undertaking an air quality review and assessment, the local authority is required to consult:

(a) the Secretary of State or the National Assembly in Wales;
(b) the Environment Agency (Scottish Environment Protection Agency in Scotland);
(c) in England and Wales, the Highways Authority;
(d) all neighbouring local authorities;
(e) the county council (where applicable);
(f) the National Park Authority (where applicable);
(g) any other public authority which the local authority considers appropriate; and
(h) bodies representing local business interests and such other bodies as the local authority considers appropriate (schedule 11, paragraph 1);
(i) and in Greater London, the London Borough Council must consult the Mayor of Greater London (LAQM. PG (09)).

For the purposes of the 1995 Act, authorities must consult on their:

- air quality review and assessment;
- further air quality assessment in an air quality management area; and
- preparation or revision of an air quality action plan.

Local authorities are also expected to consult on the declaration, amendment or revocation of any air quality management areas.

Public access to information

A report of the results of the air quality review and a report of the results of any assessment must be made available to the public free of charge at all reasonable times (schedule 11, paragraph 4).

In any event, information which the local authority holds in relation to the quality of air needs to be made available (subject to exceptions) by the Environmental Information Regulations 2003.

Functions of county councils

It is expected that, while district councils will be the lead local authorities in the remaining two tier areas of England, they will fully involve counties in the review and assessment process.

The county council may make recommendations to the district in relation to any review and assessment and the district must take account of them (section 86(2)).

In addition, schedule 11, paragraph 2 makes provision for the exchange of information between district and county councils throughout the air quality management process. LAQM. PG (09) deals with the integration of local air quality plans into transport plans and planning.

Once an air quality management area has been designated, there is a further requirement that county councils should submit details of whatever actions they propose to the district council, for inclusion in an Action Plan within a relevant period (nine months) (section 86(3) and Air Quality (England) Regulations 2000, regulation 3).

AIR QUALITY MANAGEMENT AREAS (AQMAs)

References

Environment Act 1995, sections 83 and 84 and schedule 11.
The Air Quality Regulations (England) (Scotland) and (Wales) 2000 (as amended 2002).
The Air Quality Regulations 2010.
National Air Quality Strategy for England, Scotland, Wales and Northern Ireland CM 4548 (July 2007).
DEFRA Guidance Notes:

(a) LAQM. PG(09) Policy guidance;
(b) LAQM. TG(09) Technical guidance.

Scope

Where, following a review and assessment of air quality in its area, a local authority determines that prescribed air quality objectives are not likely to be achieved within the 'relevant period' as prescribed in the regulations, it must identify the areas where this is likely and, by order, designate them as an AQMA (section 83(1)).

Such orders may subsequently be varied or revoked by further orders following further assessment and, in the case of revocation, where the prescribed standards/objections are likely to be met throughout the 'relevant period' (section 83(2)).

The AQMA

While the order needs to designate the part/s of the local authority's area to which the AQMA status is to be attached, it is expected that delineation should make appropriate use of relevant physical and geographical boundaries (LAQM. PG(09) paragraph 3.2).

Action by local authority following designation of AQMA

The local authority is to undertake a further assessment of air quality within the AQMA itself to supplement information already obtained in the detailed assessment, identify likely sources and inform the preparation of the action plan (section 84(1)). These are intended to refine the outcomes of the earlier stages of

FC15 Air quality management areas (AQMAs) – Sections 83 and 84 Environment Act 1995

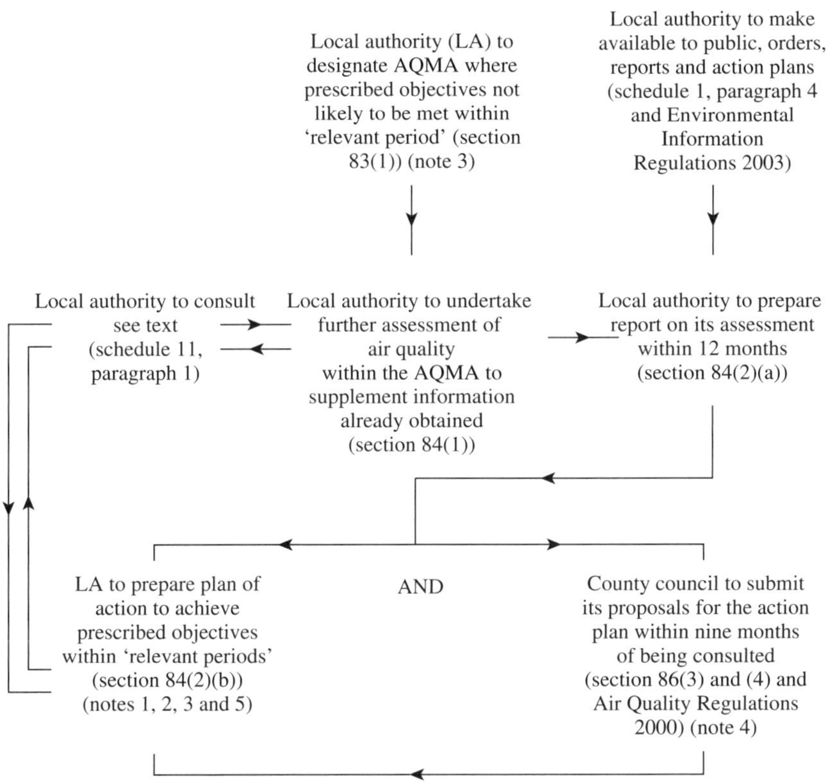

Notes

1. The local authority may revise the action plan periodically (section 84(4)) following periodic reviews and assessments of air quality.
2. There is no statutory period for the preparation of an action plan but LAQM. PG(09) paragraph 4.3 suggests the action plan should be prepared within 12–18 months from the declaration of an AQMA.
3. The Secretary of State or Scottish Environment Protection Agency has powers of direction (text).
4. In the event of disagreement between district and county, the matter is determined by the Secretary of State (section 84(5)).

review and assessments and provide a final check or clarification of earlier results from stage 3.

The local authority is required to produce a report on this assessment within 12–18 months of the operation of the order (section 84(2)(a)) and to produce a plan of action (section 84(2)(b)) normally within 12 months of the completion of the further assessment.

Air quality action plans

These should include:

(i) quantification of the source contributions to the predicted exceedences of the objectives. This will allow the action plan measures to be effectively targeted;
(ii) evidence that all available options have been considered;
(iii) how the local authority will use its powers and also work in conjunction with other organisations in pursuit of the air quality objectives;
(iv) clear timescales in which the authority and other organisations and agencies propose to implement the measures within its plan;
(v) where possible, quantification of the expected impacts of the proposed measures and, where possible, an indication as to whether the measures will be sufficient to meet the air quality objectives. Where feasible, data on emissions could be included as well as data on concentrations where possible; and
(vi) how the local authority intends to monitor and evaluate the effectiveness of the plan. (LAQM (09) paragraph 4.2)

Air quality action plans and transport plans

Given the importance of vehicle pollution in the assessment of air quality and its improvement, it is necessary for the air quality action plan to be integrated with the transport plan. Guidance on this is given in chapter 6 of LAQM. PG(09).

Single vehicle emissions in AQMAs

The Road Traffic (Vehicle Emissions) (Fixed Penalty) (England) Regulations 2002 allow local authorities who have designated AQMAs to apply to the Secretary of State to be given powers to issue fixed penalty notices to users of vehicles who fail to meet the emission requirements in the Road Vehicles (Construction and Use) Regulations 1986.

Consultation

In undertaking the further air quality assessment and in preparing its action plan, the local authority is required to consult widely. The bodies concerned are the same as those for the initial review and assessment on page 94 (schedule 11, paragraph 1).

Public access to information

The local authority is required to make available to the public free of charge copies of the order designating the AQMA, the report of the further assessment, the action plan and the proposals of the county council, where applicable (schedule 11, paragraph 4).

In any event, information which the local authority holds in relation to air quality must (with certain exceptions) be made available to the public under the Environmental Information Regulations 2004.

County councils

It is expected that, whilst the lead local authorities include district councils in the remaining two-tier areas of England, county councils will be fully involved by the district in the air quality management process generally, and schedule 11, paragraph 2 makes provision for the exchange of information between the two local authorities.

In relation to AQMAs, there is a specific requirement upon county councils that it should, within nine months from the date it is first consulted on the action plan, submit to the district details of its proposals for the action plan (sections 86(3) and (4) and regulation 3(1) of the Air Quality Regulations 2000). Disputes between districts and county councils about the content of action plans are to be settled by the Secretary of State or Scottish Environment Protection Agency (section 84(5)).

Secretary of State powers

The Act does provide for the Secretary of State (or SEPA in Scotland) to direct local authorities on the action it should take (including the designation of AQMAs) where prescribed air quality standards/objectives are unlikely to be achieved within the 'relevant periods', and the local authority has either failed to discharge a duty imposed on it or has taken inappropriate action to deal with the situation (sections 86(3) and (4)).

The Air Quality Regulations 2010

These regulations implement European Community Directive 2008/50/EC on ambient air quality and cleaner air for Europe and Directive 2004/107/EC relating to arsenic, cadmium, mercury, nickel and polycyclic aromatic hydrocarbons in ambient air.

The regulations apply mainly in England.

The regulations give the Secretary of State duties in the following areas.

- assessment of sulphur dioxide, nitrogen dioxide and oxides of nitrogen, particulate matter, lead benzene and carbon monoxide;
- assessment of ozone;
- assessment of arsenic, cadmium, nickel, mercury, benzo(a)pyrene and other polycyclic aromatic hydrocarbons;

- limit values, target values, long-term objectives, information and alert thresholds and critical levels for the protection of vegetation;
- requirements in relation to $PM_{2.5}$ in addition to the limit value and target value for this pollutant and calculation of an average exposure indicator (AEI) for the UK, the calculation of a national exposure reduction target based on the AEI, attainment of the national exposure reduction target in the UK and compliance with a limit on the AEI for 2015;
- requirements to draw up air quality plans in relation to limit values and target values and short-term action plans in relation to alert thresholds.

ENVIRONMENTAL DAMAGE

Reference

Environmental Damage (Prevention and Remediation) Regulations 2009.

Extent

These provisions apply to England only.

Scope

The regulations apply in relation to environmental damage if it is caused by an activity below:

- waste management operations;
- mining waste;
- discharges into the inland surface water that require prior authorisation;
- water abstraction and impoundment;
- dangerous substances, plant protection products and biocidal products;
- transport;
- genetically modified organisms;
- transboundary shipment of waste;
- geological storage of carbon dioxide.

The regulations also apply to environmental damage to protected species, natural habitats or a site of special scientific interest, and to environmental damage caused by any other activity if the operator intended to cause environmental damage or was negligent as to whether environmental damage would be caused (regulation 5).

There are some exemptions to the regulations, including acts of terrorism, an exceptional natural phenomenon, activities to protect from natural disasters, activities to serve national defence or international security, certain activities involving radioactivity and damage caused in the course of some commercial sea fishing (regulation 8). There are some exemptions to damage to water – for example, damage caused by new modifications to the physical characteristics of a surface water body or an alteration to the level of a body of groundwater provided certain conditions are applied.

The conditions are:

(a) all practicable steps are taken to mitigate the adverse impact on the status of the body of water;
(b) the reasons for those modifications or alterations are specifically set out and explained in the river basin management plan;
(c) the reasons for those modifications or alterations are of overriding public interest, or the result of the damage is outweighed by the benefits of the new modifications or alterations to human health, to the maintenance of human safety or to sustainable development; and
(d) the beneficial objectives served by those modifications or alterations of the water body cannot for reasons of technical feasibility or disproportionate cost be achieved by other means.

The enforcing authorities are those designated under the Environmental Permitting (England and Wales) Regulations 2007 (see page 153) where the local authority is responsible for granting the permit under part 2, and if granted under part 3, is enforced by the local authority if the damage is to land, the Environment Agency if the damage is to water and Natural England if the damage is to natural habitats, protected species or a site of special scientific interest (regulations 10 and 11).

If there is more than one type of damage, the regulations are enforced by any or all of the specified enforcing authorities. However, an enforcing authority may appoint any other enforcing authority to act on its behalf (regulation 12).

Preventing environmental damage

Where an operator of an activity that has caused, or causes, an imminent threat of environmental damage, or has caused damage where there are reasonable grounds to believe that the damage is or will become environmental damage, the enforcing authority may serve a notice on that operator that:

- describes the damage or threat;
- requires the operator to provide additional information on any damage that has occurred;
- specifies the measures required to prevent the damage or further damage; and
- requires the operator to take those measures, or measures at least equivalent to them, within the period specified in the notice.

Failure to comply to such a notice served is an offence (regulations 13 and 14).

The enforcing authority can take the actions which the operator of an activity should have taken in an emergency, if the operator cannot be ascertained, or if the operator fails to comply with a notice (regulation 15).

The operator may recover the costs when acting in accordance with the instructions of a public authority, and as a result causes or threatens to cause environmental damage, unless the instructions related to an emission or incident caused by the operator's own activities (regulation 16).

Remediation

Where damage has been caused, the enforcing authority must establish whether or not it is environmental damage (regulation 17).

If the enforcing authority decides that the damage is environmental damage, it must notify the operator of any activity or activities that caused the damage (the responsible operator) that:

(a) the damage is environmental damage;
(b) the responsible operator's activity was a cause of the environmental damage;
(c) the responsible operator must submit proposals, within a time specified, for measures that will achieve the remediation of the environmental damage; and
(d) the responsible operator has a right to appeal.

The enforcing authority may withdraw the notification if it is satisfied that the notification should not have been served or that an appeal against the notice is likely to succeed (regulation 18 and schedule 4).

Appeals

A person served with such a notification may notify the Secretary of State that they intend to appeal against that notification.

Notice of appeal must be within 28 days of service of the notification unless the time limit is extended by the Secretary of State.

The grounds of appeal are:

(a) the operator's activity was not a cause of the environmental damage;
(b) the enforcing authority has acted unreasonably in deciding that the damage is environmental damage;
(c) the environmental damage resulted from compliance with an instruction from a public authority (except an instruction relating to an emission or incident caused by the operator's own activities);
(d) the responsible operator was not at fault or negligent and the environmental damage was caused by an emission or event expressly authorised by, and fully in accordance with, the conditions of a permit;
(e) the responsible operator was not at fault or negligent and the environmental damage was caused by an emission or activity or any manner of using a product in the course of an activity that the operator demonstrates was not considered likely to cause environmental damage according to the state of scientific and technical knowledge at the time when the emission was released or the activity took place;
(f) the environmental damage was the result of an act of a third party and occurred despite the fact that the responsible operator took all appropriate safety measures.

The person deciding the appeal may confirm or quash the notice (regulation 19 and schedule 5).

Remediation notices

On receiving the proposals from the responsible operator, the enforcing authority must consult any interested party and any person on whose land the remedial measures will be carried out and may consult any other person where necessary.

Following consultation, the enforcing authority must serve a remediation notice on the responsible operator that specifies:

(a) the damage;
(b) the measures necessary for remediation of the damage, together with the reasons;
(c) the period within which those measures must be taken;
(d) any additional monitoring or investigative measures that the responsible operator must carry out during remediation; and
(e) the right of appeal against the remediation notice.

Failure to comply with a remediation notice is an offence (regulation 20).

Appeal against the remediation notice

The responsible operator may notify the Secretary of State that they intend to appeal against the remediation notice on the grounds that its contents are unreasonable. Such an appeal may only be brought against those parts of the remediation notice that are different from proposals made by the responsible operator. Notice of appeal must be served within 28 days of service of the remediation notice unless the time limit is extended by the Secretary of State.

The Secretary of State may confirm, vary or quash the notice, and must give written notification of the final decision and the reasons for it. They may add further compensatory remediation requirements necessitated by the lapse of time since the remediation notice was served. The notice is frozen until the appeal is determined unless the person hearing the appeal directs otherwise (regulation 21 and schedule 5).

An enforcing authority may serve further remediation notices at any time while remediation is being carried out or, if remediation has not been achieved, at the end of the remediation period, requiring further or different remediation (regulation 22).

The enforcing authority may carry out any reasonable works:

(a) at any time if a responsible operator cannot be identified;
(b) if a responsible operator fails to comply with a remediation notice, whether or not an appeal is pending; or
(c) if the responsible operator is not required to remediate under these regulations (regulation 23).

Costs

The enforcing authority can recover justified costs (remediation and administration) when the enforcing authority acts instead of the operator. Relevant costs are:

- service of notices;
- assessing whether the damage is environmental damage;
- establishing who is the responsible operator;
- establishing what remediation is appropriate;
- carrying out necessary consultation; and
- monitoring the remediation, both during and after the work.

No proceedings for the recovery of costs may be commenced by the enforcing authority more than five years after the completion of the measures or the identification of the operator liable to carry out the measures, whichever is the later. Costs are recoverable from the owner as a charge on premises (regulations 24 to 27).

Requests for action by interested parties

Any person who is affected or likely to be affected by environmental damage, or who otherwise has a sufficient interest, may notify the appropriate enforcing authority of any environmental damage which is being, or has been, caused or of which there is an imminent threat. A notification must be accompanied by a statement explaining the way the notifier will be affected by the damage, or the reason that the notifier has a sufficient interest; and sufficient information to enable the enforcing authority to identify the location and nature of the incident.

The enforcing authority must consider the notification and inform the notifier as to the action, if any, that it intends to take.

Before taking any decision, the enforcing authority must, if practicable, notify the operator concerned of the notification and the accompanying information, and invite that operator to submit comments on them.

The enforcing authority need not act if:

(a) the notifier is not likely to be affected or does not have a sufficient interest;
(b) in the opinion of the enforcing authority, the information provided does not disclose any environmental damage or threat of environmental damage; or
(c) as a result of the urgency of the situation, it is not practicable for the enforcing authority to comply (regulation 29).

Powers of authorised persons

The powers in section 108 of the Environment Act 1995 apply in relation to these regulations, and the powers of persons authorised by the Environment Agency in that section are exercisable by persons authorised by any enforcing authority (regulation 31).

Provision of information to the enforcing authority

An enforcing authority may require an operator to provide such information as it may reasonably require to enable the enforcing authority to carry out its functions under the regulations. Failure to provide such information is an offence (regulation 32).

FC16 Request for action by interested parties – Environmental damage – Regulation 29 Environmental Damage (Prevention and Remediation) Regulations 2009

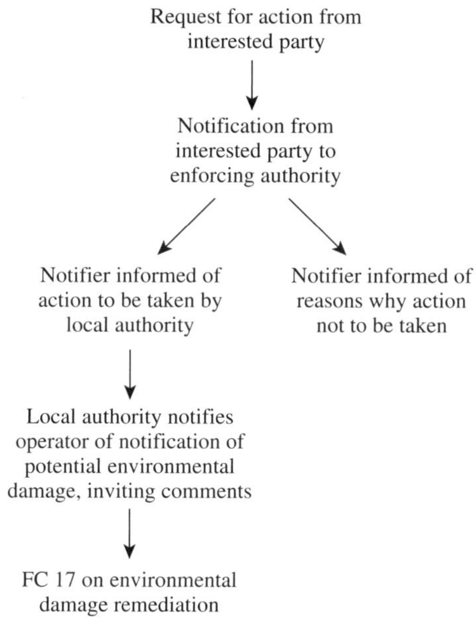

FC17 Environmental damage – Regulations 13 to 23 Environmental Damage (Prevention and Remediation) Regulations 2009

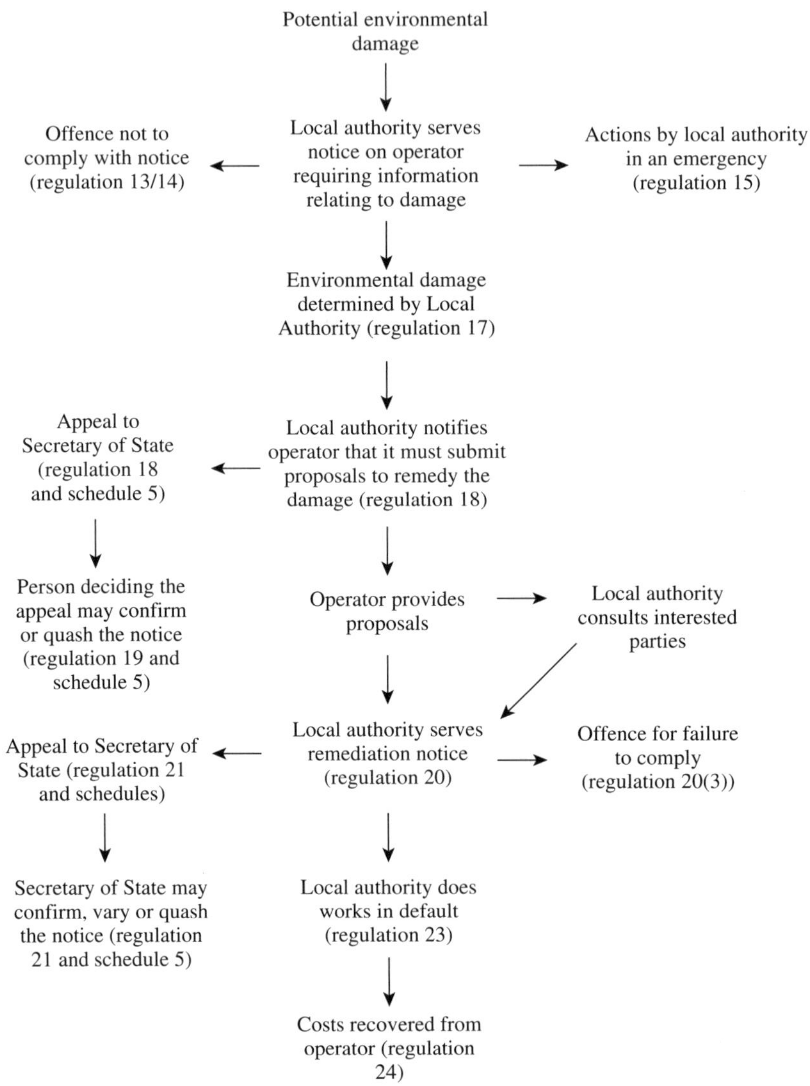

Enforcement

No enforcement action may be taken 30 years or more after the emission, event or incident concerned (regulation 33).

Penalties

A person guilty of an offence under these regulations is liable:

(a) on summary conviction, to a fine not exceeding the statutory maximum or to imprisonment for a term not exceeding three months, or both; or
(b) on conviction on indictment, to a fine or to imprisonment for a term not exceeding two years, or both.

Where a body corporate is guilty of an offence under these regulations and that offence is proved to have been committed with the consent or connivance of, or to have been attributable to, any neglect on the part of:

(a) any director, manager, secretary or other similar person of the body corporate; or
(b) any person who was purporting to act in any such capacity,

that person is guilty of the offence as well as the body corporate (regulation 34).

Definition

Environmental damage is damage to:

(a) protected species or natural habitats, or a site of special scientific interest;
(b) surface water or groundwater; or
(c) land (regulation 4).

ENVIRONMENTAL PROTECTION ACT 1990 (WITH NOISE ACT 1996 AND THE CLEAN NEIGHBOURHOODS AND ENVIRONMENT ACT 2005) GENERAL PROCEDURAL PROVISIONS

There are relatively few general provisions in the Environmental Protection Act which are applicable to all the procedures in this part, and these are shown below. Each set of procedures contains its own definitions, powers of entry, etc., and these are included in the text of each section.

Crown premises

The Act generally is applicable to Crown premises but does not create a criminal liability. The powers of entry given to local authority officers in respect of the various procedures may, however, be curtailed by the Secretary of State in the interests of national security in relation to specific premises and powers to be specified in a certificate (section 159).

Powers of entry

The general powers of entry for authorised officers for procedures under this part are either provided especially or are contained in powers relating to 'pollution control functions' in section 108 of the Environment Act 1995 (pages 87–88).

CONTAMINATED LAND

Statutory nuisance provisions do not apply where the nuisance consists of or is caused by land being in a 'contaminated state', i.e. land where there are substances in, on or under that land such that harm is being caused, or there is a possibility of its being caused, or the pollution of controlled waters is being, or is likely to be, caused (section 79(1A) and (1B) as inserted by the Environment Act 1995).

Contaminated land is dealt with by special procedures (see FC18 and FC19).

THE IDENTIFICATION AND REMEDIATION OF CONTAMINATED LAND

References

Environmental Protection Act 1990 Part 2A (inserted by section 57 of the Environment Act 1995) as amended also by the Clean Neighbourhoods and Environment Act 2005.
The Contaminated Land (England) Regulations 2006 as amended.
Environmental Protection Act 1990 Part 2A Contaminated Land Statutory Guidance (April 2012).
BS 10175:2011 Investigation of Potentially Contaminated Sites – Code of Practice.
Local Authority Guide to the Application of Part 2A of the Environmental Protection Act of 1990, www.cieh.org/uploadedFiles/Core/Policy/Publications_and_information_services/Policy_publications/Publications/LocalA2007.pdf
Contaminated land technical advice is available from the CIEH website at www.cieh.org/policy/policy2.aspx?id=13418

Extent

These provisions apply in England, Wales and Scotland with minor differences in Wales and Scotland.

Scope

These procedures provide for the identification of contaminated land and formal determination by local authorities. They also provide for what is to be done by way of remediation, which may be required for any contaminated land where the

FC18 Identification of contaminated land – Part 2A Environmental Protection Act 1990

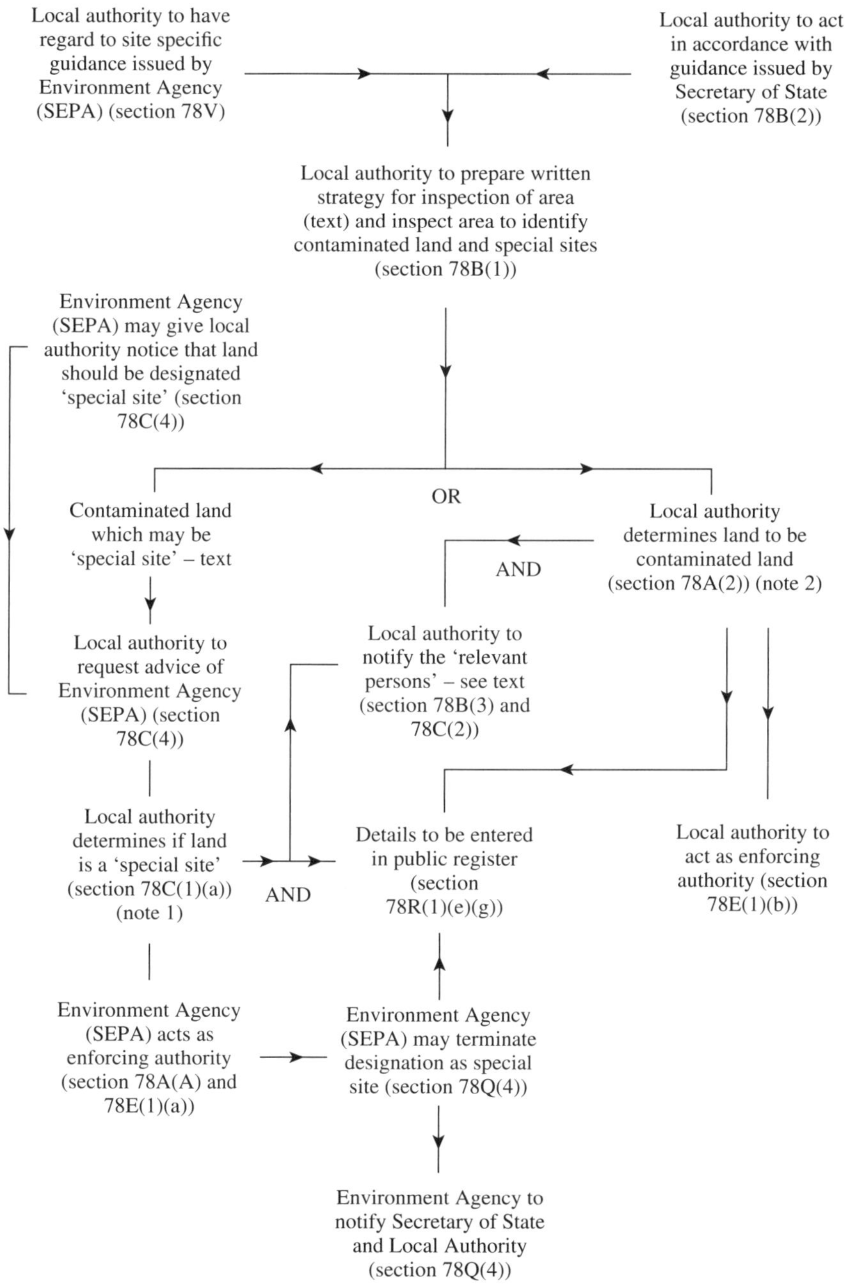

Notes

1. Disputes between the local authority and Environment Agency (SEPA) about the designation of special sites are referred to the Secretary of State, whose decision must be notified to the relevant persons as well as to the parties involved (section 78D).
2. This information is subject to the Environmental Information Regulations 2004, subject to national security or commercial confidentiality.

FC19 Remediation of contaminated land – Part 2A Environmental Protection Act 1990

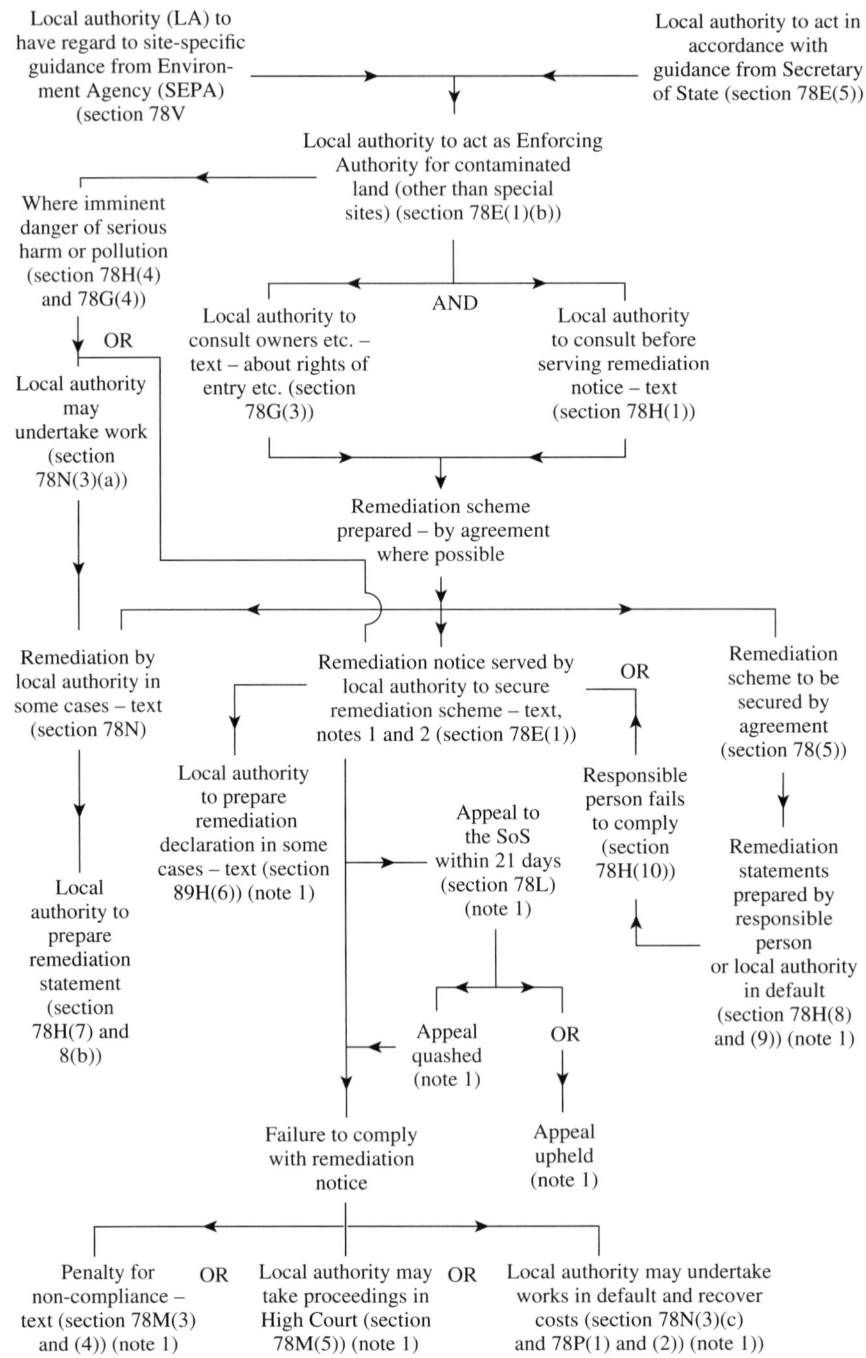

Notes

1. Details to be included in remediation register.
2. Remediation notices may be used several times on the same site if necessary to produce a staged result or to react to changing assessments and circumstances.
3. SoS = Secretary of State.

local authority is the enforcing authority, i.e. all except special sites that are the responsibility of the Environment Agency. The guidance identifies three categories of remediation action, i.e. assessment, remedial treatment and monitoring in paragraphs C64–C71 of Statutory Guidance.

Role of local authorities

The role of the local authorities is to:

(a) prepare a written strategy regarding inspection of their area (section 78B(1) and paragraph B12 of statutory guidance);
(b) identify any potentially contaminated land (section 78B(1)) and any special sites (section 78C(1) and regulations 2 and 3);
(c) determine whether land is contaminated land (section 78A(2)–(5) and paragraphs B37–51 of statutory guidance);
(d) establish who is the owner of the land (section 78A(9)), who appears to be in occupation, who is the appropriate person to bear responsibility for any necessary remediation action (sections 78E(1) and 78F);
(e) notify such persons and the Environment Agency that land is determined as contaminated and establish special sites (section 78B(3) and (4));
(f) require appropriate remediation defined in section 78A(7) and implement under sections 78E, 78H and 78N;
(g) maintain a register (section 78R and regulation 13).

In undertaking these roles, the local authorities are required to act in accordance with guidance issued by the Secretary of State (section 78B(2)) and to have regard to any site-specific advice given to it by the Environment Agency (Scottish Environment Protection Agency in Scotland) (section 78V).

Contaminated land

This is defined as:

> any land which appears to the local authority in whose area it is situated to be in such a condition by reason of substances in, on or under the land that:
>
> (a) significant harm is being caused or there is a significant possibility of such harm being caused; or
> (b) pollution of controlled waters is being, or is likely to be, caused (section 78A(2)).

In this context, 'harm' means harm to the health of living organisms or other interference with the ecological systems of which they form part and, in the case of man, includes harm to his property (section 78A(4)).

Section 78A(5) provides that the following are to be determined by the local authority in accordance with the statutory guidance (Chapter A parts 3 and 4):

(a) what harm is to be regarded as significant (part 3 paragraphs A23–A26 and table A);
(b) whether the possibility of significant harm is significant (part 3 paragraphs A27–A34 and table B);

(c) whether pollution of controlled waters is being, or is likely to be, caused (part 4 paragraphs A35–A39).

The Contaminated Land (England) Regulation 2006 and the Radioactive Contaminated Land (Modification of Enactments) (England) Regulation 2006 have brought within the scope of this definition land which is contaminated as a result of radioactive substances in or under the ground. Such sites are to be regarded as special sites (see below) and are to be dealt with by the Environment Agency.

Special sites

Wherever the local authority has determined land to be contaminated land, it must also decide whether it meets the description of special sites prescribed for the purposes of section 78C(8), regulations 2 and 3 and schedule 1 of the 2006 regulations and identify special sites for which the Environment Agency is the enforcing authority.

If the local authority considers that the land might be designated a special site, it should seek the advice of the Environment Agency (section 78C(8)). The Environment Agency also needs to consider whether any contaminated land should be designated as a special site. This might be based on information from its other pollution control functions. There is a duty on the Environment Agency to notify the local authority (section 78C(4)). Where the contaminated land meets one or more of the prescribed descriptions, the local authority must designate it a special site. Disputes between a local authority and the Environment Agency are to be settled by the Secretary of State.

Having made such a designation, the local authority must give written notice to:

(a) the Environment Agency;
(b) the owner of the land;
(c) any person who appears to be the occupier of all or part of the land; and
(d) each person who appears to be an appropriate person (see below).

Following this process, the Environment Agency and not the local authority is responsible for securing any necessary remediation (sections 78C, 78D and 78E(1)).

The identification of contaminated land

Part 3 of chapter B of the statutory guidance sets out the inspection duty. The local authority has sole responsibility for determining whether land is contaminated and it cannot delegate this except in accordance with section 101 of the Local Government Act 1972. The duty to inspect under section 78B(1) is extended by the statutory guidance to include a strategic approach. Every local authority must therefore have set out its approach to this duty in a written strategy and published it by July 2001 (chapter B paragraphs B12 and B15). This approach should enable the local authority to identify, in a rational, ordered and efficient manner, the land which merits detailed individual inspection, identifying the most pressing and serious problems first and concentrating resources on the areas where contaminated land is most likely to be found (chapter B paragraph B9).

Paragraphs B18–25 of the statutory guidance cover detailed inspections and use of powers of entry. If land has been determined as contaminated land and is likely to be designated a special site, the local authority should always make arrangements for the Environment Agency to carry out the inspection on its behalf. Section 108 of the Environment Act 1995 (see pages 87–88) can also be used to authorise Environment Agency staff.

'Contaminated Land Inspection Strategies' DETR (2001) is a Technical Advice Note for local authorities at this stage.

BS 10175:2001 provides guidance for local authorities on investigation techniques, sampling and on-site testing and laboratory analysis.

To enable a local authority to make a judgement regarding significant harm (section 78A(2)), scientific guidance was published by DEFRA in March 2002. This consists of the Contaminated Land Exposure Assessment Model CRL 10 (CLEA) together with accompanying reports C7–10. The key element of CLEA is the soil guidance values (SGVs) which represent 'intervention' levels above which unacceptable risks may exist and requires further investigation or remediation.

Appropriate persons

These are the persons responsible for any remediation of contaminated land that the local authority may require. Part 2A of the Act defines two categories of 'appropriate persons' and sets out the circumstances in which they might be liable.

The first category is created by section 78F(2) – 'any person or any persons who caused or knowingly permitted the substances, or any of the substances, by which the contaminated land in question is such land, to be in, or under that land, is an appropriate person'. These are referred to as Class A persons by the statutory guidance. Such a person will be the appropriate person only in respect of any remediation, which is referable to particular substances, which he caused or knowingly permitted to be in, on or under the land (section 78F(3)).

The second category arises where it is not possible to find a Class A person. In these circumstances, the owner (see definitions) or occupier for the time being of the land is an appropriate person, referred to as a Class B person. Occupier is not defined but DETR circular 02/2000 states that this would normally be the person in occupation, e.g. tenant or licensee.

The statutory guidance in part 3 of chapter D guides local authorities in the circumstances where two or more appropriate persons are liable for remediation. It allows a local authority to determine who should bear the liability and identifies five distinct phases in this procedure.

Remediation (see definition on page 115)

Having identified land as contaminated land, the local authority has a statutory duty to ensure that appropriate remediation is carried out (section 78E(1)(b)).

It is the government's intention that, as far as possible, remediation should be carried out by agreement rather than through the use of the formal notice procedures. In any event, the local authority is required to consult with 'appropriate persons' (see above), owners etc. about what is to be done by way of remediation before serving any remediation notices (section 78H(1)) and to produce a remediation scheme.

From a practical point of view, remediation is likely to be phased with different remediation actions being required at different times, e.g. assessment actions, remedial treatment actions and monitoring actions.

The local authority must have regard to the standard of remediation, i.e. it should result in land being 'suitable for use'. Paragraphs C17–C28 of the statutory guidance give general advice on the best practicable techniques, multiple pollution linkages, volunteered remediation and the circumstances where land is no longer contaminated.

The practicability, effectiveness and durability of remediation are governed by section 78E(5)(b), and paragraphs C44–C63 of the guidance set out the general criteria to meet these objectives.

Section 78E(5)(c) requires that regard should be had to the reasonableness of remediation. Paragraphs C29–C33 set out the criteria for this. Paragraphs C34–C37 set out the matters which the local authority should take into account when considering the costs involved and compensation that is to be paid in accordance with section 78G(5) involving compliance with the remediation notice.

Urgent remediation action

Where there is imminent danger of serious harm or serious pollution of controlled water being caused, the local authority can either:

(a) serve an urgent remediation notice without going through the normal consultation and other procedural requirements (section 78H(4)); or
(b) where an urgent remediation notice would not result in remediation happening soon enough, carry out the urgent remediation itself and recover its costs where possible (section 78H(7)).

Depending upon the circumstances, the local authority may take such action either before any remediation work has commenced by normal procedures or during the course of such work.

If the local authority carries out the work itself, it must produce a remediation statement describing the actions it has carried out (section 78H(7)).

'Imminent' and 'serious' are not defined in part 2A of the Act. The statutory guidance states that a local authority needs to judge each case on the normal meaning of the words and the facts of the case. Paragraphs C39–C43 guide local authorities with regard to the seriousness of any significant harm. A22, A34 and Tables A and B assist in the interpretation of significant harm.

Remediation notices

Where it has not been possible to secure a remediation scheme by agreement, the local authority is required to serve a remediation notice on each 'appropriate person' (see above) specifying what needs to be done by way of remediation and the time periods within which each action must be taken (section 78E(1) and regulation 4). The content of notices is specified in sections 78E(1) and (3) and in regulation 4.

Regulation 5 requires that copies of the notice are sent at the same time to those persons required to be consulted under sections 78G(3) and 78H(1) and to the Environment Agency.

Remediation notices may be served on different persons for action relating to different substances (section 78E(2)) and, where served on more than one person, the notice must state the proportion of the costs to be borne by each person (section 78E(3)).

In specifying the works to be undertaken, the local authority must consider any work to be reasonable having regard to:

(a) the costs; and
(b) the seriousness of the harm or of the pollution (section 78E(4)).

In determining what is to be done by way of remediation, the local authority must also have regard to the standard to which land is to be remediated and to what is regarded as reasonable (section 78E(5)). Remediation should not be required for the purpose of achieving any aims other than those set out in paragraphs C18–C24 of the guidance; in particular, not for dealing with matters which do not themselves form part of a significant pollution linkage or making the land suitable for any uses other than its current use.

Notices may be modified by the authority under section 78L(2)(b) and are then subject to notification and appeal provisions (regulation 11).

Remediation declarations

Where the local authority has identified works which could be carried out but is precluded from including them on a remediation notice because of the 'cost' and 'seriousness' tests above, it must produce and publish a remediation declaration which indicates:

(a) the work in question;
(b) why the local authority would have otherwise specified that work; and
(c) the grounds on which the local authority feels justified in not specifying the work on the notice (section 78H(6)).

Such declarations must be included in the remediation register (section 78R(1)(c)).

Appeals

The grounds for appeal to the Secretary of State against a remediation notice are set out in regulation 7. The person has 21 days to appeal and, where it is duly made, the notice is suspended pending the decision. In these circumstances, the local authority needs to consider whether it should carry out urgent remediation itself using its powers under section 78N. The Secretary of State must hold a hearing or local enquiry (regulation 9).

Offences

Persons failing to comply with a remediation notice are liable, on summary conviction, to a fine not exceeding level 5 and to one-fifth of level 5 for each day on which the offence continues.

Where the land concerned is industrial, trade or business premises, the maximum fine is increased to £20,000 or one-tenth daily.

As an alternative to proceeding summarily, the local authority may go to the High Court when it feels the former would not secure an effectual remedy (section 78M).

In addition, the local authority may itself undertake the works in default and recover costs (section 78N(3)(c)).

Remediation statements

Where, without the service of a remediation notice, remediation is to take place, the person who is to carry it out (including the local authority where this is the case) is required to prepare and publish a remediation statement which records:

(a) the things which have been, are being or are to be done;
(b) the name and address of the person undertaking the works; and
(c) the time which the work is expected to take (section 78H(7) and (8)).

These statements must be entered in the remediation register (section 78(1)(c)).

Where the requirement for a remediation statement is not met by the person concerned, the local authority may produce it and recover its costs in so doing (section 78H(9)).

Remediation registers

Each local authority must maintain a remediation register which will contain details specified in regulation 13 and schedule 3 (section 78R(1) and regulation 13).

Certain information is excluded from this requirement where this is in the interest of national security or it is commercially confidential (sections 78S and 78T).

The register is available for inspection at the local authority's principal offices at all reasonable times (section 78R(8)(a)) and there must be a facility for members of the public to obtain copies of entries at a reasonable charge (section 78R(8)(b)).

Schedule 3 of the regulations prescribes full particulars of the following matters to be included in the register by the local authority:

(a) remediation notices;
(b) appeals against remediation notices;
(c) remediation declarations;
(d) appeals against charging notices;
(e) designation of special sites;
(f) notification of claimed remediation;
(g) convictions for offences under section 78M;
(h) guidance issued under section 78V(1); and
(i) other environmental controls.

Definitions

Controlled waters are defined in section 78A(9) by reference to part 3, section 104 of the Water Resources Act 1991. This embraces territorial and coastal waters, inland fresh waters and groundwaters but does not include waters contained in underground strata but above the saturation zone.

Notice means a notice in writing.

Owner, in relation to any land in England and Wales, means a person (other than a mortgagee not in possession) who, whether in his or her own right or as trustee for any other person, is entitled to receive the rackrent of the land, or, where the land is not let at a rackrent, would be so entitled if it were so let.

Owner, in relation to any land in Scotland, means a person (other than a creditor in a heritable security not in possession of the security subjects) for the time being entitled to receive or who would, if the land were let, be entitled to receive the rents of the land in connection with which the work is used and includes a trustee, factor, guardian or curator and in the case of public or municipal land includes the persons to whom the management of the land is entrusted (section 78A(9)).

Remediation means:

(a) the doing of anything for the purpose of assessing the condition of:
  (i) the contaminated land in question;
  (ii) any controlled waters affected by that land; or
  (iii) any land adjoining or adjacent to that land;
(b) the doing of any works, the carrying out of any operations or the taking of any steps in relation to any such land or waters for the purpose:
  (i) of preventing or minimising, or remedying or mitigating the effects of, any significant harm, or any pollution of controlled waters, by reason of which the contaminated land is such land; or
  (ii) of restoring the land or waters to their former state; or
(c) the making of subsequent inspections from time to time for the purpose of keeping under review the condition of the land or waters (section 78A(7)).

NOISE: INTRODUCTION

Noise nuisances

Noise complaints have been dealt with under statutory nuisance provisions for a number of years and are still included in the statutory nuisances provisions of the Environmental Protection Act 1990 section 79 (see pages 471–483).

Control over 'noisy parties'

Whilst certain types of these activities may be controlled through the Licensing Act 2003 (see page 539), many noisy parties are private, domestic, held in the home and for no private gain. In such cases, the use of the statutory noise nuisance legislation is one possible means of controlling noise.

It is unlikely that these provisions were drafted with this particular use in mind; nevertheless they can form an effective remedy, and DEFRA and CIEH have issued guidance to local authorities: 'Neighbourhood Noise Policies and Practice for Local Authorities – a Management Guide'. The way in which the statutory nuisance provisions can be used in this way is shown in FC20.

However, specific powers are now provided by the provisions of the Noise Act 1996, set out in FC28, to deal with night noise from certain premises.

Noise in streets

The Noise and Statutory Nuisance Act 1993 brought the definition of a new statutory nuisance from noise arising from a vehicle, machinery or equipment in streets. This was done to allow local authorities to deal with problems that could give rise to considerable nuisance but which were not covered by previous legislation, which was restricted to noise from premises. In particular, this was the case with DIY car repairs, cooling engine noise from refrigerated lorries, misfiring of car alarms, buskers, etc.

The provisions also introduced a special procedure for dealing with noise which amounts to a statutory nuisance from vehicles, machinery or equipment in a street which are unattended and where the person responsible cannot be readily found.

The special procedure allows for the fixing of the abatement notice to the unattended vehicle but the authorised officer must spend up to one hour attempting to trace the person responsible (for appeals against these notices, see page 476).

Where he is successful in so doing, that person must then comply with the notice, there being subsequent penalties for non-compliance and the local authority having default powers to take any steps necessary to abate the nuisance. To allow sufficient time for the person responsible, having been contacted, to abate the nuisance, the local authority may indicate on the notice attached to the unattended vehicle etc. that the time limit for compliance will be extended if it is possible for the person responsible to be served with a copy of the notice. A new notice can then be served with a different period allowed. Where the person

FC20 Statutory nuisances – control over noisy parties – Section 79 Environmental Protection Act 1990

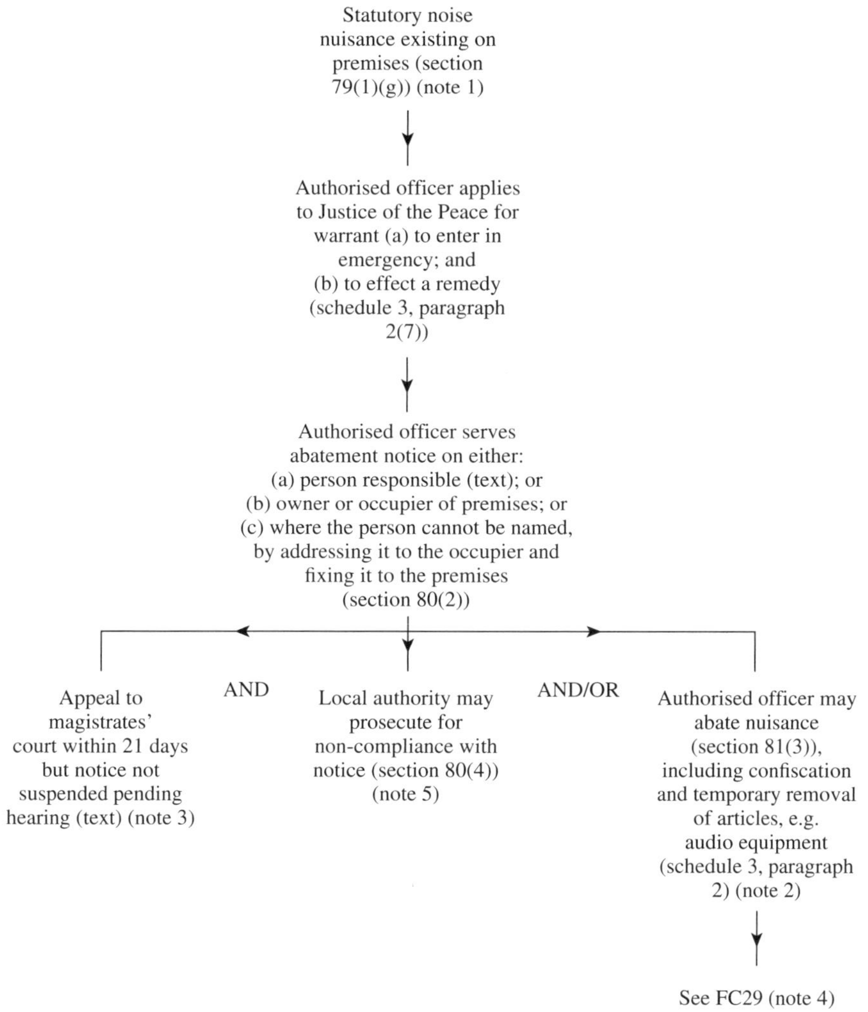

Notes

1. Anticipated noise nuisances can also be dealt with (pages 118, 475).
2. These powers must be exercised reasonably (section 81(3)).
3. The court may quash or vary the notice or dismiss the appeal (regulation 2(5)).
4. The procedure for dealing with the seizure of equipment is provided for in section 10 and the schedule to the Noise Act 1996 – see FC29.
5. For the adoptive procedure dealing with night noise from dwellings, see FC28.

FC21 Statutory nuisances – certain noise in the street – Section 80 Environmental Protection Act 1990

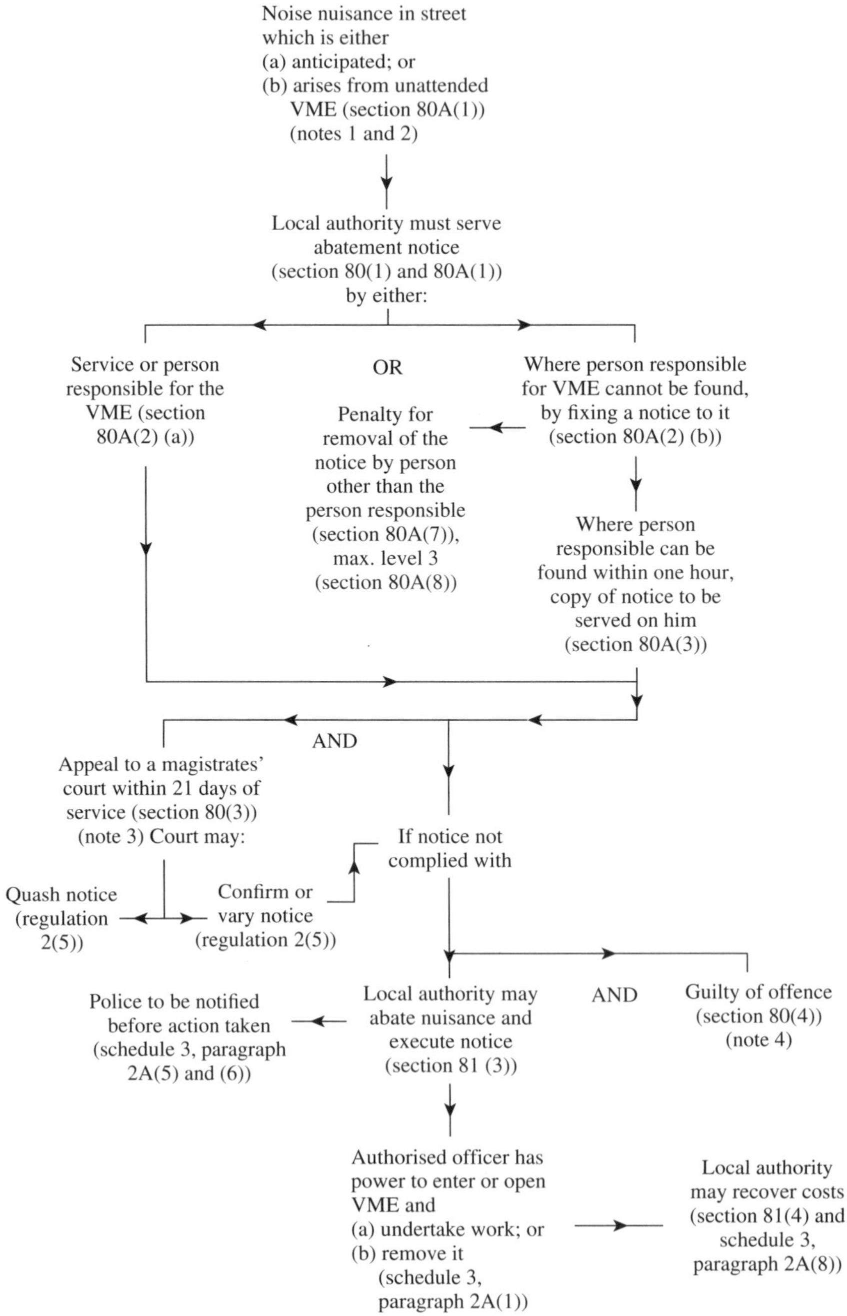

responsible cannot be traced within one hour, the local authority may then enforce the notice. Before any action is taken, the police must be informed but the authorised officer has the power to enter or open the vehicle etc., if necessary by force, or remove it to a safe place. Having secured the abatement of the nuisance, the authorised officer must secure the vehicle etc., e.g. reset the vehicle alarm. The authorised officer must not cause any more damage than is necessary to execute the notice (section 80A).

Section 4 of the Clean Neighbourhoods and Environment Act 2005 now provides another remedy to some of the situations previously dealt with under this procedure. This provides that the person carrying on works of repair, maintenance, etc. to a motor vehicle on a road, with certain exceptions, is guilty of an offence. Section 6 of that Act provides a system of fixed-penalty notices for this offence.

CONTROL OF POLLUTION ACT 1974 (NOISE CONTROL PROVISIONS) GENERAL PROCEDURAL PROVISIONS

The following provisions are generally applicable to procedures under the Control of Pollution Act 1974.

Power of entry

Any person authorised in writing by an authority charged with responsibilities under the Control of Pollution Act 1974 may enter any land or vessel at any reasonable time for the following purposes:

(a) to perform any function conferred on the authority;
(b) to determine whether such a function should be performed;
(c) to determine whether the Act or instruments made by virtue of the Act are being complied with;
(d) to carry out inspections, measurements and tests (section 91(1)).

Where:

(a) admission has been refused; or
(b) refusal is apprehended; or
(c) the land or vessel is unoccupied; or

Notes

1. For other statutory noise nuisances in streets, see page 116.
2. VME – vehicle, machinery or equipment.
3. The local authority will have indicated in the notice that it is to be suspended pending the hearing of any appeal (page 473).
4. Details of penalties are given on page 11.
5. For the control over loudspeakers in streets, see FC22.
6. A similar procedure for use in Scotland is set out in schedule 1 of the Noise and Statutory Nuisance Act 1993 in conjunction with section 58 of the Control of Pollution Act 1974.
7. Regulation numbers refer to the Statutory Nuisance (Appeals) Regulations 1995.
8. See DoETR circ. 9/97 for guidance on the operation of this procedure.

(d) the occupier is temporarily absent; or
(e) the case is one of emergency; or
(f) application for admission would defeat the object of entry *and* there is a reasonable ground for entry,

the authority may apply to a Justice of the Peace for a warrant to enter, if need be by force (section 91(2)).

Admission to land or a vessel used for residential purposes, and admission with heavy equipment to any other land or vessel, cannot be demanded of right unless at least seven days' notice of intention has been given, except in emergency or where the land or vessel is unoccupied (section 92(3)).

An authorised person may take with him such persons and equipment as necessary (section 92(2)).

After entering unoccupied land, the authorised person must leave it as secure against trespass as he found it (section 92(4)). The authority is liable to pay compensation for any damage sustained in exercising its powers and responsibilities in this respect (section 92(5)). The maximum penalty for obstruction is level 3 on the standard scale (section 92(6)).

Information

An authority is provided with a general power to obtain information necessary to perform its functions under the Control of Pollution Act 1974 by the service of notice specifying the information required and giving a time period within which or time at which it must be supplied. The maximum penalty for non-compliance or wilfully giving false information is level 5 on the standard scale (section 93).

Definitions

Best practicable means – 'practicable' means reasonably practicable, having regard among other things to local conditions and circumstances, to the current state of technical knowledge and to the financial implications. 'Means' includes design, installation, maintenance and manner and periods of operation of plant and machinery, and the design, construction and maintenance of buildings and acoustic structures. The test of best practicable means is to apply only so far as is compatible with statutory duties imposed, with safety and safe working practices and with the exigencies of any emergency or unforeseen circumstances. Regard shall be had to any relevant provisions of an approved Code of Practice (section 72).
Noise includes vibration (section 73(1)).
Notice means a notice in writing (section 105(1)).
Owner, except in Scotland, means the person for the time being receiving the rackrent of the premises in connection with which the word is used whether on his own account or as agent or trustee for another person, or who would so receive the rackrent if the premises were let at a rackrent (section 105(1)).
Person responsible, in relation to the emission of noise, means the person to whose act, default or sufferance the noise is attributable (section 73(1)).
Premises includes land (section 105(1)).

CONSENTS FOR THE USE OF LOUDSPEAKERS IN STREETS

References

Control of Pollution Act 1974, section 62.
Noise and Statutory Nuisance Act 1993, section 8 and schedule 2.
DEFRA Guidance on the Noise Act 1996 as amended – Guidance for Local Authorities, March 2008.
Code of Practice on Noise from Ice-Cream Van Chimes etc. 1982 (about to be amended).

Extent

This provision applies in England, Wales and Scotland (section 109 and section 13).

Scope

This procedure allows a local authority to consent to the operation of a loudspeaker in a street between the hours of 9 pm and 8 am, being the period during which such operation is generally prohibited under section 62(1) of the Control of Pollution Act 1974 (Noise and Statutory Nuisance Act 1993, schedule 2, paragraph 1).

Consents may not be granted in connection with any election or for advertising any entertainment, trade or business (schedule 2, paragraph 1(2)). It is thought that local authorities may wish to use a consent system – for example, to allow charity events or street events to continue beyond 9 pm.

Adoption of scheme

The consent scheme operates only if the local authority has adopted its provisions by resolution of the council. Notice that such a resolution has been made must be published in a local newspaper in two consecutive weeks before becoming operative on a date specified, which cannot be less than one month from the date of the resolution.

The notice in the newspaper must set out the effect of the resolution and explain the procedure for applying for consents (section 8(1)–(4)).

Applications

These must be made to the local authority in writing and must contain the information requested by the local authority to enable it to reach a decision. The local authority may set a reasonable fee for each application (schedule 2, paragraphs 3 and 5).

FC22 Consents for the use of loudspeakers in streets – Section 62 Control of Pollution Act 1974 and Section 8 Noise and Statutory Nuisance Act 1993

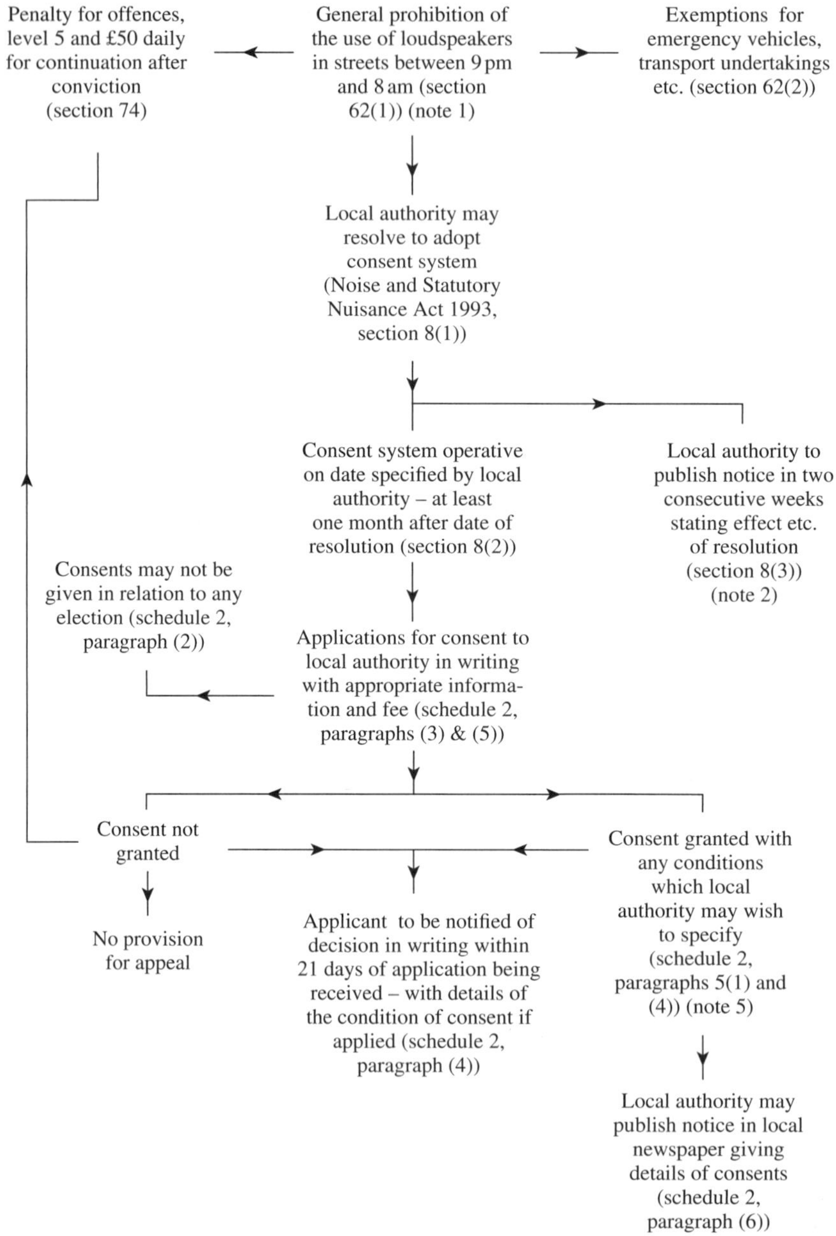

Notes

1. There is a total prohibition at any time on the use of loudspeakers in streets for advertising any entertainment, trade or business (there is an exception for advertising the presence of a vehicle selling perishable commodities, e.g. ice-cream, between noon and 7 pm). The consent procedure does not allow this prohibition to be breached.
2. There is no provision for the making or consideration of representations.
3. Where not otherwise indicated, section and schedule numbers refer to the Noise and Statutory Nuisance Act 1993.
4. For the control of statutory noise nuisances in streets, see FC21.
5. Operating a loudspeaker with a consent but in breach of its conditions would appear to create an offence under the Control of Pollution Act 1974, i.e. for breaching the general prohibition order (section 62(1)).

Considerations

The matters to be taken into account by the local authority are not specified but might reasonably include:

(a) location;
(b) adjacent activities;
(c) time;
(d) number of loudspeakers and their output power;
(e) methods of noise control;
(f) the planned route of any procession, etc.

Consents

Local authorities may decide not to issue a consent or issue one at their discretion, with or without conditions. There are no appeal provisions (schedule 2, paragraph 1). Notification of the local authority's decision must be put in writing to the applicant within 21 days of the application being made, with details of any conditions which are to be attached (schedule 2, paragraph 4).

The conditions attached to any consent could include the specification of noise levels and of any route to be taken, but could cover any issue of legitimate concern to the local authority.

The local authority may (but is not obliged to) publicise details of any consents in a local newspaper (schedule 2, paragraph 6).

Consents and statutory nuisance

The issue of a consent does not affect the operation of procedures to deal with statutory noise nuisances in streets (FC21).

If during the operation of a consent such a nuisance is created, the local authority may use its powers under that procedure.

Definitions

Local authority – see pages 8, 55.
Street means a highway and any other road, footway, square or court which is, for the time being, open to the public (Control of Pollution Act 1974 section 62(1)).

NOISE FROM CONSTRUCTION SITES

References

Control of Pollution Act 1974, sections 60 and 61.
Control of Noise (Appeals) Regulations 1975.
Control of Noise (Codes of Practice for Construction and Open Sites) (England) Order 2002.

DEFRA Guidance on the Noise Act 1996 as amended – Guidance for Local Authorities, March 2008.

Extent

These provisions apply to England, Wales and Scotland.

Scope

The first of these two procedures stems from an application for consent to the local authority, which, if operated in accordance with any conditions or with limitations, provides a defence against any proceedings taken by a local authority following the second procedure, i.e. a local authority notice specifying noise emission standards. The procedures could be operated together but only beneficially where, following application to the local authority, consent has been granted without conditions or limitations, or no consent was given.

The procedures are applicable to the following works:

(a) erection, construction, alteration, repair or maintenance of buildings, structures or roads;
(b) breaking up, opening or boring under any road or adjacent land in connection with the construction, inspection, maintenance or removal of works;
(c) demolition or dredging works;
(d) any other works of engineering construction (section 60(1)).

Criteria for consideration

In judging both applications for consent and whether or not to serve notices, local authorities must have regard to the following points:

(a) BS 5228 part 1 1997, part 3 1997, part 4 1992 and part 5 1997. Noise control on construction and open sites (see the Control of Noise (Codes of Practice for Construction and Open Sites) (England) Order 2002);
(b) the need to ensure that best practicable means are employed to minimise noise;
(c) in considering the specification of any particular methods or plant etc., the interest of the applicant or recipient of the notice in regard to alternative methods etc. which might be as effective but more acceptable to them;
(d) the need to protect any persons in the locality from the effects of noise (section 60(4)).

Local authority notices

1. **Person responsible.** This is the person carrying out, or going to carry out, the works and any other persons who appear to the local authority to be responsible for, or have control over, those works. Notices may be served on all or any such persons (section 60(5)).

FC23 Noise from construction sites – Sections 60 and 61 Control of Pollution Act 1974

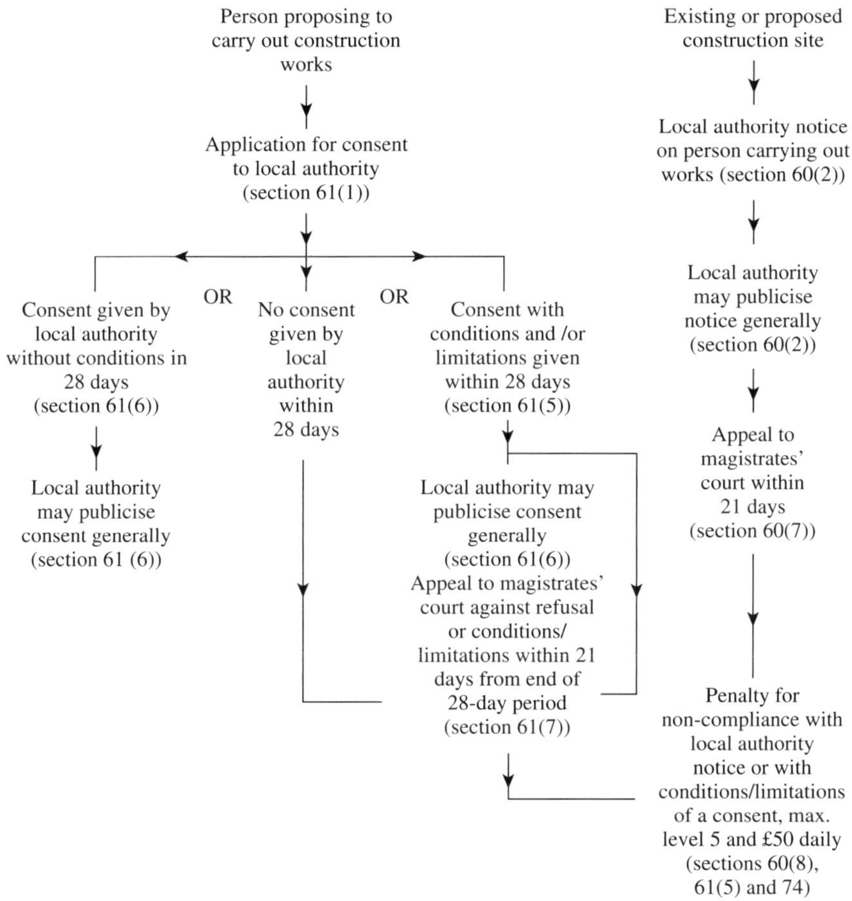

Note

1. There are no local authority powers of default following the notice procedure.

2. **Content.** Notices may include:
   (a) specification of plant and machinery which is or is not to be used;
   (b) the hours during which the works may be carried out;
   (c) specification of noise levels from the premises, from any specified point on the premises or during specified hours;
   (d) a provision for any change of circumstances (section 60(3)).

   A notice may specify a time within which the notice should be complied with, which should not be less than the appeal period, i.e. 21 days, and may require the execution of works or other steps, but the local authority has no default powers (section 60(6)).

   If the local authority intends that the notice should not be suspended in the event of an appeal, this must be stated in the notice (regulation 5 Control of Noise (Appeals) Regulations 1975).
3. **Appeals.** The grounds of appeal are set out in the Control of Noise (Appeals) Regulations 1975 as:
   (a) the notice is not justified;
   (b) there is some material informality, defect or error in the notice;
   (c) the local authority has refused unreasonably to accept compliance with alternative requirements or the requirements of the notice are unreasonable or unnecessary;
   (d) any time period allowed for compliance is not reasonably sufficient;
   (e) the notice should have been served on some other person in substitution for, or in addition to, the appellant;
   (f) the local authority has not had regard to the matters set out in section 60(4) (regulation 5).

   The court may dismiss the appeal, vary or quash the notice and/or make an order as to respective responsibilities for compliance and costs.

   In the event of an appeal, local authority notices are suspended until the appeal has been determined or abandoned where:
   (a) the noise is caused in the performance of a statutory duty imposed on the appellant; or
   (b) compliance would involve expenditure being incurred before the hearing, except in those cases where the local authority is of the opinion that:
   (a) the noise is injurious to health; or
   (b) the noise is likely to be of limited duration so that suspension would give no practical effect to the notice; or
   (c) the expenditure to be incurred would not be disproportionate to the public benefit from compliance, provided that this has been stated on the original local authority notice (regulation 10 Control of Noise (Appeals) Regulations 1975).
4. **Defence.** It is a defence in any proceedings for non-compliance with a local authority notice to prove that the alleged contravention amounted to the carrying out of the works in accordance with a consent under section 61 (section 61(8)).

Consents

1. **Applications.** These must be made at the same time as, or later than, any application for building regulations approval and may be made by anyone intending to carry out construction works. The application must contain particulars of:
   (a) the works proposed and the methods to be used to carry them out; and
   (b) the steps proposed to minimise noise (section 61(2) and (3)).

2. **Consent.** If the local authority considers that it would not serve a notice under section 60 if the works are carried out in accordance with the application, it must give consent and may:
   (a) attach conditions;
   (b) limit or qualify consent to allow for any changes in circumstances;
   (c) limit the duration of the consent (section 61(4) and (5)).

   A consent under this procedure does not of itself constitute a defence against proceedings for statutory nuisance by aggrieved persons under section 80 Environmental Protection Act 1990, and this must be stated in the consent. However, it will be a defence against proceedings by a local authority for failure to comply with an abatement notice under section 80 Environmental Protection Act 1990 to show that the alleged offence was covered by a notice under section 60 Control of Pollution Act or by a consent under sections 61 or 65 Control of Pollution Act. It will be a defence in proceedings following section 60 notices to show that the works were carried out in accordance with that consent (section 61(8) and (9) and section 80(9) Environmental Protection Act 1990).

3. **Appeals.** Grounds of appeal against no consent being granted or against attachment of conditions, limitations or qualifications are set out in Control of Noise (Appeals) Regulations 1975 as being:
   (a) that any condition, limitation or qualification is not justified;
   (b) that there is some material informality, defect or error in connection with the consent;
   (c) that the requirements of any condition are unreasonable in character or extent, or are unnecessary;
   (d) that the time within which any conditions etc. are to be complied with is not reasonably sufficient (regulation 6).

Publication

In order to bring either notices under section 60 or consents under section 61 to the attention of the public who might be affected, the local authority is able to publicise the contents of notices and consents in any ways which it thinks appropriate (sections 60(2) and 61(6)).

Definition

Work of engineering construction means the construction, structural alteration, maintenance or repair of any railway line or siding or any dock, harbour or inland navigation, tunnel, bridge, viaduct, waterworks, reservoir, pipeline, aqueduct, sewer, sewage works or gas holder (section 73(1)).

Practical issues

Some local authorities use the spirit of the legislation to work with contractors to limit noise complaints rather than serve notices, prior consents, etc. Responsible contractors approach local authorities for prior consent. They prefer an informal approach rather than notices and consents. Informal action works well with contacts in organisations through other areas of work, e.g. Integrated Pollution Prevention and Control (IPPC).

It is worth noting that local authorities need to coordinate their approaches to these organisations in licensing, planning, building control and noise control. Duplicating conditions in planning permission and in licensing may cause problems for licensing and noise control enforcement.

NOISE ABATEMENT ZONES: DESIGNATION BY LOCAL AUTHORITIES

Reference

Control of Pollution Act 1974, section 63 and schedule 1.

Extent

These provisions apply in England, Wales and Scotland.

Local authority duty to inspect district

Each local authority is under a statutory obligation to inspect its district from time to time to decide how to exercise its powers relating to noise abatement zones (NAZ), but this does not need to precede the designation of each zone (section 57(b)).

Noise abatement zones

The order designating a NAZ may relate to all or part only of the local authority district and may subsequently be revoked or varied by the local authority. The order must detail the classes of premises which will be covered by the NAZ (section 63).

Publication of intention

Before declaration of the NAZ, or before revoking or varying it, the local authority is required to:

(a) serve notice on every owner, lessee and occupier, other than tenants of one month or less, of each premises in the area of a class to which the order relates; and

FC24 Noise abatement zones: designation by local authorities – Section 63 and Schedule 1 Control of Pollution Act 1974

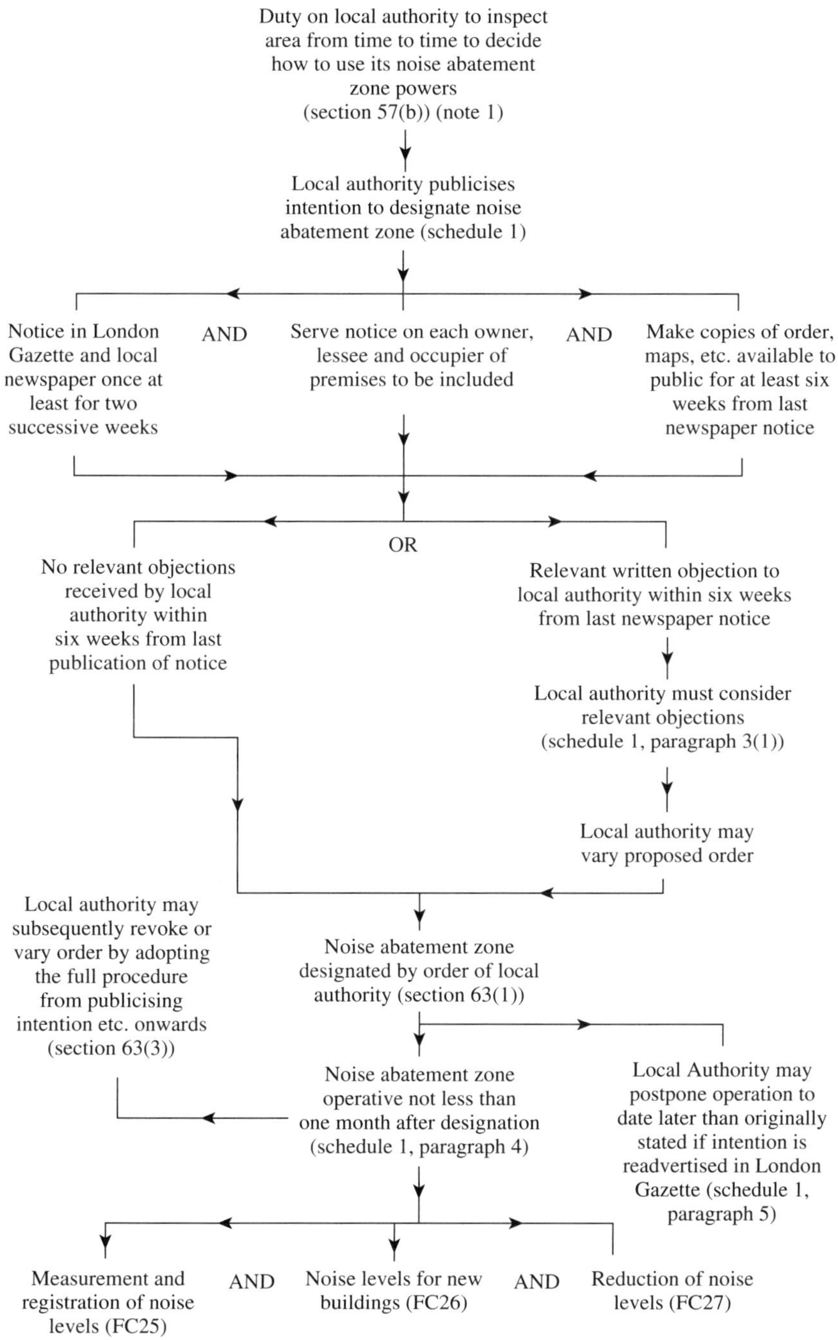

Note

1. This is not a necessary prerequisite to the declaration of each noise abatement zone.

(b) publish a notice in the *London Gazette* and at least once in each of two successive weeks in a local newspaper.

These notices referred to in (a) and (b) above must:

(a) contain a statement that the local authority intends to make that order;
(b) contain the general effect of the order;
(c) specify where copies of the order and maps etc. may be seen by the public free of charge for at least six weeks from the last newspaper publication of the notice;
(d) contain a statement that objections may be made in writing to a local authority within a six-week period (schedule 1, paragraphs 1 and 2).

Objections

Before making an order designating a NAZ, the local authority must consider any objections received during the six-week period unless it is satisfied that it need not do so, having regard to:

(a) the nature of the premises to which the order will relate;
(b) the nature of the interests of the person making the objection;
(c) in the case of an order revoking or varying a previous order, the fact that the substance of the objection is the same as an objection made to the initial order (schedule 1, paragraphs 3 and 4).

Operation

A NAZ becomes operative on a date specified in the order by the local authority, which, except for orders revoking or modifying previous orders, must be at least one month from the date on which the order was made (schedule 1, paragraph 4).

NOISE ABATEMENT ZONES: MEASUREMENT AND REGISTRATION OF NOISE LEVELS AND CONSENT TO EXCEED THEM

References

Control of Pollution Act 1974, sections 64 and 65.
Control of Noise (Appeals) Regulations 1975.
Control of Noise (Measurements and Registers) Regulations 1976.

Extent

These provisions apply to England, Wales and Scotland.

Measurement

Once a noise abatement zone becomes operative, the local authority is required to measure noise levels emanating from premises within the area which are subject to the noise abatement order (section 64(1)).

FC25 Noise abatement zones: measurement and registration of noise levels and consent to exceed them – Sections 63 to 67 Control of Pollution Act 1974

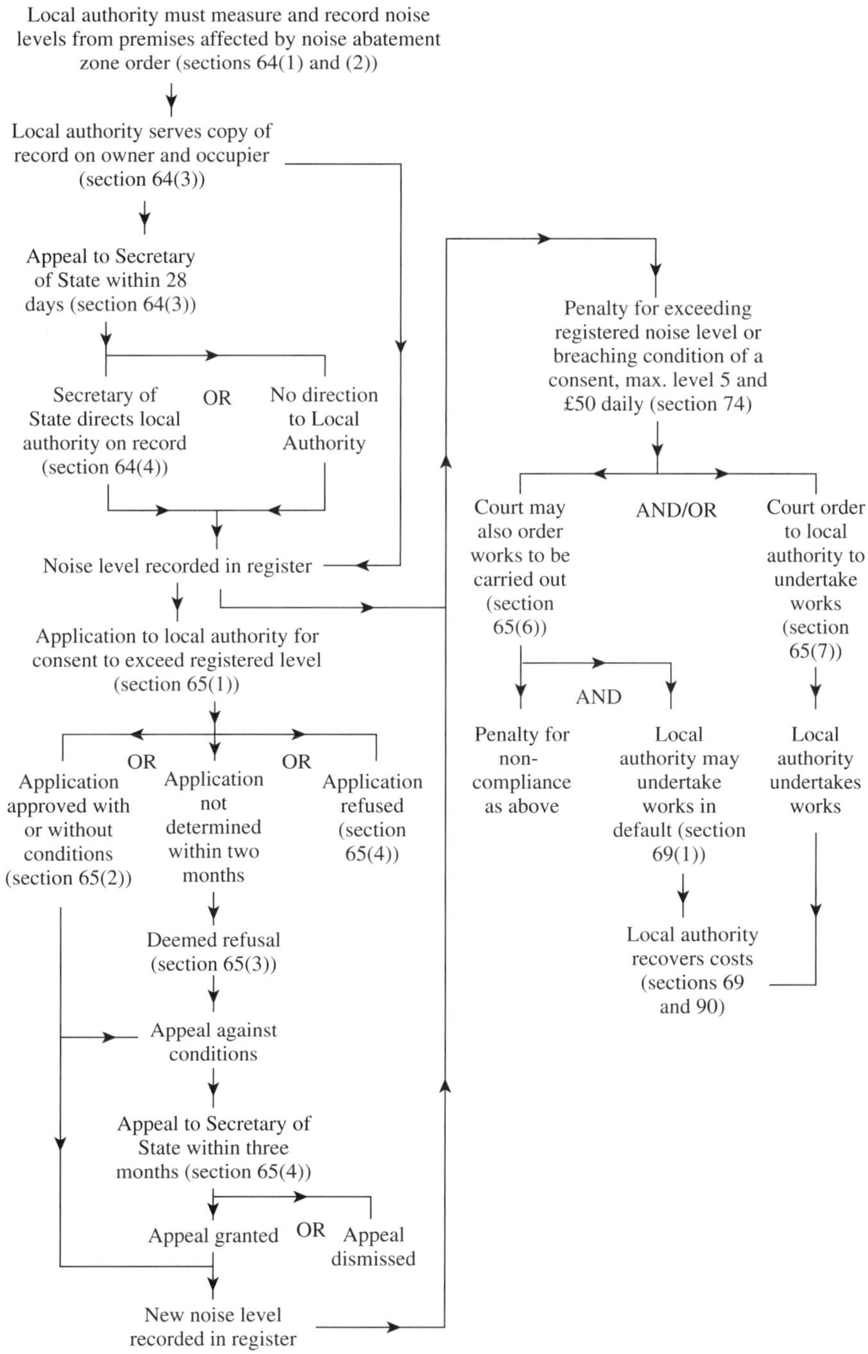

Notes

1. For noise reduction notices in noise abatement zones, see FC27.
2. For provisions relating to new buildings in noise abatement zones, see FC26.

Detailed methods of measurement and calculation are laid down in the Control of Noise (Measurement and Registers) Regulations 1976.

Recording of measurements

The local authority is required to record the measured levels in a register, the form and content of which is detailed in schedule 1, paragraph 6 Control of Noise (Measurement and Registers) Regulations 1976. No time period is stipulated within which these measurements must be taken. Copies of the entries are to be sent to the owner and occupier of each premises and they may appeal to the Secretary of State (section 64(3)).

The register must be open to public inspection free of charge and copies of entries provided at reasonable charge (section 64(7)).

Consents to exceed registered level

Application for such consent may be made to the local authority and these may be granted, with or without conditions, or refused. If granted by the local authority or subsequently on appeal, the consent must be recorded in the noise level register (section 65(2)).

Appeals

The procedure relating to appeals to the Secretary of State against the recorded noise level for a premises, refusal to consent to exceed the registered noise level or against conditions attached to a consent is laid down in regulation 9 Control of Noise (Appeals) Regulations 1975. Within seven days of lodging the appeal, the appellant must send to the Secretary of State:

(a) the application made to the local authority (if any);
(b) any relevant plans etc. submitted to the local authority;
(c) any relevant records, consents, notices, etc. issued by the local authority;
(d) copies of correspondence with the local authority;
(e) a plan of the premises concerned.

The Secretary of State may require further written evidence from the appellant and from the local authority and may hold a local inquiry at his discretion. No specific grounds for appeal are laid down.

Offences

An offence is committed if the registered noise level is exceeded, a condition attached to a consent is breached or a court order to undertake works is not complied with (sections 65(5) and (6)).

No specific defences are laid down. A consent to exceed the registered noise level does not of itself provide a defence against proceedings relating to noise nuisance by aggrieved persons under section 82 Environmental Protection Act 1990, but is a defence against proceedings for failure to comply with a local authority abatement notice under section 80 Environmental Protection Act 1990 (section 65(8) and Environmental Protection Act 1990 section 80(9)(a)).

NOISE ABATEMENT ZONES: NOISE LEVEL DETERMINATIONS FOR NEW BUILDINGS

References

Control of Pollution Act 1974, section 67.
Control of Noise (Appeals) Regulations 1975.

Extent

These provisions apply in England, Wales and Scotland.

New buildings

This procedure applies in noise abatement zones (NAZ) where:

(a) a new building is to be erected which will be used for a purpose classified by the noise abatement order as one to which the order will apply; and
(b) existing buildings which, when works are completed, will be used for a purpose covered by the order (section 67).

Applications to local authority

Local authorities may decide to determine a registered noise level on their initiative but applications to them to do so may be made by:

(a) the owner or occupier; or
(b) any person negotiating to acquire an interest in the premises (section 67 (1), (2), and (3)).

Appeals

The provisions covering appeals to the Secretary of State are laid down in part 3, regulation 7 Control of Noise (Appeals) Regulations 1975 and are the same as those relating to registration of noise levels generally in NAZs, as set out on page 132.

Defences

It will be a defence in proceedings for non-compliance with a local authority abatement notice under section 80 Environmental Protection Act 1990 to show that the noise did not exceed the level registered under this procedure (Environmental Protection Act 1990 section 80(9)(c)).

FC26 Noise abatement zones: noise level determinations for new buildings – Section 67 Control of Pollution Act 1974

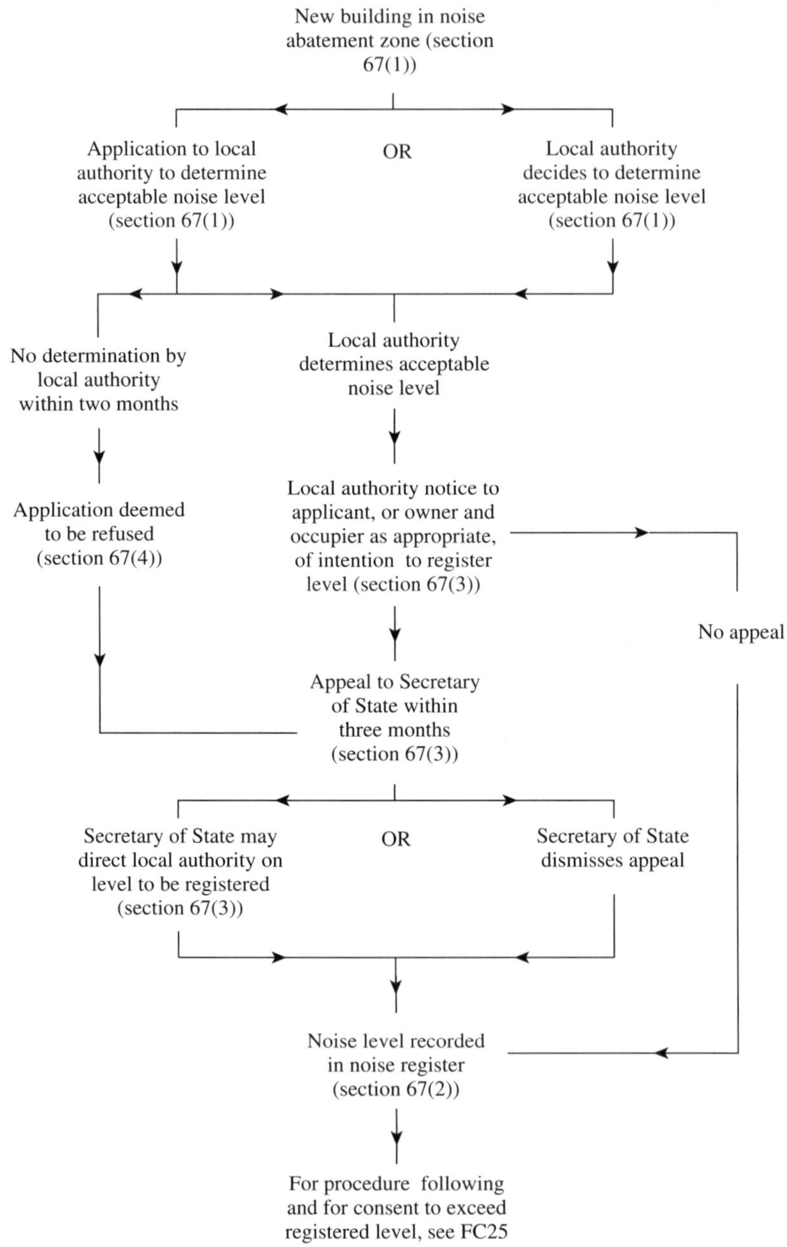

Notes

1. For measurement of noise levels generally in noise abatement zones, see FC25.
2. For noise reduction notices for new buildings, see FC27.
3. For noise nuisances, see FC20, FC28 and FC121.

NOISE ABATEMENT ZONES: NOISE REDUCTION NOTICES

References

Control of Pollution Act 1974, section 66.
Control of Noise (Appeals) Regulations 1975.

Extent

These provisions apply to England, Wales and Scotland.

Criteria for serving notices

Before serving a noise reduction notice, the local authority must be satisfied that:

(a) the premises concerned are of a class to which the noise abatement order designating the NAZ applies;
(b) the noise level emanating from the premises is not acceptable having regard to the purposes of the NAZ;
(c) a reduction in noise level is practicable at reasonable cost and would afford a public benefit (section 66(1)).

Noise reduction notices

The notice must include:

(a) a requirement to reduce the level of noise to one specified in the notice;
(b) a prohibition on any subsequent increase in the level without the consent of the local authority;
(c) the specification of such works as may be necessary to achieve (a) and (b) above (section 66(2)).

In addition, the notice may specify particular times or days during which the level is to be reduced and different noise levels may be set for different times or days (section 66(4)).

The local authority must specify in the notice a time within which the noise level is to be reduced and within which any required works are to be completed. This time must be not less than six months or, for new buildings, three months (sections 66(3) and 67(5)).

A noise reduction notice may be served even if a consent under section 65 has been given (page 130) (section 66(5)).

Notices are served on 'the persons responsible' (definition on page 124) (section 66(1)).

Appeals

The grounds for appeal are set out in the Control of Noise (Appeals) Regulations 1975 as including:

FC27 Noise abatement zones: noise reduction notices – Section 66 Control of Pollution Act 1974

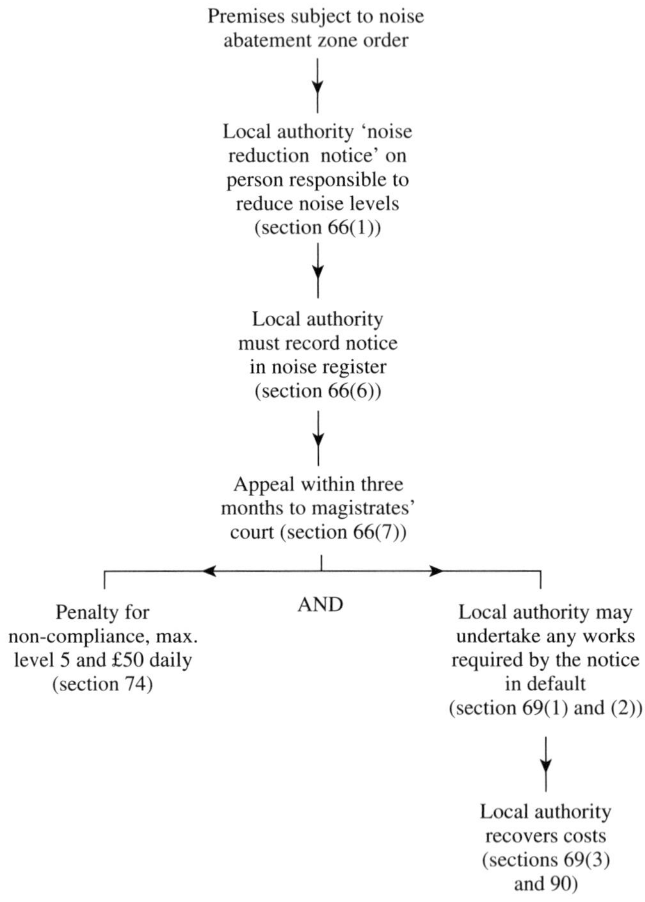

Notes

1. For declaration of NAZ, see FC24.
2. For measurement and registration of noise levels in NAZs and consent to exceed, see FC25.
3. NAZ = noise abatement zone.

(a) the notice is not justified;
(b) there is some material informality, defect or error in connection with the notice;
(c) the local authority has unreasonably refused compliance with alternative requirements or the requirements of the notice are unreasonable in character or extent or are unnecessary;
(d) the time allowed is not reasonably sufficient;
(e) where the noise emanates from a trade or business, the best practicable means have been used;
(f) the notice should have been served on some other person in addition to, or in substitution for, the appellant (regulation 7(2)).

The court may vary or quash the notice or dismiss the appeal (regulation 7(5)).

The provisions relating to the suspension of notices pending the determination of appeal are as set out for noise from construction sites on page 126.

Defences

In proceedings for contravention of a noise reduction notice relating to a trade or business, it is a defence to prove that best practicable means have been used (section 66(9)).

In respect of any proceedings for non-compliance with a local authority abatement notice under the Environmental Protection Act 1990 section 80, it will be a defence to show that the noise level at the time did not contravene the requirements of any noise reduction notice under this procedure (Environmental Protection Act 1990 section 80(9)(b)).

New buildings

These provisions relating to noise reduction notices also apply to buildings which:

(a) are to be constructed and are a class to which the noise abatement order applies; and
(b) will be of such a class when works are completed except that:
    (i) before serving a notice the local authority need not be satisfied that noise reduction is practicable at reasonable cost and would afford a public benefit;
    (ii) the minimum period allowed for compliance is three months and not six months;
    (iii) the defence of best practicable means for trade or business is not available (section 67(5)).

NOISE FROM CERTAIN PREMISES AT NIGHT

References

Noise Act 1996 as amended by the Anti-Social Behaviour Act 2003 and the Clean Neighbourhoods and Environment Act 2005.

The Environmental Offences (Fixed Penalties) (Miscellaneous Provisions) Regulations 2007.
DEFRA Guidance on the CNEA 2005 – Noise 2006 and Fixed Penalty Notices 2006.
DEFRA Guidance on the Noise Act 1996 as amended – Guidance for Local Authorities, March 2008.
DEFRA Circular NN/31/03/2004.
CIEH Neighbourhood Noise Policies and Practice for Local Authorities – A Management Guide 2006
NB. The general provisions of the Environmental Protection Act 1990 do not apply to this procedure.

Extent

This procedure applies in England and Wales (section 1).

Scope

The procedure enables a local authority to deal with complaints from any individual present in a dwelling (definition below) during night hours that excessive noise is being emitted from another dwelling or a premises which is the subject of a premises licence or a temporary event notice under the Licensing Act 2003 (see FC143 and FC147).

Night hours are defined as from 11 pm to 7 am the following morning (sections 2(2) and (6)).

Complaints

The local authority may investigate complaints of night noise which appear to be covered by the provisions. Such investigations must be undertaken by an officer of the local authority (section 2(1)).

Warning notices

Where upon investigation the officer is satisfied that:

(a) noise is being emitted from the offending premises during night hours; and
(b) the noise if measured within the complainants' dwelling would or might exceed the permitted level,

he may serve a warning notice (section 2(4)).

At this stage, actual measurement of the noise level is not obligatory. Noise above the permitted level does not necessarily create a statutory nuisance.

The permitted levels in relation to licensed premises and dwellings will be set by the Secretary of State in directions made under section 5. See DEFRA Guidance on the Noise Act 1996 as amended – Guidance for Local Authorities.

FC28 Noise from certain premises at night – Noise Act 1996

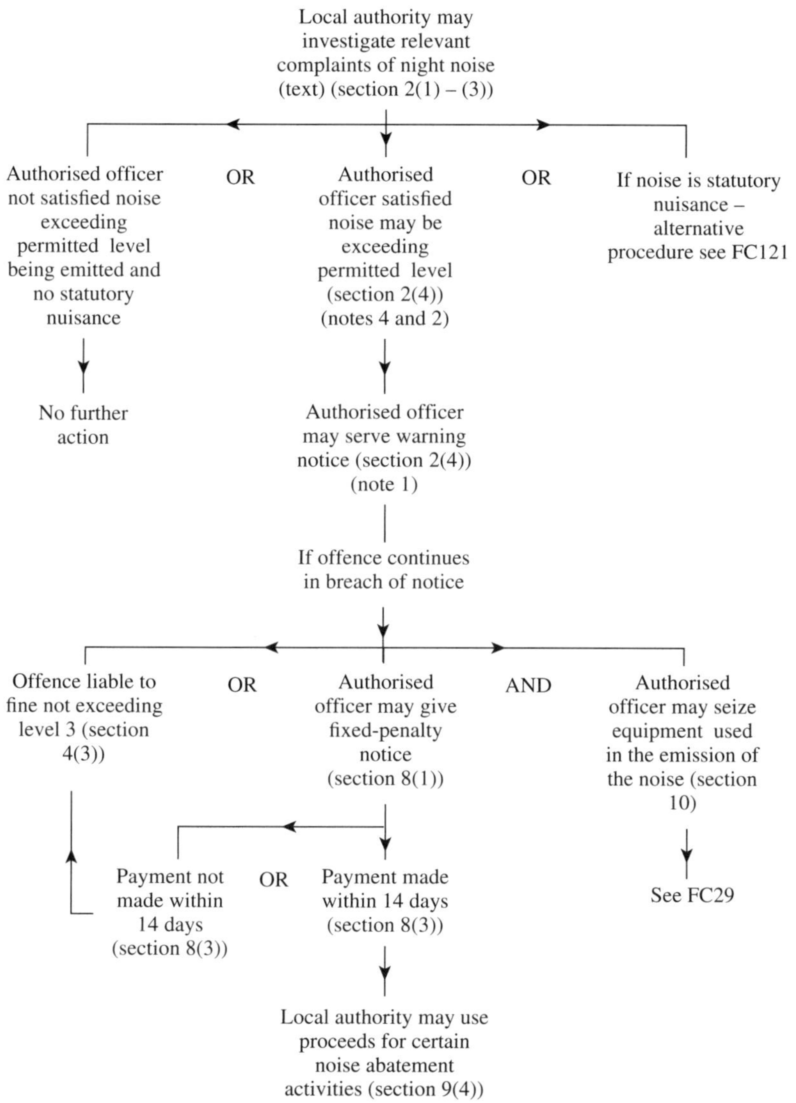

Notes
1. There is no prescribed form for a warning notice but see CIEH guidance.
2. Action may be taken against a premises within an adjoining local authority area (section 2(7)).
3. For noise which is a statutory nuisance, see FC121.
4. This may or may not be a statutory nuisance.

Complaints made by a resident of one local authority area about noise arising from an adjoining local authority area may be pursued by the first local authority and warning notices served (section 2(7)).

In judging whether or not the noise complained about is, would or might exceed the permitted level, the officer has discretion as to whether or not measurements are taken and whether the assessment of the noise is from within or outside the complainants' dwelling (section 2(5)).

Content of warning notices

The notice must:

(a) state that the officer considers that there is night noise from the offending premises which exceeds, or may exceed, the permitted level as measured within the complainants' dwelling; and
(b) give warning that any person responsible for noise emitted above the permitted level in the period specified in the notice may be guilty of an offence (section 3(1)).

The period specified must begin not earlier than ten minutes from the notice being served and end at the following 7 am. Officers may use discretion as to how much 'grace' should be given before the notice comes into effect given the circumstances of the situation. The notice must also state the time at which it is served (section 3(2) and (4)).

Service of notices

Service of warning notices is effected either by delivering to any person present at or near the offending dwelling and appearing to be responsible for the noise or, where identification of such a person is not reasonably practicable, by leaving it at the offending dwelling (section 3(3)).

A person is deemed as responsible for the noise if he is a person by whose act, default or sufferance the emission of the noise is wholly or partly attributable (section 3(5)).

Offences

Once a warning notice has been served, if noise is emitted from the offending dwelling in excess of the permitted level as measured from within the complainants' dwelling, any person responsible (see above) is guilty of an offence and a fine not exceeding level 3 (section 4).

It should be noted that, at this stage, the noise must be measured and the measurement taken from within the dwelling (as distinct from the situation at the investigative stage).

Evidence of measurements taken will not be admissible in court proceedings unless they were taken using a device approved by the Secretary of State and used in accordance with conditions prescribed by him (section 6).

There are detailed provisions about the way in which evidence must be gathered and presented (section 7).

It is a defence to show that there was a reasonable excuse for the act, default or sufferance in question (section 4(2)).

Fixed-penalty notices

Where the authorised officer believes that an offence has been committed by the infringement of a warning notice, he may serve a fixed-penalty notice as an alternative to prosecution.

The notice offers the person the opportunity of discharging liability to conviction by payment within 14 days. The notice is served either by giving it to the person or by leaving it, addressed to him, at the offending dwelling or licensed premises. In the latter case, the notice should be given to the most senior person present. The notice must state:

(a) details of the offence;
(b) the period during which proceedings will not be taken, i.e. 14 days;
(c) the amount of the fixed penalty (£100); and
(d) the ways in which the payment can be made (these are not 'on the spot' fines).

The fixed penalty in respect of dwellings is, at the discretion of the local authority, between £75 and £110 (the Environmental Offences (Fixed Penalties) (Miscellaneous Provisions) Regulations 2007). For licensed premises, the fine is fixed at £500 (Clean Neighbourhoods and Environment Act 2005 schedule 1 and section 6 Noise Act 1996).

Where a payment is not received during the period specified (if by post the date of payment is deemed to be the time at which the letter would be delivered in the ordinary course of post), the local authority may proceed by way of prosecution for the offence (section 8).

Only one fixed-penalty notice may be given to the same person for a particular premises the same night (section 9(2)).

Seizure of equipment/powers of entry

Where noise exceeding the permitted level has been emitted from a dwelling during the currency of a warning notice, officers may enter and seize or remove any equipment which is being or has been used in the emission of the noise (also see FC29).

In exercising this power of entry, the authorised officer must, if requested, produce his authority. There are provisions for application to a Justice of the Peace for a warrant authorising entry, if need be by force, where:

(a) a Warning Notice has been served;
(b) noise has exceeded the permitted level in contravention of the notice; and
(c) entry has been refused, refusal is apprehended or a request for admission would defeat the object of entry.

Warrants are effective until their purpose has been satisfied. Persons executing the entry may take other persons or equipment as necessary and must leave unoccupied premises secured. The penalty for obstructing an authorised officer is a fine not exceeding level 3.

This power also applies to offences under the statutory noise nuisance procedure under the Environmental Protection Act 1990 (see FC121) (section 10).

Definition

Dwelling means any building, or part of a building, used or intended for use as a dwelling and references to noise emitted from a dwelling include noise emitted from any garden, yard, outhouse or other appurtenance belonging to or enjoyed within the dwelling (section 11(2)).

SEIZURE OF EQUIPMENT USED TO MAKE NOISE UNLAWFULLY

References

Noise Act 1996, section 10 and schedule 1.
Environmental Protection Act 1990, sections 79(1)(g) and 81(3).
DEFRA Guidance on the Noise Act 1996 as amended – Guidance for Local Authorities, March 2008.
CIEH Neighbourhood Noise Policies and Practice for Local Authorities – A Management Guide, 2006.

Extent

In addition to England and Wales, the Act applies, with some minor amendments, in Northern Ireland but not in Scotland (section 14(4)).

Scope

This procedure is available to authorised officers of local authorities where:

(a) a warning notice under the Noise Act 1996 has been served (FC28) and noise emitted from the offending dwelling has exceeded the permitted level as measured from within the complainants' dwelling; (section 10) or
(b) the local authority is using its powers under section 81(3) of the Environmental Protection Act 1990 to itself abate or prevent a statutory noise nuisance under section 79(1)(g) following the service of an abatement notice (FC121).

Seizure of equipment

An officer of the local authority or someone authorised by the local authority has the power to enter the dwelling from which the noise is being or has been

FC29 Seizure of equipment used to make unlawful noise – Section 10 Noise Act 1996

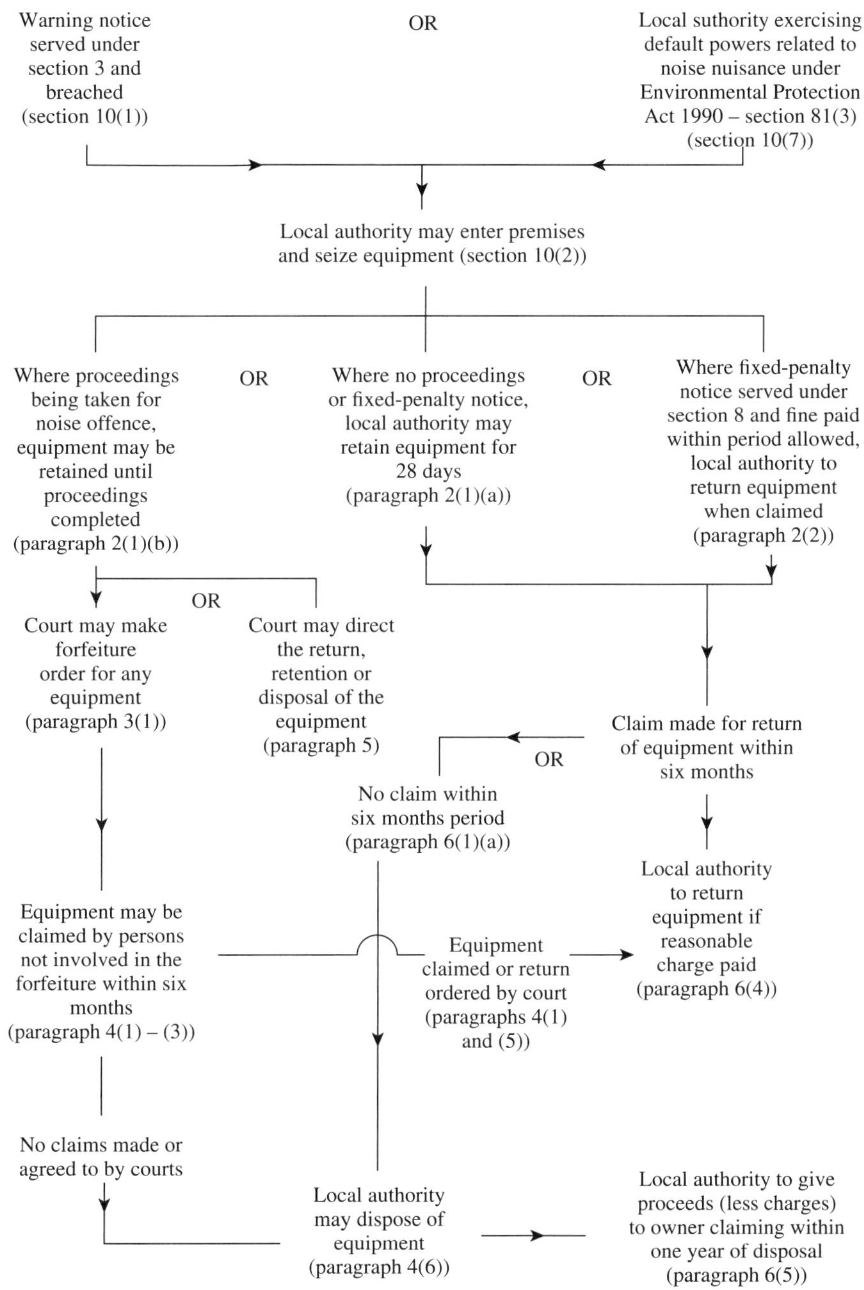

Note

1. Paragraph numbers refer to schedule 1 of the Noise Act 1996.

emitted and may seize and remove any equipment which is being or has been used to make the noise. Any person exercising these powers must produce his or her authority, if required to do so (sections 10(2) and (3)).

Note – if entry is in connection with enforcement under the 1990 Act, then 24 hours' notice of intention to enter must be given.

Warrants

Where a Justice of the Peace is satisfied on sworn information that:

(a) a warning notice has been served;
(b) noise has been emitted in breach of the requirements of the notice; and
(c) entry to the dwelling has been refused, such refusal is apprehended or a request for entry would defeat the object of admission,

the Justice of the Peace may by warrant authorise the local authority to enter, if necessary by force. Persons authorised by the local authority to effect the warrant may take with them any other person or equipment as may be necessary but must ensure when leaving that any unoccupied premises are left as effectively secured against trespass as they found them.

Warrants continue in force until the purpose for entry has been satisfied (sections 10(2)–(6)).

Offences

Any person wilfully obstructing someone exercising the powers of this procedure is liable to a fine not exceeding level 3 (section 10(8)).

Retention of equipment

Equipment seized under these powers may be retained by the local authority for 28 days unless it was:

(a) the subject to proceedings for a noise offence – in which case it may be retained until the proceedings have been dealt with; or
(b) equipment used to emit noise which has been the subject of a fixed-penalty notice under section 8 – in which case it must be returned on request (schedule 1, paragraph 2).

Forfeiture

When a person is convicted by a magistrates' court of a noise offence (either the breach of a warning notice under the Noise Act or an abatement notice for a statutory noise nuisance under the Environmental Protection Act), the court may order the forfeiture of any seized equipment. In these cases, third parties (e.g. hiring companies) may make application to the court within six months for its return. Persons subject to the forfeiture order have no further rights to the equipment named (paragraph 3 and 4).

The court may also give directions (where a forfeiture order is not made) as to the return, retention or disposal of the equipment (paragraph 5).

Return of seized equipment

Equipment which is claimed within the required periods or is ordered to be returned to its owner by a magistrates' court must be returned by the local authority once a reasonable charge has been paid to it (paragraph 6(5)).

Notification of claim rights

Local authorities have a duty to take reasonable steps to bring to the attention of persons who may be entitled to do so their rights to make claims for the return of seized equipment (paragraphs 4(4) and 6(3)).

AUDIBLE ALARM NOTIFICATION AREAS

References

The Clean Neighbourhoods and Environment Act 2005, sections 69 to 81.
The Environmental Offences (Fixed Penalties) (Miscellaneous Provisions) Regulations 2006.
DEFRA Guidance on the Clean Neighbourhoods and Environment Act 2005 – Noise 2006 and Fixed Penalty Notices 2006.
DEFRA Guidance on the Noise Act 1996 as amended – Guidance for Local Authorities, March 2008.

Extent

This provision applies in England, Wales and Scotland.

Scope

This is a procedure that enables local authorities to deal with annoyance caused by audible intruder alarms. It allows a local authority to designate any or all of its area as an alarm notification area (ANA). The effect is to require all intruder alarms, residential and non-residential, to be notified to the local authority with the identification of a key-holder (section 69(1)).

Designation

A notice of the proposal must be published in a local newspaper, stating:

(a) that representations can be made;
(b) the date by which these must be received being at least 28 days after the publication of the notice.

FC30 Audible alarm notification areas (ANAs) – Sections 69 to 80 Clean Neighbourhoods and Environment Act 2005

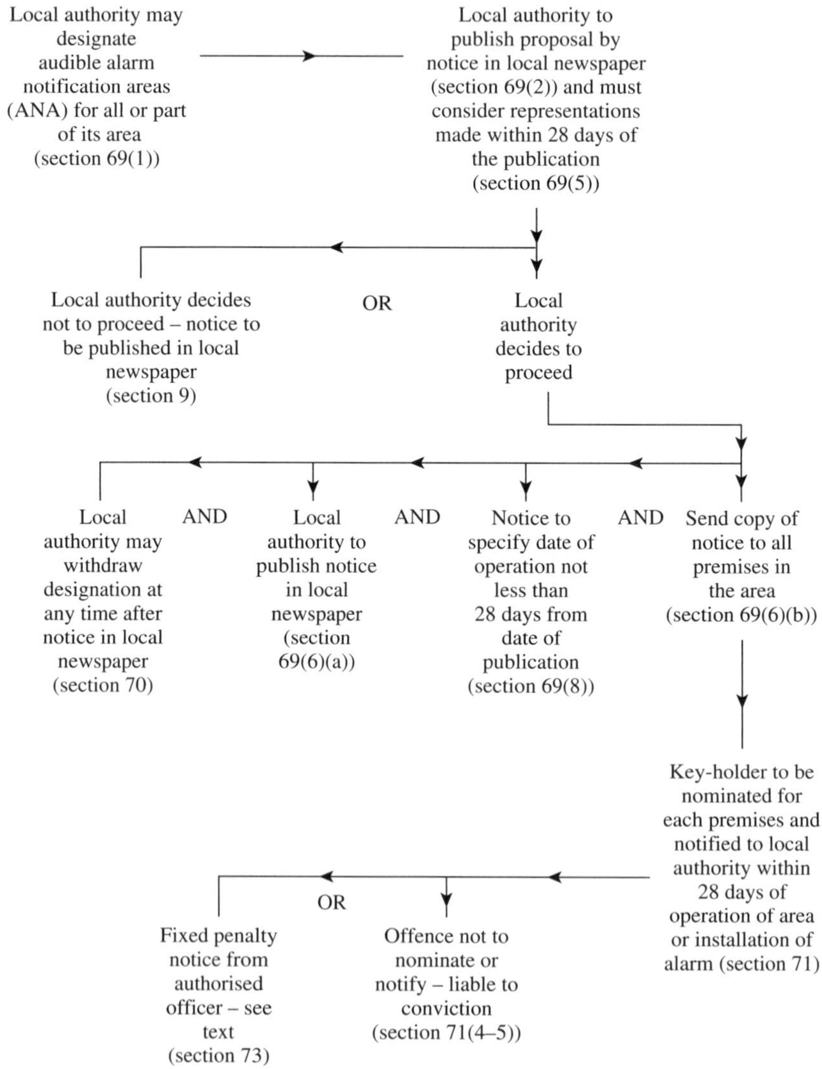

The local authority must consider all representations made within the time limit (section 69(2–5)).

If the local authority determines to go ahead, it must:

(i) arrange for the decision to be published in a local newspaper;
(ii) send a copy to all premises in the area (NB 'premises' does not include vehicles in this procedure);
(iii) specify a date of operation, which must be at least 28 days after the notice has been published (section 69(6–8)).

If the local authority decides not to go ahead, it must publish a notice to that effect (section 69(9)).

Withdrawal of designation

The designation may be withdrawn by publishing the proposal with an implementation date and notifying all of the premises in the area (section 70).

Notification of nominated key-holders

In a designated area, the occupier, or owner where there is no occupier, must nominate a key-holder and then notify the local authority in writing within 28 days of the operation of the designation, or 28 days from installation, of that holder's name, address and telephone number.

A person eligible to be the key-holder:

(a) must hold sufficient keys to gain access;
(b) normally resides or is situated in the vicinity;
(c) has information that will enable him to silence the alarm;
(d) has agreed to be the nominated key-holder;
(e) for residential premises, is not the occupier or a key-holding company;
(f) for non-residential premises, is the responsible person or acts on their behalf or it is a key-holding company (sections 71 and 72(3–8)).

Offences

It is an offence to fail to nominate or notify details of a key-holder and the penalty on conviction is a fine not exceeding level 3 (section 71(4–5)).

Fixed penalties

Where it appears that an offence has been committed, an authorised officer may issue a fixed-penalty notice and this, if paid within 14 days, removes the liability to conviction (section 73). The amount of the fixed penalty is set by the 2006 regulations at between £50 and £80 at the discretion of the local authority and with discounts for early payment. Sums collected by the local authority may be used to offset the costs of this procedure and other identified functions under the Noise Act 1996 and in dealing with statutory noise nuisances (section 75).

Authorised officers are given the power to request the name and address of offenders where he proposes to issue a fixed-penalty notice (section 76).

Power of entry

An authorised officer is able to enter a premises, other than by force, in order to silence an intruder alarm in any part of the district and not just in a designated area provided that:

- the alarm has been sounding for more than 20 minutes or intermittently for more than one hour; and
- the sounding of the alarm is likely to give reasonable cause for annoyance to persons living or working in the vicinity; and
- if the premises is in a notification area and reasonable steps have been taken to get the nominated key-holder to silence it (sections 77 and 79).

An authorised officer may apply to a Justice of the Peace for a warrant to enter a premises to silence an alarm but the Justice of the Peace must be satisfied that it has not been possible to enter without the use of force (section 78). Local authorities may recover costs of entry (section 79).

ANTI-SOCIAL BEHAVIOUR ORDERS (ASBOs)

References

The Crime and Disorder Act 1998, sections 1 and 4, as amended by the Police Reform Act 2002, the Anti-Social Behaviour Act 2003, the Serious Organised Crime and Police Act 2005 and the Crime and Security Act 2010.

The Magistrates' Courts (Sex Offenders and Anti-Social Behaviour Orders) Rules 1998.

A Guide to Anti-Social Behaviour Orders, Home Office, August 2006.

Taking Action: Tackling Anti-Social Behaviour – a toolkit for environmental health practitioners, available at www.cieh.org/uploadedFiles/Core/Policy/Publications_and_information_services/Policy_publications/Publications/AntiSocialBehaviourToolkitSeptember2006.pdf

Extent

This procedure applies in England and Wales but the Act makes a similar provision for Scotland.

Scope

An ASBO may be issued by a magistrates' court or county court, both sitting in a civil capacity, against an individual aged ten or over:

(a) who has acted in an anti-social manner – that is, in a manner that caused or was likely to cause harassment, alarm or distress to one or more persons not of the same household; and

FC31 Anti-social behaviour orders (ASBOs) – Sections 1 to 4 Crime and Disorder Act 1998

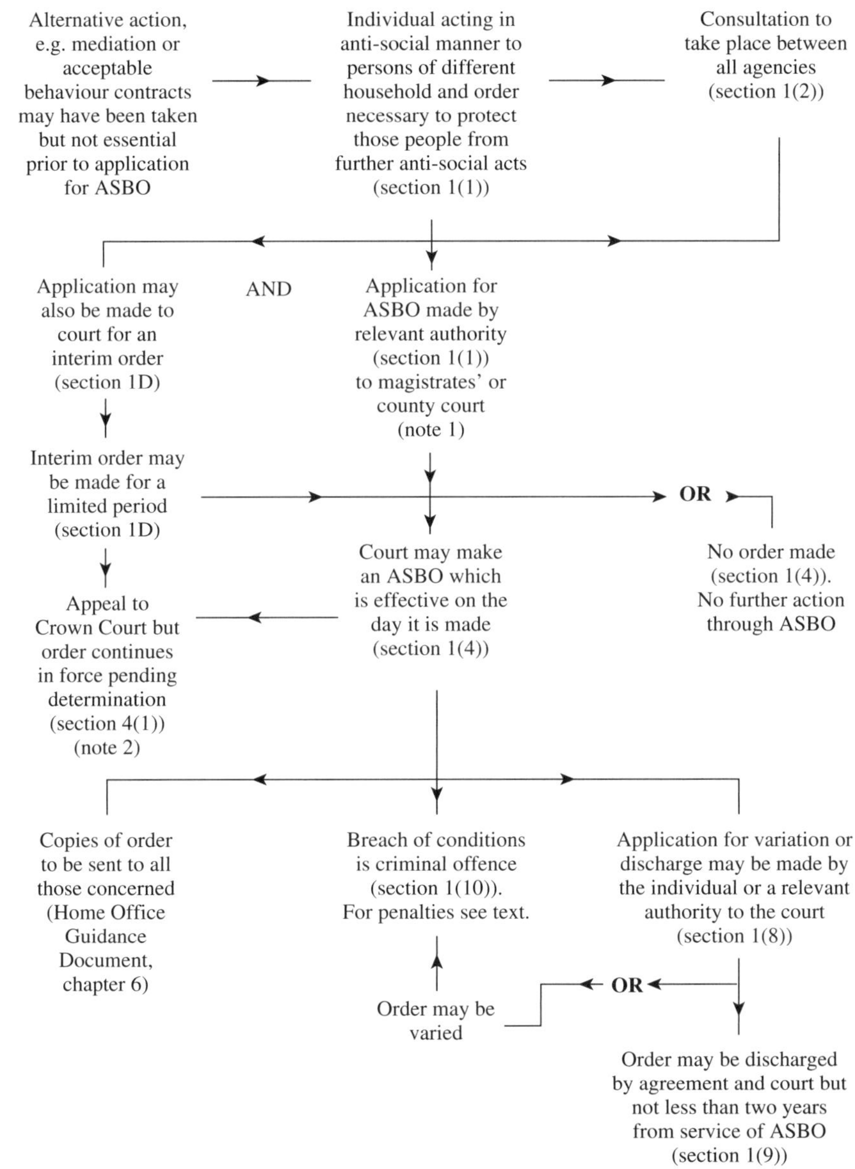

Notes

1. The form of application is set out in schedule 4 of the 1998 Rules.
2. There may be an appeal to the High Court against the decision of the Crown Court or an application for judicial review.

(b) the order is necessary to protect persons against whom the harassment etc. was directed from further anti-social acts by that person (section 1).

The purpose of an ASBO is to restrain serious anti-social behaviour caused to individuals who cannot be expected to tolerate the behaviour unchecked. This could include situations, e.g. noise, that cannot be dealt with effectively or appropriately through other remedies, e.g. as statutory nuisances under section 80 of the Environmental Protection Act 1990.

Usage

Types of anti-social behaviour where an ASBO may be appropriate include:

(i) where individuals intimidate neighbours and others through threats or violence or a mixture of unpleasant actions;
(ii) persistently unruly behaviour by a small group of individuals on a housing estate or other local area, who may dominate others and use minor damage to property and fear of retaliation, possibly at unsociable hours, as a means of intimidating other people;
(iii) where there are families whose anti-social behaviour, when challenged, leads to verbal abuse, vandalism, threats and graffiti, sometimes using children as the vehicle for action against neighbouring families;
(iv) persistent abusive behaviour towards elderly people or towards mentally ill or disabled people, causing them fear and distress;
(v) serious and persistent bullying of children on an organised basis in public recreation grounds or on the way to school or within the school grounds if normal disciplinary procedures do not stop the behaviour;
(vi) persistent racial harassment or homophobic behaviour;
(vii) persistent anti-social behaviour as a result of drugs or alcohol abuse.

Amongst the options which may be considered before application for an ASBO are:

- mediation;
- acceptable behaviour contracts – see below;
- injunctions.

However, there is no requirement to demonstrate that every other remedy has been exhausted before applying for an ASBO. The key is that one should be used where it is the most appropriate remedy, is necessary to protect an individual, a group of people or a community and is the proportionate step to take.

Partnership

Action on ASBOs should be taken within the framework of partnership arrangements under section 5 of the Crime and Disorder Act 1998 between the police, local authority, health authority and the probation service and be subject to a local protocol for ASBOs, which also involves social landlords, organisations representing local residents and agencies such as social services and the youth offending teams.

The protocol should set out agreed procedures for considering and making applications for ASBOs, for their monitoring and applying for the variation and discharge of them. As part of this protocol, an ASBO coordinator should be identified for each local authority district.

Consultation

Before making application for an ASBO, the authority making the application must consult with other relevant authorities – see below (section 1E). This consultation should involve all the relevant departments and also include other agencies as appropriate, e.g. social services, health service, landlords, etc., and be conducted through a consultative group established as part of the local protocol (section 1E).

Agreement between the police and local authority is not a requirement but, in the absence of it, the court must be informed when the application is made.

Applications

Applications for orders are made by way of complaint to the magistrates' or county court for the area where the harassment etc. is taking place, although there have been cases where applications have been made to neighbouring courts (section 1(3)).

Applications for an ASBO may be made by:

(a) the council for a local government area;
(b) a county council (in England);
(c) the police;
(d) British Transport Police (BTP);
(e) registered social landlords;
(f) a housing trust.

These are called 'relevant authorities'.

Local authorities and the police may apply to protect people within their area irrespective of whether or not the original anti-social behaviour took place in their area. The BTP may apply to protect people on or in the vicinity of premises that are policed by them whilst registered social landlords may look to protect people on or in the vicinity of premises owned by them (section 1B).

Proceedings to apply for an ASBO are civil and the case needs to be proved according to the rules of civil evidence.

Orders

If it is shown that the conditions in 'Scope' above exist, the court may make an ASBO against a named individual which prohibits that person from doing anything described in the order (section 1 (4)). Such prohibitions are to be those necessary to prevent further anti-social behaviour by that individual and may relate to more than one local government area and may even relate to the whole of England and Wales (section 1(5) and (6)).

Prohibitions should be specific in time and place so that it is readily apparent both to the defendant and to those enforcing the order what does or does not constitute a breach. Orders should contain a prohibition on inciting/encouraging the commission of specified anti-social acts within the meaning of the Act, including where appropriate by minors in the household, and by others over whom the person subject to the order has control. Any requirements in the order must be negative. There is no power to compel an individual to do anything, only not to take particular actions.

Orders are effective for the period specified in it, with a minimum of two years, unless subsequently varied or discharged – see below (section 1 B(7)).

Interim orders

Relevant authorities may make applications to a magistrates' or county court for an interim order before the full application process is complete. Such an order, if granted, can prohibit the defendant from doing anything detailed in the order. An interim order is for a fixed period but can be varied, renewed or discharged. It will cease to have effect when the main application is decided. They may be effective for the protection of persons anywhere in England and Wales and there are the same provisions relating to appeals as for ASBOs (section 1D).

Orders on conviction

Criminal courts are able, without application being made, to make an ASBO where the defendant has been convicted of an offence. The court needs to be satisfied that the offender has acted in an anti-social manner and the order is necessary to protect people anywhere in England and Wales from further anti-social acts (section 1C).

Appeals

Appeals against the making of an ASBO or an interim order are made to the Crown Court, which may make such orders as necessary to give effect to its decision, including consequential or incidental orders (section 4).

Challenge to a decision of the Crown Court is to the High Court by way of case stated or by an application for judicial review.

Effect of an ASBO

Breach of the prohibitions in the order is a criminal offence (section 1(10)) and the standard of proof required is therefore that of 'beyond reasonable doubt'. Prosecutions may be taken by the Crown Prosecution Service or a relevant authority.

Cases are triable either way with penalties on summary conviction of imprisonment for up to six months and a fine not exceeding the statutory maximum, or both, or on indictment imprisonment for up to five years or a fine, or both. There is a defence of 'reasonable excuse' (section 1(10)).

Variation and discharge of orders

Relevant authorities or the individual against whom the order was made may apply for these by way of complaint to the court. The order cannot be discharged less than two years after its service (not making) without the agreement of both the police and the local authority (section 1(8) and (9)).

If successful, the variation could, for example, change the prohibitions.

Acceptable behaviour contracts (ABCs)

These are informal arrangements that may be used as an alternative to ASBOs where the latter are not considered to be justified. An ABC is an individual written agreement by a person with a partner agency and the police not to carry on with certain identifiable acts that could be construed as anti-social behaviour. Whilst not a legally binding document, a breach of it could be cited in any application for an ASBO.

ENVIRONMENTAL PERMITTING

Introduction

The Environmental Permitting system implements the requirements of the Integrated Pollution Prevention and Control (IPPC) Directive, which applies to the more complex and potentially polluting industries and deals with any likely significant pollutant emission to air, water or land. It also addresses noise, energy efficiency, waste minimisation, prevention of accidental emissions and site protection and restoration.

The Environmental Permitting system is designed to allow regulators to focus on medium- and high-risk operations whilst continuing to protect the environment and human health.

The Environmental Permitting (England and Wales) Regulations 2010 provide industry, regulators and others with a permitting and compliance system, including systems for discharge, consenting, groundwater authorisations and radioactive substances regulation.

Larger industrial installations are subject to integrated controls on all likely significant emissions. Most are regulated by the Environment Agency and are often referred to as 'Part A(1)' activities. These are regulated for emissions to air, land, water and other environmental considerations.

Local authorities regulate premises with 'Part A(2)' (comparatively less polluting) installations for air, land, water and other environmental considerations and 'Part B' activities (smaller installations), which are generally limited to air pollution control.

The Part B regime is known as Local Air Pollution Prevention and Control (LAPPC) and the A2 regime as Local Authority Industrial Pollution Prevention and Control (LA-IPPC). The port health authority may be the regulator in port areas (regulation 32).

Part A(1) and A(2) installations

The IPPC system as enforced by both the Environment Agency and local authorities for the installations allocated to them applies an integrated approach to ensure a high level of protection to the environment as a whole and to prevent and, where that is not practicable, reduce emissions to acceptable levels. The conditions set in granting permits are to be based on the use of 'best available techniques' (BAT) – see below. The regulations also include provisions relating to energy efficiency, site restoration, noise, odour, waste minimisation, accident prevention and heat and vibrations.

Part B installations

Regulated by local authorities under LAPPC, these are also subject to conditions based on BAT but only so far as these are related to emissions to air.

Definitions

Activities 'are industrial activities forming part of an installation' and are defined as 'activities of any nature, whether:

(a) industrial or commercial or other activities; or
(b) carried on on particular premises or otherwise,

and includes (with or without other activities) the depositing, keeping or disposal of any substance'(Pollution Prevention and Control Act 1999 section 1(1)).

Change in operation – a change in the nature or functioning, or an extension, of the installation which may have consequences for the environment. A change in operation could entail either technical alterations or modifications in operational or management practices, including changes to raw materials or fuels used and to the installation throughput (paragraph 5, schedule 5).

Emission means:

(a) in relation to a Part A installation, the direct or indirect release of substances, vibrations, heat or noise from individual or diffuse sources in the installation into the air, water or land;
(b) in relation to a Part B installation, the direct release of substances or heat from individual or diffuse sources in the installation into the air;
(c) in relation to Part A mobile plant, the direct or indirect release of substances, vibrations, heat or noise from the mobile plant into the air, water or land;
(d) in relation to Part B mobile plant, the direct release of substances or heat from the mobile plant into the air;
(e) in relation to a waste operation, the direct or indirect release of substances, vibrations, heat or noise from individual or diffuse sources related to the operation into the air, water or land;
(f) in relation to a mining waste operation, the direct or indirect release of substances, vibrations, heat or noise from individual or diffuse sources related to the operation into the air, water or land; and

(g) in relation to a radioactive substances activity, the direct or indirect release of radioactive material or radioactive waste.

Emission limit value means the mass, expressed in terms of specific parameters, concentration or level of an emission, which may not be exceeded during one or more periods of time.

Environmental pollution means pollution of the air, water or land which may give rise to any harm; and for the purposes of this definition (…)

(a) pollution includes pollution caused by noise, heat or vibrations or any other kind of release of energy; and
(b) air includes air within buildings and air within other natural or man-made structures above or below ground.

In this definition, harm means:

(a) harm to the health of human beings or other living organisms;
(b) harm to the quality of the environment, including:
  (i) harm to the quality of the environment taken as a whole;
  (ii) harm to the quality of the air, water or land; and
  (iii) other impairment of, or interference with, the ecological systems of which any living organisms form part;
(c) offence to the senses of human beings;
(d) damage to property; or
(e) impairment of, or interference with, amenities or other legitimate uses of the environment (expressions used in this paragraph having the same meaning as in Council Directive 96/61/EC) (Pollution Prevention and Control Act 1999 section 1(1)).

Installation means (except where used in the definition of 'excluded plant'):

(a) a stationary technical unit where one or more activities are carried on; and
(b) any other location on the same site where any other directly associated activities are carried on, and references to an installation include references to part of an installation.

Mobile plant means any of the following:

(a) Part A mobile plant;
(b) Part B mobile plant;
(c) waste mobile plant (regulation 2).

Operator, in relation to a regulated facility, means:

(a) the person who has control over the operation of the regulated facility;
(b) if the regulated facility has not yet been put into operation, the person who will have control over the regulated facility when it is put into operation; or
(c) if a regulated facility authorised by an environmental permit ceases to be in operation, the person who holds the environmental permit (regulation 7).

Pollutant means any substance liable to cause pollution (regulation 2).

Pollution, in relation to a water discharge activity or groundwater activity, means the direct or indirect introduction, as a result of human activity, of substances or heat into the air, water or land which may:

(a) be harmful to human health or the quality of aquatic ecosystems or terrestrial ecosystems directly depending on aquatic ecosystems;
(b) result in damage to material property; or
(c) impair or interfere with amenities or other legitimate uses of the environment.

Pollution, other than in relation to a water discharge activity or groundwater activity, means any emission as a result of human activity which may:

(a) be harmful to human health or the quality of the environment;
(b) cause offence to a human sense;
(c) result in damage to material property; or
(d) impair or interfere with amenities or other legitimate uses of the environment.

Regulated facility means any of the following:

(a) an installation;
(b) mobile plant;
(c) a waste operation;
(d) a mining waste operation;
(e) a radioactive substances activity;
(f) a water discharge activity;
(g) a groundwater activity (regulation 8).

Substantial change means a change in operation of an installation which in the regulator's opinion may have significant negative effects on human beings or the environment and includes:

(a) in relation to a Part A installation, a change in operation which in itself meets the thresholds, if any, set out in part 2 of schedule 1; and
(b) in relation to an incineration plant or co-incineration plant for non-hazardous waste, a change in operation which would involve the incineration or co-incineration of hazardous waste.

Best available technique (BAT)

Operators of Part A(1) installations are required to use BAT to achieve a high level of protection of the environment taken as a whole and, for Part B, in relation to emissions into the air. BAT is defined in the IPPC directive as:

> the most effective and advanced stage in the development of activities and their methods of operation which indicates the practical suitability of particular techniques for providing in principle the basis for emission limit values designed to prevent and, where that is not practicable, generally to reduce emissions and the impact on the environment as a whole.

In essence, BAT is a technique that balances the costs to the operator with the benefits to the environment, with a consideration of local circumstances and provides the main basis for the setting of emission limit values – see definition on page 162.

Guidance

1. *Integrated Pollution Prevention and Control – a Practical Guide, 4th ed. 2005 – DEFRA*
   This a general guide to the whole of the IPPC system and covers all aspects. It is primarily intended for Environment Agency regulators in relation to A(1) installations and persons affected by it.
2. Environmental permitting guidance relevant to the Local Authority Pollution Control regime, in particular the *LAPPC General Guidance Manual* and *Controlling Pollution from Industry: Regulation by Local Authorities – A Short Guide*.
   These manuals are the principal guidance to local authorities in exercising their functions under the Environmental Permitting Regulations, i.e. LA-IPPC and LAPPC, and are applicable to all procedures in this chapter.
3. *Process Guidance Notes (PGNs) and Sector Guidance Notes (SGNs)*
   These are issued by the Secretary of State under regulation 64 and form statutory guidance on what constitutes BAT for LA-IPPC installations for each of the main sectors regulated.
4. *AQ Notes*
   These are issued by DEFRA and form additional guidance to local authorities on a wide range of both technical and administrative issues.
5. *Explanatory Memorandum to Environmental Permitting (England and Wales) Regulations 2010*
6. *Environmental Permitting Guidance: The IPPC Directive Part A(1) Installations and Part A(1) Mobile Plant for the Environmental Permitting (England and Wales) Regulations 2010*
7. *Environmental Permitting Guidance: Core Guidance for the Environmental Permitting (England and Wales) Regulations 2010*
8. *Environmental Permitting General Guidance Manual on Policy and Procedures for A2 and B Installations Local Authority Integrated Pollution Prevention and Control (LA-IPPC) and Local Authority Pollution Prevention and Control (LAPPC)*
9. *General Guidance Manual on Policy and Procedures for A2 and B Installations, Part B of Manual Annexes and Parts C and D*

All of the publications above are available on the DEFRA website at www.defra.gov.uk/industrial-emissions/las-regulations/guidance/

Instruments

An instrument means a notice, notification, certificate, direction or form under these regulations. An instrument must be in writing.

Instruments may be served on or given to a person by personal delivery, leaving it at the person's proper address or sending it by post or electronic means to the person's proper address.

In the case of a body corporate, an instrument may be served on or given to the secretary or clerk. In the case of a partnership, an instrument may be served on or given to a partner or a person having control or management of the partnership business.

The proper address is the last known address except:

(a) in the case of a body corporate or their secretary or clerk, the registered or principal office of that body, or the e-mail address of the secretary or clerk;
(b) in the case of a partnership or a partner or person having control or management of the partnership business, the principal office of the partnership, or the e-mail address of a partner or a person having that control or management (regulation 10).

Powers of entry

These are provided for in the Environment Act 1995 and are detailed on pages 87–88.

Application to the Crown

Whilst the Crown is bound by the Environmental Permitting Regulations, it is not criminally responsible and cannot be prosecuted for non-compliance with notices etc. However, local authorities may apply to the High Court to have the Crown's actions, or lack of them, declared unlawful (regulation 11 and schedule 4).

Obtaining of information

By notice served on any person, local authorities may require to be given such information as is specified in the notice and in such form and time/period as is specified (regulation 60(1)).

PERMITTING OF ACTIVITIES BY LOCAL AUTHORITIES

References

Environment Act 1995.
Pollution Prevention and Control Act 1999.
Environmental Permitting Regulations (England and Wales) Regulations 2010 as amended.
Environmental Permitting: General Guidance Manual on Policy and Procedures for A2 and B Installations, March 2010.
Environmental Permitting Regulations 2010: Local Authority Toolkit for Inspections (ENTEC/DEFRA), September 2010.

Extent

These provisions apply to England and Wales.

Scope

These procedures apply to activities, installations and mobile plant, prescribed in parts A(2) and B of schedule 1 to the Environmental Permitting Regulations 2010. These activities etc. are controlled and permitted by the local authority. In each case, the operator (definitions on page 155) is required to obtain a permit from the local authority before operating the installation or plant. All installations etc. existing before March 2007 should now have permits so that these procedures relate to the supervision of those plants and to the permitting of new ones.

Applications (FC32)

Applications are to be made in writing to the local authority, with the prescribed fee (see page 173), and must contain all of the information specified in regulation 14 and paragraph 2 of part 1 of schedule 5. Application forms are in part C of the Secretary of State Guidance Manual.

For novel or complex installations, staged applications are possible by agreement between the operator and the local authority.

Local authorities are empowered to request by notice additional information sufficient for them to determine the application. If the information requested has not been provided within the time specified in the notice, the application is deemed to have been withdrawn (schedule 5, part 1, paragraph 4).

Consultations

Subject to the exceptions in schedule 4 (see FC33), the local authority must send copies of the application within the communication consultation period to the public consultees (a person whom the regulator considers is affected by, is likely to be affected by or has an interest in an application). The consultation period is 30 working days starting on the day the regulator receives a duly-made application, unless national security or confidentiality is an issue, in which case the 30 days starts on the determination date.

The local authority must have regard to make representations from the public consultees before determining the application (schedule 5, paragraph 11).

The Environment Agency should always be a public consultee as should be the local authority when the Environment Agency is dealing with applications for part A(1). There is a specimen consultation letter in part C of the Guidance Manual.

FC32 Permitting of scheduled activities by local authorities – Regulation 14 and Schedule 5 Environmental Permitting Regulations 2010

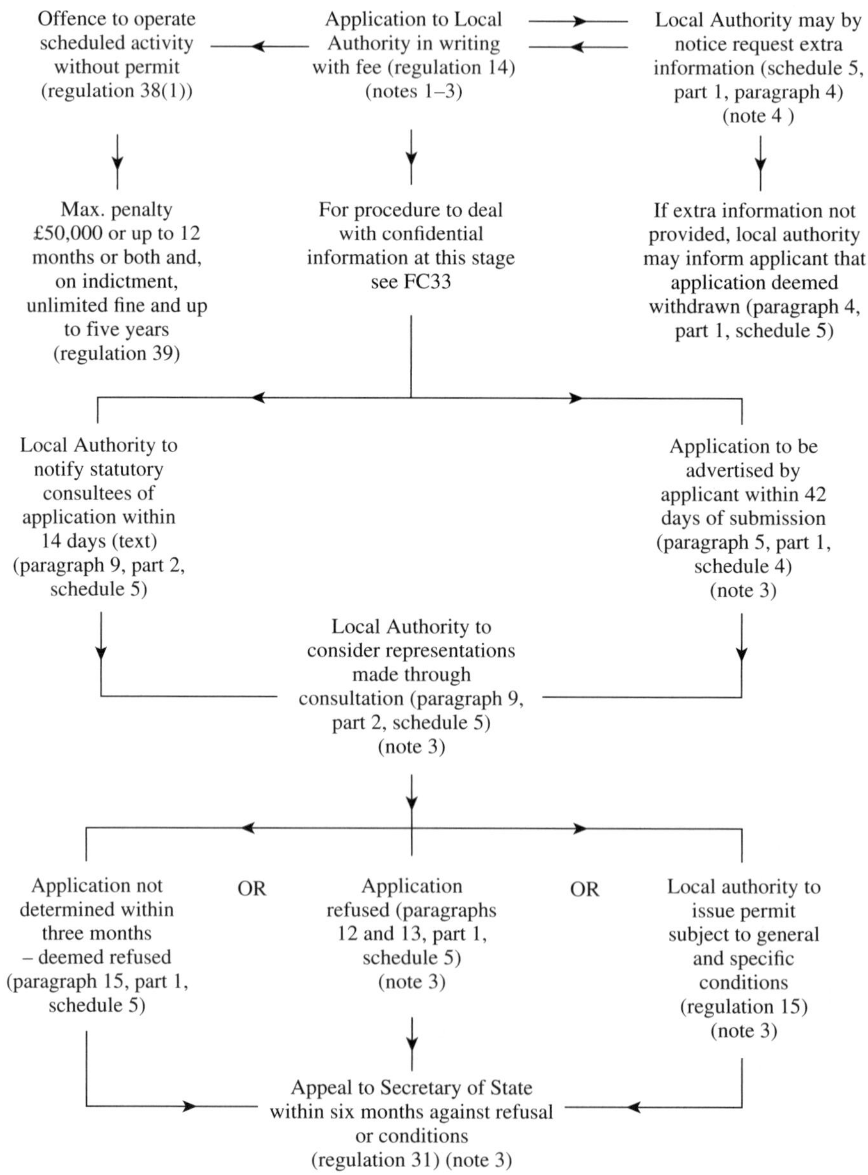

Notes

1. For details to be included, see regulation 14 and part 1, schedule 5.
2. For consolidation of a number of regulated facilities with a single permit (regulation 18).
3. Details of these events to be included in public register (regulation 46 and paragraph 1, schedule 24).
4. Request for information notice form in part D of General Manual.

FC33 Handling of commercially confidential and national security information by local authorities – Regulations 47 to 55 Environmental Permitting Regulations 2010

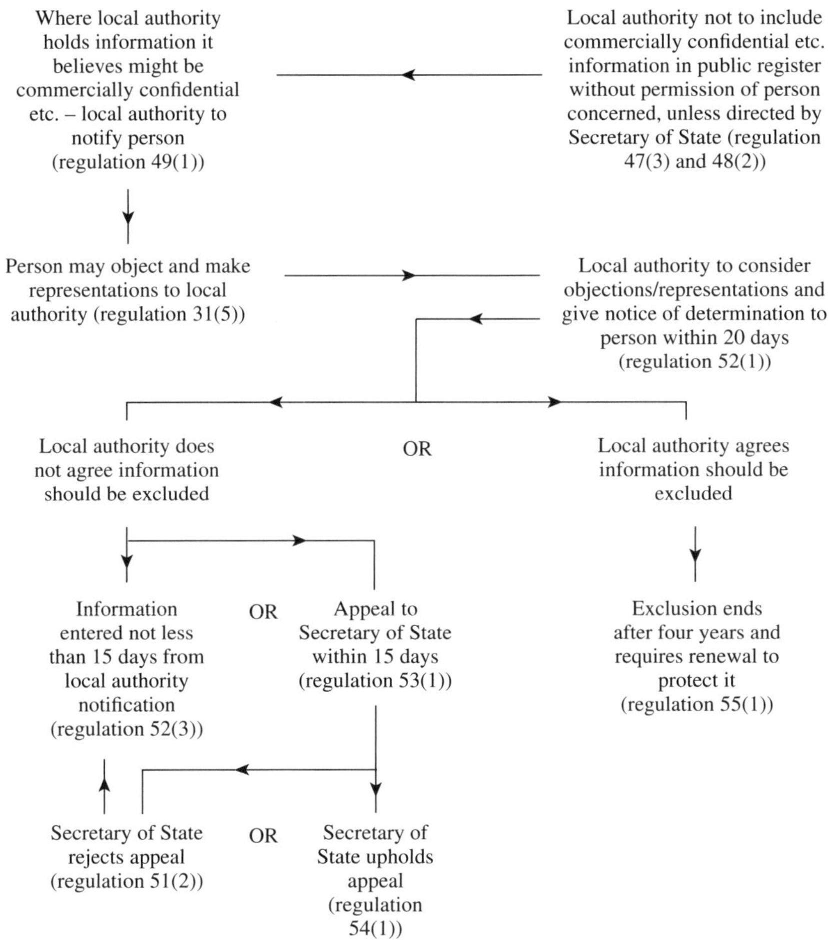

Determination of applications

The general principles against which all applications, parts A and B, are to be considered are:

(i) that all appropriate preventative measures must be taken against pollution and in particular the application of BAT (see page 156);
(ii) that no significant pollution will be caused.

The additional general principles to be considered for part A activities are:

(a) that satisfactory methods will be in place to avoid waste production and, where it is produced, for its recovery and disposal;
(b) that there is efficient use of energy;
(c) that measures are taken to prevent accidents and to limit their consequences;
(d) that, on the cessation of operation of the installation, satisfactory measures will be taken to avoid pollution and return the site to a satisfactory state.

The local authority must determine applications generally within three months of submission unless a longer period is agreed. In the case of transfer, it is two months and, where national security and confidentiality is an issue, it is four months. Failure to do so is deemed to be a refusal (paragraph 15, part 1, schedule 5).

Content and form of the permit

An environmental permit must specify the operator and the regulated facility whose operation it authorises. It can be in electronic form and must include a map, plan or other description of the site showing the geographical extent of the site of the regulated facility, but the map etc. does not apply to mobile plant or radioactive substances activities (regulation 14)

Conditions may be attached to the permit requiring the operator to carry out works or do other things on land which the operator is not entitled to do without obtaining the consent of another person (regulation 15).

A regulator cannot authorise the operation of more than one regulated facility under a single environmental permit but can authorise a single environmental permit for the operation by the same operator of more than one mobile plant, standard facility, regulated facility on the same site or more than one radioactive substances activity (regulation 17).

If the local authority decides to issue a permit, it may attach conditions (paragraph 12, part 1, schedule 5) appropriate to the following issues and in accordance with sectoral technical guidance. For all activities, these are:

(a) emission limit values (EMLs) or equivalent parameters for pollutants, in particular those likely to be emitted in significant quantities. These will normally be based on BAT (see page 156), taking account of the particular characteristics and the local environment of the plant;
(b) best available technique for preventing or reducing emissions.

It is suggested that content and conditions should cover:

(a) long-distance and transboundary pollution;
(b) the protection of soil and groundwater and the management of waste;
(c) precautions to protect the environment when the installation is not operating normally, e.g. during start-up;
(d) site monitoring and remediation;
(e) the on-going monitoring of emissions and the submission of reports to the local authority;
(f) notification procedures to deal with incidents or accidents;
(g) taking account of conditions for emissions to water specified by the Environment Agency; and
(h) avoiding conflict with other legislation prescribing release levels.

Standard rules

The Secretary of State is empowered to make standard rules (generally known as process and sector guidance notes) for certain types of installation, which can be used by a local authority instead of site-specific conditions. Standard rules will, by their nature, be suitable for industry sectors where installations share similar characteristics (regulations 26 to 30).

Refusals

A permit must be granted or refused following an application. The permit cannot be granted by the local authority unless it considers that the applicant will be the person who will have control after the permit is issued and will ensure that the installation or mobile plant is operated so as to comply with the environmental permit (paragraphs 12 and 13, part 1, schedule 5). This may be, for example, where there is reason to believe that the operator lacks the management systems or competence to run the installation according to the application or any permit conditions.

There is an appeal procedure against refusal (see below).

Transfers (FC34)

Transfers of permits between operators must be the subject of a joint application to the local authority, who may agree to the transfer unless it considers that the conditions attached will not be complied with (regulation 21). There is an appeal against refusal (regulation 31) (see below).

Permit reviews

The local authority must periodically review the environmental permits and make periodic inspections of regulated facilities (regulation 34).

It is suggested that the reviews be carried out where:

(a) the installation causes such significant pollution that the local authority must change the emission limit values;

FC34 Transfer of permits – Regulation 21 and Schedule 5 Environmental Permitting Regulations 2010

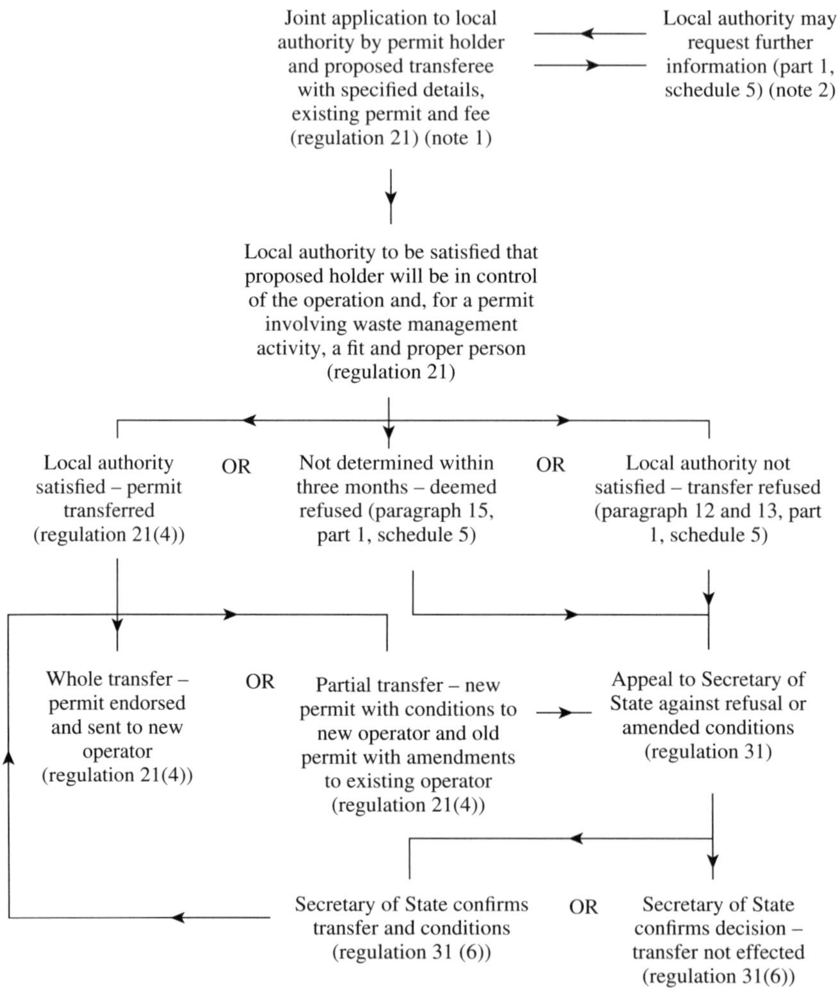

Notes

1. Application form in part C of General Manual.
2. Specimen request for information notice in part D of the General Manual.
3. Information to be included in public register (regulation 46 and paragraph 1, schedule 24).

(b) substantial changes in best available technique make it possible to reduce emissions significantly without excessive costs; and
(c) operators must change techniques for reasons of safety.

Variation of conditions (FC35)

Operators of permitted installations and mobile plant must make an application for variation to the local authority of their intention to make a 'substantial change' and the rules for consultation and public participation apply (form in part C of the General Guidance Manual) (schedule 5).

The local authority may vary the conditions at any time on its own initiative. The prescribed fee must accompany applications (see 'Charges' on page 173). The application or local authority-led variation is subject to the same procedures as an initial application, including a period of public participation and consultation before determination.

There is an appeal against the change of conditions and against a refusal to grant an application for change (see below) (regulation 20).

Surrender of permits (FC36 and FC37)

An environmental permit may be surrendered by the operator of an activity where he intends to cease operation. The operator must notify the local authority about a Part B installation, except a waste operation, mobile plant or stand-alone water discharge activity or stand-alone groundwater activity. In all other cases, an application must be made to the local authority. The local authority may accept the surrender of the permit if it is satisfied that the steps to be taken to implement the closure are appropriate to avoid risk of pollution and will return the site to a satisfactory state.

In the case of partial surrenders, the regulator must serve a notice on the operator specifying that it is necessary to vary the conditions, the variation and the date the variation takes effect.

Applications must be determined in the same way as initial applications, variations and transfers and failure to do so is deemed to be a refusal. There is an appeal against refusal to accept surrender (regulations 24 and 25).

Enforcement notices (FC38)

If the local authority is of the opinion that an environmental permit has been is or is likely to be contravened, they may serve an enforcement notice on the operator. The notice must specify:

(a) the contravention;
(b) the steps necessary to remedy it; and
(c) the time period allowed.

Non-compliance is an offence – see below (regulation 36).

FC35 Variations to permit conditions – Regulation 20 and Schedule 5 Environmental Permitting Regulations 2010

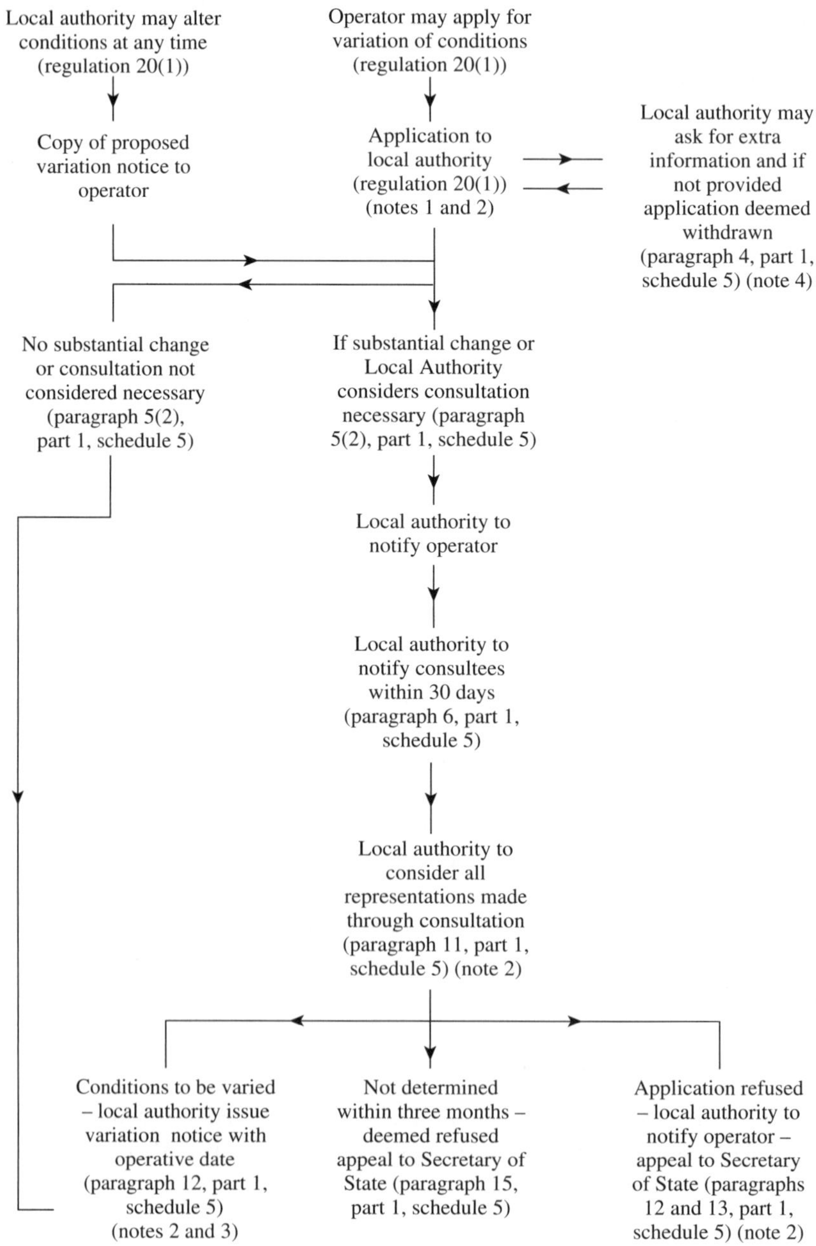

Notes

1. The details to be included in the application are set out in regulation 14 and paragraph 2 of part 1 of schedule 5.
2. Information to be included in public register (regulation 46 and paragraph 1, schedule 24).
3. Model form in part C of General Manual.
4. Specimen request for information notice in part D of General Manual.

FC36 Application for surrender of an environmental permit (Part A processes) – Regulation 25 Environmental Permitting Regulations 2010

Operator ceases or intends to cease operating whole or part of an installation other than part B installation, mobile plant, stand-alone water discharge or groundwater activity (regulation 25(1))

↓

Operator makes application to Local Authority with specified details as if an initial application is being made(text) and a date of surrender not less than 20 days from notification (regulation 24(3)) (note 1)

↓

Permit ceases to be effective on date specified in notification (regulation 24(4))

↓

For partial surrenders Local Authority may vary conditions for remaining permit by serving a variation notice (see FC35) (regulation 24(6))

Notes

1. Application form in part C of General Manual.
2. Part 1 of schedule 5 of the Environmental Permitting Regulations 2010 applies to surrender of an environmental permit.
3 The local authority can:
   - request more information (paragraph 4, part 1, schedule 5);
   - refuse an application (paragraph 12, part 1, schedule 5).
4. If application is refused or deemed to be refused, the operator can appeal to the Secretary of State (regulation 31).

FC37 Notification of surrender of an environmental permit (Part B processes) – Regulation 24 Environmental Permitting Regulations 2010

Operator ceases or intends to cease operating whole or part of a part B installation, mobile plant, stand-alone water discharge or groundwater activity (regulation 24(1))

Operator to notify local authority with specified details as if an initial application is being made(text)
and a date of surrender not less than 20 days from notification (regulation 24(3)) (note 1)

Permit ceases to be effective on date specified in notification (regulation 24(4))

For partial surrenders local authority may vary conditions for remaining permit by serving a variation notice (see FC35) (regulation 24(6))

Notes

1. Application form in part C of General Manual.
2. Part 1 of schedule 5 of the Environmental Permitting Regulations 2010 applies to surrender of an environmental permit.

FC38 Enforcement and suspension notices – Regulations 36 and 37 Environmental Permitting Regulations 2010

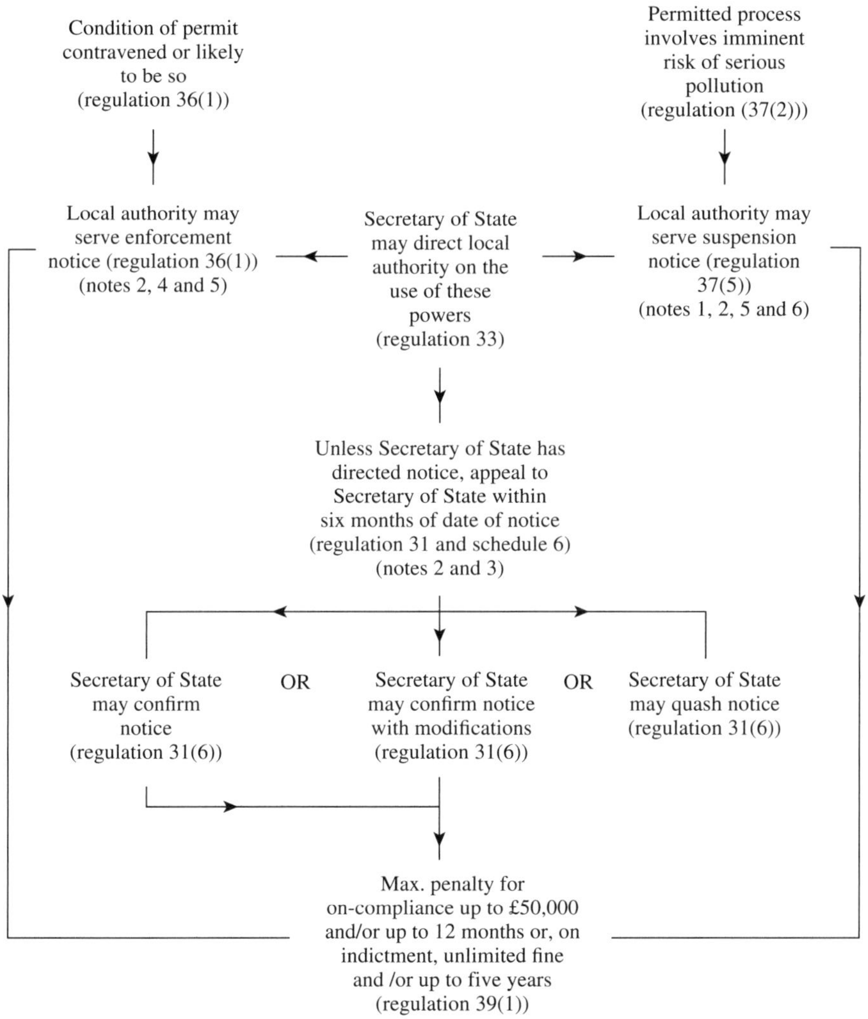

Notes

1. Where these circumstances exist, the local authority may serve a suspension notice unless taking action itself under regulation 57 – see FC40.
2. Details to be entered in public register (regulation 46 and paragraph 1, schedule 24).
3. The notices are not suspended during the appeal process (regulation 31(9)).
4. Instead of serving an enforcement notice, the local authority may prosecute for the contravention (regulation 38).
5. An enforcement or suspension notice may be withdrawn at any time (regulations 36(4) and 37(8)).
6. Specimen Suspension Notice in part D of the General Manual.

Suspension notices

Where the local authority believes that the continued operation of the activities will involve a risk of serious pollution, it may serve a suspension notice on the operator. This applies whether or not the activities are permitted and may deal with a failure to comply with conditions of a permit. Alternatively, the situation may be dealt with by use of its powers under regulation 57 – see below.

The suspension notice must:

(a) specify what the risk of serious pollution is;
(b) specify the steps to be taken;
(c) specify the period within which the steps must be taken;
(d) state that any permit shall cease to have effect to the extent specified in the notice until the notice is withdrawn;
(e) where the activities will continue in part, state any additional measures to (b) above that must be taken.

The local authority may withdraw the notice at any time and must do so when the risk has been removed (regulation 37).

Prevention or remedying of pollution (FC39)

Where the local authority believes that the operation of an activity where an environmental permit is in place involves a risk of serious pollution or there has been the commission of an offence by the operator that causes pollution, it may arrange itself for steps to be taken to remove that risk or pollution.

Before taking any steps, the local authority must give the operator five days' notice.

The costs of the necessary works may be recovered from the operator unless the operator can show that there was no risk or the costs, or part of them, were incurred unnecessarily (regulation 57).

Revocation notices (FC40)

The local authority may revoke a permit at any time, in whole or in part, by the service of a revocation notice. This is a wide power that can be used whenever the local authority considers it to be appropriate.

The notice must be served on the operator and specify:

(a) the reasons for the revocation;
(b) in the case of a partial revocation:
    (i) the extent to which the permit is being revoked;
    (ii) any variation to the conditions of the permit; and
(c) the date on which the revocation will take place, at least 20 working days after the notice is served (regulations 22 and 23).

Unless the local authority withdraws a revocation notice, an environmental permit ceases to have effect on the date specified in the notice:

(a) in the case of a revocation in whole, entirely; or
(b) in the case of a partial revocation, to the extent of the part revoked.

FC39 Local authority powers to prevent or remedy pollution – Regulation 57 Environmental Permitting Regulations 2010

There is in relation to an installation or mobile plant an imminent risk of serious pollution (regulation 57(1)) (note 3)

↓

Pollution caused by installation or mobile plant where:
(a) no permit has been issued;
(b) there is a contravention of conditions; or
(c) there is a failure to comply with a variation or suspension notice
(regulation 57 (2)) (note 1)

↓

Local authority to give operator five days' notice of steps to be taken (regulation 57(3))

←

Local authority may take steps to remove that imminent risk or the effects of pollution (regulation 57(1) and (2)) (note 2)

↓

Local authority may recover costs of the actions unless operator shows that:
(a) there was no risk of serious pollution; or
(b) costs were unnecessarily incurred
(regulation 57(4) and (5))

Notes

1. This power may be implemented in addition to the prosecution of committed offences.
2. There is no provision for appeal against these actions, only a challenge to the recovery of costs.
3. Where these circumstances exist, the local authority may implement this procedure or serve a suspension notice – see FC39.

FC40 Revocation of permits by local authority – Regulations 22 and 23 Environmental Permitting Regulations 2010

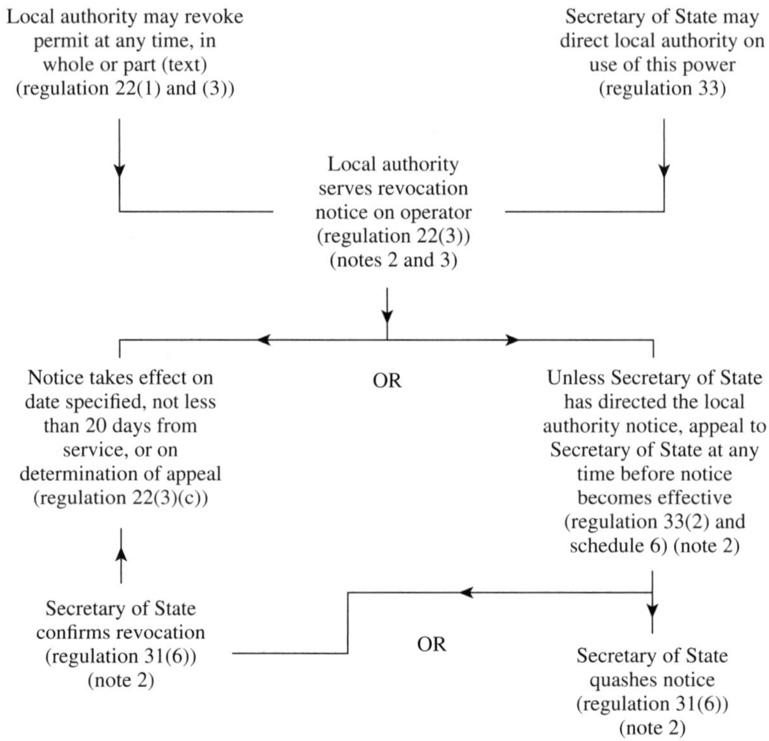

Notes

1. The revocation is suspended pending determination of the appeal or withdrawal of the notice (regulation 31(9)(b)).
2. Details to be entered in public register (regulation 46 and paragraph 1, schedule 24).
3. Specimen Revocation Notice in part D of the General Manual.

Public registers

Local authorities must maintain public registers containing information on all Part A(2) and Part B permits issued by them for installations in their areas and also details of installations in their area regulated by the Environment Agency (regulation 46). The content of the register is prescribed by schedule 24, paragraph 1. The register may be in any form and the necessary content of registers is set out in annex 13 of the General Guidance Manual.

Commercial confidentiality

The local authority must exclude what it considers commercially confidential information or if it receives a notice from the operator that the information is confidential (see FC33 and FC41). Information is commercially confidential if the information is commercially or industrially confidential in relation to any person (regulation 48).

If the local authority considers that information may be confidential but has not received an objection notice, it must give notice of that view to the operator. The operator must, within 15 working days after the date of the notice, give consent to the regulator to include the information on the register or give an objection notice to the regulator (regulation 49).

Charges

The Secretary of State, under the powers of regulation 65, determines the type and level of charges that may be made by local authorities in relation to their activities for environmental permits. This scheme is reviewed annually and is set out in the local authority charging schemes: the Local Authority Permits for Part B Installations and Mobile Plant (Fees and Charges) (England) Scheme 2012 and the Local Authority Permits for Part A(2) Installations and Mobile Plant (Fees and Charges) (England) Scheme 2012.

The charges may be raised relating to:

(a) permit applications and 'substantial change';
(b) late applications;
(c) variations;
(d) transfers;
(e) surrender;
(f) subsistence charges on an annual basis.

The fees are based on a risk-based approach in line with the 'polluter pays' principle. All processes, except waste oil burners and petrol stations, are required to be risk-rated to establish the inspection frequency and hours that the local authority needs to spend on them.

FC41 Applications to local authority to exclude commercially confidential information from public register – Regulations 48 to 55 Environmental Permitting Regulations 2010

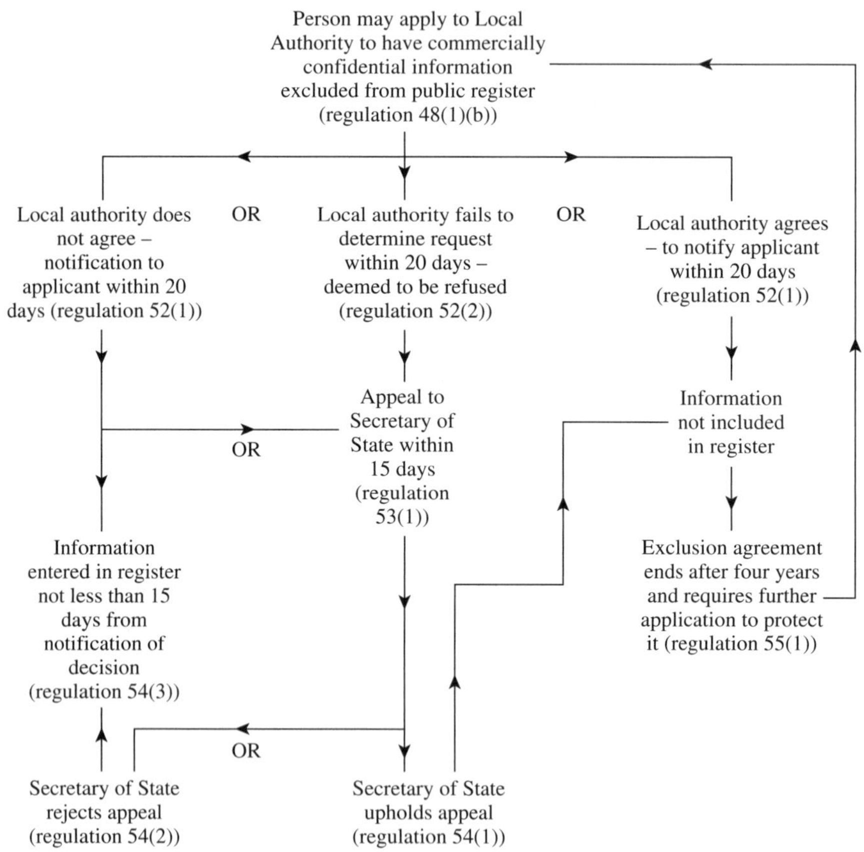

Appeals

Unusually for environmental health enforcement work, rights of appeal are to the Secretary of State and not to a court of law. The operator may appeal against local authority decisions relating to:

(a) refusal of an application for a permit (regulation 13);
(b) refusal of an application for the variation of a permit (regulation 20);
(c) the service of a revocation, enforcement or suspension notice (regulations 22, 36 and 37);
(d) determination that information is not commercially confidential (regulation 53);
(e) refusal of an application to transfer or surrender a permit (regulations 21, 24 and 25);
(f) the service of a variation notice on the local authority's initiative (regulation 20).

The full procedure with timescales is set out in schedule 6 to the Environmental Permitting Regulations. The Secretary of State has the power to affirm or quash the local authority decisions and to alter the terms of any conditions (regulation 31(6) and schedule 6).

Offences

Offences committed against any of the requirements of the Environmental Permitting Regulations are punishable, on summary conviction, of a fine up to £50,000 and/or up to 12 months' imprisonment. Conviction in the Crown Court may lead to a fine and/or to imprisonment for up to five years (regulations 38 and 39).

Chapter 3

FOOD SAFETY

FOOD CONTROL AND HYGIENE LEGISLATION

Legal structure

The current structure of food safety legislation for the UK took effect on 1 January 2006. This stems from two principal EC regulations:

(a) EC Regulation 852/2004 on the hygiene of foodstuffs; and
(b) EC Regulation 853/2004, laying down specific hygiene rules for food of animal origin.

Several others that contain important procedural matters relating to enforcement, including EC Regulation 854/2004, laying down specific rules for the organisation of official controls on products of animal origin intended for human consumption, support these.

These regulations have both a direct effect on UK enforcement, in that some of the provisions are directly enforceable, e.g. the registration and approval of food premises, and an indirect effect that has required new legislation in the UK to implement them, e.g. the use of hygiene improvement notices. In the light of UK devolution, the legislation dealing with these issues is a matter for the devolved governments. This work is based on the provisions in England but similar legislation is to be found in Northern Ireland, Scotland and Wales. The principal English regulations affecting this work are:

(a) the Food Hygiene (England) Regulations 2006 (FH (E) Regulations); and
(b) the Official Feed and Food Controls (England) Regulations 2009 (OF and FC (E) Regulations).

Whilst the Food Safety Act 1990 remains with some amendments, its enforcement provisions are likely to be used only in respect of food standards matters and not for food hygiene, where the provisions of the 2006 regulations will take effect.

Codes of Practice (CoP) and Practice Guidance (PG) form a very important part of the procedural requirements. The Food Law Code of Practice (England) 2006 was issued by the Secretary of State under section 40 of the Food Safety Act 1990 and regulation 24 of the Food Hygiene (England) Regulations 2006. Food authorities are therefore required to follow and implement it, and food authorities who do not do so may find their decisions or actions successfully challenged or evidence ruled inadmissible.

The Practical Guidance (England) 2006 has a different status, being issued by the Food Standards Authority to assist food authorities with the enforcement of food law, and is non-statutory.

AUTHORISED OFFICERS

An enforcement officer of the food authority enforces most of the procedures dealt with in this part. This is a person (whether or not an officer of the authority) who is authorised in writing, either generally or specifically, to act in matters arising under the Food Safety Act and the Food Hygiene (England) Regulations 2006.

There are specific requirements for the levels of qualifications and experience held by authorised officers carrying out inspections or other enforcement matters under food law. These are set out in the Code of Practice, chapter 1.2 and are noted in the procedures here.

Definitions

The following definitions are relevant to all procedures in this part:

1. Food

The definition in the Food Safety Act was changed by the Food Safety Act 1990 (Amendment) Regulations 2004 in order to bring it into line with EC Regulation 178/2002. It is as follows:

> Food means any substance or product, whether processed, partially processed or unprocessed, intended to be, or reasonably expected to be, ingested by humans.

Food includes drink, chewing gum and any substance, including water, intentionally incorporated into the food during manufacture, preparation or treatment.

Food shall not include:

- (a) feed;
- (b) live animals unless they are prepared for placing on the market for human consumption;
- (c) plants prior to harvesting;
- (d) medicinal products within the meaning of Council Directives 65/65/EEC and 92/73/EEC;
- (e) cosmetics within the meaning of Council Directive 76/768/EEC;
- (f) tobacco and tobacco products within the meaning of Council Directive 89/622/EEC;
- (g) narcotic or psychotropic substances;
- (h) residues and contaminants (section 1(1)).

This definition applies to all procedures in this part.

2. Food business

This is 'any undertaking, whether for profit or not and whether public or private, carrying out any of the activities related to any stage of production, processing and distribution of food' (EC 178/2002, chapter 1, article 3).

3. Food business operator

This is 'the natural or legal person responsible for ensuring that the requirements of food law are met within the food business under their control' (EC 178/2002, chapter 1, article 3).

4. Specified community provision

This means 'any provision of the Community Regulations that is listed in Column 1 of Schedule 2 and whose subject-matter is described in Column 2 of that Schedule' (Food Hygiene (England) Regulations 2006, regulation 2).

5. The Community Regulations

This means 'Regulation 852/2004, Regulation 853/2004, Regulation 854/2004, Regulation 2073/2005 and Regulation 2075/2005' (Food Hygiene (England) Regulations 2006, regulation 2).

6. The Hygiene Regulations

This means 'these (the Food Hygiene (England) Regulations 2006) and the Community Regulations' (Food Hygiene (England) Regulations 2006, regulation 2).

GENERAL PROCEDURAL PROVISIONS

1. Notices

In most cases, the forms and notices to be used in the various procedures are specified in the Code of Practice at annexes 7, 8 and 9.

The requirements relating to notices are:

(a) all notices should be in writing;
(b) they must be signed (including a facsimile signature) by the proper officer or any officer authorised in writing;
(c) service may be effected by either:
  (i) delivering it to the person;
  (ii) by leaving it or sending it in a pre-paid letter addressed to him at the usual or last known residence;
  (iii) for an incorporated company or body, by delivering it to the secretary or clerk at its registered office or by sending it in a pre-paid letter addressed to him there;
  (iv) where it has not been possible to ascertain the name and address of an owner or occupier or if the premises are unoccupied, by addressing it 'the owner' or 'the occupier' and delivering it to someone on the premises or affixing it to the premises.

(Food Safety Act 1990, section 50 and similar provisions in the Food Hygiene (England) Regulations 2006, regulation 28)

For some of the procedures here, the Code of Practice indicates acceptable, alternative methods of service, e.g. courier service, and these are indicated as appropriate.

2. Powers of entry

Similar provisions are made by the Food Safety Act 1990 at section 32 and the Food Hygiene (England) Regulations 2006 at regulation 14, except that in the latter there is a power to enter premises in the area of another food authority. The provisions under the 2006 regulations are that an authorised officer, upon producing some duly authenticated document if requested, has a right at all reasonable hours to enter any premises:

(a) within his own food authority area to check for contraventions of the Hygiene Regulations;
(b) inside or outside of his food authority area to see if there is evidence of such contravention;
(c) to perform the functions of the food authority under the Hygiene Regulations.

Admission to a private dwelling may not be demanded unless 24 hours' notice of entry is given. There are provisions for application to a Justice of the Peace for a warrant to enter if necessary (regulation 14).

3. Obstruction

Persons who obstruct anyone executing the 'Hygiene Regulations' or, without reasonable cause, fail to give assistance or information, or give false or misleading information are guilty of an offence (Food Safety Act 1990, section 33 and Food Hygiene (England) Regulations 2006, regulation 15).

4. Appeals

Where provision is made for appeals against the actions of authorised officers in these procedures, these are made to the magistrates' court within one month and may be followed by a further appeal to the Crown Court (Food Hygiene (England) Regulations 2006, regulations 20–22 and, in relation to approvals of premises dealing with foods of animal origin, the Official Feed and Food Control (England) Regulations 2009, regulations 12–13).

5. Due diligence

In any proceedings, it will be a defence to show that the person took all reasonable precautions and exercised due diligence to avoid the commission of the offence either by himself or by a person under his control (Food Safety Act 1990, section 21(1) and Food Hygiene (England) Regulations 2006, regulation 11).

6. Offences due to the fault of another person

Where the commission of an offence by a person is due to the act or default of another person, that other person is also guilty of an offence (Food Hygiene (England) Regulations 2006, regulation 10).

7. Time limit for prosecutions

No prosecution should take place under these procedures after the expiry of:

(a) three years from the commission of the offence; or
(b) one year from its discovery by the prosecutor,

whichever is the earlier (Food Safety Act 1990, section 34 and Food Hygiene (England) Regulations 2006, regulation 16).

8. Offences and penalties
Any person who contravenes any of the 'specified community provisions' listed in schedule 2 of the 2006 regulations is guilty of an offence. These provisions deal with the various relevant EC regulations. In addition, it is an offence not to comply with the 'Hygiene Regulations', which are the 2006 regulations and EC Regulations 852/2004, 853/2004, 854/2004, 2073/2005 and 2075/2005.

The penalties for these offences are:

(a) on summary conviction, a fine not exceeding the statutory maximum; or
(b) on conviction on indictment, imprisonment for not exceeding two years, a fine or both (Food Hygiene (England) Regulations 2006, regulation 17).

REGISTRATION OF FOOD BUSINESS ESTABLISHMENTS

References

EC Regulation 852/2004 on the Hygiene of Foodstuffs, 29 April 2004.
The Food Hygiene (England) Regulations 2006.
Food Law Code of Practice (England), April 2012, DEFRA.

Extent

These provisions apply in England only but there are similar provisions in Wales.

Scope

Regulation 852/2004 requires that each food business operator must supply the competent authority, in this case the food authority, with details to enable each establishment used for any of the stages of production, processing and distribution of food to be registered (A6.2). An establishment is defined as including any unit of a food business and will therefore include those on land and water and those which are mobile (A2 paragraph 1(c)). The purpose is to allow the food authority to perform official controls efficiently (recital paragraph (19)).

Each separate unit must be registered, except:

(a) those which require to be approved – see FC43;
(b) those which are not to be regarded as food businesses for the purpose of this procedure. These are:
   (i) primary production for private domestic use;
   (ii) domestic preparation, handling or storage for private domestic consumption;

(iii) the direct supply by the producer of small quantities of primary products to the final consumer or to a local retail establishment directly supplying the final consumer;
(iv) collection centres and tanneries;

(c) ocean-going ships, aircraft, trains and long-distance coaches, which are not expected to register for practical reasons, but shore-based vessels, e.g. floating restaurants, and boats plying on inland waterways have to register;
(d) individual vending machines, although distribution centres at which food is stored or transported to vending machines should be registered;
(e) in relation to markets, vehicles and stalls used for transporting or preparing food or the sale of food within the market, which are to be registered even if owned by the market operator. Operators working from permanent stalls within a market must register;
(f) moveable establishments other than those which form part of a market or operate within the area of a market, e.g. ice-cream vans, hot dog vendors etc., which should be registered by the food business operator with the food authority in the area in which they are ordinarily kept (Code of Practice 1.5.1.2 and 8).

Registration

Operators must register their premises with the food authority at least 28 days before food operations commence and it is an offence not to do so. The food authority is required to provide a registration form for this to be undertaken and a model form is contained in annex 8 (Code of Practice 1.5.3.2 and 3).

On receiving the form, the local authority must record the date of receipt on it and, if there are any activities noted that are outside its enforcement role, send it to the appropriate authority. If there are any parts not completed, the food authority should make contact with the operator or, if the missing information is substantial, return it to him for final completion (Code of Practice 1.5.5).

Registration is then effected by entering the details on a database and keeping the form in the file for those premises. Certificates of registration are not to be issued to the operator but the food authority can confirm the receipt of the form and the entry of the premises in the database (Code of Practice 1.5.6). In two-tiered areas, the food authority is required to send a copy of the information to the county council within 28 days (Code of Practice 1.1.5).

Food authorities are asked to consider an inspection of the premises when the form is received but registration is not dependent upon the results of that inspection. The registration process is informatory (Code of Practice 1.5.5).

Database

Food authorities must draw up a list of food establishments registered with them and will include within it premises previously registered under the 1991 regulations (EC Regulation 882/2004, A31 (1)(b)). The list is to be made available to the general public at all reasonable times and should contain the following details for each business:

FC42 Registration of food business establishments – EC Regulation 852/2004 on the Hygiene of Foodstuffs

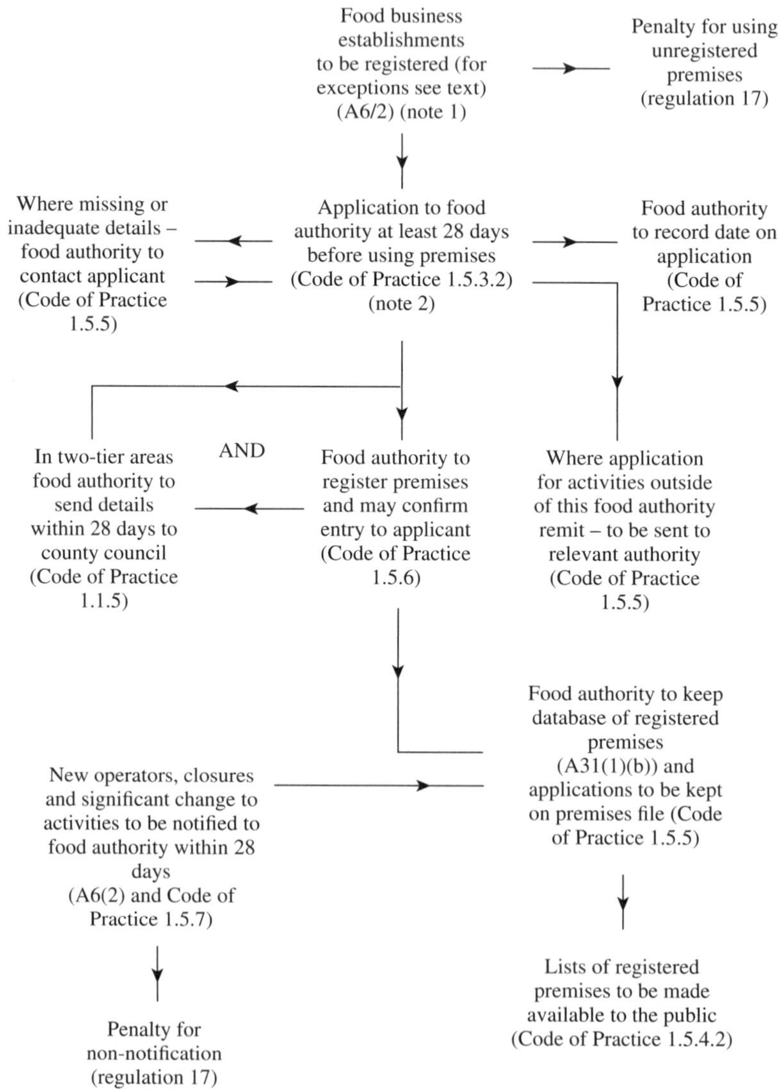

Notes

1. 'A' references are to EC Regulation 852/2004; regulation refers to the 2006 England regulations and Code of Practice to the 2012 Code of Practice.
2. A model form is at Annex 8 of the Code of Practice and should be made available by the food authority.
3. For approval of product-specific establishments, see FC43.

(i) name of the food business;
(ii) address;
(iii) particulars and nature of the business.

Details of food premises records are to be made available to the Health Protection Agency and the Consultant in Communicable Disease Control to facilitate the investigation of outbreaks of disease (Code of Practice 1.4.3). These records are subject to both the Data Protection Act 1998 and the Freedom of Information Act 2000.

Changes to activities etc.

Food business operators are required to ensure that the food authority always has up-to-date information on establishments by notifying them of any significant changes in activities or of any closure within 21 days of the changes happening (A6.2 of 852/2004 and Code of Practice 1.5.7). The new operator should make notification of a change of operator. The database should be updated accordingly and the details placed on the premises file (Code of Practice 1.5.7).

APPROVAL OF PRODUCT-SPECIFIC ESTABLISHMENTS

References

EC Regulation 852/2004 on the hygiene of foodstuffs, 29 April 2004.
EC Regulation 853/2004 on specific hygiene rules for food of animal origin, 29 April 2004.
EC Regulation 854/2004 laying down specific rules for the organisation of official controls on products of animal origin intended for human consumption.
EC Regulation 882/2004 on the official controls performed to ensure the verification of compliance with feed and food law, animal health and animal welfare rules.
The Food Hygiene (England) Regulations 2006.
The Official Feed and Food Controls (England) Regulations 2009.
Food Law Code of Practice (England), April 2012, DEFRA (Code of Practice).
Food Law Practice Guidance (England), April 2012, Food Standards Agency (Practice Guidance).

Extent

These provisions apply in England only but there are similar provisions in Wales.

Scope

The competent authority must approve establishments handling products of animal origin. These are unprocessed and processed products of animal origin but not foods consisting partly of products of plant origin (A4 853/2004). 'Establishment' is defined in A2 paragraph 1(c) of EC852 as being 'any unit of a food business' and will therefore include both mobile units and those on water, including fishery vessels. The food authority is responsible for approvals of premises where any of the following products are produced alone or in combination:

(a) minced meat;
(b) meat preparations;
(c) mechanically separated meat;
(d) meat products;
(e) live bivalve molluscs, but not the classification of beds, which is done by the Food Standards Agency;
(f) fishery products;
(g) raw milk other than cows' milk;
(h) dairy products;
(i) eggs but not from primary production;
(j) egg products;
(k) frogs' legs and snails;
(l) rendered animal fats and greaves;
(m) treated stomachs, bladders and intestines;
(n) gelatine and collagen – collection centres and tanneries supplying raw material for these premises do not require to be approved but do need authorisation by the food authority;
(o) certain cold stores and wholesale markets.

The Food Standards Agency is the competent authority for the approval of:

(i) raw cows' milk;
(ii) slaughterhouses;
(iii) game handling establishments;
(iv) cutting plants putting fresh meat on the market; and
(v) activities co-located with these where approval would otherwise be dealt with by the food authority, e.g. meat products establishments (regulation 5, Food Hygiene (England) Regulations 2006).

Exemptions

The following are exempted from the need to seek approval:

1. Primary production for private domestic use;
2. Domestic preparation, handling or storage of food for private domestic consumption;
3. The direct supply by the producer of small quantities of primary products to the final consumer or to local retail establishments directly supplying the final consumer;
4. Hunters who supply small quantities of wild game or wild game meat directly to the final consumer or to local retail establishments supplying the final consumer;

FC43 Approval of product-specific establishments by food authorities – EC Regulation 882/2004

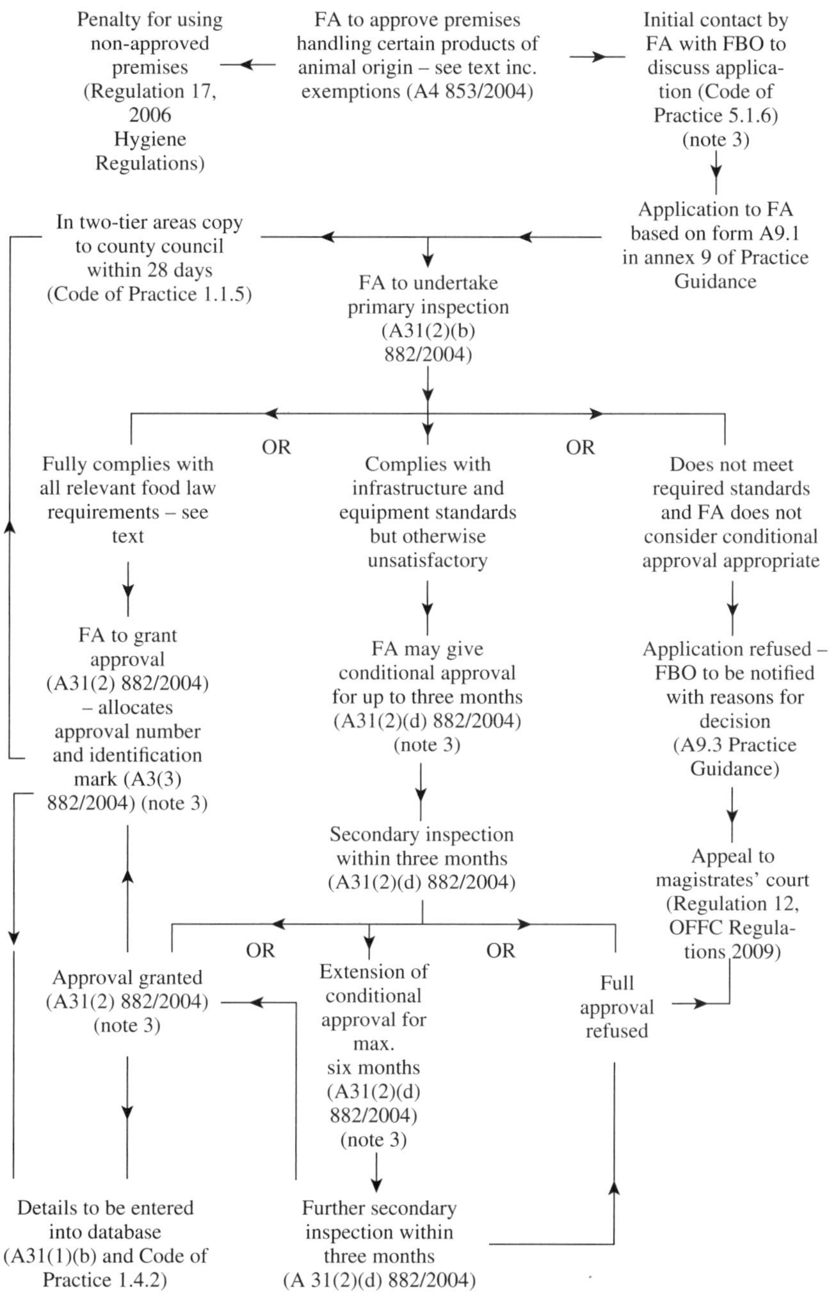

Notes

1. FBO = food business operator.
2. FA = food authority.
3. FSA to be notified of these decisions.
4. OFFC Regulations = Official Feed and Food Controls Regulations 2009.

5. Food containing both products of plant origin and processed products of animal origin;
6. The production, transportation or storage of products not requiring temperature-controlled storage conditions (EC Regulation 853/2004 A1).

Activities such as the occasional preparation of food by individuals or groups for gatherings or for sale at charitable events are also exempt (recital 9 of 852/2004).

A detailed appraisal of these exemptions is given in the Code of Practice 5.1.4.

Applications for approval

There should be initial contact and discussion between the food authority and the food business operator prior to the application being submitted (Code of Practice 5.1.6). The form of application is at A.9.1 in annex 9 of the Practice Guidance.

Having ensured that all the necessary information has been obtained, the local authority is required to undertake a primary inspection (A31(2)(b) of 882/2004). This is to check that the premises meets all of the requirements for that type of premises as set out in detail in annex 2 of 853/2004. A sectoral approach is taken in determining the specific hygiene provisions for each foodstuff but the food authority may grant special conditions to take account of traditional production methods.

Where the application is for a live bivalve mollusc purification centre, or a modification to an existing centre, a copy must be sent to the Food Standards Agency for consultation with the Centre for Environment, Fisheries and Aquaculture Science (CEFAS). The food authority cannot determine such applications until they have received a response from CEFAS and must include any operating conditions set by them in the approval document (Code of Practice 5.1.10).

Approvals

Approval can only be given if the premises complies with all of the food law requirements relevant to that business (EC 882/2004 A31 (2)). Some conditions are specific for each type of food and are to be found in the annexes to EC Regulation 853/2004, whilst other general requirements are laid down in 852/2004. When approval is given, the food authority must notify the food business operator and the Food Standards Agency – see A.9.2 in annex 9 of the Practice Guidance. Detailed guidance to food authorities on each product dealt with by them is contained in annexes 3–8 of the Practice Guidance.

Each establishment must be allocated a three-digit approval number unique to that authority, which should consist of the food authority's two-letter code followed by the approval number (A3 (3) 854/2004 and Code of Practice 5.1.12). The approval code is then to be incorporated into an identification mark to be applied to the products from that establishment, the form of which is contained in annex 2, section 1B of 853/2004 and the Practice Guidance, section 5.

Conditional approval

The food authority may grant a conditional approval where the premise does not fully comply but only if the infrastructure and equipment requirements are met. In such cases, a further inspection must be carried out within three months and the conditional approval may be extended for a further six months (A 31(2)(d) EC882/2004 and Code of Practice 5.1.8). When granting a conditional approval, the food business operator and the Food Standards Agency must be notified – see A9.2 of annex 9 of the Practice Guidance. At the end of the time limit, the application must be granted or refused – see A9.6 in annex 9 of the Practice Guidance.

Refusals to approve

If the requirements of 853/2004 are not met and the food authority has decided not to issue a conditional approval, the food authority will refuse the application. The applicant must be notified in writing of the refusal (see A9.3 in annex 9 of the Practice Guidance) with the reasons for that decision and the rights of appeal (Code of Practice 5.1.14).

Suspension and withdrawal of full and conditional approvals

Before deciding to use these options, the food authority must first have considered the use of the other enforcement options – see FCs 44–47. The withdrawal of an approval should only be implemented if the food authority is satisfied that it does not have a reasonable expectation that the deficiencies will be rectified and acceptable standards will be maintained in the future (EC Regulation 882/2004 A 31(2)(e)).

Decisions to suspend or withdraw approvals must be notified to the food business operator in writing – see A 9.4 and A 9.5 of annex 9 of the Practice Guidance – together with the reasons for that decision and the rights of appeal. The Food Standards Agency must also be informed. Pending the outcome of an appeal, the establishment is able to continue business subject to any conditions imposed by the food authority to protect public health, although the food authority will still have available the other enforcement options in that period (Code of Practice 5.2.5 and 6).

Changes of details or activities

Food business operators must ensure that the food authority has up-to-date information and, in particular, notify the food authority of significant changes of operation and closures (A 6(2) 852/2004 and Code of Practice 5.1.16).

Appeals

Appeals against decisions of the food authority to refuse full or conditional approval or to withdraw or suspend approval are dealt with by regulations 12–13 of the Official Feed and Food Controls (England) Regulations 2009. They are made initially to the magistrates' court with further appeal rights to the Crown Court.

Database

Comprehensive details of premises subject to approvals and conditional approvals are to be kept by the food authority and this should be made available to the public (Code of Practice 1.4.2). Information on individual premises must also be supplied to the Health Protection Agency and the Consultant for Communicable Diseases Control to aid investigations of outbreaks of infectious diseases (Code of Practice 1.4.3). The database should be updated accordingly.

Penalties

Offences of operating non-approved premises and other offences related to this procedure are dealt with by regulation 17 of the Food Hygiene (England) Regulations 2006. On conviction, persons are subject to a fine not exceeding the statutory maximum on summary conviction and, on indictment, imprisonment for up to two years and/or a fine.

Further guidance

See www.food.gov.uk/enforcement

HYGIENE IMPROVEMENT NOTICES

References

The Food Hygiene (England) Regulations 2006.
Food Law Code of Practice (England), April 2012, DEFRA (CoP).
Food Law Practice Guidance (England), April 2012, FSA (PG).

Extent

These provisions apply in England only but there are similar provisions in Wales.

Scope

If an authorised officer of the food authority has reasonable grounds for believing that a food business operator is failing to comply with the 'Hygiene Regulations', he may serve a hygiene improvement notice on that person (regulation 6(1)). The 'Hygiene Regulations' are the 2006 Regulations and Community Regulations 852/2004, 853/2004, 854/2004, 2073/2005 and 2075/2005 (regulation 2(1)).

The service of a hygiene improvement notice is considered to be appropriate where:

(a) formal action is proportionate to the risk; and/or
(b) there is a record of non-compliance by that food business operator; and/or
(c) the authorised officer has reason to believe that an informal approach will not be successful (Code of Practice 3.2.2.1).

They are not considered to be appropriate where:

(a) the contravention may be a continuing one;
(b) in transient situations, swift enforcement action is needed;
(c) there is a breach of good hygiene practice but no breach of regulations (Code of Practice 3.2.2.2).

The hygiene improvement notice may only be signed by an officer authorised to do so by the food authority, who must be either:

(i) an environmental health officer enforcing food hygiene or food processing regulations;
(ii) a holder of the Higher Certificate in Food Premises Inspection, who is authorised to carry out inspections; or
(iii) a holder of the Ordinary Certificate in Food Premises Inspection in relation to the premises they are authorised to inspect (Code of Practice 1.2.9.1.5).

Content

The hygiene improvement notice should:

(a) identify the grounds for believing that the food business operator is failing to comply;
(b) specify the matters which constitute that failure;
(c) indicate the measures to be taken to secure compliance; and
(d) specify a period of at least 14 days during which the measures must be taken (regulation 6(1)).

A model form of notice is found at A7.1/A7.2 of annex 7 of the Code of Practice.

Time limits

The hygiene improvement notice should indicate that, should he wish to do so, the food business operator may apply in writing for the time limit to be extended. The authorised officer will consider such request and, if feeling that an extension is justified, should withdraw the hygiene improvement notice and serve a new one with a different time limit (Practice Guidance 3.2.6).

FC44 Hygiene improvement notice (HIN) – Regulation 6 Food Hygiene (England) Regulations 2006

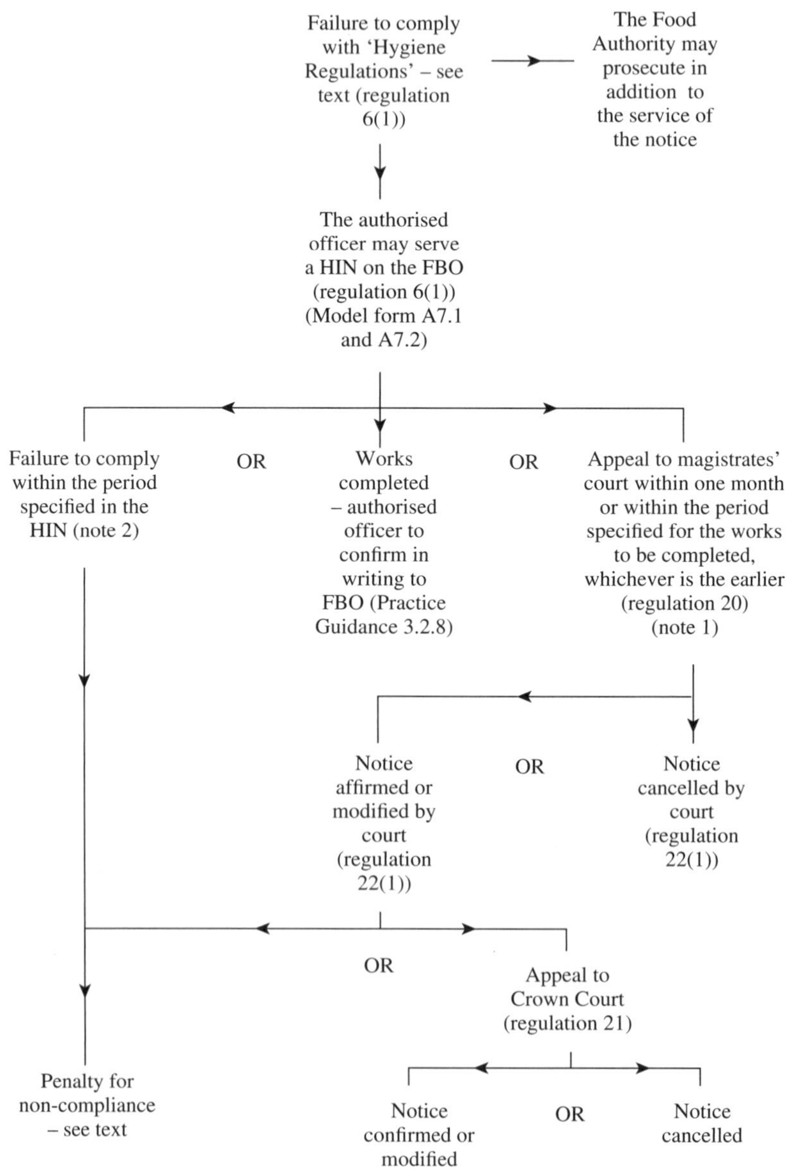

Notes

1. An appeal suspends the period for compliance until it has been disposed of (regulation 22).
2. Although there is no provision for the authorised officer to extend the period, the Code of Practice considers it reasonable to do so in appropriate cases by withdrawing the notice and serving a fresh one with a new period (Practice Guidance 3.2.6).
3. FBO = food business operator.

Appeals

The recipient of the hygiene improvement notice may appeal to a magistrates' court within one month of service or, if it is shorter, the period specified in the notice for compliance (regulation 20). The court may cancel or affirm the notice with or without amendment. The notice is not enforceable until the appeal is disposed of or withdrawn (regulation 22). There is provision for a subsequent appeal to the Crown Court (regulation 21).

Penalty

A food business operator failing to comply with the notice is guilty of an offence (regulation 6(2)) and is liable to a penalty on summary conviction of a fine up to the statutory maximum or, on indictment, a fine and/or imprisonment for up to two years (regulation 17(2)). The prosecutor must commence action for prosecution before three years from the commission of the offence or one year from its discovery (regulation 16).

HYGIENE PROHIBITION ORDERS

References

Food Hygiene (England) Regulations 2006.
Food Law Code of Practice, April 2012, DEFRA (CoP).
Food Law Practice Guidance, April 2012, FSA (PG).

Extent

These provisions apply in England only but there are similar provisions in Wales.

Scope

Hygiene prohibition orders are made by a magistrates' court following the conviction of the food business operator of offences under the 2006 regulations (regulation 7(1)). There is no requirement for application to be made by the food authority, although they may bring the procedure to the attention of the court.

Health risk condition

Before making an order, the court must be satisfied that there is a risk of injury to health from:

(a) the use of any process or treatment;
(b) the construction of the premises or the use of equipment; or
(c) the condition of the premises or equipment (regulation 7(2)).

FC45 Hygiene prohibition orders (HPOs) – Regulation 7 Food Hygiene (England) Regulations 2006

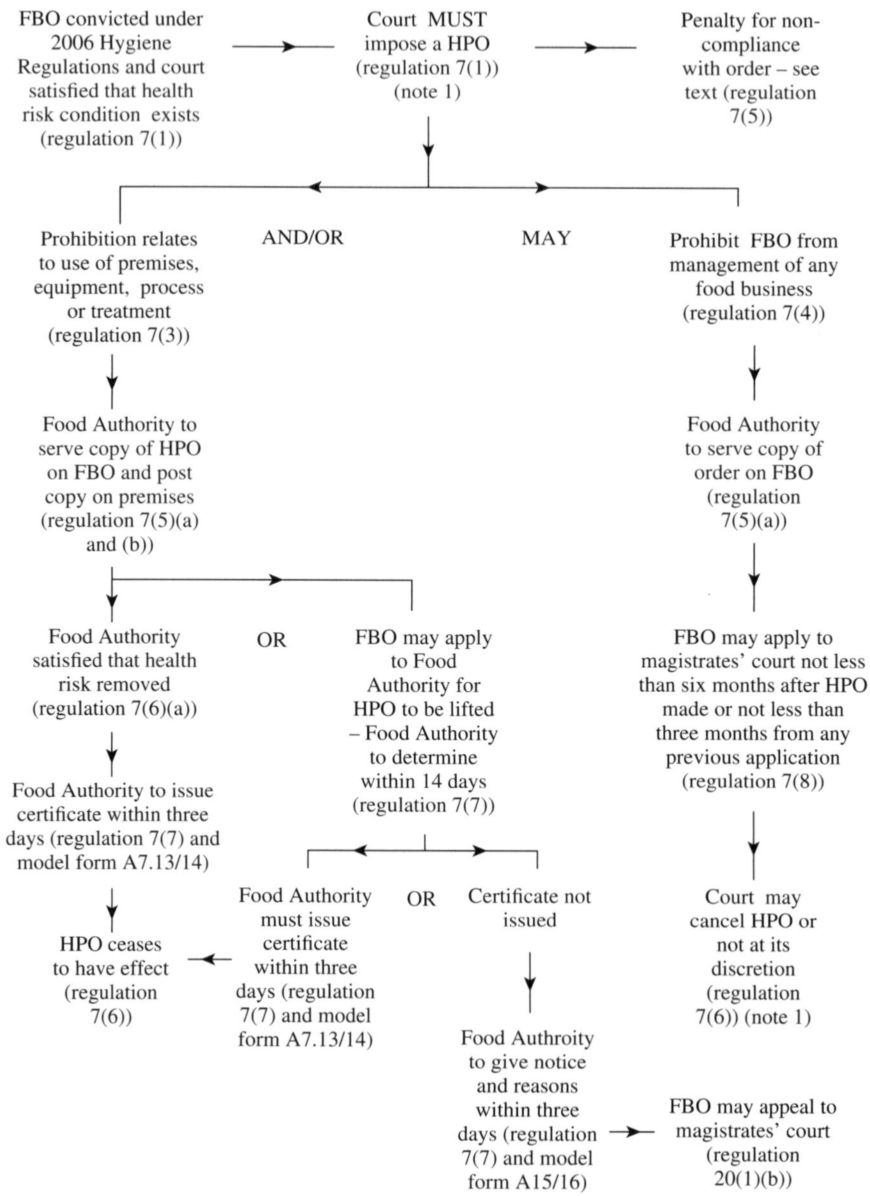

Notes

1. The food authority should notify other food authorities through the Chartered Institute of Environmental Health (Code of Practice, paragraph 3.3.2.5).
2. FBO = food business operator.

Where any of these conditions exist and the food business operator is convicted of an offence under the 2006 regulations, the court must make a hygiene prohibition order, although the prohibition of a person from the management of a business is discretionary (regulation 7(4)).

Types of prohibition

The type of prohibition specified in the order must be appropriate to the circumstances and may either:

(a) prohibit the use of a process or treatment for the purposes of the business;
(b) prohibit the use of the premises or equipment for the food business or similar food businesses;
(c) prohibit the use of the premises or equipment for the purposes of any food business (regulation 7(3)).

In the case of the prohibition of a person, the food business operator may be prohibited from participation in the management of any food business or businesses of a particular type (regulation 7(4)). This type of prohibition may also be applied to any other person convicted as a result of the commission of offences by the food business operator (regulation 7(10)).

Guidance on the appropriate use of these prohibitions is given in the Code of Practice at paragraph 3.3.2.

Determination of hygiene prohibition orders

Orders prohibiting a person from the management of premises are determined by the court following application by the person. The food authority deals with all other hygiene prohibition orders made with other prohibitions, either on its own initiative or on application by the food business operator.

Appeals

Decisions of the food authority not to determine the hygiene prohibition order are to be communicated to the food business operator together with reasons for that decision and are subject to appeal to the magistrates' court (regulation 7(7)(b) and regulation 20). Details of appeals are to be given with the notice of reasons for the decision (Practice Guidance 3.3.22).

Penalties

A food business operator failing to comply with a hygiene prohibition order is guilty of an offence (regulation 7(5)) and is liable to a penalty on summary conviction of a fine up to the statutory maximum or, on indictment, a fine and/or imprisonment for up to two years (regulation 17(2)).

HYGIENE EMERGENCY PROHIBITION NOTICES AND ORDERS

References

Food Safety (England) Regulations 2006.
Food Law Code of Practice (England), April 2012, DEFRA (CoP).
Food Law Practice Guidance (England), April 2012, (FSA).

Extent

These provisions apply in England only but similar provisions apply in Wales.

Scope

Unless the use of a voluntary procedure is appropriate (see below), an authorised officer may instigate emergency action to secure a hygiene emergency prohibition order (HEPO) where there is an *imminent* risk of injury to health in respect of any food business (regulation 8(1) and (2)).

Hygiene emergency prohibition notices (HEPNs)

This is served by an authorised officer on the food business operator where there is an *imminent* risk of injury to health. The parameters for the determination of this health risk condition are the same as for HEPOs (see below), except that the risk to health must be *imminent* (regulation 7(2) and 8(1)). Guidance on the health risk condition is given in the Code of Practice at 3.3.2.1–3.

The effect of the service of a HEPN is to immediately close the premises, or prevent the use of equipment or a process/treatment. Unlike a HEPO, this procedure may not be used to prohibit a person from management of a business.

HEPNs may only be signed by environmental health officers who have at least two years' post-qualification experience and are currently involved in food safety enforcement. They must also be adequately trained, competent and duly authorised (Code of Practice 1.2.9.1.7).

Hygiene emergency prohibition orders (HEPOs)

The HEPN ceases to have effect after three days (or before if the conditions which led to it are removed) unless the authorised officer applies to the magistrates' court for the making of an HEPO. The food business operator must be given at least one day's notice of the intention to apply for an order (regulation 8(3), (7) and (8)). If the court is satisfied that the imminent health risk condition exists, it must make an HEPO (regulation 8(2)).

As soon as possible after the order is made, an authorised officer must fix a copy of it on the premises and serve a copy on the food business operator (regulation 8(6)).

FC46 Hygiene emergency prohibition notices and orders (HEPNs and HEPOs) – Regulation 8 Food Hygiene (England) Regulations 2006

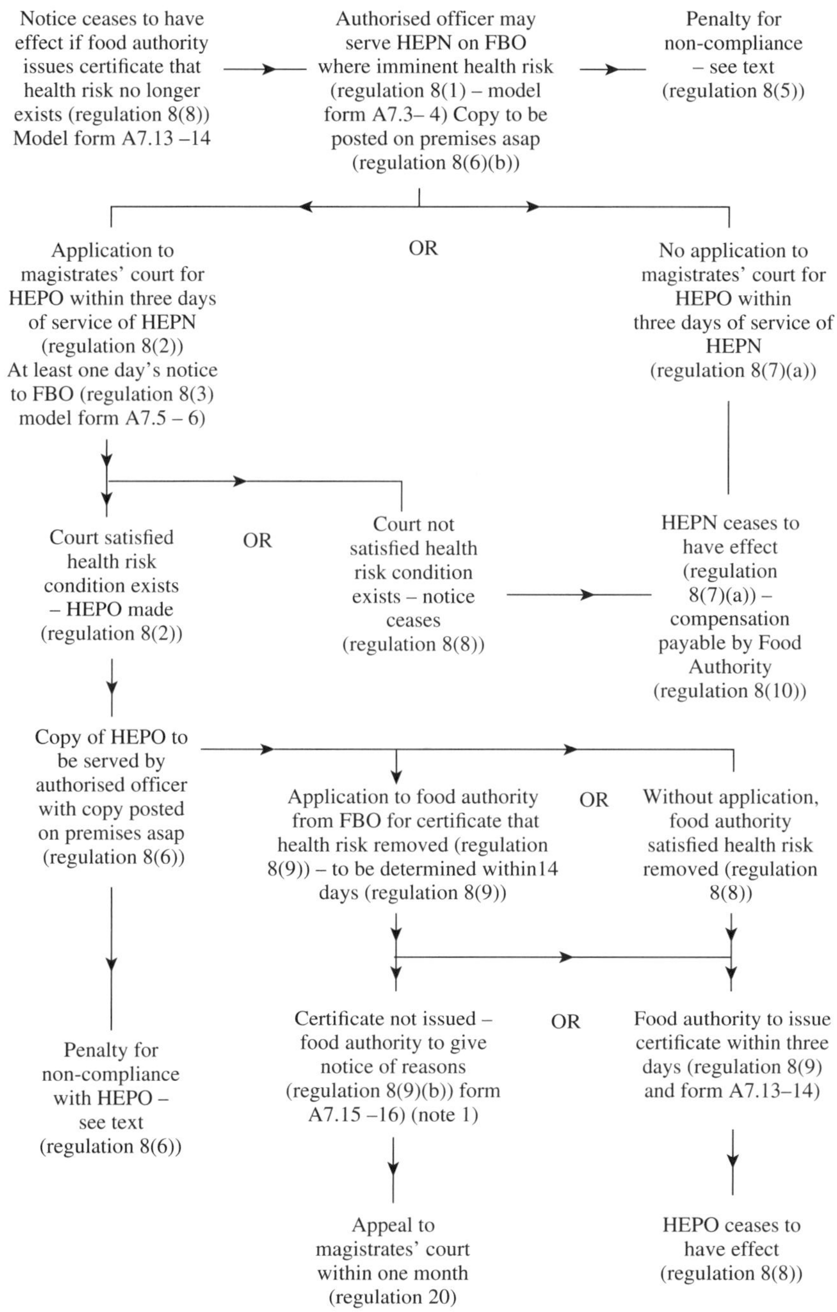

Notes

1. No time period is indicated within which this must be done.
2. The model forms are found in annex 7 of the Code of Practice.
3. FBO = food business operator.

Determination of HEPOs

The food authority effects this when satisfied that the health risk condition has been removed and it may be undertaken on its own initiative or on written application from the food business operator (regulation 8(8) and (9)). In the latter case, the decision must be made within 14 days (regulation 8(9)). A certificate that the HEPO has been removed must be issued within three days of the determination but, if refused, the food business operator must be notified with reasons for the decision (regulation 8(9)(b)). There is a right of appeal to the magistrates' court, which must be made within one month of the notice of the decision (regulation 20).

Compensation

The food authority must compensate the food business operator for losses arising from the service of an HEPN if an HEPO is not applied for within three days or if the court, in dealing with the application, considers that the health risk condition did not exist at the time the notice was served (regulation 8(10)).

Offences

Persons contravening an HEPN or an HEPO are guilty of an offence (regulation 8(6)) and are subject to a penalty on summary conviction of a fine not exceeding the statutory maximum or, on indictment, a fine and/or imprisonment for up to two years (regulation 17(2)).

Voluntary procedure

The removal of a situation where an imminent health risk condition exists may be dealt with voluntarily by the closure of the premise, where the food business operator agrees that the condition exists. Such agreements are to be made in writing (Code of Practice 3.3.2.4).

REMEDIAL ACTION NOTICES AND DETENTION NOTICES

References

EC Regulation 853/2004 on specific hygiene rules for food of animal origin, 29 April 2004.
The Food Safety (England) Regulations 2006.
Food Law Code of Practice, April 2012, DEFRA (CoP).

Extent

These provisions apply in England only but there are similar provisions in Wales.

Scope

This procedure allows an authorised officer of the food authority to deal with a situation at premises handling food of animal origin and approved by them (FC43) where:

(a) any of the 'Hygiene Regulations' are being breached; or
(b) inspection under those regulations is being hampered (regulation 9(1)).

The 'Hygiene Regulations' are the 2006 Regulations and Community Regulations 852/2004, 853/2004, 854/2004, 2073/2005 and 2075/2005 (regulation 6(1)).

The Code of Practice indicates that they should only be used after other enforcement options have been considered, including those in the 2006 regulations dealing with hygiene improvement notices, hygiene emergency prohibition notices and hygiene emergency prohibition orders (see FCs 44 and 46) and the suspension or withdrawal of the approval (FC43) (Code of Practice 3.5.2).

Examples of the circumstances in which these actions might be considered are at paragraph 3.5.3 of the Code of Practice.

Remedial action notices (RANs)

Where it is considered to be the most appropriate form of action, an authorised officer may serve a remedial action notice in writing on the food business operator or on his duly authorised representative. The notice may:

(i) prohibit the use of any equipment or any part of the establishment;
(ii) impose conditions on or prohibit the carrying out of any process;
(iii) require the rate of operation to be reduced to such an extent as specified or to be stopped completely.

The notice must be served as soon as is practicable and must state why it is being served. If served under (i) above, it must state the breach and the action needed to be taken (regulation 9(1–3)). Note – Remedial action notices are only available in Wales. Their availability for use in all food businesses in England has been delayed.

Detention notices (DNs)

These notices may be served by an authorised officer where, at approved premises, he wishes to detain any animal or food for examination, including the taking of samples (regulation 9(5)). The action may be taken either alone or in combination with the service of a remedial action notice.

Withdrawal of RANs and DNs

As soon as such action as required has been taken to remedy the issue or there is no longer a need for the animal or food to be detained, an authorised officer of the same food authority may withdraw the notice in writing served on the food business operator or his authorised representative (regulation 9(4) to (6)).

FC47 Remedial action notices (RANs) and detention notices (DNs) – Regulation 9 Food Hygiene (England) Regulations 2006

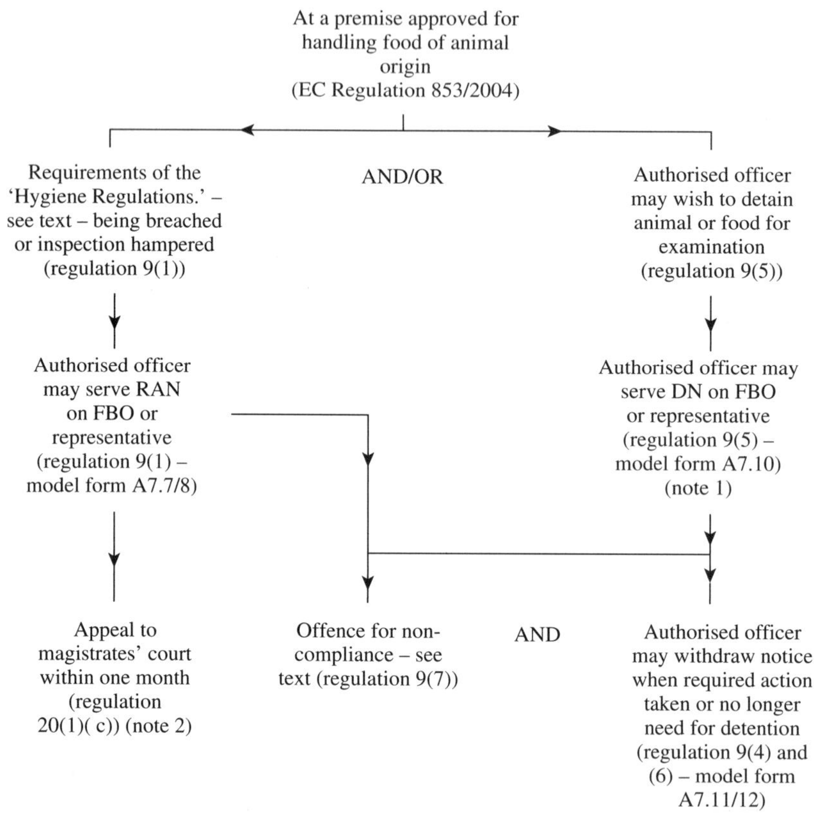

Notes

1. There is no provision for appeal against a detention notice.
2. The court may cancel, affirm or modify the notice (regulation 22(1)).
3. FBO = food business operator.

Qualifications of authorised officers for the service of RANs and DNs

These may be served only by environmental health officers who have at least two years' post-qualification experience in food safety matters, are currently involved in food law enforcement and who are properly trained, competent and duly authorised (Code of Practice 1.2.9.1.8).

Appeals

There is a right of appeal to a magistrates' court against a remedial action notice but not a detention notice. The appeal must be brought within one month of service. The remedial action notice is not suspended pending appeal. The court may cancel, approve or modify the notice and there is a further right of appeal to the Crown Court (regulations 20(1), 21 and 22).

Penalties

It is an offence not to comply with a remedial action notice or a detention notice (regulation 9(7)), the penalty for which on summary conviction is a fine not exceeding the statutory maximum or, on indictment, a fine and/or imprisonment for up to two years (regulation 17(2)).

DETENTION AND SEIZURE OF FOOD

References

The Food Safety (England) Regulations 2006.
Food Safety Act 1990, section 9.
Detention of Food (Prescribed Forms) Regulations 1990.
The Meat (Enhanced Enforcement Powers) Regulations 2000.
Food Law Code of Practice 2012, DEFRA (Code of Practice).
Food Law Practice Guidance 2012, FSA (Practice Guidance).

Extent

These provisions apply in England, Wales and Scotland. This procedure is based on both the Food Safety Act 1990 and the Food Hygiene (England) Regulations 2006.

Power of inspection of food

An authorised officer of a food authority may at all reasonable times inspect any food intended for human consumption which:

(a) has been sold;
(b) is offered or exposed for sale;

FC48 Detention and seizure of food notices – Section 9 Food Safety Act 1990

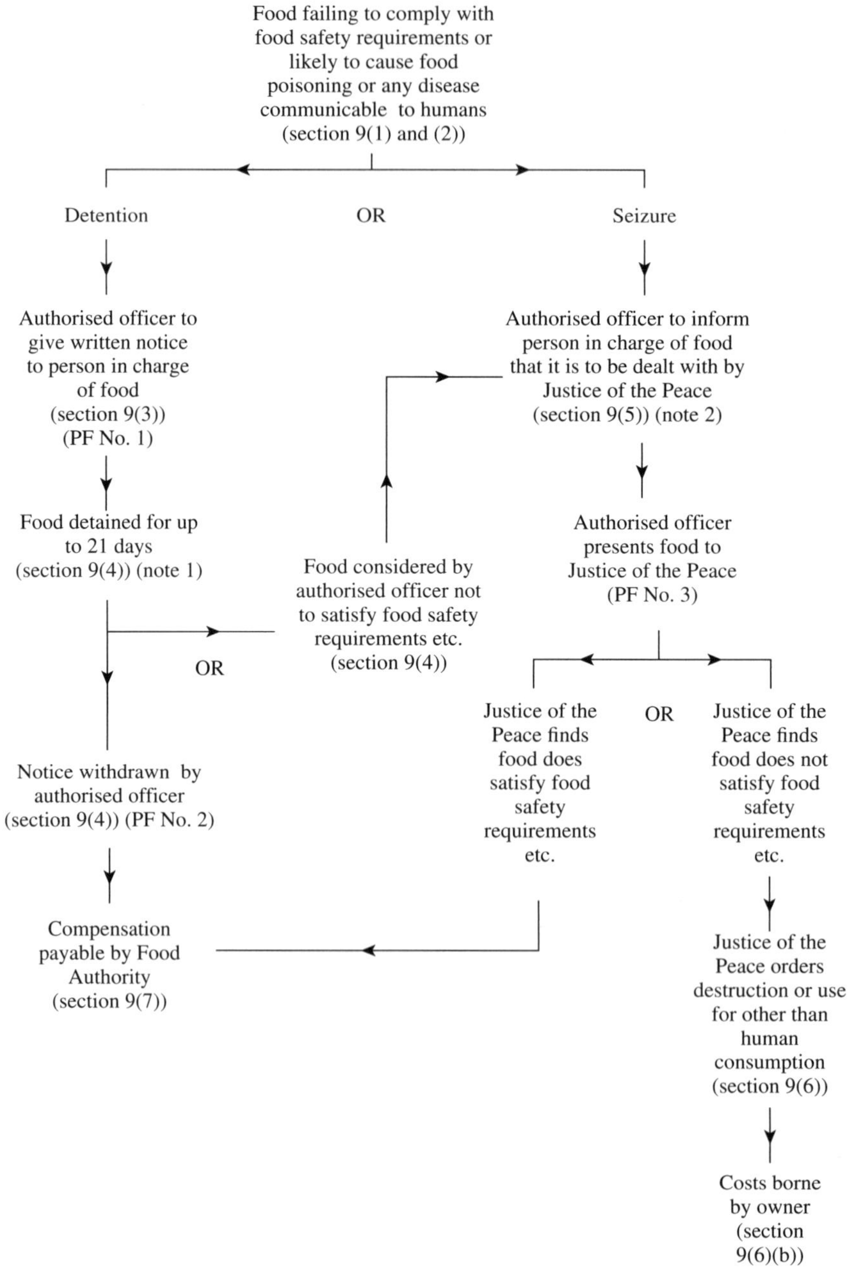

Notes

1. It is an offence to breach the requirements of the notice (section 9(3)).
2. Persons liable to prosecution have a right to attend the hearing and present evidence and witnesses (section 9(5)(a)).
3. The prescribed forms are detailed in the Detention of Food (Prescribed Forms) Regulation 1990.

(c) is in the possession of, has been deposited with or consigned to any person for the purpose of sale or preparation for sale (section 9(1)).

Whilst the 1990 Act defines 'human consumption' as 'includes use in the preparation of food for human consumption' (section 53(1)), the phrase 'intended for human consumption' is defined in greater detail in regulation 3 of the Food Safety (Hygiene) Regulations 2006 in relation to procedures under those regulations as:

(a) any food commonly used for human consumption if placed on the market or offered, exposed or kept for placing on the market; and
(b) any food commonly used for human consumption and found on premises used for preparation, storage, etc. of that food; and
(c) any articles or substances commonly used in the manufacture of food for human consumption found on such premises shall, until the contrary is proved, be presumed as intended for human consumption.

For the definition of 'food', see page 177.

Voluntary procedures

It is possible for an authorised officer to use a voluntary procedure to remove food not suitable for human consumption from the food chain, at the instigation of either the authorised officer or the owner. However, such a procedure must be well documented in accordance with Code of Practice, paragraph 3.4.9.

Detention and seizure of food

Where either:

(a) on inspection, food is found to fail to comply with food safety requirements; or
(b) it appears to the authorised officer that any food is likely to cause food poisoning or any disease communicable to humans,

the authorised officer may either:

(i) detain the food; or
(ii) seize it and remove it in order to have it dealt with by a Justice of the Peace (section 9(1)–(3)).

Food is deemed not to comply with food safety requirements if:

(a) it is unfit for human consumption; or
(b) it has been rendered injurious to health by the addition of articles or substances, by the use of an article or substance as an ingredient in its preparation, by the abstraction of any constituent or by the application of any process or treatment; or
(c) it is so contaminated (by extraneous matters or otherwise) that it would not be reasonable to expect it to be used for human consumption in that state (section 8(2));

(d) an authorised officer certifies that the food has not been produced, processed or distributed in compliance with the 'Hygiene Regulations' – see page 176. Such certification should use the model form at A7.17/18 of annex 7 of the Code of Practice (regulations 23 and 27).

The Meat (Enhanced Enforcement Powers) (England) Regulations 2000 extended the use of section 9 powers dealing with the inspection and seizure of food to include illegally produced meat which does not have the required health mark.

The Code of Practice requires that the procedures dealing with the inspection, detention and seizure of food under section 9 should only be exercised by appropriately qualified officers, i.e. environmental health officers, and, where appropriate, official veterinary surgeons, and in relation to meat, only officers qualified under the Authorised Officers (Meat Inspection) Regulations 1987 (paragraph 1.2.9.3.1).

Detention of food notice

The detention of the food is effected by a notice (form prescribed by the Detention of Food (Prescribed Forms) Regulations 1990) to be served on the person in charge of the food requiring that:

(a) the food is not to be used for human consumption; and
(b) it is not to be removed, except to a place specified in the notice.

Failure to comply with the notice is an offence for which the penalty on indictment is an unlimited fine and/or imprisonment for up to two years or on summary conviction to a fine up to the statutory maximum and/or imprisonment for not exceeding six months (sections 9(3) and 35(2)).

Having served such notice, the authorised officer is required to determine whether or not the food complies with food safety requirements as soon as is reasonably practicable and in any event within 21 days. If satisfied of compliance, the food is released and the notice withdrawn using a withdrawal of detention of food notice (PF No. 2); if satisfied of non-compliance, the food must be seized (section 9(4)).

Seizure

Seizure of food, with or without prior detention, is effected by the authorised officer removing it to have it dealt with by a Justice of the Peace. The person in charge of the food must be informed by the service of a Food Condemnation Warning (PF No. 3) and any person who may be liable to prosecution for the condition of the food may attend before the Justice of the Peace and is entitled to be heard and call witnesses (section 9(5)).

If the Justice of the Peace considers, on the basis of the evidence, that the food fails to comply with food safety requirements or is likely to cause food poisoning or a human communicable disease, he must condemn the food and order it to be destroyed or disposed of so as to prevent it being used for human consumption. In this case, the owner of the food is required to defray the costs incurred in the destruction or disposal of the food (section 9(6)).

The Code of Practice (paragraph 3.4.5) requires that, if possible, food which is seized should be presented to a Justice of the Peace within two days and, where the food is highly perishable, as soon as possible.

Compensation

If the Justice of the Peace is not satisfied that the food fails to comply, the food authority is required to compensate the owner for any depreciation in its value. This is also the case if a detention notice is withdrawn by the authorised officer (section 9(7)).

Disposal of food

Once a Justice of the Peace has condemned food, it is the responsibility of the food authority to ensure its adequate destruction by supervision throughout the process. If necessary, the food should be rendered unusable for a return to the food chain (Code of Practice 3.4.10).

Dealing with batches, lots or consignments

Article 14(6) of EC Regulation 178/2002 states that where

> any food that is unsafe forms part of a batch, lot or consignment of food of the same class or description, it shall be presumed that all the food in that batch etc. is also unsafe, unless following a detailed assessment there is no evidence that the rest of the batch etc. is unsafe.

This definition is applicable to the detention and seizure of food using these procedures.

IMPORTED FOOD – OFFICIAL CONTROLS ON FEED AND FOOD OF NON-ANIMAL ORIGIN FROM THIRD COUNTRIES

References

Official Food and Feed Control (England) Regulations 2009 (in this procedure, regulation numbers refer to these regulations).
Official Controls (Animals, Feed and Food) (England) Regulations 2006.
Guidance and Regulatory Advice on Import Legislation (GRAIL) database, www.food.gov.uk/business-industry/imports/enforce_authorities/grail
Inland enforcement of imported feed and food controls – resource pack, www.food.gov.uk/business-industry/imports/enforce_authorities/resourcepack#

Extent

These provisions apply in England but similar provisions apply in Wales, Scotland and Northern Ireland.

Enforcement authorities

Port health authorities are local authority bodies at seaports set up under the Public Health (Control of Diseases) Act 1984. At UK airports and some seaports, the local authority in whose area the port is located is responsible for enforcing the food import controls.

Outside of port areas, local authorities ('inland' local authorities) are responsible for enforcing food safety controls on foodstuffs, including imported food.

Deferred execution and enforcement

The enforcement authority which is the ultimate destination of imported food becomes responsible for enforcing and executing the import provisions with respect to that product once it arrives there, in the following circumstances:

- a product from a third country has entered England;
- customs examination of that product has been completed or has been deferred until it reaches its place of destination elsewhere in the United Kingdom;
- an authorised officer of the enforcement authority for the place of entry has on reasonable grounds issued an authorisation confirming that:
  - (i) examination of the product for the purposes of the import provisions should be deferred until the product arrives at its destination elsewhere in England; or
  - (ii) such examination should take place when the product arrives at its destination elsewhere in the United Kingdom under legislation with respect to imported products in force there; and
- a person importing the product gives that authorised officer an undertaking in writing which states:
  - (i) the destination of the product; and
  - (ii) confirms that the container containing the product has been sealed and will not be opened until it has reached that destination, the opening of the container has been authorised by the enforcement authority for the place in which the destination is located, and the container will be available at that destination for examination under the import provisions.

Where an authorised officer issues an authorisation, he shall notify the enforcement authority where the imported food is destined for that the product has not been examined under the import provisions, if customs examination of the product has been deferred and a copy of any undertaking (regulation 27).

Checks on products

The importer must allow an authorised officer to carry out checks in relation to the product. The importer must provide the facilities and assistance which the

FC49 Imported food – Regulations 27 to 42 Official Food and Feed Control (England) Regulations 2009

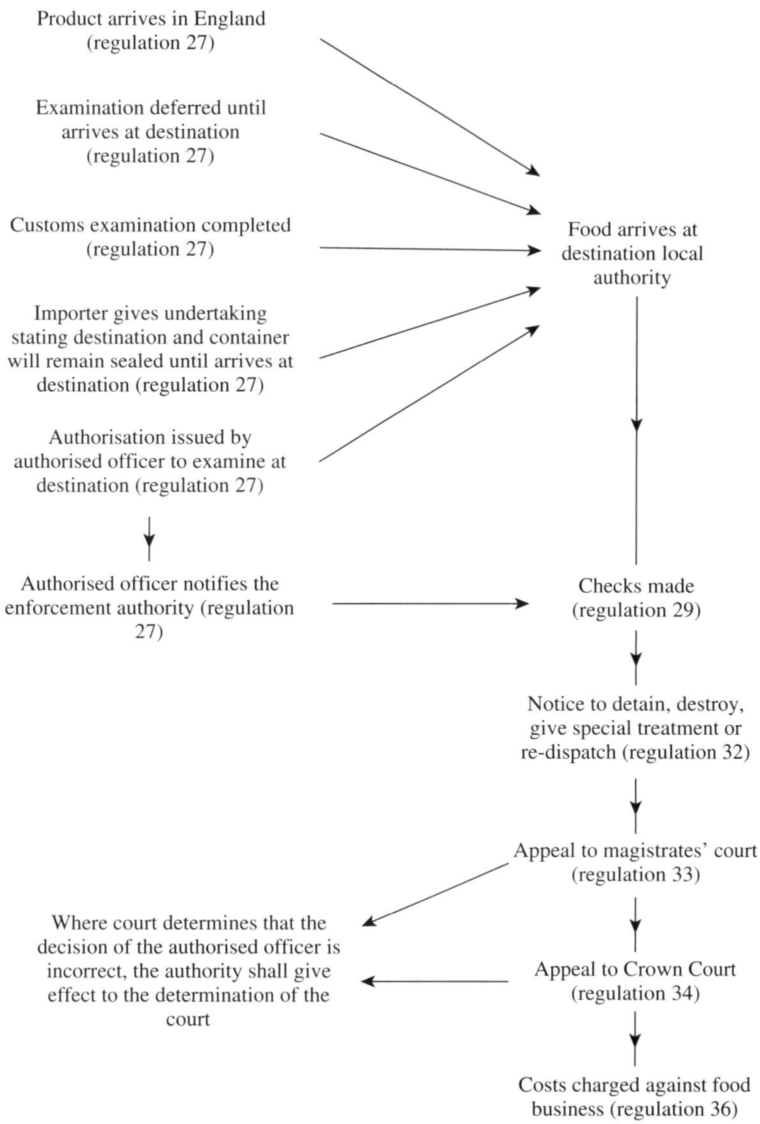

authorised officer reasonably requires to carry them out at a specified place (regulation 29).

Checks include a systematic documentary check, a random identity check and, as appropriate, a physical check (Article 16 of Regulation 882/2004).

An enforcement authority has the powers of detention, destruction, special treatment, re-dispatch and other appropriate measures and charging for costs (regulation 31).

Notices under Articles 18 and 19 of Regulation 882/2004 (imports of feed and food from third countries)

An authorised officer must serve a notice on the feed or food business operator responsible for it, where he proposes:

- to place a consignment of feed or food under official detention;
- to destroy the feed or food;
- to subject it to special treatment;
- other appropriate measures;
- where already on the market, to monitor, recall or withdraw it before taking the measures above (regulation 32).

Right of appeal

The person receiving the notice can appeal against the notice to the magistrates' court by way of complaint for an order within one month from the date of the notice. Where the magistrates' court determines that the decision of the authorised officer is incorrect, the authority must follow the instructions of the court (regulation 33). A person aggrieved by the dismissal of an appeal by a magistrates' court may appeal to the Crown Court (regulation 34).

Costs and fees

The enforcement authority may claim costs for its actions from the feed or food business operator or its representative (regulation 36).

Procurement by authorised officers of samples with regard to food

An authorised officer may:

(a) purchase a sample of any food, or any substance capable of being used in the preparation of food;

(b) take a sample of any food, or any such substance, which:

  (i) appears to him to be intended for placing on the market or to have been placed on the market, for human consumption; or

  (ii) is found by him on or in any premises which he is authorised to enter;

(c) take a sample from any food source, or a sample of any contact material, which is found by him on or in any such premises; and

(d) take a sample of any article or substance which is found by him on or in any such premises and which he has reason to believe may be required as evidence in proceedings (regulation 37).

Analysis etc. of samples

An authorised officer can submit the sample for analysis by a public analyst or examination by a food examiner. A certificate specifying the result of the analysis or examination must be provided and signed by the food analyst or examiner (regulation 38).

Powers of entry of authorised officers of a food authority

An authorised officer, on producing some duly authenticated document showing his authority, has a right at all reasonable hours to enter any premises within the authority's area for the purpose of:

- ascertaining whether there is or has been on the premises any contravention of the provisions of the import provisions in relation to food;
- ascertaining whether there is on the premises any evidence of any such contravention within that area (the officer has the power to enter premises outside his enforcement authority area); and
- ascertaining the performance by the authority of their functions under the import provisions.

Admission to a private dwelling-house can only be obtained after 24 hours' notice of the intended entry has been given to the occupier.

A warrant for entry can be obtained from a Justice of the Peace, on sworn information in writing, if satisfied that there is reasonable ground for entry onto any premises and either that admission to the premises has been refused, or a refusal is apprehended, and that notice of the intention to apply for a warrant has been given to the occupier; or that an application for admission, or the giving of such a notice, would defeat the object of the entry; or that the case is one of urgency; or that the premises are unoccupied or the occupier is temporarily absent. The warrant continues in force for one month.

The authorised officer may take with him such other persons as he considers necessary, and on leaving any unoccupied premises must leave them as effectively secured against unauthorised entry as he found them.

The authorised officer may inspect any records, including computer records, relating to a food business. The officer may also seize and detain any records which he has reason to believe may be required as evidence (regulation 39).

Obstruction etc. of officers

It is an offence to obstruct any person acting in the execution of the import provisions or to fail to give any assistance or information which that person may reasonably require of him for the performance of his functions. It is also an offence to recklessly give information that is known to be false or misleading in a material particular (regulation 40).

Offences and penalties

It is an offence to contravene or fail to comply with any of the specified import provisions and, if a person is guilty of an offence, they are liable on summary conviction to a fine not exceeding the statutory maximum or, on conviction on indictment, to imprisonment for a term not exceeding two years, to a fine or to both.

A person guilty of obstruction is liable on summary conviction to a fine not exceeding level 5 on the standard scale or to imprisonment for a term not exceeding three months, or to both (regulation 41).

Time limit for prosecutions

Prosecutions must be taken within three years of the offence or one year from its discovery by the prosecutor, whichever is the earlier (regulation 42).

SAMPLING OF FOOD FOR ANALYSIS AND EXAMINATION

References

Food Safety Act 1990, sections 29–30.
Food Safety (Sampling and Qualifications) Regulations 1990.
Food Hygiene (England) Regulations 2006.
Quick Frozen Foodstuffs (England) Regulations 2007.
Materials and Articles in Contact with Food (England) Regulations 2012.
Food Law Code of Practice, April 2012, DEFRA (CoP).
Food Law Practice Guidance, April 2012, FSA (PG).
Guidance on Food Sampling for Microbiological Examination, January 2006, LACORS.
EU Directive 92/2 – sampling procedure and community method of analysis for the official control of temperatures of quick frozen foods intended for human consumption.

Extent

Most provisions apply in England, Wales and Scotland.

Scope

Powers of sampling for analysis (chemical) and examination (microbiological) by an authorised officer are provided in both the Food Safety Act 1990, section 29 and the 2006 regulations (regulation 12). In practical terms, examination for microbiological condition will relate to the hygienic conditions in a premises and should therefore be undertaken using the 2006 regulations. The powers under the 1990 Act should be used to check food standards requirements (chemical).

Both sampling regimes are, however, covered by the Food Safety (Sampling and Qualifications) Regulations 1990.

Generally, the authorised officer may purchase or take (without payment) a sample of food or substance capable of being used in food that appears to be intended for human consumption or is found on any premises. Samples may also be taken from any food sources or of contact materials (regulation 12).

Sampling policy

The formal sampling procedures as identified here should form only part of a sampling policy, which itself should be an integral part of the food authority's enforcement policy. The sampling policy should be drawn up in liaison with the public analyst and/or the food examiner and should take account of the home authority arrangements (Code of Practice 6.1.1/2).

Analysis and examination

The Food Safety Act 1990 recognises that the microbiological examination of food is a separate activity from its analysis.

Analysis: '…includes microbiological assay and any technique for establishing the composition of the food' (section 53(1)).

Examination is defined as: 'a microbiological examination' (section 28(2)).

Samples taken by the authorised officer are to be submitted either to a public analyst for analysis or, if for examination, to a food examiner. The necessary qualifications for each are specified in the Food Safety (Sampling and Qualifications) Regulations 1990 (section 30(1) and regulations 3 and 4). Code of Practice, paragraph 6.1.10 requires that samples for chemical analysis must be submitted to a laboratory accredited for that purpose and which appears on the list of official food control laboratories.

Food authorities, other than remaining two-tier district councils in England, who are not generally involved in matters of food composition, are required to appoint one or more public analysts for their areas (section 27).

Food authorities including district councils may provide their own facilities for microbiological examination (section 28) or, alternatively, may submit samples to a food examiner.

Analysts' and examiners' certificates

Following analysis or examination, the analyst or examiner must submit a certificate to the authorised officer specifying the result (section 30(6) and regulation 13(6)). Each certificate must be signed (section 30(7) and regulation 13(7)). The form of certificate to be used is specified in schedule 3 of the Food Safety (Sampling and Qualifications) Regulations 1990.

In any legal proceedings taken following the analysis or examination, the certificate or a copy of it is sufficient evidence of the facts stated therein, unless either party wishes the food analyst or examiner to be called as a witness (section 30(8) and regulation 13(8)).

FC50 Sampling of food for analysis and examination – Food Safety Act 1990 and Food Safety (Sampling and Qualifications) Regulations 1990

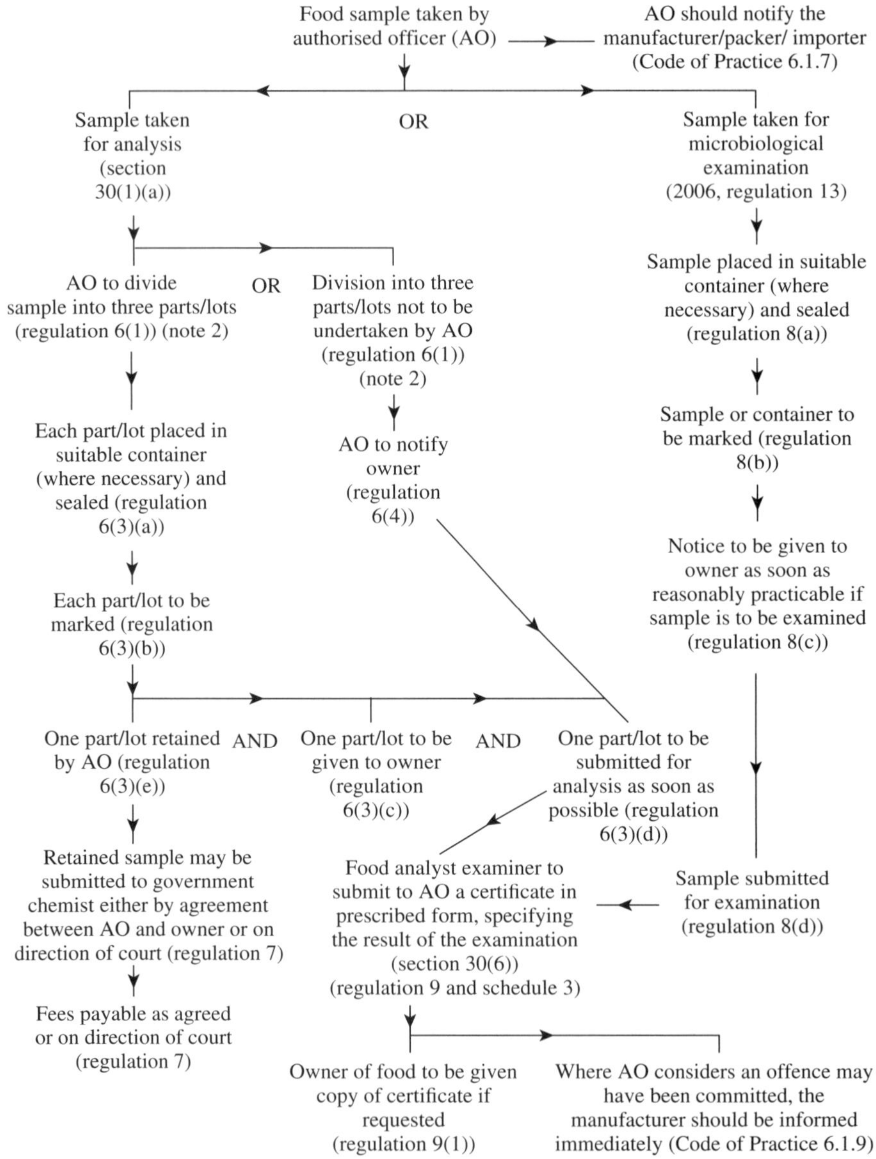

Notes

1. Regulation numbers refer to the Food Safety (Sampling and Qualifications) Regulations 1990, except where 2006 is shown, in which case they refer to the Food Hygiene (England) Regulation 2006.
2. Division into three parts or lots is not required if this is not reasonably practicable or may impede analysis (regulation 6(1)).
3. This sampling procedure does not apply to the enforcement of certain food regulations (page 211).
4. Sometimes there are difficulties in establishing who the 'owner' is in certain multiple retailer situations – in these cases, the registered head office of the company should be notified.

Sampling procedure

Following sampling, the sample must be dealt with in accordance with the Food Safety (Sampling and Qualifications) Regulations 1990 and the procedure is indicated in FC50. More detailed guidance will be found in the Code of Practice, section 6 and the Practice Guidance, section 6.

This procedure is *not* applicable to sampling under the following food regulations, in which particular sampling procedures are laid down:

- Poultrymeat (England) Regulations 2011;
- Natural Mineral Water, Spring Water and Bottled Drinking Water (England) Regulations 2007;
- The Materials and Articles in Contact with Food Regulations 2012;
- The Animals and Animal Products (Examinations for Residues and Maximum Residue Limits) Regulations 1997.

The sampling procedure detailed in FC50 covers the sampling of foods in all other situations.

Division of samples

Samples intended for analysis (but not for examination) are to be divided into three parts unless either:

(a) this is not reasonably practicable; or
(b) it is likely to impede analysis (regulation 6(1) and (4)).

If the sample itself consists of sealed containers and the authorised officer believes that the opening of them would impede proper analysis, he may proceed by dividing the sample into parts by putting the containers into three lots (regulation 6(2)).

Sampling and analysis under other provisions

Quick Frozen Foodstuffs (England) Regulations 2007 (regulation 7)

If, on inspection, an authorised officer has reasonable grounds to believe that the temperatures that are being or have been maintained in respect of any quick-frozen foodstuff are not the temperatures prescribed, he must inspect the quick-frozen foodstuff and their temperatures in accordance with the provisions of EU Directive 92/2.

Materials and Articles in Contact with Food (England) Regulations 2012

An authorised officer who has procured a sample of materials or articles in contact with food under the Food Safety Act 1990 and who considers it should be analysed shall divide the sample into three parts.

If the sample consists of sealed containers and opening them would, in the opinion of the authorised officer, impede a proper analysis, the authorised officer shall divide the sample into parts by putting the containers into three lots, and each lot shall be treated as being a part.

The authorised officer shall:

(a) if necessary, place each part in a suitable container and seal it;
(b) mark each part or container;
(c) as soon as is reasonably practicable, give one part to the owner and notify the owner in writing that the sample will be analysed;
(d) submit one part for analysis; and
(e) retain one part for future submission (regulation 25).

Secondary analysis by the government chemist

Where a sample has been retained and proceedings are intended to be or have been commenced against a person for an offence under the regulations, and the prosecution intends to present as evidence the result of the analysis, the authorised officer should send the retained part of the sample to the government chemist for analysis:

- of the officer's own volition;
- if requested by the prosecutor (if a person other than the authorised officer);
- if the court so orders; or
- if requested by the accused.

Where a request is made by the accused, the authorised officer may give notice in writing to the accused requesting payment of a fee specified in the notice to defray some or all of the government chemist's charges for performing the analysis and, in the absence of agreement by the accused to pay the fee specified in the notice, the authorised officer may refuse to comply with the request (regulation 26(6)).

The government chemist shall analyse the part and send the authorised officer a certificate specifying the results of the analysis.

The authorised officer shall immediately on receipt supply the prosecutor and the accused with a copy of the government chemist's certificate of analysis (regulation 26).

CLOSURE NOTICES ON MOLLUSC HARVESTING AREAS

References

EC Regulation 852/2004.
EC Regulation 853/2004.
EC Regulation 854/2004.
Food Hygiene (England) Regulations 2006.
Food Law Code of Practice, April 2012, DEFRA (CoP).
Food Law Practice Guidance, April 2006, FSA (PG).

Extent

These provisions are made under EC regulations and apply in England.

Scope

Where, based on sampling, either:

(a) the health standards set out in Regulation 853/2004, section 7, chapter 5 are exceeded (these standards incorporate the microbiological standards in EC Regulation 852/2004); or
(b) there is otherwise a risk to human health,

the food authority must either:

(a) close the production area by the issue of a closure notice (CN); or
(b) if appropriate, re-classify the area as class B or C if it meets the criteria in part A of annex 2 of EC854/2004 (see below).

Consultation

The food authority should liaise with the Food Standards Agency in considering its actions, including on possible action to withdraw any molluscs from that area which may have been distributed (Code of Practice 5.3.5). They should also consider consulting appropriate experts, such as the Consultant Communicable Disease Control or the consultant microbiologist at the Health Protection Agency (Practice Guidance A4.14).

Re-classification

Classification of harvesting areas is normally undertaken by the Food Standards Agency but, in the urgent situation faced by this procedure, re-classification is by the food authority, in consultation with the food business operator and the Food Standards Agency. The classifications are laid out in EC 854/2004 at annex 2, chapter 2A and are:

(a) Class A areas from which molluscs may be collected for direct human consumption – these must meet the standards in annex 3, section 7, chapter 5 of EC 853/2004;
(b) Class B areas where molluscs may be collected but only placed on the market after treatment in a purification centre or after relaying so as to meet the standards in (a) above;
(c) Class C from where molluscs may only be placed on the market after relaying for a long period so as to meet the standards in (a) above.

Closure notices

The issue of these notices by the food authority has the effect of temporarily closing the harvesting area, although no time limit is set. The form of notice to be used is at annex 9 of the Code of Practice.

FC51 Closure notices on mollusc harvesting areas

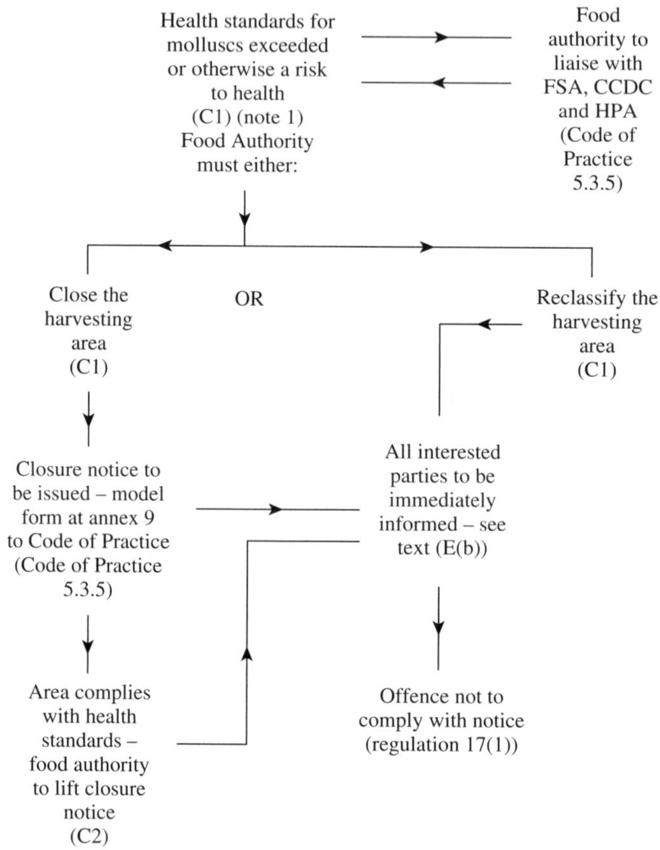

Notes

1. C and E numbers refer to EC Regulation 854/2004, annex 2, chapter 2.
2. FSA = Food Standards Agency.
3. CCDC = Consultant in Communicable Disease Control.
4. HPA = Health Protection Agency.

Information

The food authority must immediately inform all interested parties, such as producers, gatherers and operators of purification and dispatch centres, of any re-classification or issue of a closure notice (854/2004, annex 2, chapter 2 E(b)). The Code of Practice suggests that this is best done in relation to closure notices by sending them copies. Where there is re-classification, notices should also be posted prominently around the area (Code of Practice 5.3.5).

Re-opening of production areas

The food authority should remove the closure notice immediately upon being satisfied that harvesting in accordance with the EC regulations may be resumed. This will follow a period of sampling (854/2004, annex 2, chapter 2 C2 and Code of Practice 5.3.5). Again, interested parties must be informed as above.

Offences

Non-compliance with a closure notice constitutes an offence (regulation 17(1), Food Hygiene (England) Regulations 2006).

Chapter 4

HEALTH AND SAFETY

GENERAL PROCEDURAL PROVISIONS

Unless otherwise indicated, the following provisions are applicable to procedures in this chapter.

References

Health and safety at Work etc Act 1974 as amended.
The Health and Safety (Enforcing Authority) Regulations 1998.

Extent and application

The procedures in this chapter apply in Wales and Scotland but not in Northern Ireland (section 84).

Enforcing authority

Section 18(7)(a) of the Health and Safety at Work etc. Act 1974 defines enforcing authority as 'the Executive (Health and Safety Executive (HSE)) or any other authority which is by relevant statutory provision or by regulations … made responsible for the enforcement of any of the provisions to any extent'.

The Health and Safety (Enforcing Authority) Regulations 1998 make local authorities (as defined below) responsible for enforcement where the following are the main activities undertaken at any non-domestic premises.

1. The sale or storage of goods for retail or wholesale distribution except:
   (a) at container depots where the main activity is the storage of goods in the course of transit to or from dock premises, an airport or a railway;
   (b) where the main activity is the sale or storage for wholesale distribution of any substance or preparation dangerous for supply;
   (c) where the main activity is the sale or storage of water or sewage or their by-products or natural or town gas;

   and for the purposes of this paragraph where the main activity carried on in premises is the sale and fitting of motor car tyres, exhausts, windscreens or sunroofs, the main activity shall be deemed to be the sale of goods.
2. The display or demonstration of goods at an exhibition for the purposes of offer or advertisement for sale.

3. Office activities.
4. Catering services.
5. The provision of permanent or temporary residential accommodation, including the provision of a site for caravans or campers.
6. Consumer services provided in a shop, except dry-cleaning or radio and television repairs, and in this paragraph 'consumer services' means services of a type ordinarily supplied to persons who receive them otherwise than in the course of a trade, business or other undertaking carried on by them (whether for profit or not).
7. Cleaning (wet or dry) in coin-operated units in launderettes and similar premises.
8. The use of a bath, sauna or solarium, massaging, hair transplanting, skin piercing, manicuring or other cosmetic services and therapeutic treatments, except where they are carried out under the supervision or control of a registered medical practitioner, a dentist registered under the Dentists Act 1984, a physiotherapist, an osteopath or a chiropractor.
9. The practice or presentation of the arts, sports, games, entertainment or other cultural or recreational activities, except where the main activity is the exhibition of a cave to the public.
10. The hiring out of pleasure craft for use on inland waters.
11. The care, treatment, accommodation or exhibition of animals, birds or other creatures, except where the main activity is horse breeding or horse training at a stable, or is an agricultural activity or veterinary surgery.
12. The activities of an undertaker, except where the main activity is embalming or the making of coffins.
13. Church worship or religious meetings.
14. The provision of car parking facilities within the perimeter of an airport.
15. The provision of childcare, or playgroup or nursery facilities (schedule 1).

Schedule 2 of the Health and Safety (Enforcing Authority) Regulations 1998 allocate the following premises to the HSE for enforcement:

1. Any activity in a mine or quarry other than a quarry in respect of which notice of abandonment has been given under section 139(2) of the Mines and Quarries Act 1954.
2. Any activity in a fairground.
3. Any activity in premises occupied by a radio, television or film undertaking in which the activity of broadcasting, recording or filming is carried on, and the activity of broadcasting, recording or filming wherever carried on and, for this purpose, 'film' includes video.
4. The following activities carried on at any premises by persons who do not normally work in the premises:
   (a) construction work if:
      (i) regulation 7(1) of the Construction (Design and Management) Regulations 1994 (which requires projects which include or are intended to include construction work to be notified to the Executive) applies to the project which includes the work;

(ii) the whole or part of the work contracted to be undertaken by the contractor at the premises is to the external fabric or other external part of a building or structure; or
(iii) it is carried out in a physically segregated area of the premises, the activities normally carried out in that area have been suspended for the purpose of enabling the construction work to be carried out, the contractor has authority to exclude from that area persons who are not attending in connection with the carrying out of the work and the work is not the maintenance of insulation on pipes, boilers or other parts of heating or water systems or its removal from them;

(b) the installation, maintenance or repair of any gas system, or any work in relation to a gas fitting;
(c) the installation, maintenance or repair of electricity systems;
(d) work with ionizing radiations except work in one or more of the categories set out in schedule 3 to the Ionizing Radiations Regulations 1985.

5. The use of ionizing radiations for medical exposure (within the meaning of regulation 2(1) of the Ionizing Radiations Regulations 1985).
6. Any activity in premises occupied by a radiography undertaking in which there is carried on any work with ionizing radiations.
7. Agricultural activities, and any activity at an agricultural show which involves the handling of livestock or the working of agricultural equipment.
8. Any activity on board a sea-going ship.
9. Any activity in relation to a ski slope, ski lift, ski tow or cable car.
10. Fish maggot and game breeding, except in a zoo.
11. Any activity in relation to a pipeline within the meaning of regulation 3 of the Pipelines Safety Act 1996.
12. The operation of a railway (schedule 2).

The 'main activity' test does not apply to the following premises, all of which are allocated to the HSE:

(a) the tunnel system within the meaning of section 1(7) of the Channel Tunnel Act 1987;
(b) an offshore installation within the meaning of regulation 3 of the Offshore Installations and Pipeline Works (Management and Administration) Regulations 1995;
(c) a building or construction site, that is to say, premises where the only activities being undertaken are construction work and activities for the purpose of or in connection with such work;
(d) the campus of a university, polytechnic, college, school or similar educational establishment;
(e) a hospital (regulation 3(5)).

In addition, all premises occupied by the following are reserved for HSE enforcement:

(a) local authority;
(b) parish and community councils;

(c) police authorities;
(d) fire authorities;
(e) organisations under the International Headquarters and Defence Organisations Act 1964, including authorities of visiting forces;
(f) United Kingdom Atomic Energy Authority;
(g) the Crown.

Where premises are mainly occupied by one of these bodies but also occupied by someone else, the Health and Safety Executive assumes responsibility for the whole premises (regulation 4).

Regulation 5 allows for a transfer of enforcement responsibility between Health and Safety Executive and local authorities in either direction by agreement.

Guidance on the Enforcing Authority Regulations is given in LAC 23/15 (Rev. 1) February 2011, HSE, www.hse.gov.uk/lau/lacs/23-15.htm

Appointment of inspectors

An enforcing authority may appoint persons having suitable qualifications as inspectors. Each appointment must be in writing and specify which of the powers given to inspectors by the Act (below) are to be exercisable. In exercising these duties, the inspector must be able to produce if required his instrument of appointment or a duly authenticated copy of it (section 19).

Powers of inspectors

An enforcing authority may confer on an inspector all or any of the following powers in connection with his appointment:

(a) enter any premises at any reasonable time or, where a situation may be dangerous, at any time;
(b) take with him a police constable;
(c) take with him any other person duly authorised by the enforcing authority and any equipment or materials he may require;
(d) make such examination and investigation as necessary;
(e) direct that a premises or part of it remain undisturbed for as long as reasonably necessary for examination or investigation;
(f) take such measurements, photographs or records as necessary;
(g) take samples of articles, or substances in the premises or the atmosphere in, or in the vicinity of, the premises;
(h) dismantle, treat or test any article or substance likely to cause danger to health or safety, if requested, in the presence of a responsible person at the premises;
(i) take possession of or detain an article or substance likely to cause danger to health or safety in order to examine it, prevent it being tampered with or to keep it available for evidence, having previously taken a sample, where possible, and left notice of his action in the premises;
(j) require any person to answer and sign a declaration of truth;
(k) require the production of, inspect and take copies of any books or documents;
(l) require such facilities or assistance as necessary;

(m) exercise any other power necessary to carry out his responsibilities (section 20).

Notices

(a) **Form.** Notices should be in writing.
(b) **Authentication.** Since improvement and prohibition notices are served by an inspector, they should be signed by that person.
(c) **Service.** Notices to be served by an inspector may:
  (i) be served by delivering it to the person, or leaving it at his proper address or sending it by post to him at that address;
  (ii) in the case of a body corporate, be served or given to the secretary or clerk of that body;
  (iii) in the case of a partnership, be served on or given to a partner or person having control or management of the partnership.

The proper address is:

(a) his last known address; or
(b) in the case of a body corporate on their secretary or clerk, at the registered or principal office; or
(c) in the case of a partnership, the principal office of the partnership.

If a person has specified an address in the UK, other than his proper address, to which notices should be sent, this will be acceptable as proper service. Notices to be served on an owner or occupier of any premises may be served by sending it by post to him at that premises or by addressing it to him by name but delivering it to some responsible person, resident or employee on the premises. Where, after reasonable enquiry, the name and address of the owner or occupier cannot be found, notice may be served by addressing it to 'the owner' or 'the occupier' of the premises (describing them) and delivering it to a responsible person, resident or employee at the premises, or, if no such person is available, by fixing it, or a copy, to some conspicuous part of the premises (section 46).

Offences

Failure to discharge a duty, to contravene any of the provisions of the act or its regulations or to obstruct an inspector will result on summary conviction in a penalty of imprisonment for a term not exceeding 12 months, or a fine not exceeding £20,000, or both, or in some cases, a fine not exceeding level 5 on the standard scale. The penalty on conviction on indictment is imprisonment for a term not exceeding two years, or a fine, or both (section 33 and schedule 3A).

Definitions

Unless otherwise indicated, the following definitions are applicable to procedures in this chapter.
Employee means an individual who works under a contract of employment.

Personal injury includes any disease and any impairment of a person's physical or mental condition.

Premises includes any place and, in particular, includes:

(a) any vehicle, vessel, aircraft or hovercraft;
(b) any installation on land (including the foreshore and other land intermittently covered by water), any offshore installation, and any other installation (whether floating, or resting on the seabed or the subsoil thereof, or resting on other land covered with water or the subsoil thereof); and
(c) any tent or moveable structure.

Substance means any natural or artificial substance, including micro-organisms, whether in solid or liquid form or in the form of a gas or vapour (section 53).

Work means work as an employee or as a self-employed person. The employee is at work throughout the time when he is in the course of his employment, but not otherwise and a self-employed person is at work throughout such time as he devotes to work as a self-employed person (section 52(1)).

The meaning of 'work' has been extended by regulation to include any activity involving genetic manipulation or dangerous pathogens and the duty of care under section 3(2) of the Act extends to such an activity undertaken by research students etc. 'Work' has also been extended to include participants on government training schemes, school-age pupils on work experience and college students on 'sandwich course' external training (Health and Safety (Training for Employment) Regulations 1990).

In relation to premises outside of Great Britain but controlled under the Health and Safety at Work etc. Act 1974, e.g. oil rigs, the definition of 'at work' has also been extended so that an employee or a self-employed person is deemed to be at work throughout the time that he is present at the premises (regulation 23(2), the Management of Health and Safety Regulations 1999).

IMPROVEMENT AND PROHIBITION NOTICES

References

Health and Safety at Work etc. Act 1974, sections 21 to 24 inclusive.
Employment Tribunals (Constitution and Rules of Procedure) Regulations 2004 as amended 2008, 2010 and 2012.
HELA Circular LAC 22/1 Rev., September 2000 – Choice of appropriate Enforcement Procedure.
HSC Health and Safety in the Local Authority Enforced Sectors, section 18.
Guidance to LAs 2001 (Rev.), which includes at Annex I the Statement of Enforcement Policy, January 2002.
Section 18 Standard, www.hse.gov.uk/section18/index.htm

Inspector

Improvement and prohibition notices may be served by persons appointed by an enforcing authority as an inspector under the Act and no further authority is required for them to exercise these powers (section 19).

Relevant statutory provisions

Both notice procedures relate to the application of 'relevant statutory provisions', which are:

(a) Part 1 Health and Safety at Work etc. Act 1974;
(b) health and safety regulations; and
(c) enactments and regulations made under them in force before the passing of the Health and Safety at Work etc. Act 1974 and now listed in schedule 1 of that Act (section 53(1)).

Enforcement procedures

In using the powers to serve improvement and prohibition notices, local authorities must have regard to the detailed guidance from the Health and Safety Executive contained in LAC 22/1 and to the HSC Statement of Enforcement Policy. Guidance is also to be found in the HSE Enforcement Guide 2003 and the Section 18 Standard 2008. One aspect of this is to be sure that the service of an improvement notice or prohibition notice is the most appropriate form of action and, in this context, attention is drawn to the enforcement option of a formal caution, which was introduced in the 2008 HSC Enforcement Policy Statement, www.hse.gov.uk/foi/internalops/ocs/100-199/130_6.htm.

Improvement notices

(a) **Scope.** Improvement notices may be served when a relevant statutory provision is either being contravened or, having been contravened, likely to be continued or repeated (section 21).
(b) **Content.** The notice must include:
  (i) the inspector's opinion concerning the contravention;
  (ii) specification of the provision or provisions being contravened;
  (iii) reasons for him being of that opinion;
  (iv) a requirement to remedy the contravention; and
  (v) a period within which the contravention must be remedied, which must be not less than 21 days (section 21).

The notice need not specify the actual measures to be taken to remedy the contravention but, if it does, it may refer to any approved Code of Practice and allow a choice of remedies (section 23(2)).

Prohibition notices

(a) **Scope.** These notices may be used by an inspector where there is a risk of serious personal injury from an activity which is either being carried on or is likely to be carried on, and to which a relevant statutory provision applies (section 22(1) and (2)).
(b) **Content.** The notice must:
  (i) state the inspector's opinion relating to the risk of serious personal injury;

FC52 Improvement notices – Section 21 Health and Safety at Work Act 1974

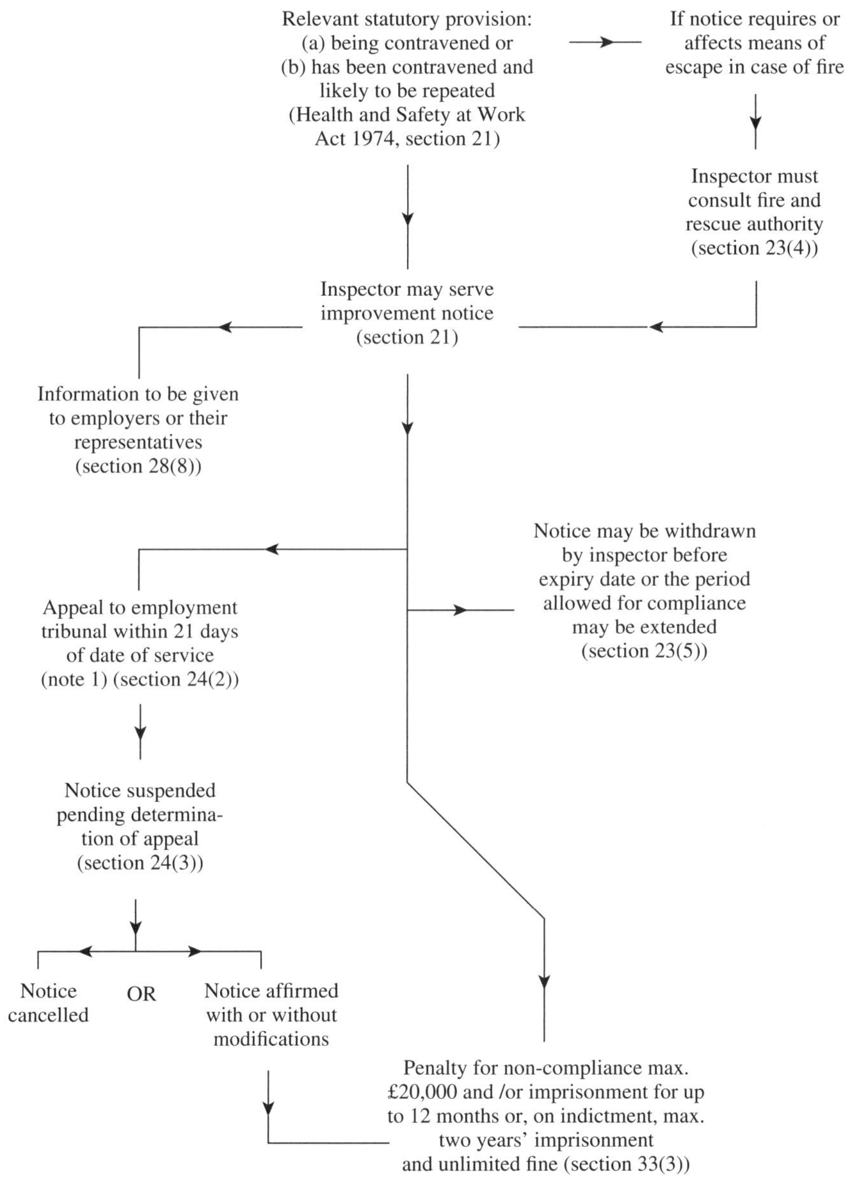

Notes

1. The period allowed for appeal may be extended by an employment tribunal (Employment Tribunals (Constitution and Rules of Procedure) Regulations 2004).
2. This procedure cannot be applied to the Crown (Health and Safety at Work Act 1974, section 48).
3. Unless the notice imposes requirements solely relating to the protection of persons at work, it must be included in a public register under the provisions of the Environment and Safety Information Act 1988.

FC53 Prohibition notices – Section 22 Health and Safety at Work Act 1974

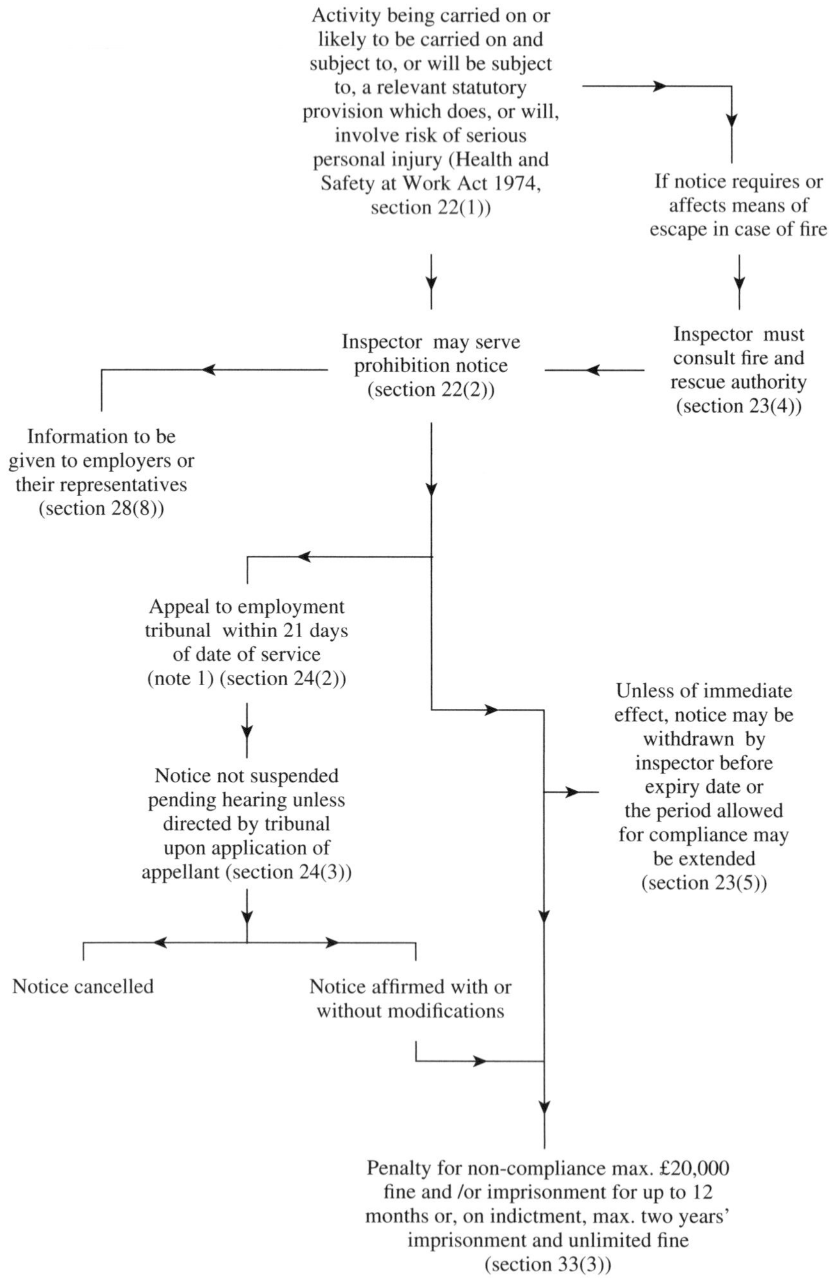

Notes

1. The period allowed for appeal may be extended by an employment tribunal (Employment Tribunals (Constitution and Rules of Procedure) Regulations 2004 as amended 2008, 2010, 2012).
2. This procedure cannot be applied to the Crown (Health and Safety at Work Act 1974, section 48).
3. Unless the prohibition notice imposes restrictions solely relating to the protection of persons at work, it must be included in a public register under the provisions of the Environment and Safety Information Act 1988.

(ii) specify the matters which give rise to that risk;
(iii) where a relevant statutory provision is being contravened, specify the provision concerned and give reasons for that opinion;
(iv) direct that the activity concerned shall not be carried on unless the matters specified in the notice have been remedied.

The notice will normally be given immediate effect if the inspector is of the opinion that there is a risk of serious personal injury. Alternatively, a period must be specified following which the notice comes into effect. There is no minimum period to be allowed and, since in the event of an appeal the notice is not automatically suspended, the period allowed may be shorter than the appeal period. The notice can only be suspended on application to the tribunal and the direction of the tribunal (sections 22 and 24).

The notice need not specify the actual measure to be taken but, if it does, it may refer to any approved Code of Practice and allow a choice of remedies (section 23(2)).

Standard of requirements

Unless a relevant statutory provision lays down a specific requirement, requirements in improvement notices relating to buildings must be no more onerous than that necessary to secure compliance with any building regulations in force as if the building was being newly erected (section 23(3)).

Person responsible

Improvement notices are to be served on the person who is contravening the relevant statutory provision or who has contravened it and is likely to continue or repeat it (section 21). Prohibition notices must be served on the person by or under the control of whom the particular activity is being or is about to be carried on (section 22(2)).

Appeals

Appeal against either notice must be made to an employment tribunal within 21 days of the date of service, although the tribunal may allow an extension of this period on application from the appellant, either before or after the expiration of the 21 days, if satisfied that it was not reasonably practicable to have brought the appeal within 21 days.

There are no specific grounds in either the Act or the regulations for lodging an appeal. The procedure to be adopted by the employment tribunal for receiving and hearing appeals is laid down in schedule 4 to the regulations (section 24 and Employment Tribunals (Constitution and Rules of Procedure) Regulations 2004 as amended 2008, 2010 and 2012).

Where an appeal is made against an improvement notice, the notice stands suspended until the appeal is determined or the appeal is withdrawn but appeals against prohibition notices do not automatically suspend the notice. Any suspension of a prohibition notice may be ordered only by the tribunal and then only from the date of the direction (section 24(3)).

Note – some employment tribunals call themselves industrial tribunals, particularly in Northern Ireland.

Defences

Where the duty imposed on a person by a relevant statutory provision is qualified by either:

(a) so far as is practicable; or
(b) as far as is reasonably practicable; or
(c) to use the best practicable means,

it is for the defendant to prove that it was not practicable or not reasonably practicable to do more than was in fact done to satisfy the duty or requirement, or that there was no better practicable means than was in fact used to satisfy the duty or requirement (section 40).

REPORTING OF INJURIES, DISEASES AND DANGEROUS OCCURRENCES

References

The Reporting of Injuries, Diseases and Dangerous Occurrences Regulations 1995 as amended 2012.
A Guide to the Reporting of Injuries, Diseases and Dangerous Occurrences Regulations 1995 (RIDDOR) L73 (Third Edition) HSE 2008.

Scope

The following four categories of incident are to be notified or reported.

Notifiable injuries

(a) any person (i.e. at work or not) dying as a result of an accident arising out of or in connection with work (regulation 3(1)(a));
(b) a major injury (see below) to any person at work as a result of an accident arising out of or in connection with work (regulation 3(1)(b));
(c) a person not at work suffering any injury as a result of an accident arising out of or in connection with work and that person is taken from the scene to a hospital for treatment (regulation 3(1)(c));
(d) a person not at work suffering a major injury as a result of an accident arising out of or in connection with work at a hospital (regulation 3(1)(d));
(e) an injury to a person at work causing incapacity for his normal work or work of a similar nature for more than seven consecutive days (excluding the day of the accident but including any days which would not have been working days) (regulation 3(2)).

In addition, where an employee has suffered a reportable injury as a result of an accident at work and dies within one year from that cause, the employer must inform the enforcement agency in writing as soon as the death comes to his knowledge (regulation 4).

Major injuries are defined in schedule 1 as:

1. any fracture, other than to the fingers, thumbs or toes;
2. any amputation;
3. dislocation of the shoulder, hip, knee or spine;
4. loss of sight (whether temporary or permanent);
5. a chemical or hot metal burn to the eye or any penetrating injury to the eye;
6. any injury resulting from an electric shock or electrical burn (including any electrical burn caused by arcing or arcing products) leading to unconsciousness or requiring resuscitation or admittance to hospital for more than 24 hours;
7. any other injury:
   (a) leading to hypothermia, heat-induced illness or to unconsciousness;
   (b) requiring resuscitation; or
   (c) requiring admittance to hospital for more than 24 hours;
8. loss of consciousness caused by asphyxia or by exposure to a harmful substance or biological agent;
9. either of the following conditions which result from the absorption of any substance by inhalation, ingestion or through the skin:
   (a) acute illness requiring medical treatment; or
   (b) loss of consciousness;
10. acute illness which requires medical treatment where there is reason to believe that this resulted from exposure to a biological agent or its toxins or infected material.

Dangerous occurrences

Dangerous occurrences as detailed in schedule 2 (below) are notifiable. There are also lists detailing specific dangerous occurrences in offshore workplaces, mines, quarries and transport systems. Those which apply generally are as follows:

1. The collapse of, the overturning of, or the failure of any load-bearing part of any lifting machinery or equipment.
2. The failure of any closed vessel (pressure systems) (including a boiler or boiler tube) or of any associated pipework, in which the internal pressure was above or below atmospheric pressure, where the failure has the potential to cause the death of any person.
3. The failure of any freight container in any of its load-bearing parts while it is being raised, lowered or suspended.
4. Any unintentional incident in which plant or equipment either:
   (a) comes into contact with an uninsulated overhead electric line in which the voltage exceeds 200 volts; or
   (b) causes an electrical discharge from such an electric line by coming into close proximity to it.

5. Electrical short circuit or overload attended by fire or explosion which results in the stoppage of the plant involved for more than 24 hours or which has the potential to cause the death of any person.
6. Various incidents involving explosives.
7. Any accident or incident which resulted or could have resulted in the release or escape of a biological agent likely to cause severe human infection or illness.
8. Incidents involving malfunction of a radiation generator or its ancillary equipment used in fixed or mobile industrial radiography, the irradiation of food or the processing of products by irradiation.
9. Any incident in which breathing apparatus malfunctions.
10. Any incidents in relation to a diving operation, such as the failure or the endangering of any lifting equipment associated with the diving operation, or life support equipment, including control panels, hoses and breathing apparatus, which puts a diver at risk; any damage to, or endangering of, the dive platform; the trapping of a diver; any explosion in the vicinity of a diver; or any uncontrolled ascent or any omitted decompression which puts a diver at risk.
11. The complete or partial collapse of any scaffold which is more than five metres in height.
12. Any unintended collision of a train with any other train or vehicle, which caused, or might have caused, the death of, or major injury to, any person.
13. Various incidents in relation to a well (other than a well sunk for the purpose of the abstraction of water).
14. Various incidents in respect of a pipeline or pipeline works, including the uncontrolled or accidental escape of anything from, or inrush of anything into, a pipeline; the unintentional ignition of anything in a pipeline; any damage to any part of a pipeline; any substantial and unintentional change in the position of a pipeline; any unintentional change in the subsoil or seabed in the vicinity of a pipeline; any failure of any pipeline isolation device, equipment or system which has the potential to cause the death of, major injury or damage to the health of any person or which results in the pipeline being shut down for more than 24 hours.
15. The following incidents on fairground equipment in use or under test:
    (a) the failure of any load-bearing part;
    (b) the failure of any part designed to support or restrain passengers; or
    (c) the derailment or the unintended collision of cars or trains.
16. Any incident involving a road tanker or tank container used for the carriage of a dangerous substance.
17. Any unintended collapse or partial collapse of:
    (a) any building or structure (whether above or below ground) under construction, reconstruction, alteration or demolition which involves a fall of more than five tonnes of material;
    (b) any floor or wall of any building (whether above or below ground) used as a place of work; or
    (c) any false-work.
18. An explosion or fire occurring in any plant or premises which results in the stoppage of that plant or, as the case may be, the suspension of normal work

in those premises for more than 24 hours, where the explosion or fire was due to the ignition of any material.

19. The sudden, uncontrolled release of flammable liquid.
20. The accidental release or escape of any substance in a quantity sufficient to cause the death, major injury or any other damage to the health of any person.

Diseases

Where a person at work suffers from any of the listed occupational diseases and his work involves a corresponding work activity, that situation is reportable (regulation 5(1)). To become reportable, the disease must have been confirmed by a registered medical practitioner (regulation 5(2)).

The diseases and activities are set out in schedule 3 to the regulations and include:

- Conditions due to physical agents and the physical demands of work
  1. Inflammation, ulceration or malignant disease of the skin due to ionising radiation.
  2. Malignant disease of the bones due to ionising radiation.
  3. Blood dyscrasia due to ionising radiation.
  4. Cataract due to electromagnetic radiation.
  5. Decompression illness.
  6. Barotrauma resulting in lung or other organ damage.
  7. Dysbaric osteonecrosis
  8. Cramp of the hand or forearm due to repetitive movements.
  9. Subcutaneous cellulitis of the hand (*beat hand*).
  10. Bursitis or subcutaneous cellulitis arising at or about the knee due to severe or prolonged external friction or pressure at or about the knee (*beat knee*).
  11. Bursitis or subcutaneous cellulitis arising at or about the elbow due to severe or prolonged external friction or pressure at or about the elbow (*beat elbow*).
  12. Traumatic inflammation of the tendons of the hand or forearm or of the associated tendon sheaths.
  13. Carpal tunnel syndrome.
  14. Hand-arm vibration syndrome.
- Infections due to biological agents
  15. Anthrax.
  16. Brucellosis.
  17. (a) Avian chlamydiosis.
      (b) Ovine chlamydiosis.
  18. Hepatitis.
  19. Legionellosis.
  20. Leptospirosis.
  21. Lyme disease.
  22. Q fever.
  23. Rabies.
  24. Streptococcus suis.
  25. Tetanus.

26. Tuberculosis.
27. Any infection reliably attributable to the performance of the work.

- Conditions due to substances

28. Poisonings by any of the following:
    (a) acrylamide monomer;
    (b) arsenic or one of its compounds;
    (c) benzene or a homologue of benzene;
    (d) beryllium or one of its compounds;
    (e) cadmium or one of its compounds;
    (f) carbon disulphide;
    (g) diethylene dioxide (dioxan);
    (h) ethylene oxide,
    (i) lead or one of its compounds;
    (j) manganese or one of its compounds;
    (k) mercury or one of its compounds;
    (l) methyl bromide;
    (m) nitrochlorobenzene, or a nitro- or amino- or chloro-derivative of benzene or of a homologue of benzene;
    (n) oxides of nitrogen;
    (o) phosphorus or one of its compounds.
29. Cancer of a bronchus or lung.
30. Primary carcinoma of the lung where there is accompanying evidence of silicosis.
31. Cancer of the urinary tract.
32. Bladder cancer.
33. Angiosarcoma of the liver.
34. Peripheral neuropathy.
35. Chrome ulceration of:
    (a) the nose or throat; or
    (b) the skin of the hands or forearm.
36. Folliculitis.
37. Acne.
38. Skin cancer.
39. Pneumoconiosis (excluding asbestosis).
40. Byssinosis.
41. Mesothelioma.
42. Lung cancer.
43. Asbestosis.
44. Cancer of the nasal cavity or associated air sinuses.
45. Occupational dermatitis.
46. Extrinsic alveolitis (including farmer's lung).
47. Occupational asthma.

Gas incidents

Death or major injury arising out of gas being distributed, filled, imported or supplied are notifiable to the Health and Safety Executive (regulation 6).

FC54 The reporting of injuries, diseases and dangerous occurrences – RIDDOR 1995

(a) Fatalities
(b) Major injuries
(c) Dangerous occurrences
(regulation 3(1))

↓

Reportable to Incident Contact Centre or enforcing authority without delay (regulation 3(1))

↓

To be confirmed in writing within ten days (regulation 3(1)) Form F2508

Notifiable diseases (regulation 5(1))

↓

Reportable to Incident Contact Centre or enforcing authority forthwith on form F2508A (regulation 5(1))

Incapacitation from normal work for more than seven days (regulation 3(2))

↓

Reportable to Incident Contact Centre or enforcing authority in writing as soon as practicable but within 15 days (regulation 3(2)) Form F2508

↓

Notification is to the enforcing authority or the Incident Contact Centre (see text on methods)

AND

Records (see text) to be kept for at least three years and enforcing authority may require extracts (regulation 7)

Exemptions

Road traffic accidents causing death or injury are not notifiable under these provisions unless they involved the exposure of substances carried on the vehicle or another work activity, e.g. roadworks adjoining the carriageway. None of the requirements relates to the Armed Forces (regulation 10).

Notifications

The responsible person must notify deaths, major injuries, accidents to people who are not at work and dangerous occurrences without delay. Notifications of other reportable injuries are to be reported as soon as practicable and in any event within 15 days of the accident (regulation 3(2)), whilst notifiable diseases are reportable forthwith using form F2508A (regulation 5(1)).

All of these accidents may be notified to the Incident Contact Centre (ICC) in any of the following ways:

1. by telephone on Monday to Friday from 8.30 am to 5.00 pm to 0845 300 9923;
2. by fax to 0845 300 9924;
3. by internet at www.riddor.gov.uk or www.hse.gov.uk/riddor/report.htm;
4. by e-mail to riddor@natbrit.com;
5. by post to Incident Contact Centre, Caerphilly Business Park, Caerphilly, CF83 3GG, using form F2508 or F2508A as appropriate.

Reporting of reportable diseases is to the ICC or enforcing authority.

Written notifications and confirmations are to be made on form F2508A for occupation-related diseases or F2508 for all other incidents.

Notification to the local enforcing agency is also acceptable by phone or post and confirmation by the appropriate form. These reports will be sent on to the ICC for processing.

Records

In respect of deaths, major injuries and dangerous occurrences, the following particulars must be kept by the responsible person:

1. Date and time of the accident or dangerous occurrence.
2. In the event of an accident suffered by a person at work, the following particulars of that person:
   (a) full name;
   (b) occupation;
   (c) nature of injury.
3. In the event of an accident suffered by a person not at work, the following particulars of that person (unless they are not known and it is not reasonably practicable to ascertain them):
   (a) full name;
   (b) status (for example, ‘passenger’, ‘customer’, ‘visitor’ or ‘bystander’);
   (c) nature of injury.

4. Place where the accident or dangerous occurrence happened.
5. A brief description of the circumstances in which the accident or dangerous occurrence happened.
6. The date on which the event was first reported to the relevant enforcing authority.
7. The method by which the disease was reported.

In relation to reportable diseases the following information must be recorded:

1. Date of diagnosis of the disease.
2. Name of the person affected.
3. Occupation of the person affected.
4. Name or nature of the disease.
5. The date on which the disease was first reported to the relevant enforcing authority.
6. The method by which the disease was reported.

In addition, records must be kept of any injury resulting from an accident where a person is incapacitated for more than three consecutive days (excluding the day of the accident, but including any days which would not have been working days).

The records must be kept either at the place where the work to which it relates is carried on or at the usual place of business of the responsible person, and must be kept for at least three years.

Enforcing authorities may require extracts from the records to be sent to them (regulation 7).

Defences

In proceedings for offences under these provisions, it will be a defence to prove that the person was not aware of the event and had taken all reasonable steps to have such events brought to his attention (regulation 11).

Definitions

Accident includes:

(a) an act of non-consensual physical violence done to a person at work; and
(b) an act of suicide which occurs on, or in the course of the operation of, a relevant transport system.

Disease includes a medical condition.
Responsible person means:

(a) in the case of:
   (i) a mine, the manager of that mine;
   (ii) a quarry, the operator of that quarry;
   (iii) a closed tip, the owner of the mine with which that tip is associated;
   (iv) an offshore installation (otherwise than in the case of a disease

reportable under regulation 5), the duty holder for the purposes of the Offshore Installations and Pipeline Works (Management and Administration) Regulations 1995 provided that for the purposes of this provision regulation 3(2)(c) of those regulations shall be deemed not to apply;

(v) a dangerous occurrence at a pipeline (being an incident to which paragraph 14(a)–(f) of part I of schedule 2 applies), the owner of that pipeline;

(vi) a dangerous occurrence at a well, the person appointed by a licensee to execute the function of organising and supervising the drilling of, and all operations to be carried out by means of, that well or, where no such person has been appointed, the licensee (and, for this purpose, 'licensee' means any person to whom a licence to search and bore for and get petroleum in respect of any area within relevant waters is granted pursuant to section 3 of the Petroleum Act 1998);

(vii) a diving project (otherwise than in the case of a disease reportable under regulation 5), the diving contractor;

(viii) a vehicle to which paragraph 16 or 17 of part I of schedule 2 applies, the operator of the vehicle;

(b) (where sub-paragraph (a) above does not apply) in the case of the death of or other injury to an employee reportable under regulation 3 or of a disease suffered by an employee reportable under regulation 5, his employer; and

(c) in any other case, the person for the time being having control of the premises in connection with the carrying on by him of any trade, business or other undertaking (whether for profit or not) at which, or in connection with the work at which, the accident or dangerous occurrence reportable under regulation 3, or case of disease reportable under regulation 5, happened (regulation 2).

NOTIFICATION OF COOLING TOWERS AND EVAPORATIVE CONDENSERS

References

The Notification of Cooling Towers and Evaporative Condensers Regulations 1992.

Approved Code of Practice and Guidance – L8. Legionnaires' disease: the control of legionella bacteria in water systems, HSE 2001.

LAC 46/1 September 2000, Legionnaire's disease, HELA.

LAC 46/2 January 2001, Control of legionella, HELA.

LAC 46/3 February 2001, Control of legionella in wet cooling systems, HELA.

LAC 46/4 February 2002, Investigation of outbreaks and single cases of Legionnaire's disease involving cooling towers and evaporative condensers, HELA.

Scope

Notification is required to be made to the local authority in whose area the premises are situated in respect of all cooling towers and evaporative condensers except:

(a) where it contains no water exposed to air;
(b) its water supply is not connected; and
(c) its electrical supply is not connected (regulation 2 and 3(1)),

and where the device is situated in premises other than domestic premises, which are used for, or in connection with, the carrying on of a trade, business or undertaking (whether for profit or not).

The purpose of the notification is to allow the local authority to monitor the environmental control of the devices with particular reference to the avoidance of Legionnaire's disease and to have an awareness of the location of them in order to investigate outbreaks of such disease.

Person responsible

Notification is to be made by each person who has, to any extent, control of the premises and must be made before the device is situated there. Where the premises fall to any extent under the control of the manufacturer of the device, e.g. during installation, that person is required to effect the notification (regulation 3(1) and (2)).

Notification

This is to be done in writing using the form prescribed by the Health and Safety Executive for the purpose. The detail to be supplied is:

(a) the address of the premises where the notifiable device is to be situated;
(b) the name, address and telephone number of a person who has, to any extent, control of the premises referred to in (a) above;
(c) the number of notifiable devices at the premises referred to in (a) above;
(d) the location on the premises of each notifiable device referred to in (c) above.

Where the notification is being made by the manufacturer, the details required relate only to (a) and (b) above (regulation 3(1) and (2)).

Subsequent change

Any changes to the notified details must be notified to the local authority by either the person having control of the premises or the manufacturer in writing within one month of the change (regulation 3(3)).

The local authority must be informed in writing as soon as reasonably practicable where the device ceases to be used, other than for seasonal shutdown or maintenance (regulation 3(4)).

FC55 Notification of cooling towers and evaporative condensers – Notification of cooling towers and evaporative condensers Regulations 1992

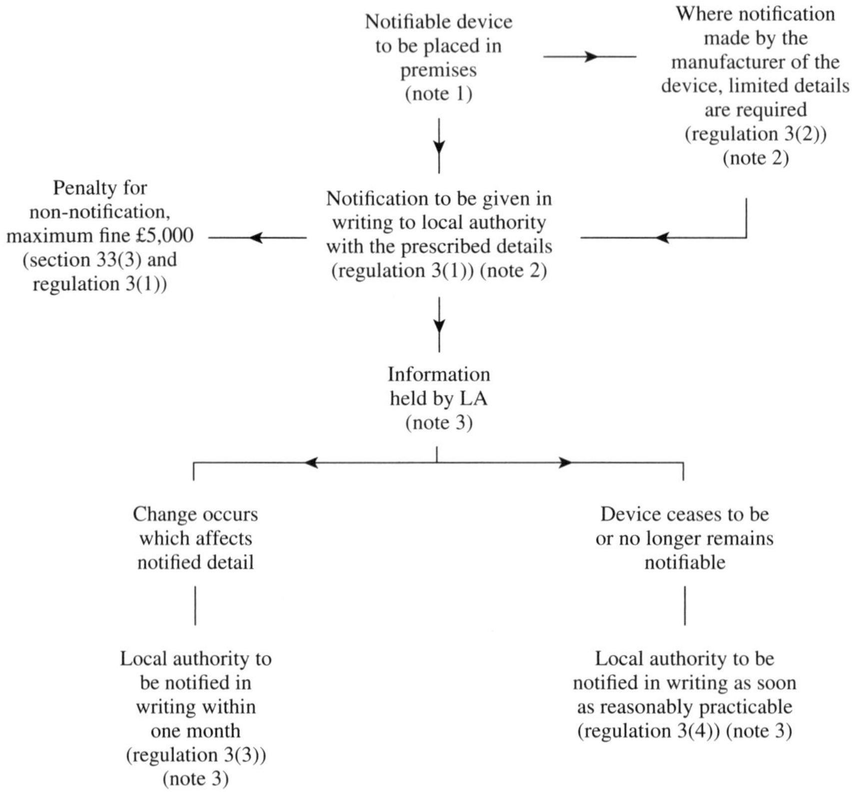

Notes

1. Notification must be made before the device is placed in the premises, otherwise an offence is committed (regulation 3(1)).
2. The form to be used is prescribed by the Health and Safety Executive.
3. Local authorities should make available to the Health and Safety Executive information from notifications relating to equipment on premises for which the Health and Safety Executive has enforcement responsibility.

Definitions

Cooling tower means a device whose main purpose is to cool water by direct contact between that water and a stream of air.
Evaporative condenser means a device whose main purpose is to cool a fluid by passing that fluid through a heat exchanger, which is itself cooled by contact with water passing through a stream of air.
Heat exchanger means a device for transferring heat between fluids which are not in direct contact with one another.

SEIZURE OF DANGEROUS ARTICLES OR SUBSTANCES

Reference

Health and Safety at Work etc. Act 1974, section 25.

Scope

An inspector appointed under the Act by an enforcing authority may seize any article or substance on any premises which he has power to enter if he believes that, in the circumstances in which he found it, the article or substance is a cause of imminent danger of serious personal injury to either employees or other persons (section 25(1)).

Customs officers are given the power to assist inspectors by seizing and detaining for up to two working days any imported article or substance (section 25A).

Sampling

Before seizing either:

(a) any article which forms part of a batch of similar articles; or
(b) any substance,

the inspector is required, where it is practicable for him to do so, to take a sample and give a portion of it, marked in such a way as to identify it, to a responsible person at the premises where it was found (section 25(2)).

Action following seizure

Having seized the article or substance, the inspector is to render it harmless and this may be by any method, including destruction (section 25(1)).

Having dealt appropriately with the article or substance, the inspector must:

(a) prepare and sign a written report of the matter, including the circumstances in which the article or substance was seized and dealt with; and
(b) give a copy of that report to a responsible person at the premises where the article or substance was found; and

FC56 Seizure of dangerous articles or substances – Section 25 Health and Safety at Work etc. Act 1974

Articles or substances on any premises causing imminent danger of serious personal injury (Health and Safety at Work Act 1974, section 25(1))

If
(a) article forms part of batch of similar articles or
(b) a substance is involved

→ Inspector must take sample and give portion to a responsible person at the premises (section 25(2))

OR

If
(a) article does not form part of batch of similar articles or
(b) sampling is not practicable

↓

Inspector may seize article or substance and cause it to be rendered harmless (section 25(1))

↓

Inspector must prepare and sign written report (section 25(3))

↓

Give copy of report to a responsible person at the premises where the article or substance was found (Health and Safety at Work Act 1974, section 25(3))

AND

Serve copy of report on owner of the substance or article (Health and Safety at Work Act 1974, section 25(3))

Note

1. These procedures may not be applied to Crown premises (Health and Safety at Work Act 1974, section 48).

(c) serve a copy of the notice on the owner of the article or substance where that is a different person from (b) above.

If, after reasonable enquiry, the inspector is unable to ascertain the name and address of the owner, a copy may be served on the owner by giving it to the responsible person (section 25(3)).

Chapter 5

HOUSING AND HEALTH

STRUCTURE OF THE LAW DEALING WITH HOUSING STANDARDS AND THEIR ENFORCEMENT

The main provisions are found in the Housing Act 2004, which contains:

(a) a system for assessing housing conditions based on the use of the Housing Health and Safety Rating System (HHSRS);
(b) enforcement options; and
(c) licensing of certain houses in multiple occupation and selective licensing for other residential accommodation.

The Housing Act 1985 provisions relating to demolition orders and clearance areas together with those relating to overcrowding continue in force with amendments. The ability for local authorities to declare Renewal Areas is found in the Local Government and Housing Act 1989, and the Housing Grants, Construction, Development and Regeneration Act 1996 and the Regulatory Reform (Housing Assistance) (England and Wales) Order 2002 have provisions for grants for the renewal of private sector housing.

GENERAL PROCEDURAL PROVISIONS

Extent

The provisions in this part apply to England and Wales (Housing Act 1986, section 625(5), Housing Grants, Construction and Regeneration Act 1996, section 148 and Housing Act 2004, section 270) but, whilst the power to make orders and regulations under the Housing Act 2004 lies with the Secretary of State, he must consult the National Assembly for Wales before acting in relation to any residential properties in Wales.

Definitions

Local housing authorities

In relation to England, these are:

- unitary councils;
- district councils;
- London borough councils;
- the common Council of the City of London;
- the Sub-Treasurer of the Inner Temple and the Under-Treasurer of the Middle Temple; and
- the Council of the Isles of Scilly.

For Wales, they are:

- county councils or county borough councils (section 261).

Throughout this part, local housing authority is abbreviated to local authority.

Residential premises

In relation to Part 1 of the 2004 Act, this means:

(a) a dwelling;
(b) a house in multiple occupation;
(c) unoccupied accommodation;
(d) any common parts of a building containing one or more flats (section 1(4)).

This definition is further clarified:

(a) 'a building containing one or more flats' (in (d) above) does not include a house in multiple occupation;
(b) 'common parts' in relation to a building containing one or more flats includes the structure and exterior of the building and common facilities provided (whether or not in the building) for persons who include the occupiers of one or more of the flats;
(c) 'dwelling' means a building or part of a building occupied or intended to be occupied as a separate dwelling;
(d) 'external common parts' in relation to a building containing one or more flats means common parts of the building which are outside it;
(e) 'flat' means a separate set of premises (whether or not on the same floor) which forms part of a building, is constructed or is adapted for use for the purposes of a dwelling and either the whole or a material part of which lies above or below some other part of the building (section 1(5));
(f) house in multiple occupation is as defined in sections 254–259 – see page 313.

'Unoccupied house in multiple occupation accommodation' means a building or part of a building constructed or adapted for use as a house in multiple occupation but for the time being is either unoccupied or only occupied by persons who form a single household.

References to a dwelling, house in multiple occupation or a building containing one or more flats include (where the context permits) any yard, garden, outhouses and appurtenances belonging to or usually enjoyed with the dwelling, house in multiple occupation or building (or any part of it) (section 1(6)).

Person having control

This is the person who receives, either directly or as agent or trustee, the market rent from the tenants or is entitled to receive the rent if the premises were let, i.e. an owner (section 263).

Person managing

This is a person who receives the rent directly from the occupier. Rent includes ground rent, so this person may be a managing agent (section 263).

Obtaining information

The local authority may require by notice the production of documentation that may be necessary to carry out its enforcement functions. Such notices may be served on the following 'relevant persons':

- holders of house in multiple occupation licences or those proposing to hold such a licence;
- any person with an estate or interest in the premises;
- a person who is or is proposing to be managing or having control;
- anyone otherwise involved, or so proposing, in the management of the premises; and
- an occupier.

Persons failing to comply with the notice are guilty of an offence and subject to a fine not exceeding level 5 (sections 235 and 236).

Service of notices

The form, authentication and methods of service of notices or other documents arising from the Housing Act 2004 are as set out in sections 233–234 of the Local Government Act 1972 and the provisions of section 16 of the Local Government (Miscellaneous Provisions) Act 1976 relating to the requisition of information regarding ownership etc. are also made available for enforcement of this Act (see page 28) (Housing Act 2004, section 246).

Where the recipient has given prior approval, a notice or licence may be served electronically and detailed provisions for this are to be set out in the regulations (sections 247–248).

The Secretary of State has powers by regulations to dispense with the need for notices etc. where there is no prescribed form (section 244).

Appeals

Appeals against the various actions of the local authority in these procedures are to be made to a Residential Property Tribunal whose functions are exercised by

Rent Assessment Committees, which become a Residential Property Tribunal when dealing with these procedures. This applies not only to procedures in the 2005 Act but also to demolition notices under the 1985 Act (section 229). The powers of the Residential Property Tribunal are contained in section 230 and these include a general power to give such directions as it considers necessary or desirable for securing the just, expeditious and economical disposal of the proceedings. The more detailed provisions for appeals are noted in each procedure below.

With the approval of the Residential Property Tribunal, appeals against its decisions are to the Lands Tribunal within 28 days of notification of the decision (section 231).

Detailed rules for the operation of Residential Property Tribunals are contained in the Residential Tribunal Procedures and Fees (England) Regulations 2011 and with similar regulations for Wales. Further information on the working of Residential Property Tribunals can be found at www.justice.gov.uk/tribunals/residential-property

Charging for enforcement action

Local authorities are given the power to make a reasonable charge for recovering certain costs in connection with:

1. the service of improvement notices;
2. making prohibition orders;
3. serving hazard awareness notices;
4. taking emergency remedial action;
5. making emergency prohibition orders;
6. making demolition orders.

The costs that may be recovered relate to:

1. inspection of the premises;
2. consideration of the action to be taken;
3. the service of the notice (section 49).

The local authorities should take account of personal circumstances of the person to be charged (Enforcement Guidance, paragraph 5.52).

There is an appeal to the Residential Property Tribunal against the charges raised. The demand for payment becomes operative 21 days after service and is, until recovered, a charge on the property and is a local land charge. After one month, the debt may be recovered by the appointment of a receiver (section 50).

The Act provides for maximum charges to be set by the Secretary of State but these powers have not yet been used.

Powers of entry etc.

An authorised officer may enter premises where it is necessary to carry out an inspection and either:

(a) the inspection is required to see if a category 1 or 2 hazard exists or to check if any of their powers under parts 1–4 of the 2004 Act need to be used; or
(b) where the premises are specified in an improvement notice or prohibition order; or
(c) there is a management order in force.

The authorised officer must have written authorisation and give at least 24 hours' notice of intention to enter. His authorisation must be given by an officer of at least deputy chief officer status whose duties consist of or include the exercise of functions relevant to the authorisation (section 239).

Where:

(a) admission has been refused; or
(b) the premises are unoccupied or temporarily unoccupied and waiting for the occupier's return may defeat the object of entry; or
(c) application for admission to the occupier would defeat the object of entry,

the local authority may apply to a magistrates' court for a warrant to enter, if necessary by force (section 240).

A person obstructing a person in performance of duties under parts 1–4 commits an offence and is liable to penalty not exceeding level 4 (section 241).

Powers of entry relating to demolition orders remain in section 319 of the Housing Act 1985 and require at least seven days' notice.

HOUSING STANDARDS – THE HOUSING HEALTH AND SAFETY RATING SYSTEM (HHSRS)

References

Housing Act 2004.
Housing Health and Safety Rating System Operating Guidance, Department for Communities and Local Government (DCLG), 2006.

Background

The Housing Health and Safety Rating System of assessing housing conditions was introduced in the Housing Act 2004 and replaced the former system based on the test of fitness for human habitation. The system operates by reference to category 1 and category 2 hazards in residential premises. The system and the way in which it should be operated are described in Housing Health and Safety Rating System Operating Guidance, Department for Communities and Local Government (DCLG), 2006. The system and the options for remedial actions apply to residential premises (definition on page 241).

Assessment process

The assessment process is:

(a) full inspection of the residential premises to identify any deficiencies;
(b) inspectors judge whether the deficiencies mean that there are any hazards that are significantly worse than the average for residential premises of that age and type; and if so:
  (i) the likelihood of an occurrence that could cause harm over the next 12 months; and
  (ii) the severity of the outcomes from such an occurrence.

The Housing Health and Safety Rating System Operating Guidance issued by the DCLG in February 2006 under section 9 of the Act gives detailed technical guidance on the way in which the system is to be used, and local authorities must have regard to this in carrying out their functions.

Inspection

Local authorities are required to keep under review the housing conditions in their area with a view to identifying what courses of action are necessary under the Housing Acts 1985–2004 (section 3 and Enforcement Guidance 2.1–2.3).

Where the local authority considers, from information gained from its duties under section 3, or for any other reason, that it would be appropriate for a residential premises to be inspected with a view to determining if category 1 or 2 hazards exist, the local authority must arrange for an inspection. An inspection obligation also follows an official complaint about the condition of the premises from a Justice of the Peace or from a parish or community council (section 4).

Inspections are to be carried out in accordance with regulation 5 of the Housing Health and Safety Rating System (England) Regulations 2005 and an inspector must:

- have regard to any guidance issued under section 9 (Enforcement Guidance 3.2);
- inspect with a view to preparing an accurate record of the condition of the property; and
- prepare and keep such a record in written or electronic form.

Where an inspection reveals that category 1 or 2 hazards exist, the proper officer must report in writing to the local authority, who must consider the report as soon as possible (section 4(6–7)).

Hazards

A hazard is a situation where a risk of harm (see below) is associated with any of the matters or circumstances listed in schedule 1 of the Housing Health and Safety Rating System (England) Regulations 2005, being as follows:

Damp and mould growth

1. Exposure to house dust mites, damp, mould or fungal growths.

Excess cold

2. Exposure to low temperatures.

Excess heat

3. Exposure to high temperatures.

Asbestos and MMF

4. Exposure to asbestos fibres or manufactured mineral fibres.

Biocides

5. Exposure to chemicals used to treat timber and mould growth.

Carbon monoxide and fuel combustion products

6. Exposure to:
   (a) carbon monoxide;
   (b) nitrogen dioxide;
   (c) sulphur dioxide and smoke.

Lead

7. The ingestion of lead.

Radiation

8. Exposure to radiation.

Uncombusted fuel gas

9. Exposure to uncombusted fuel gas.

Volatile organic compounds

10. Exposure to volatile organic compounds.

Crowding and space

11. A lack of adequate space for living and sleeping.

Entry by intruders

12. Difficulties in keeping the dwelling or house in multiple occupation secure against unauthorised entry.

Lighting

13. A lack of adequate lighting.

Noise

14. Exposure to noise.

Domestic hygiene, pests and refuse

15. (a) Poor design, layout or construction such that the dwelling or house in multiple occupation cannot readily be kept clean.
    (b) Exposure to pests.
    (c) An inadequate provision for the hygienic storage and disposal of household waste.

Food safety

16. An inadequate provision of facilities for the storage, preparation and cooking of food.

Personal hygiene, sanitation and drainage

17. An inadequate provision of:
    (a) facilities for maintaining good personal hygiene;
    (b) sanitation and drainage.

Water supply

18. An inadequate supply of water free from contamination, for drinking and other domestic purposes.

Falls associated with baths etc.

19. Falls associated with toilets, baths, showers or other washing facilities.

Falling on level surfaces etc.

20. Falling on any level surface or falling between surfaces where the change in level is less than 300 millimetres.

Falling on stairs etc.

21. Falling on stairs, steps or ramps where the change in level is 300 millimetres or more.

Falling between levels

22. Falling between levels where the difference in levels is 300 millimetres or more.

Electrical hazards

23. Exposure to electricity.

Fire

24. Exposure to uncontrolled fire and associated smoke.

Flames, hot surfaces, etc.

25. Contact with:
    (a) controlled fire or flames;
    (b) hot objects, liquid or vapours.

Collision and entrapment

26. Collision with, or entrapment of body parts in doors, windows or other architectural features.

Explosions

27. An explosion at the dwelling or house in multiple occupation.

Position and operability of amenities etc.

28. The position, location and operability of amenities, fittings and equipment.

Structural collapse and falling elements

29. The collapse of the whole or part of the dwelling or house in multiple occupation.

Assessment of hazards

The assessment of hazards identified is a two-stage process, which first considers the likelihood of an occurrence and second the range of harm outcomes. The calculations are based on the risk to the most vulnerable potential occupant, whether or not anyone is resident at the time of inspection. The system relates poor housing conditions to the kinds of harm attributable to such conditions. It does not try to assess a specific health outcome in relation to the current occupant.

'Likeliness' is considered within a period of 12 months, beginning on the date of the assessment, during which the relevant occupier may suffer harm from identified hazards. Likelihood is expressed as ratio from Table 1 in regulation 6 of the 2005 regulations.

In this connection, 'harm' is defined by the 2005 regulations by reference to four classes contained in schedule 2. These classes are grouped on harm as is reasonably foreseeable as a result of the hazard and being extreme, severe, serious or moderate harm. Each class has a list of specific conditions that may be suffered by the occupiers. The inspector must assess which of the four classes of harm a relevant occupier is most likely to suffer during the 12-month period.

To allow for the comparison of the significance of the widely differing hazards, the Housing Health and Safety Rating System uses a formula to generate hazard scores. This formula uses the likelihood expressed as a ratio in Table 1 of regulation 6, the spread of possible harm outcomes expressed as a percentage in Table 2 (regulation 6) and the weightings given to the four Housing Health and Safety Rating System Classes of Harm (reflecting the degree of incapacity suffered by each class). The hazard score is:

	Class of harm weighting	*Likelihood*	*Spread of harm (%)*
S1 =	10,000 x	1/L x	O1
S2 =	1,000 x	1/L x	O2
S3 =	300 x	1/L x	O3
S4 =	10 x	1/L x	O4

For the purposes of this formula:

L is the representative scale point in column 2 of Table 1 of regulation 6.
O1 is the representative scale point of the percentage range (RSPPR) recorded under paragraph (4) in relation to Class 1 harm.
O2 is the RSPPR for Class 2 harm.
O3 is the RSPPR for Class 3 harm.
O4 is the RSPPR for Class 4 harm.

The numerical hazard scores are banded into ten bands in Table 3 of regulation 7 and these are:

Band	*Numerical score range*
A	5,000 or more
B	2,000–4,999
C	1,000–1,999
D	500–999
E	200–499
F	100–199
G	50–99
H	20–49
I	10–19
J	9 or less

Categories of hazard

For the enforcement of the procedures to remedy unsatisfactory conditions, hazards are classified as either category 1 or category 2. Category 1 are those hazards banded as A, B or C in the table above and category 2 are those banded as D–J (regulation 8).

Application of Housing Health and Safety Rating System in houses in multiple occupation

The Part 1 functions are available for houses in multiple occupation but are not part of the licensing procedure for those premises – see below. There is a requirement in section 55(5)(c) and 55(6) that, not later than five years after an application for a house in multiple occupation licence, there should be no Part 1 enforcement functions (see below) that ought to be carried out by the local authority.

HOUSING STANDARDS FOR ACCOMMODATION OF THE HOMELESS

Reference

Homelessness (Suitability Of Accommodation) (England) Order 2012.

Extent

These standards apply in England only but there are similar provisions in Wales.

Where a local authority wish to secure accommodation for homeless the accommodation must comply with suitable standards.

Article 2 of the order makes provision for the matters to be taken into account in determining whether accommodation is suitable for a person.

Article 3 sets out circumstances where accommodation is not to be regarded as suitable.

Matters to be taken into account in determining whether accommodation is suitable for a person.

In determining whether accommodation is suitable for a person, the local authority must take into account:

- the location of the accommodation;
- any disruption which would be caused by the location of the accommodation to the employment, caring responsibilities or education of the person or members of the person's household;
- the proximity and accessibility of the accommodation to medical facilities and other support which are currently used by or provided to the person or members of the person's household, and are essential to the well-being of the person or members of the person's household; and
- the proximity and accessibility of the accommodation to local services, amenities and transport (article 2).

Circumstances in which accommodation is not to be regarded as suitable for a person

Accommodation in the private rented sector shall not be regarded as suitable where in the view of the Local Authority, one or more of the following apply–

(a) the accommodation is not in a reasonable physical condition;
(b) any electrical equipment supplied with the accommodation is unsuitable
(c) the landlord has not taken reasonable fire safety precautions with the accommodation and any furnishings supplied with it;
(d) the landlord has not taken reasonable precautions to prevent the possibility of carbon monoxide poisoning in the accommodation;
(e) that the landlord is not a fit and proper person to act in the capacity of landlord, having considered if the person has committed certain sexual, discrimination or housing related offences,
(f) the accommodation an unlicensed house in multiple occupation,
(h) the accommodation does not have a valid energy performance certificate
(i) the accommodation does not have a current gas safety record
(j) the landlord has not provided the Local Authority a written tenancy agreement, which the landlord proposes to use, and the Local Authority considers to be adequate (article 3).

ENFORCEMENT

General approach to enforcement

The Enforcement Guidance Office of the Deputy Prime Minister (ODPM), February 2006, Part 2 sets out the policy context in which it is expected that local authorities will conduct their enforcement responsibilities. The outline of this is as follows:

- Local authorities are required to consider the housing conditions in the district and consider what action under the 2004 Act needs to be taken.
- Local authorities need to take a view of the spread of hazards and prioritise their actions on those with the more serious impact on health or safety.
- Where practicable, local authorities should consult adjoining local authorities about actions in estates that straddle boundaries.
- Enforcement policies should take account of the circumstances and views of tenants, landlords and owners and should provide for consultation with social services, tenancy support, housing needs and housing management officers.
- Local authorities are encouraged to adopt the Enforcement Concordat (page 3), which provides a basis for fair, practicable and consistent enforcement. The essence of this is that anyone likely to be the subject of enforcement action should receive clear explanations of what they are required to do and have an opportunity to resolve difficulties before formal action is taken. This allows local authorities to take informal action where that appears to be the most satisfactory course of action in the circumstances.

Part 1 enforcement options

The process of identifying and assessing hazards is the first part of the process that may lead to enforcement action. The Enforcement Guidance indicates a three-stage process:

1. The hazard score as calculated above.
2. The local authority judgement as to the most appropriate action to deal with the hazard in all the circumstances, including taking account of both potential and actual vulnerable occupants.
3. Whether the local authority, in the light of the score, has a discretion or duty to act.

The enforcement provisions that follow the Housing Health and Safety Rating System assessments are contained in Part 1 of the 2004 Act and are known as the Part 1 functions. In reaching a decision in each case on the appropriate action to be taken, the local authority must have regard to the Enforcement Guidance 2006.

Section 5 of the Housing Act 2004 *imposes a duty* on a local authority to take appropriate (best course) enforcement action where there is a category 1 hazard. The options available to local authorities in respect of category 1 hazards are:

- improvement notices (Housing Act 2004, section 11 – FC66);
- prohibition orders (Housing Act 2004, section 20 – FC68);
- hazard awareness notices (Housing Act 2004, section 28 – FC65);
- emergency remedial action notices (Housing Act 2004, section 40 – FC67);
- emergency prohibition orders (Housing Act 2004, section 43 – FC69);
- demolition orders (Housing Act 1985, section 265(1) or (2) – FC63);
- clearance area declarations (Housing Act 1985, section 289(2ZB) – FC64); and
- in relation to prohibition and demolition orders, local authorities also have the alternative of making a determination to purchase under section 300(1) or (2) of the Housing Act 1985.

Where two or more courses of action are available to deal with a category 1 hazard, the local authority must take that which they consider to be the most appropriate (section 5(4)).

Section 7 lists the discretionary enforcement powers of local authorities where a category 2 hazard exists and these are:

- improvement notices (Housing Act 2004 section 12 – FC66);
- prohibition orders (Housing Act 2004, section 21 – FC68);
- hazard awareness notices (Housing Act 2004, section 29 – FC65);
- demolition orders (Housing Act 1985, section 265(3) or (4) but only in prescribed circumstances – FC63); and
- clearance area declarations (Housing Act 1985, section 289(2ZB) but only in prescribed circumstances – FC64).

Reasons for decision

In each case where the local authority decides to take a Part 1 action, whether in relation to category 1 or category 2 hazards, a statement of reasons as to why the particular course of action is being taken in preference to others must be prepared. This is to be served, along with the notice or order, on those in receipt of the notice or order or a copy of them. In addition, statements that relate to the inclusion of premises in a clearance area must be published as soon as possible after the resolution declaring the clearance area has been passed (section 8).

DEFECTIVE PREMISES

Reference

Building Act 1984, section 76.

Extent

These provisions apply to England and Wales.

Scope

This procedure is available to deal with any premises which are in such a state as to be prejudicial to health (i.e. 'defective state'), but where unreasonable delay in dealing with the situation would occur if the normal provisions for dealing with statutory nuisances in section 80 of the Environmental Protection Act 1990 were used – see FC120 (section 76(1)).

Person responsible

Notices are to be served either:

(a) on the person responsible for the nuisance; or
(b) where the person in (a) cannot be found or the nuisance has not yet occurred, on the owner or occupier of the premises on which the nuisance arises; or
(c) where the nuisance arises from a structural defect, on the owner of the premises (section 27(1) and Environmental Protection Act 1990, section 80(2)).

Notices

The provisions of section 99 relating to the content and enforcement of notices do not apply here.

Notices must be in writing and, in addition to giving the local authority's intention to remedy the defects, must specify the actual defects which it intends to remedy (section 76(1) and section 92). If a counter-notice is served by the recipient of the local authority notice but works are not begun within a reasonable time or progress towards completion is not reasonable, the local authority may execute the works itself. 'Reasonable' is not defined and the period allowed by the local authority will therefore need to take account of all the relevant circumstances (section 76(2) and (3)).

FC57 Defective premises – Section 76 Building Act 1984

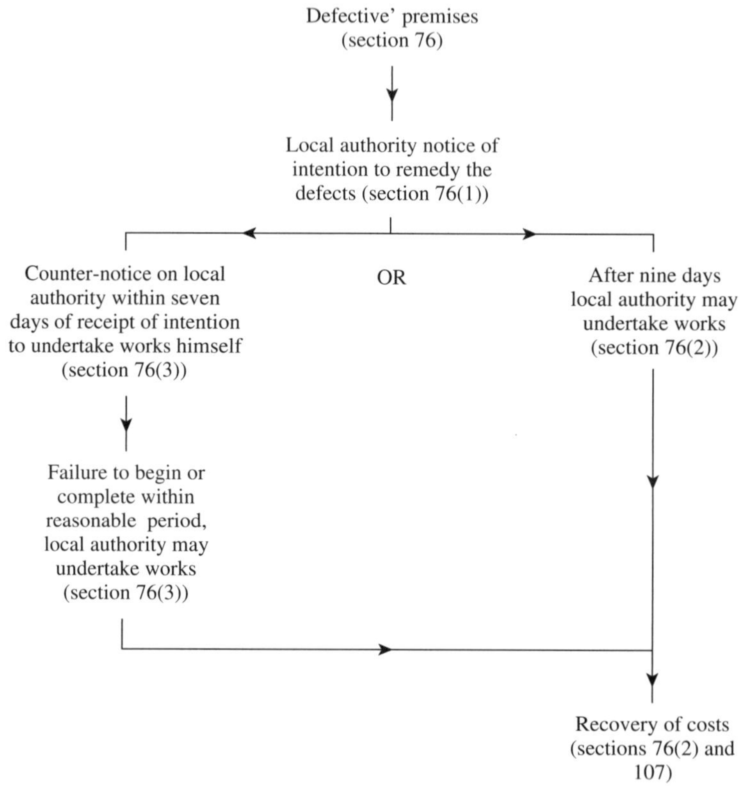

Notes

1. For normal statutory nuisance procedure relating to defective premises, see FC120.
2. There is no provision for appeal against notices served under section 76 and there is no penalty for non-compliance with the notice.

Recovery of costs

In any proceedings by the local authority to recover costs of carrying out works in default, the court must have regard to:

(a) whether the local authority was right in concluding that the premises were in such a state as to be prejudicial to health or a nuisance;
(b) whether unreasonable delay would have occurred if the normal procedures for dealing with statutory nuisances had been followed;
(c) following a counter-notice, whether the defendant had been given a reasonable time to begin or complete the works; and
(d) whether any costs ought to be borne wholly or partly by someone other than the defendant (section 76(4)).

RUINOUS AND DILAPIDATED BUILDINGS

Reference

Building Act 1984, section 79.

Extent

These provisions apply to England and Wales.

Scope

The procedure is applicable to any buildings or structures which are seriously detrimental to the amenities of the neighbourhood because of their ruinous or dilapidated condition (section 79(1)), and to rubbish resulting from the demolition of a building which renders the site seriously detrimental to the amenities of the neighbourhood (section 79(2)).

Notices

The local authority must serve notices in writing on the owner of the building or structure concerned and may require either:

(a) repair or restoration; or
(b) if the owner so elects, demolition of the building or structure or parts thereof and the removal of rubbish resulting from demolition (section 79(1)); or
(c) removal of the rubbish already existing (section 79(2)).

Notices must indicate both the nature of works of repair or restoration and the works of demolition and removal of rubbish or material (section 79(3)). The provisions of sections 99 and 102 apply with regard to appeals against and the enforcement of these notices (pages 447–449).

FC58 Ruinous and dilapidated buildings – Section 79 Building Act 1984

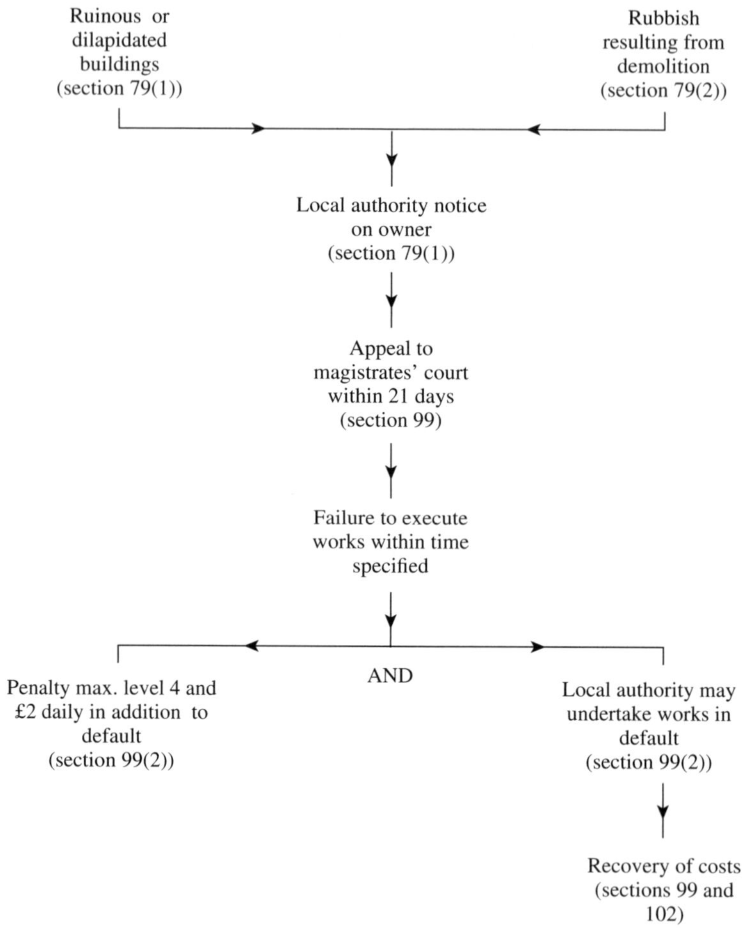

DANGEROUS BUILDINGS

Reference

Building Act 1984, sections 77 and 78.

Extent

These provisions apply in England and Wales.

Scope

The procedure may be used where a building or structure, or part thereof, is:

(a) in such a state; or
(b) is used to carry such loads

as to be dangerous (sections 77(1) and 78(1)).

Emergency measures

These may be taken where immediate action is required in the opinion of the local authority to remove the danger (section 78(1)).

The measures which may be taken are those necessary to remove the danger (section 78(1)). Where practicable, prior notice must be given to the owner and occupier (section 78(2)).

Court order

Where there is no call for immediate action, the local authority may apply to a magistrates' court and the court may order the owner to:

(a) execute necessary works to remove the danger; or
(b) if the owner elects, to demolish the building or structure and remove resulting rubble or rubbish; or
(c) where the danger arises from the overloading of the building or structure, restrict the use until the court withdraws or modifies that restriction following works having been carried out (section 77(1)).

Recovery of expenses

The local authority's expenses in executing a court order are recoverable from the owner (section 77(2)). Where the emergency provisions have been used, the local authority may recover its costs, but in any dispute as to the recovery of these costs, the court may consider whether the local authority was justified in taking emergency action as opposed to an application to the court for an order under section 77 (section 78(5)).

FC59 Dangerous buildings – Sections 77 and 78 Building Act 1984

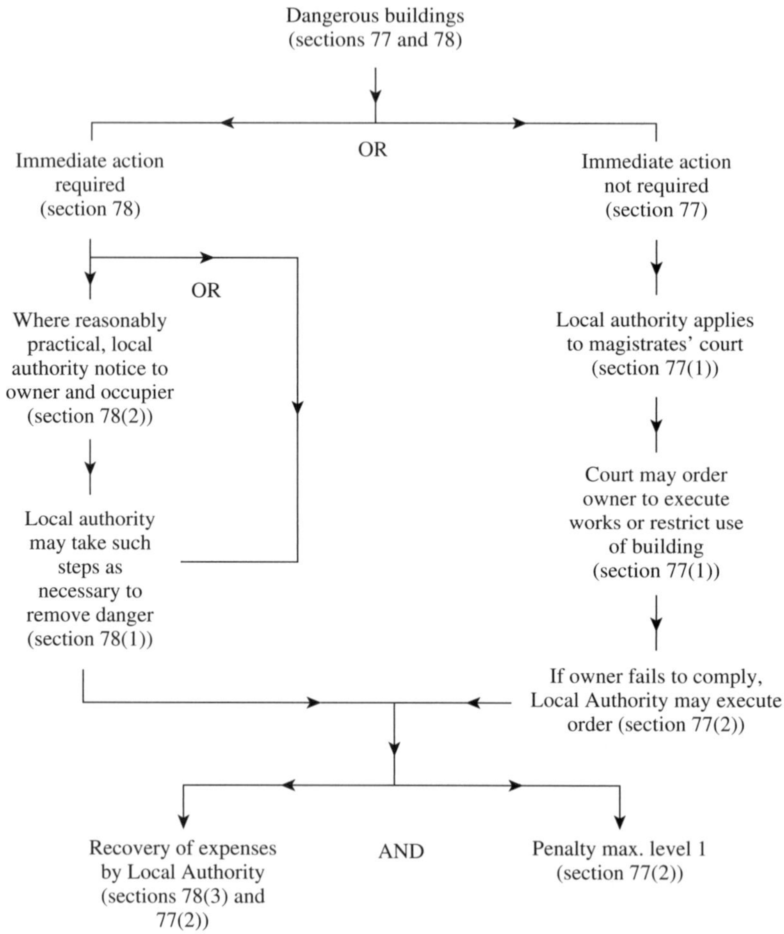

Note

1. This procedure is subject to the provisions of the Planning (Listed Buildings and Conservation Areas) Act 1990 in relation to listed buildings etc.

Recovery of expenses connected with the fencing off of a building or structure or arranging for it to be watched are not recoverable after:

(a) the danger has been removed by emergency action by the local authority; or
(b) after works required under section 77 have been carried out (section 78(4)).

PROTECTION OF BUILDINGS

Reference

Local Government (Miscellaneous Provisions) Act 1982, sections 29–32 inclusive.

Extent

This procedure applies in England and Wales.

Scope

This procedure applies to any building:

(a) which is unoccupied; or
(b) where the occupier is temporarily absent;

and which is:

(a) not effectively secured against unauthorised entry; or
(b) likely to become a danger to public health (section 29(1)).

Works

The local authority may undertake any works necessary to prevent unauthorised entry or to prevent the building becoming a danger to public health. If either:

(a) works are immediately necessary to prevent it becoming a danger to public health; or
(b) the name and address of the owner, or the whereabouts of the occupier are not reasonably practical to obtain,

a notice of intention is not required. In all other cases, at least 48 hours' written notice must be given (section 29(6) and (8)).

Local authority notices

(a) **Person.** The notice is to be served on either the owner or the occupier (section 29(6)) or on British Railways Board or any statutory undertaker in respect of their operational land (section 30(2)).

FC60 Protection of buildings – Sections 29 to 32 Local Government (Miscellaneous Provisions) Act 1982

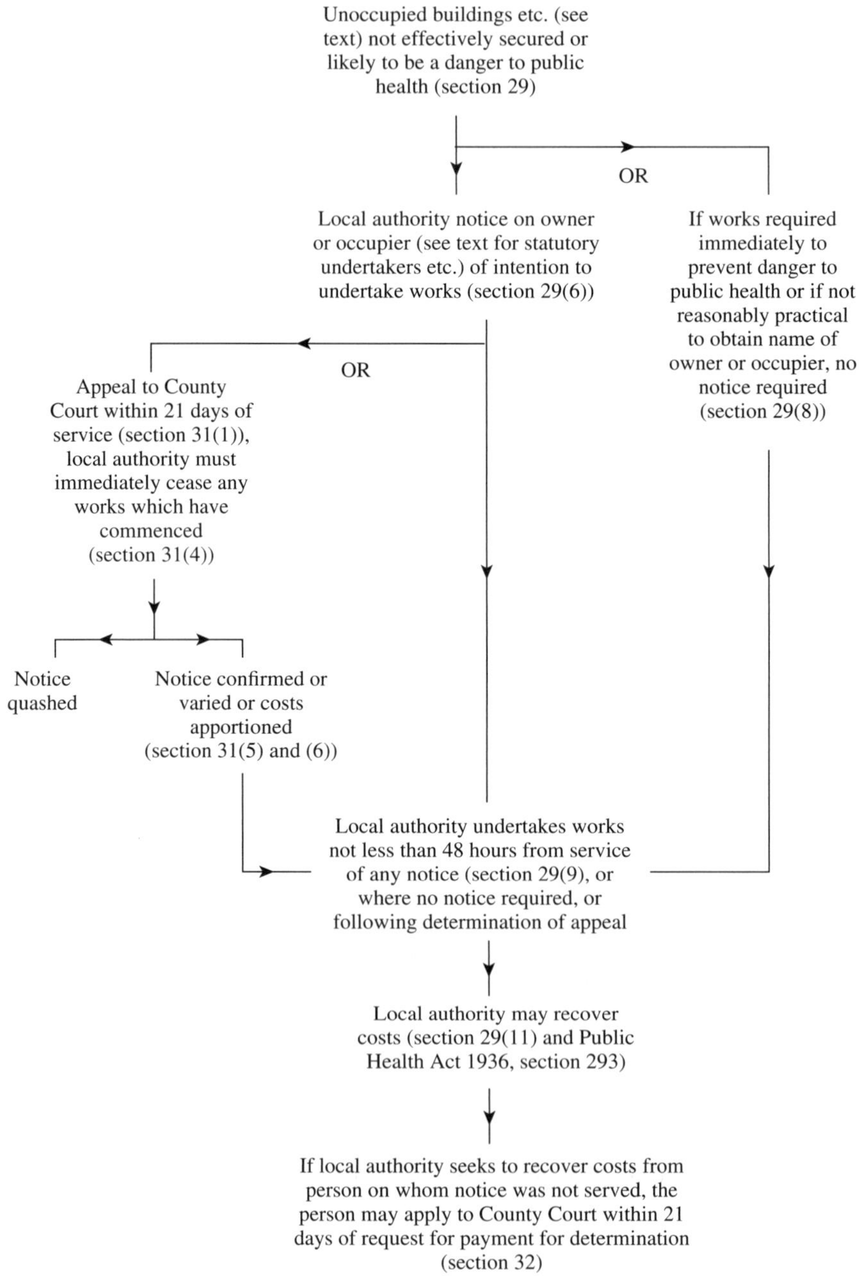

Notes

1. No offences are committed under this procedure.
2. For operational land of British Railways Board or any statutory undertaker, in carrying out works the local authority must comply with any reasonable request which British Railways Board or the statutory undertakers may impose for the protection or safety of their undertaking (section 30).
3. The procedure can be applied to any building in the specified condition, including ones subject to orders under the Housing Acts.

(b) **Content.** The notice must specify the works which the local authority intends to carry out (sections 29(7) and 30(3)).
(c) **Service.** This procedure is not incorporated in the provisions of either the Public Health Acts or the Housing Acts and therefore service should be by one of the methods specified in section 233 of the Local Government Act 1972 (page 25).

Appeals

A person in receipt of the local authority notice of intention may appeal to the County Court within 21 days of service on any of the following grounds:

(a) the works specified are not authorised by the section 29 procedure;
(b) the works are unnecessary;
(c) it was unreasonable for the local authority to undertake the works (section 31(1), (2) and (3)).

In the event of an appeal, the local authority must stop any works which it has commenced and await the determination of the court. If the works had been completed, the court may still determine the position regarding recovering of costs (section 31(4)).

The court may confirm, vary or quash the notice and make provision in its order relating to recovery of costs by the local authority (section 31(5) and (6)).

Persons not in receipt of a local authority notice but from whom costs are being recovered may appeal to the County Court within 21 days of the request for payment and ask the court for a declaration that either:

(a) the works were unnecessary; or
(b) it was unreasonable for the local authority to undertake them (section 32).

Power of entry

A person authorised by the local authority may enter:

(a) the building on which works are to be undertaken;
(b) any land appurtenant to that building;
(c) any other land if it appears to be unoccupied and where it would be impossible to undertake works without entering it (section 29(10)).

There does not appear to be a power of entry in respect of the assessment of the building before a decision about the carrying out of works is made.

Definitions

Building includes structure (section 29(3)).

EMPTY DWELLING MANAGEMENT ORDERS

References

Housing Act 2004, sections 132 to 138.
Housing (Empty Dwelling Management Orders) (Prescribed Exceptions and Requirements) (England) Order 2006 as amended.
Housing (Management Orders and Empty Dwelling Management Orders) (Supplemental Provisions) (England) Regulations 2006.

Extent

These provisions apply to England only but there are similar provisions for Wales.

Interim empty dwelling management orders

A local authority can apply to the Residential Property Tribunal to issue an interim empty dwelling management order (EDMO) with the consent of the relevant proprietor to ensure the dwelling becomes and continues to be occupied (section 132 (2)). The local authority must give the relevant proprietor three months' notice before applying to the Residential Property Tribunal.

The tribunal can confirm or vary the order. It must be satisfied that the dwelling is wholly unoccupied, and the relevant proprietor is not a public sector body.

The local authority must make reasonable efforts:

(a) to notify the relevant proprietor that they are considering making an interim EDMO in respect of the dwelling; and
(b) to ascertain what steps (if any) he is taking, or is intending to take, to secure that the dwelling is occupied.

The local authority must also take into account the rights of the relevant proprietor of the dwelling and the interests of the wider community.

The local authority may make an interim EDMO despite any pending appeal against the order of the tribunal. Before making an interim EDMO, the local authority must:

(a) serve a copy of the proposed order on each relevant person; and
(b) consider any representations.

The notice must state that the local authority is proposing to make an interim EDMO and set out:

(a) the reasons for making the order;
(b) the main terms of the proposed order (including those of the management scheme to be contained in it); and
(c) the end of the consultation period.

The local authority must consider any representation made about the management order before they consider whether to apply to the Residential Property Tribunal.

Following representations, the notice must set out:

(a) the proposed modifications;
(b) the reasons for them; and
(c) the end of the consultation period (section 133).

Details of the evidence that local housing authorities must provide for the Residential Property Tribunal are contained in the Housing (Empty Dwelling Management Orders) (Prescribed Exceptions and Requirements) (England) Order 2006 as amended.

A Residential Property Tribunal may authorise a local authority to make an interim EDMO if it is satisfied:

(a) that the dwelling has been wholly unoccupied for at least two years or such longer period as may be prescribed;
(b) that there is no reasonable prospect that the dwelling will become occupied in the near future;
(c) that, if an interim order is made, there is a reasonable prospect that the dwelling will become occupied;
(d) that the authority has made reasonable efforts to notify the relevant proprietor of the making of the order and what the proprietor intends to do with the property;
(e) that any prescribed requirements have been complied with; and
(f) that the case does not fall within one of the prescribed exceptions.

In deciding whether to authorise a local authority to make an interim EDMO, the tribunal must take account of:

(a) the interests of the community; and
(b) the effect that the order will have on the rights of the relevant proprietor and may have on the rights of third parties (section 134).

Local authority's duties once interim EDMO in force

The local authority must take such steps as they consider appropriate for securing that the dwelling becomes and continues to be occupied and making proper management arrangements, including insurance, for the dwelling pending the making of a final EDMO or the revocation of the interim EDMO.

If the local authority concludes that there are no steps which they could appropriately take, the authority must either:

(a) make a final EDMO in respect of the dwelling; or
(b) revoke the order without taking any further action.

The owner would receive the rental minus the council's costs and benefit from any repairs when they retake control.

If none of these exemptions apply and the tribunal rules in favour of the council, it could take control of the property for seven years.

Exemptions

Many properties will be exempt – for example, those from where someone has gone temporarily abroad or is in hospital; where it is a holiday home; where the proprietor is a member of the armed forces and is serving abroad; or within six months of probate being granted where the home owner has died (Housing (Empty Dwelling Management Orders) (Prescribed Exceptions and Requirements) (England) Order 2006 as amended).

Final empty dwelling management orders

A final EDMO is an order made, in succession to an interim EDMO or a previous final EDMO, for the purpose of securing that a dwelling is occupied (section 132(3)).

A local authority may make a final EDMO to replace an interim EDMO if:

(a) they consider that, unless a final EDMO is made in respect of the dwelling, the dwelling is likely to become or remain unoccupied;
(b) where the dwelling is unoccupied, they have taken all such steps as it was appropriate for them to take under the interim EDMO with a view to securing the occupation of the dwelling.

The local authority can make another final EDMO if the first one is not working.

The same criteria and provisions for making an interim EDMO apply to a final EDMO.

Local housing authority's duties once final EDMO in force

The duties are the same as for interim EDMO but the authority must from time to time review:

(a) the operation of the order and in particular the management scheme contained in it;
(b) whether, if the dwelling is unoccupied, there are any steps which they could appropriately take under the order for the purpose of securing that the dwelling becomes occupied; and
(c) whether keeping the order in force in relation to the dwelling is necessary to secure that the dwelling becomes or remains occupied.

If, on a review, the local authority considers that any variations should be made, they must proceed to make those variations. The review can be done at the local authority's own volition or on application from a relevant proprietor.

If the dwelling is unoccupied and on a review the local authority concludes that either:

(a) there are no steps which they could appropriately take; or
(b) keeping the order in force is not necessary,

they must proceed to revoke the order (section 137).

FC61 Interim and final empty property management orders – Sections 132 to 138 Housing Act 2004

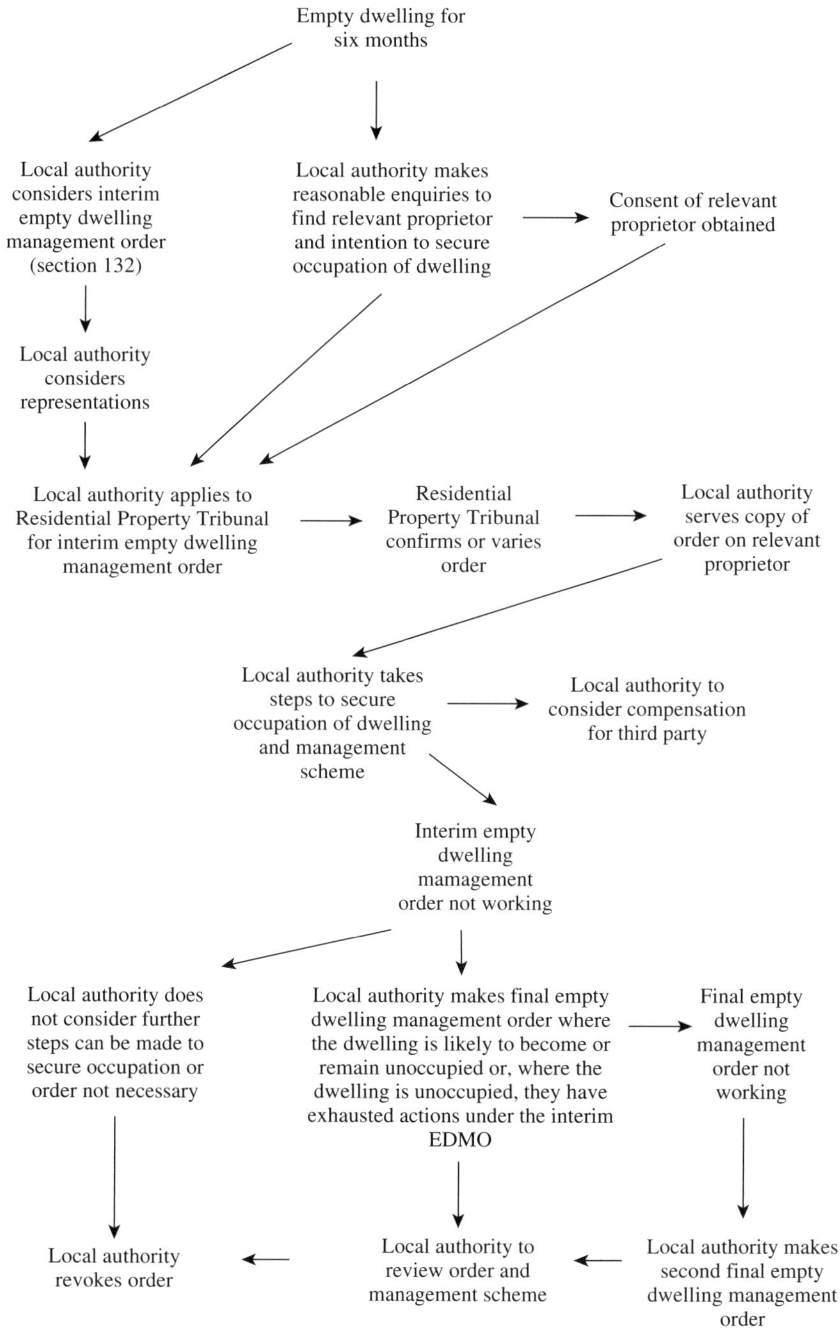

Compensation payable to third parties

Before making an interim or final EDMO, the tribunal or local authority must consider whether compensation should be paid by them to any third party in respect of any interference in consequence of the order with the rights of the third party (section 134).

A third party may apply to a Residential Property Tribunal for an order requiring the local authority to pay to him compensation in respect of any interference in consequence of the order with his rights in respect of the dwelling.

The tribunal may make an order requiring the authority to pay to the third party an amount by way of compensation for such interference.

If a third party requests them to do so at any time, the local authority must consider whether an amount by way of compensation should be paid to him in respect of any interference in consequence of a final EDMO with his rights.

The authority must notify the third party of their decision as soon as practicable.

Where the local authority decides that compensation ought to be paid to a third party, they must vary the management scheme contained in the order to specify the amount of the compensation to be paid and to make provision for its payment.

Definitions

Dwelling means:

(i) a building intended to be occupied as a separate dwelling; or
(ii) a part of a building intended to be occupied as a separate dwelling which may be entered otherwise than through any non-residential accommodation in the building.

Relevant proprietor, in relation to a dwelling, means:

(i) if the dwelling is let under one or more leases with an unexpired term of seven years or more, the lessee under whichever of those leases has the shortest unexpired term; or
(ii) in any other case, the person who has the freehold estate in the dwelling.

Third party, in relation to a dwelling, means any person who has an estate or interest in the dwelling, other than the relevant proprietor and any person who is a tenant under a lease (section 132(4)).

Wholly unoccupied means that no part is occupied, whether lawfully or unlawfully (section 133)

Other powers in respect of empty dwellings

Local authorities have a range of powers it can use to deal with empty properties.

Section 77 of the Building Act 1984 deals with making the building safe.

Section 79 of the Building Act 1984 enables councils to deal with ruinous and dilapidated buildings and neglected sites 'in the interest of amenity'.

Section 215 of the Town and County Planning Act 1990 allows councils to take action over unsightly land or the external appearance of a property.

Section 29 of the Local Government (Miscellaneous Provisions) Act 1982 allows councils to board up or make secure empty properties to stop people getting into the property or to stop the property being a danger to public health.

Additional powers that the council can use include those given by the Law of Property Act 1925, which enables councils to force the sale of empty properties where the local authority is owed money by the owner.

Empty dwelling management orders can be used on properties empty for six months or more. Here the council can take over the management of a property, essentially acting as the landlord, bringing the property back into use. These powers can only be used with the agreement of the owner.

The strongest power that the local authority can use is that of compulsory purchase where the local authority buys the property without the agreement of the owner. The local authority needs to apply to the Secretary of State in order to carry out such a purchase. They have to show that there is a public interest in compulsorily buying the property and that the other methods of bringing the property into use have not worked.

DEMOLITION OF BUILDINGS

Reference

Building Act 1984, sections 80–83 inclusive.

Extent

These provisions apply to England and Wales.

Scope

The notification procedure applies to the demolition of the whole or part of any building except:

(a) of an internal part of an occupied building which is to continue to be occupied;
(b) where the cubic content by external measurement does not exceed 1,750 cubic feet;
(c) of greenhouses, conservatories, sheds or prefabricated garages forming part of a larger building;
(d) agricultural buildings, unless it is contiguous to a non-agricultural building;
(e) demolition of a house subject to a demolition order under the Housing Act 1985 (section 80(1)).

FC62 Demolition of buildings – Sections 80 to 83 Building Act 1984

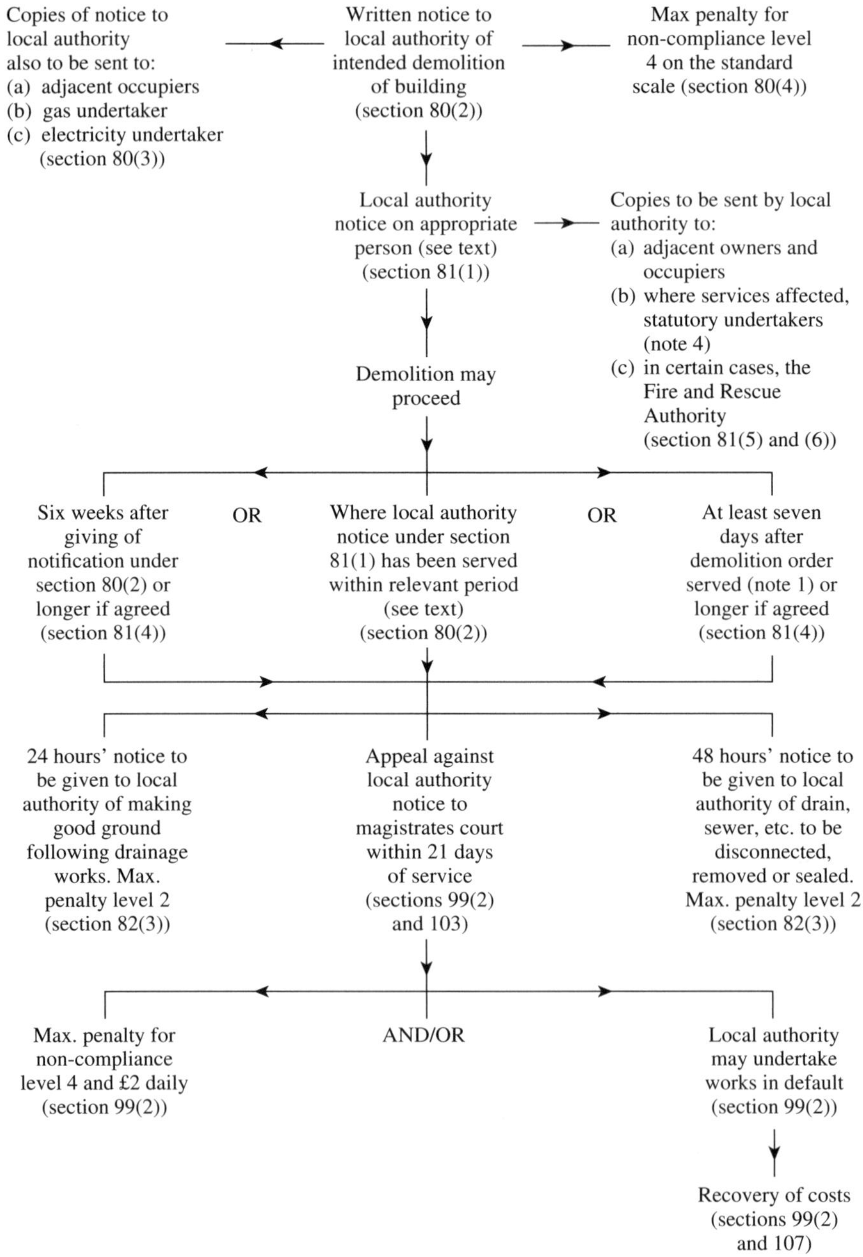

Notes

1. For demolition orders, see FC63.
2. For further procedure relating to rubbish following demolition, see FC58.
3. The section 99 procedure applies here for appeals against and the enforcement of local authority notices (pages 447–449).
4. The statutory undertakers include the water undertakers.

Notification to a local authority

The person undertaking the demolition of any building within the scope of this procedure must give prior notification to the local authority in writing indicating the building concerned and the works of demolition which are intended. Copies of this notice must be sent or given to:

(a) occupiers of adjacent buildings;
(b) the gas supplier(s);
(c) the electricity supplier(s).

Demolition must not commence until the local authority has served its notice under section 81 or the 'relevant period' (below) has expired (section 80(2) and (3)).

Local authority notices

In addition to those situations where notification must be made to the local authority, notices may also be served by the local authority in the situations identified in (i), (ii) and (iii) under (a) below.

(a) **Person.** These notices are to be served on either:
    (i) a person upon whom a demolition order under the Housing Act 1985 has been served; or
    (ii) a person not to be intending to comply with an order under section 77 of the Building Act 1984 (dangerous structure etc.) or a notice under section 79 of the Building Act 1984 (ruinous and dilapidated buildings, etc.); or
    (iii) a person who is intending, or has begun, a demolition to which section 80 of the Building Act 1984 applies ('Scope' above) (section 81(1)).
(b) **Relevant period.** This is the period within which the local authority is required to serve its notice and this is:
    (i) where a person has served a notification on the local authority under section 80, within six weeks from the giving of the notice to the local authority, or such longer period as may be agreed by the notifier in writing; and
    (ii) in relation to a demolition order under the Housing Act 1985, not less than seven days after service of the order, or such longer period as the person on whom the order was served may allow in writing (section 81(4)).
(c) **Contents.** The local authority notice may require all or any of the following:
    (i) shoring up of the adjacent buildings;
    (ii) weather-proofing of exposed surfaces of adjacent buildings and repairing or making good damage;
    (iii) removal of material from site;
    (iv) disconnection, sealing and removal of drains and sewers;
    (v) making good ground surface following disconnection or removal of drains and sewers;

  - (vi) arrangements with the statutory undertakers (including water undertakers) for the disconnection of gas, electricity and water supplies;
  - (vii) arrangements for the burning of structures and materials on site with the Fire and Rescue Authority;
  - (viii) any other steps considered necessary for the protection of the public and the preservation of public amenity (section 82(1)).
- (d) **Copies.** The local authority is required to send copies of its notices to:
  - (i) owners *and* occupiers of adjacent buildings;
  - (ii) if the notice requires disconnection of gas, electricity or water supplies, the statutory undertaker (including water undertakers) concerned;
  - (iii) if the notice relates to the burning of structures or materials on site, the Fire and Rescue Authority (section 81(5) and (6)).
- (e) **Enforcement of local authority notices.** These notices are subject to section 99 procedures (pages 447–449) and the local authority is, therefore, able to undertake works in default and recover costs as well as prosecuting for non-compliance (section 82(6)).
- (f) **Appeals.** Appeal against local authority notices served under section 81 may be made under the grounds set out in Building Act 1984, section 102 (page 447) and in addition on the following grounds:
  - (i) in relation to the shoring up of an adjacent building, that the owner is not entitled to support by the building to be demolished and he, therefore, should pay for, or contribute towards, the costs; and
  - (ii) in relation to the weather-proofing of an adjacent building, that the owner should pay for, or contribute towards the costs (section 83(2)).

  In both cases, the appellant must serve copies of his notices of appeal on the other persons concerned (section 83(3)).

DEMOLITION ORDERS

References

Housing Act 1985, part 9, sections 265 to 275 (as amended by the Housing Act 2004, section 46 and schedule 15, paragraphs 13–21).
Enforcement Guidance, ODPM February 2006, part 5, paragraphs 5.44–5.46.

Extent

These provisions apply to England and Wales.

Scope

Where a category 1 hazard exists on any residential premises, the making of a demolition order is one of the courses of action available to a local authority in complying with its duty under section 5 to take the most appropriate course of action (section 265(1) and (2)).

FC63 Demolition orders (DOs) – Sections 265 to 275 Housing Act 1985

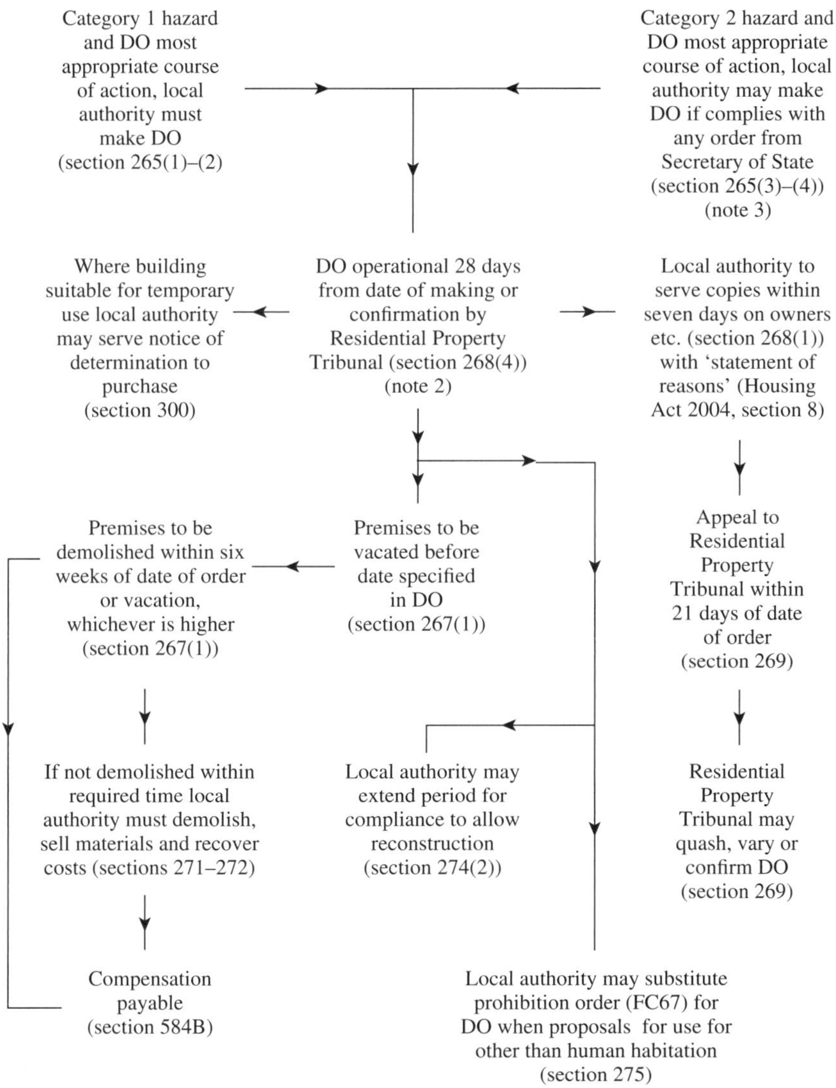

Notes

1. Section numbers relate to the Housing Act 1985 unless indicated otherwise.
2. Local authority may undertake cleansing of vermin – see text (section 273).
3. No order has yet been made by the Secretary of State.

In relation to a category 2 hazard, the local authority may use its discretion to make a demolition order as the most appropriate action but only where the circumstances are specified in an order made by the Secretary of State. No such orders have yet been made (section 265(3) and (4)).

Demolition orders are not to be made where the premises are subject to a management order or are listed buildings (sections 265(5) and 304).

In deciding whether to make a demolition order, the local authority should:

- take into account the availability of local accommodation for rehousing the occupants;
- take into account the demand for, and sustainability of, the accommodation if the hazard was remedied;
- consider the prospective use of the cleared site;
- consider the local environment, the sustainability of the area for continued residential use and the impact of a cleared site on the appearance and character of the neighbourhood (Enforcement Guidance, paragraph 5.44).

Content of demolition orders

Demolition orders must require:

- vacation within a period specified in the order but not less than 28 days from the date of operation of the order,
- the demolition of the premises within six weeks of the date of operation or the date of vacation, whichever is the longer. The local authority may agree a longer period (section 267(1)).

Service of demolition orders

Copies of the demolition order must be served within seven days of the date on which it was made on:

(a) owners and occupiers;
(b) persons authorised to permit occupation of it; and
(c) mortgagees (section 268(1)).

The service of these copies of the demolition order are subject to the provisions of section 246 of the Housing Act 2004 and not section 617 of the Housing Act 1985. Service on occupiers is deemed to have been properly made if the copy is posted on a conspicuous part of the premise (section 268(3) and (5)).

Service of copies of demolition orders is to be accompanied by a 'statement of reasons' for the local authority decision to make a demolition order (Housing Act 2004, section 8 – see pages 253, 283).

Operation of a demolition order

A demolition order becomes operative 28 days from the date on which it was made or the date when the appeal is determined by a modification or confirmation (section 268(4)).

Appeals

Any person aggrieved by a demolition order, other than an occupier under a lease or agreement with an unexpired term of three years or less, may appeal to the Residential Property Tribunal within 28 days of the order being made. The Residential Property Tribunal may vary, quash or confirm the order (section 269).

The grounds of appeal include that the demolition order was not the most appropriate course of action, which would have been an improvement notice, a hazard awareness notice, a prohibition order or declaring the area to be a clearance area (section 269A).

Cleansing before demolition

Before a demolition order becomes operative, the local authority may give notice to the owner that they intend to cleanse the property from vermin after operation of the order and before demolition takes place. Demolition may not proceed until this work is completed and the local authority has served a notice to this effect. However, the owner may, after vacation and by notice, request the local authority to carry out the treatment within 14 days from this notice and may demolish after the expiry of that period, whether or not the work has been done (section 273).

Enforcement

If the premises are not demolished within the time allowed, the local authority is required to do so, sell the materials and recover their expenses (sections 271 and 272).

If it is intended to carry out works that will result in the elimination of the category 1 hazard, the local authority may extend the period within which demolition is required to allow for the carrying out of the works (section 274(2)).

It is possible to substitute a demolition order with a prohibition order (see FC68), where proposals are made for the use of the premises other than for human habitation (section 275).

Compensation

Compensation is payable by the local authority to every owner of premises subject to a demolition order (section 584A) but is repayable where the demolition order is revoked to permit reconstruction under section 274 (section 584B).

Determinations to purchase

In circumstances where the local authority would be required to make a demolition order and it appears to them that the building can be rendered capable of providing accommodation which is adequate for the time being, the local authority may purchase the premises, by agreement or compulsory purchase order, instead of making a demolition order (section 300).

CLEARANCE AREAS (CAs)

References

Housing Act 1985, sections 289 to 298, as amended by the Housing Act 2004.
Land Compensation Act 1973, section 39, as amended.
Housing (Prescribed Forms) Regulations 1990 as amended.
Enforcement Guidance, ODPM February 2006, part 5, paragraphs 5.47–5.49.

Extent

These provisions apply to England and Wales.

Scope

A clearance area is an area which is to be cleared of all buildings in accordance with this procedure (section 289 (1)).

A local authority may declare a clearance area if either:

1. a category 1 hazard exists in each of the residential buildings in an area and if any other buildings in that area are dangerous or harmful to the health or safety of the inhabitants. This is one of the options available to local authorities in undertaking their duty to take the most appropriate action under section 5 of the Housing Act 2004; or
2. the local authority is satisfied that the residential buildings in the area are dangerous or harmful to the health or safety of the inhabitants because of their bad arrangement, or the narrowness or bad arrangement of the streets, and that any other buildings are dangerous or harmful to health. This is a discretionary power; or
3. a category 2 hazard exists in each of the residential premises and any other buildings to be included are dangerous or harmful to health and safety. This is also a discretionary power (section 289 as amended).

Before declaration of a clearance area, the local authority should consider the desirability of clearance in the context of the proposals for the wider neighbourhood of which the area forms part. In deciding to make a clearance area for an area in which hazardous dwellings are situated, the local authority should have regard to:

- the likely long-term demand for residential accommodation;
- the degree of concentration of dwellings containing serious and intractable hazards within the area;
- the density of the buildings and street pattern around them;
- the overall availability of housing accommodation in the wider neighbourhood in relation to housing needs and demands;
- the proportion of dwellings free of hazards and other, non-residential, premises in sound condition which would also need to be cleared to produce a suitable site;

FC64 Clearance areas (CAs) – Sections 289 to 298 Housing Act 1985 as amended

All residential buildings are dangerous or harmful to health or safety because of bad arrangement etc. and other buildings are dangerous or harmful to health or safety – local authority have discretion to make clearance area (section 289(2ZA))

OR

Category 1 hazards exist in each dwelling and other buildings are dangerous or harmful to health or safety – clearance area option to comply with duty under section 5 of the (Housing Act 2004 (section 289(2))

OR

Category 2 hazard exists in each dwelling and other buildings are dangerous or harmful to health or safety and meets criteria set out in Secretary of State Order – local authority have discretion to make Clearance Area (section 289(2ZB)) (note 1)

Local authority to invite representations within at least 28 days by:

Local authority to serve notice of intention (using Prescribed Form 20) on all with interest in dwellings and buildings to be included (section 289(2B)(a) and (2C))

AND

Local authority to inform owner occupiers (section 289(2B)(b) and (2D))

AND

Proposal to be published in at least two local newspapers (section 289(2B)(c) and (2E))

Local authority to consider representations (section 289(2F)) (note 3)

Local authority satisfied there is sufficient, suitable accommodation available for displaced persons (section 289(4)(a))

AND

Local authority to take account of matters laid out in the Enforcement Guidance – see text (paragraph 5.48)

AND

Local authority satisfied that sufficient resources are available to deal with clearance area (section 289(4)(b))

Area to be defined on a map (section 289(3)(a))

(continued on next page)

Notes

1. No order has yet been made.
2. The Secretary of State has no powers of intervention at this stage; these powers come with any subsequent compulsory purchase order.
3. After consideration of representations, the local authority may declare a clearance area as proposed, or with the exclusion of some residential buildings, or may decide not to proceed and take other action.

FC64 Clearance areas (CAs) – continued

(Continued from previous page)

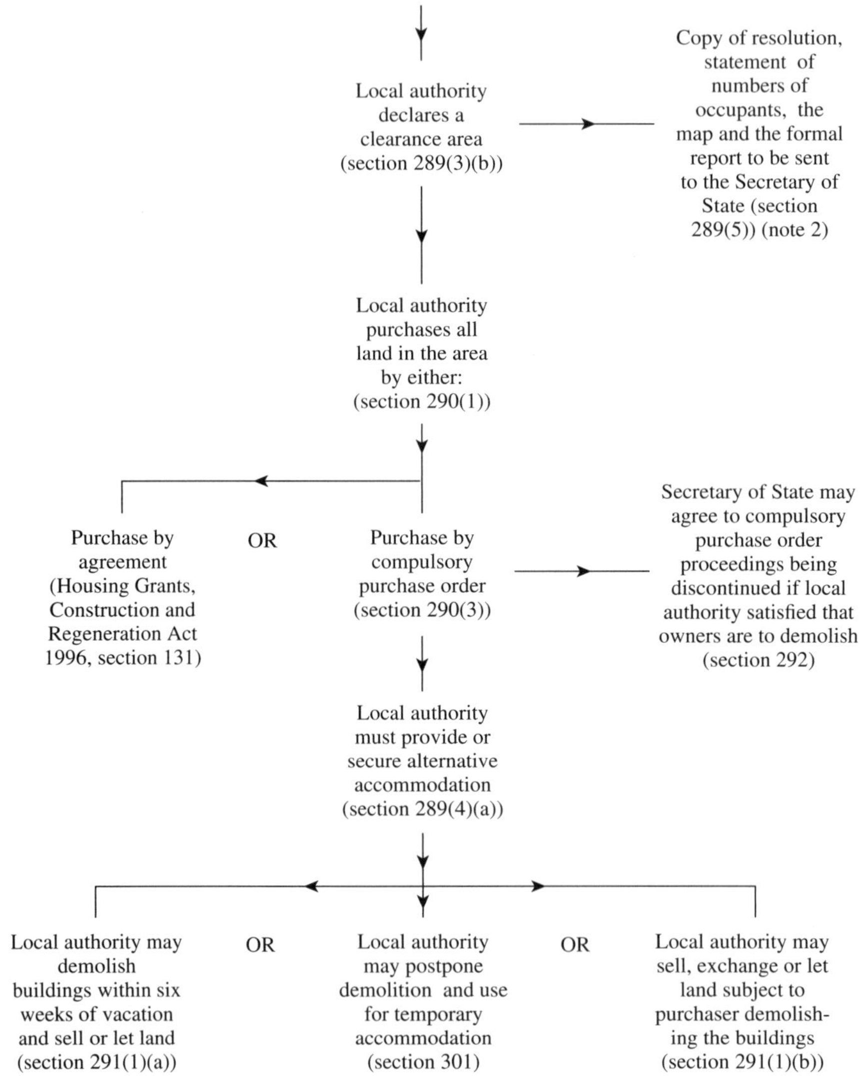

- whether it would be necessary to acquire land surrounding or adjoining the area and whether added land can be acquired by agreement;
- the existence of any listed buildings – listed buildings should only be included in a clearance area in exceptional circumstances and only when building consent has been given;
- the results of statutory consultation – see below;
- the arrangements necessary for the rehousing of displaced occupants and the extent to which occupants are satisfied with those arrangements;
- the impact of clearance on, and the scope for relocating, commercial premises;
- the suitability of the proposed after-use of the site, the needs of the wider neighbourhood and the socio-economic benefits which after-use would bring;
- the degree of support from local residents;
- the extent to which such after-use would create investment in the area (Enforcement Guidance, paragraph 5.48).

Consultation

Before making a clearance area, the local authority must consult with those people who will be directly affected. This is to be done by service of a notice on every person who has an interest, including freeholders and sub-lease holders, and on all residential occupiers. A notice must also be placed in at least two local newspapers. In both cases, the local authority is to invite representations within a period of not less than 28 days (section 289).

Options after consultation

The local authority must consider all representations made during the stated period and must then take one of the following options:

1. Declare the area to be a clearance area.
2. Declare a clearance area with the exclusion of certain residential buildings.
3. Not declare a clearance area and consider other options (in relation to category 1 hazards, the local authority is still required to take action under section 5 of the Housing Act 2004) (section 289(2F)).

A map defining the proposed clearance area must be prepared before the resolution declaring the clearance area is passed (section 289(3)(a)).

Implementation

A copy of the resolution together with a statement of numbers of occupants, the map and the formal report to the council are to be sent to the Secretary of State, although there is no power for him to intervene at this stage.

The clearance area is implemented by the purchase of all land in the area by the local authority, either by agreement or by the use of a compulsory purchase order. The local authority may also acquire land that surrounds the area and is needed to secure an area of convenient shape and dimension and land necessary for satisfactory development.

Compulsory purchase orders are made under section 290 and are regulated by the Acquisition of Land Act 1981 and regulations made under it. The compulsory purchase order will require Secretary of State confirmation or modification before proceeding (sections 290– 291).

Temporary use of residential buildings

Although purchased for demolition, if the local authority considers that any of the residential buildings can provide adequate temporary accommodation, with or without the carrying out of works, it may postpone demolition for such period as may be determined. Other buildings in the area may have demolition postponed where:

- the demolition of the residential buildings has been postponed;
- they are required to support a residential building;
- there is some other special reason connected with the provision of temporary accommodation in the area (section 301).

Demolition

Having secured the land by agreement or compulsory purchase order, the local authority must arrange for the vacation and demolition of buildings within six weeks of vacation or such other period as the local authority determines to be reasonable. The land may subsequently be sold, let or exchanged. Alternatively, the local authority may sell or exchange the land with a condition that the purchaser undertakes the demolition (section 291).

Compensation

Open market value is payable for all dwellings purchased via a compulsory purchase order. Home loss and disturbance payments are also payable (Land Compensation Act 1973, sections 29 and 37). The Regulatory Reform (Housing Assistance) Order 2002 makes provision for a local authority to provide financial assistance where there has been demolition.

HAZARD AWARENESS NOTICES

References

Housing Act 2004, sections 28 and 29 and schedule 1, part 2.
Enforcement Guidance, ODPM February 2006, part 5.

Extent

These provisions apply to England and Wales.

FC65 Hazard awareness notices (HANs) – Sections 28 and 29 Housing Act 2009

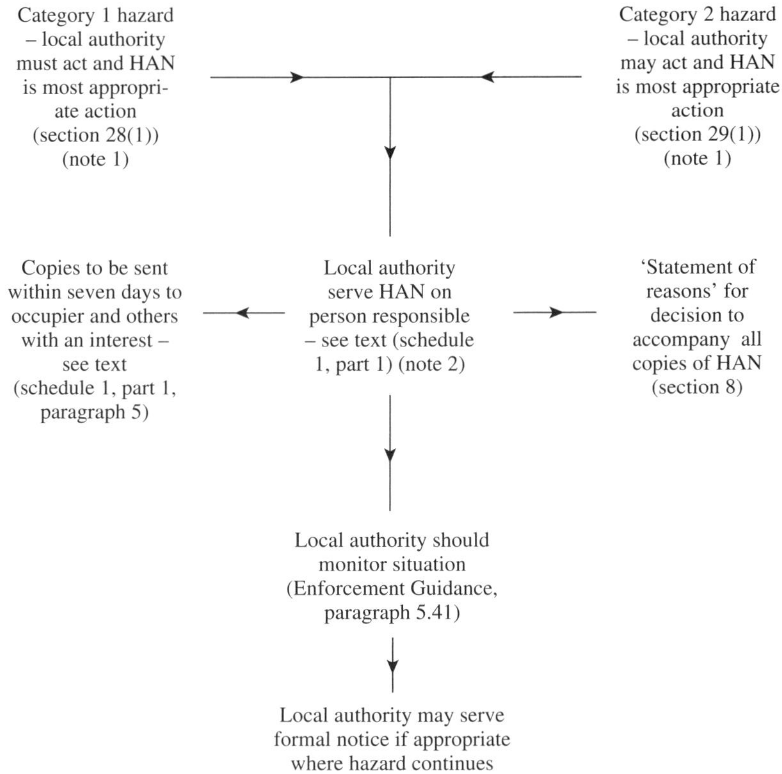

Notes

1. Where both category 1 and category 2 hazards exist, the notices under sections 28 and 29 may be combined (section 29(6)).
2. There is no appeal since there is no enforcement process.
3. There is no requirement for the HAN to be registered as a local land charge.

Scope

Hazard awareness notices may be served where they are considered to be the most appropriate action in response to both the local authority duty for category 1 (section 28(1)) and its discretionary powers for category 2 (section 29(1)) hazards, in both cases where there is no management order in place – see FC78 and FC79.

The Enforcement Guidance considers that section 29 would be a reasonable response to a less serious hazard, whilst section 28 might be used where works of improvement or prohibition of use may not be practicable or reasonable. It also suggests that a hazard awareness notice might be used instead of an improvement notice where the owner or landlord has agreed to take remedial action and the local authority is confident that the work will be completed within a reasonable time (Enforcement Guidance, paragraphs 5.38–5.39 and 5.43).

These notices are not enforceable and require no further action from the recipient. There is therefore no provision for appeal. They are simply advisory, although the Enforcement Guidance suggests that the local authority should consider monitoring the position at that premises (Enforcement Guidance, paragraph 5.41). However, the service of a hazard awareness notice does not prevent the use of a formal notice later if an unacceptable hazard remains.

Content of hazard awareness notice

The notice must specify:

1. the nature of the hazard and the premises on which it exists;
2. the deficiency giving rise to the hazard;
3. the premises on which the deficiency exists;
4. the reasons for deciding upon the action as the most appropriate course;
5. details of the remedial action required which the local authority considers would be practicable and reasonable (Enforcement Guidance, paragraph 5.40).

A hazard awareness notice served under section 28 may only contain category 1 hazards but, where there are also category 2 hazards, the notices under sections 28 and 29 may be combined (section 29(6)).

Service of hazard awareness notice

The provisions relating to the service etc. of hazard awareness notices is the same as for improvement notices in schedule 1, part 1 – see below.

IMPROVEMENT NOTICES

References

Housing Act 2004, sections 11–18 and schedules 1 and 2.
Enforcement Guidance, ODPM February 2006, part 5, paragraphs 5.4–5.14.

FC66 Improvement notices (INs) – Sections 11 to 19 Housing Act 2004

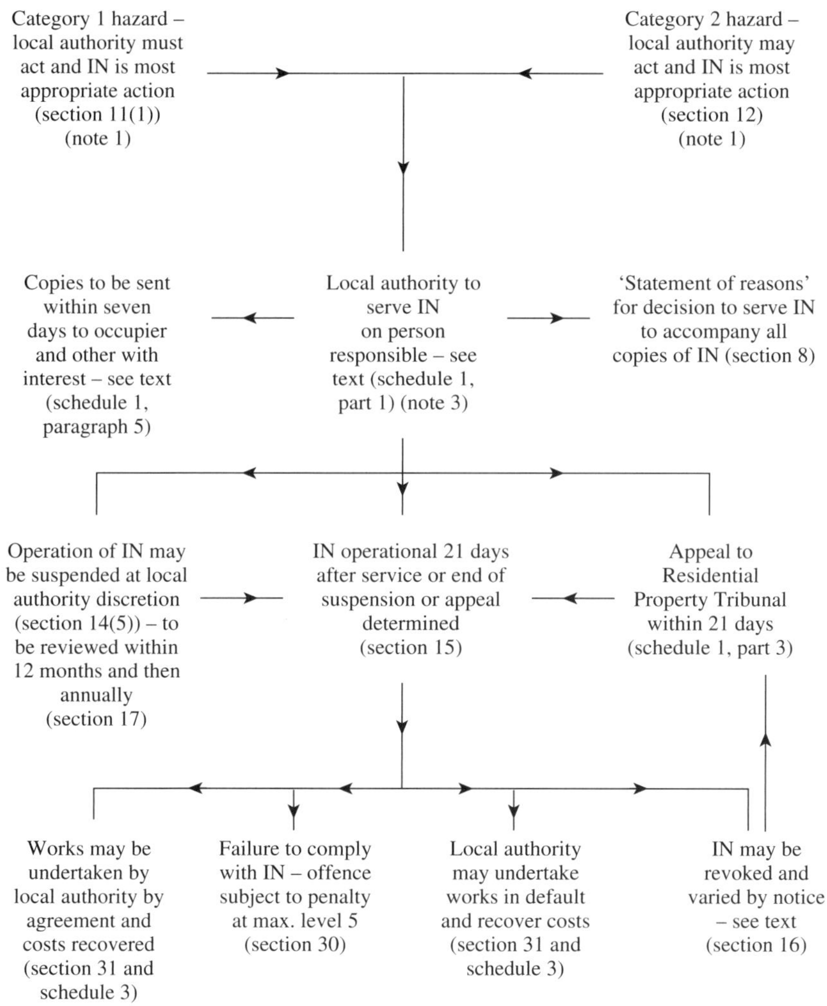

Notes

1. Where there are multiple hazards including category 2, the IN can require action to deal with both category 1 and category 2.
2. For emergency remedial action, see FC67.
3. The IN must be registered as a local land charge (section 37).

Extent

These provisions apply to England and Wales.

Scope

Improvement notices are a possible response where the local authority considers that this action is the most appropriate in all the circumstances to deal with one or more category 1 and/or category 2 hazards at residential premises (sections 11 and 12). They are notices that require the person on whom they are served to take the specified remedial action to remove the hazard.

Decision to serve an improvement notice

The local authority must be satisfied that the service of an improvement notice is the appropriate action having regard to the Enforcement Guidance. In addition, the circumstances necessary for the service of such a notice are:

(a) that the local authority is satisfied that a category 1 or category 2 hazard exists;
(b) that there is no interim or final management order relating to the premises (sections 11 and 12);
(c) that, if re-development of the land by the owner has been approved and is being proceeded with to a specified time limit, an improvement notice may not be used (section 39(5)).

Persons to be served with an improvement notice

The improvement notice is to be served on the person responsible. Identifying this depends on the type of premises (section 18 and schedule 1).

(a) In the case of houses in multiple occupation etc. licensed under Part 2 or 3 of the 2004 Act – the holder of the licence.
(b) In the case of premises not licensed under either Part 2 or 3:
    (i) where the premises is not a flat – the person having control;
    (ii) where the premises is not a flat, but is a house in multiple occupation – the person having control or the person managing the house in multiple occupation;
    (iii) where the premises is a flat and is a dwelling and not a house in multiple occupation – an owner who, in the local authority's opinion, ought to take the action specified in the notice;
    (iv) where the premises is a flat and is a house in multiple occupation – an owner who, in the local authority's opinion, ought to take the action specified in the notice, or the person managing the flat.
(c) In the case of common parts or any non-residential part of a building containing flats – an owner who, in the local authority's opinion, ought to take the action specified in the notice.

In addition to the person responsible, the local authority is required to serve a copy of the improvement notice on the occupier of the premises, and on every person who, to the knowledge of the local authority, is a freeholder, mortgagee or lessee of the premises (schedule 1, paragraph 5).

Content of improvement notice

An improvement notice can deal with one or more hazards of both categories (section 12(5)) and must, in relation to each hazard, specify:

(a) whether the notice has been served under section 11 or 12;
(b) the nature of the hazard and the residential premises to which the notice relates;
(c) the deficiency giving rise to the hazard;
(d) the premises in relation to which remedial action is to be taken and the nature of the remedial action;
(e) the date by which that remedial action is to be started (which cannot be earlier than the twenty-eighth day after the date the notice was served);
(f) the period within which the remedial action is to be completed, or the periods within which each part of that work is to be completed;
(g) information on the right to appeal under Part 3 of schedule 1;
(h) the period within which an appeal may be made.

In relation to category 1 hazards, the notice must require at least the removal of that hazard but may go further. Requirements must be reasonable (section 13).

Accompanying the improvement notice must be a statement of reasons (section 8(3) and (4)). This statement must explain why the local authority decided that the appropriate action was the service of an improvement notice rather than any of the other enforcement options (section 8(3) and (4)).

Operation of improvement notice

An improvement notice becomes operative 21 days from the date on which it was served. Where an appeal is brought, the notice does not become operative until the appeal is determined (or withdrawn) and confirmed and the period for further appeal expires (section 15).

Suspension of an improvement notice

The local authority has the discretion to suspend the operation of an improvement notice either:

(a) until a time, which may be when a person of a particular description begins or ceases to occupy the premises;
(b) until a specified event occurs, which may be a breach of an undertaking, which breach has been made known to the person on whom the notice was served (section 14).

The local authority has power to review a suspended improvement notice at any time, but must review it not later than one year after the date of service and on the anniversary of that date. The local authority's decision on review must be served on all those on whom the notice and copies were served (section 17).

The time or events triggering an end to suspension may be varied by the local authority at their own volition or as a result of the application of the person on whom it was served (section 16(4) and (8)).

Where an improvement notice has been suspended, it becomes operative again on the twenty-first day after the suspension ends and the minimum 28 days for starting work becomes 21 days (section 14(5)).

Revocation and variation of an improvement notice

Review of whether an improvement notice should or could be revoked or varied can be at the local authority's own initiative or on the application of the person on whom the notice was served (section 16(8)).

The local authority must revoke the notice where they are satisfied the requirements have been complied with (section 16(1)). This applies whether the notice relates to one or to more than one hazard.

The local authority is given a power to revoke a notice where:

(a) in the case of a notice served in respect of a category 1 hazard, they consider that there are special circumstances making revocation appropriate; or
(b) in the case of a notice served in respect of a category 2 hazard, they consider revocation appropriate (section 16(2)).

Where the notice relates to more than one hazard, the local authority has the power to revoke a part of the notice; they also have the power to vary the remainder of the notice (section 16(3)).

An improvement notice may be varied with the agreement of the person on whom it was served. It may also be varied without such agreement to alter the time or events triggering the end of a suspension (section 16(4)).

Change of person liable to comply

Where the person served with an improvement notice transfers their interest to a third party while the notice is still active, then the new person is deemed to be in the same position as the original person. However, the new person is not held subject to any liability incurred by the original recipient of the notice (section 19). It is important to note that such a transfer of interest has no effect on the period given for compliance in the notice.

Appeals

The persons on whom the improvement notice was served are given a general right to appeal against the notice to the Residential Property Tribunal within 21 days from the date of service. In addition to the open right of appeal, the schedule specifies the following grounds:

(i) that one or more other persons, as owner(s) of the premises, ought to take the specified action;
(ii) that one or more other persons, as owner(s) of the premises, ought to pay all or part of the cost of taking the specified action;
(iii) that, rather than an improvement notice, a hazard awareness notice, a prohibition order or a demolition order would have been the best course of action.

There is also an appeal process against decisions of the local authority relating to revocation and variation within 21 days from service of that notice.

The appeal is by way of a re-hearing, and the Residential Property Tribunal may have regard to matters of which the local authority was unaware. On hearing the appeal, the Residential Property Tribunal may confirm, vary or quash the notice (schedule 1, part 3).

Prosecution for non-compliance with an improvement notice

It is a criminal offence to fail to comply with the requirements of an improvement notice within the period or period(s) specified. The local authority may take proceedings against the person on whom the notice was served in the magistrates' court. The maximum fine on summary conviction is not to exceed level 5 on the standard scale. It is a defence for the person to satisfy the court that there was a reasonable excuse for failing to comply with the notice. Non-compliance is deemed to be not carrying out the specified actions within the periods specified in the notice or the actions and period(s) as determined on appeal. It should be noted that the obligation to comply with the notice continues even when the period for completion has expired (section 30).

Default powers

The Act enables a local authority to carry out the actions specified in a notice in the following circumstances:

(a) with the agreement of the person on whom the notice was served;
(b) where, in the local authority's opinion, insufficient progress is made to complete the actions required, even though the period(s) specified in the notice have not expired;
(c) where the notice has not been complied with.

Where the local authority agrees to carry out the action, or decides to do so, they have all the rights that the person served with the notice had. Seven days before invoking their right to take action, the local authority must serve notice on the person on whom the improvement notice was served and serve a copy on the occupier and on any owner of the relevant premises.

Once the local authority has taken on the right to carry out the actions required, it is an offence for the person on whom the notice was served, or for any workman or contractor of that person, to be on the premises for the purpose of carrying out any works.

Any expenses, including administrative expenses, incurred by the local authority in taking action without agreement are recoverable. Such expenses are a charge on the premises and the local authority has the powers under the Law of Property Act 1925 as if the local authority were mortgagee by deed.

Where the local authority considers that the expense incurred (and any interest) is not likely to be recovered, and that rents or other payments are being made to a person, then the local authority may apply to the Residential Property Tribunal for an order that that person makes such payments as the Tribunal thinks just.

There is a right of appeal against a demand for payment of expenses (and any interest) made by the local authority. The appeal is to the Residential Property Tribunal and must be made within 21 days of the demand (section 31 and schedule 3).

NOTICE OF EMERGENCY REMEDIAL ACTION

References

Housing Act 2004, sections 40–42 and 45 and schedule 3, part 1.
Enforcement Guidance, ODPM February 2006, part 5, paragraphs 5.31–5.34.

Extent

These provisions apply to England and Wales.

Scope

Where the local authority is satisfied that:

1. a category 1 hazard exists in a residential premises;
2. the hazard involves an imminent risk of serious harm to the health or safety of any of the occupants of that or other residential premises;
3. no management order is in force; and
4. action could be taken by way of an improvement notice under section 11,

they may take emergency remedial action as one of the options available to comply with their duty to act (section 40(1)– (3)).

'Emergency remedial action' means such remedial action in respect of the hazard concerned as the authority considers immediately necessary in order to remove the imminent risk of serious harm to the health or safety of any of the occupants (section 40(2)). Action may be taken in respect of more than one category 1 hazard on the same premises or in the same building containing one or more flats (section 40(4)).

Commencement of work

In relation to such hazards, the local authority may commence remedial work immediately without waiting for a notice of emergency remedial action to be served (paragraphs 3 to 5 of schedule 3).

FC67 Emergency remedial action (ERA) notices – Sections 40 to 42 and 45 Housing Act 2004

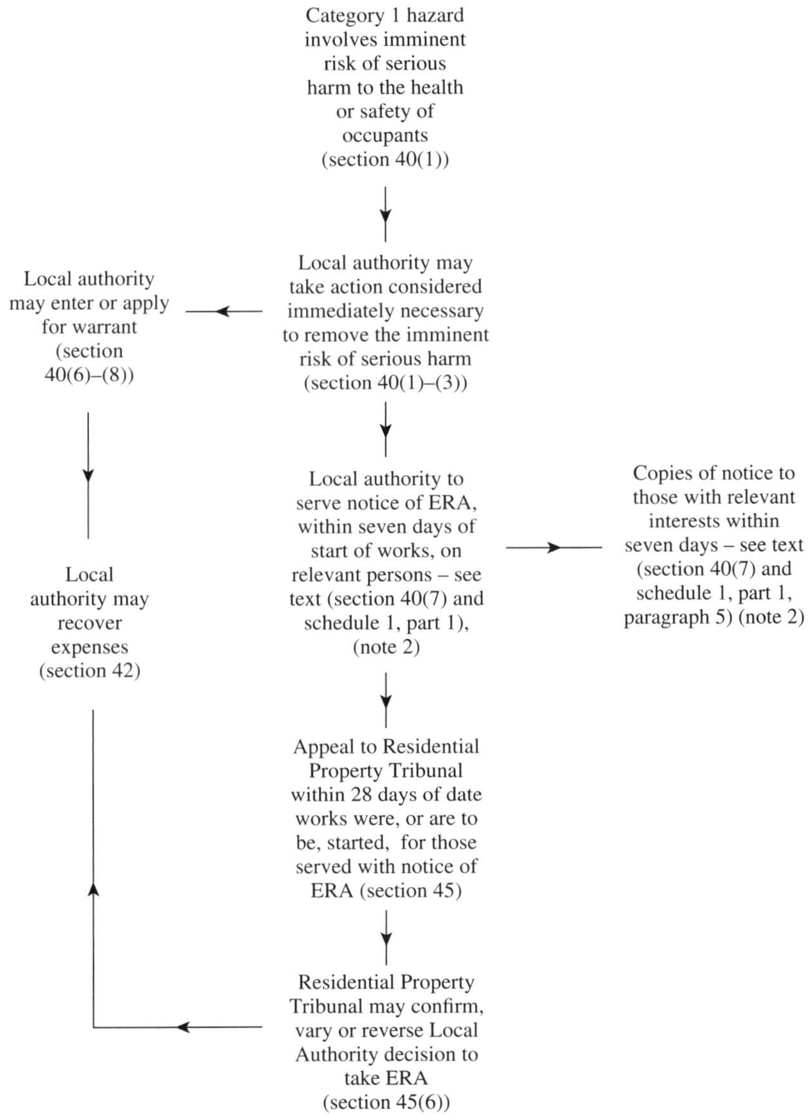

Notes

1. For the procedure for using emergency prohibition orders, see FC69.
2. Notice and copies to be accompanied by a 'statement of reasons' for the decision (section 8).

Power of entry

Paragraphs 3–5 of schedule 3, which relate to the enforcement of improvement notices, also apply to the carrying out of emergency remedial action, except that:

(a) the rights of entry may be exercised at any time; and
(b) the notice is served on occupiers before entering (section 40(5) and (6)).

If necessary, the local authority may apply to a Justice of the Peace for a warrant to enter under section 240 but a warrant can only be granted where the Justice of the Peace considers that the local authority would not gain entry without it (section 40(8)).

Notice of emergency remedial action

Within seven days of commencing work, the local authority must serve a notice of emergency remedial action and copies of it on the same persons as would be the case for an improvement notice – see page 282 (section 40(7)).

The notice must specify:

- the nature of the hazard;
- the residential premises where it exists;
- the deficiency giving rise to the hazard;
- the premises where the remedial action has been or is to be taken;
- the nature of the action;
- the power under which the action is taken;
- the date on which the action was or is to be taken; and
- information about rights of appeal.

Appeals

Persons on whom the notice of emergency remedial action was served may appeal to a Residential Property Tribunal within 28 days of the date specified in the notice when the work was, or was to be, started. The Residential Property Tribunal may allow a longer period for appeal where it feels there is good reason to do so. An appeal will not prevent the action from being put into effect and this may have been started and finished before a notice has been served. There are no specified grounds of appeal. The tribunal may confirm, reverse or vary the local authority decision to take the action (section 45). Since works may have been completed at this time, the effect of such a decision would relate to the local authority's ability to recover expenses.

Recovery of expenses

The provisions of paragraphs 6–14 of schedule 3, which relate to enforcement of improvement notices, apply to the recovery of costs under this.

PROHIBITION ORDERS

References

Housing Act 2004, sections 20–27 and schedule 2.
Enforcement Guidance, ODPM, February 2006, part 5.

Extent

These provisions apply to England and Wales.

Scope

Prohibition orders are a possible response available to the local authority to perform its duty to deal with category 1 hazards and is a discretionary power which may be used to deal with category 2 hazards, in both cases where it is considered to be the most appropriate form of action (section 20(1) and 21(1)). They may not be used when the premises is the subject of a management order – see FC78 and FC79.

A prohibition order may prohibit the use of all or part of a residential premises (definition on page 241) (sections 20(1) and 21(1)). The non-residential part of premises may also be prohibited if the defect giving rise to a hazard is located there and the prohibition is necessary for the health and safety of actual or potential occupants (section 20(4)).

The Enforcement Guidance gives examples of when the use of a prohibition order might be appropriate.

These include:

- where the conditions present a serious threat to health or safety and remedial action would be unreasonable or impracticable;
- to specify maximum numbers of occupants where it is too small for the household's needs;
- to control the number of occupants where there are insufficient facilities;
- to prohibit use by a specified group or where conditions are hazardous to some people but relatively safe for others;
- in a house in multiple occupation, to prohibit the use of specified dwelling units or common parts.

The Enforcement Guidance also indicates additional matters to be considered before deciding upon the use of a prohibition order:

- Have regard to the exclusion of vulnerable people.
- Consider if the building is listed or protected.
- Take account of the relationship of the building to neighbouring buildings, which may mean that prohibition is preferable to, for example, demolition.
- Consider alternative uses other than those proposed by the owner.
- Take account of the existence of a conservation or renewal area.
- Consider effects of complete prohibition on the community and the appearance of the area.

FC68 Prohibition orders (POs) – Sections 20 to 27 Housing Act 2004

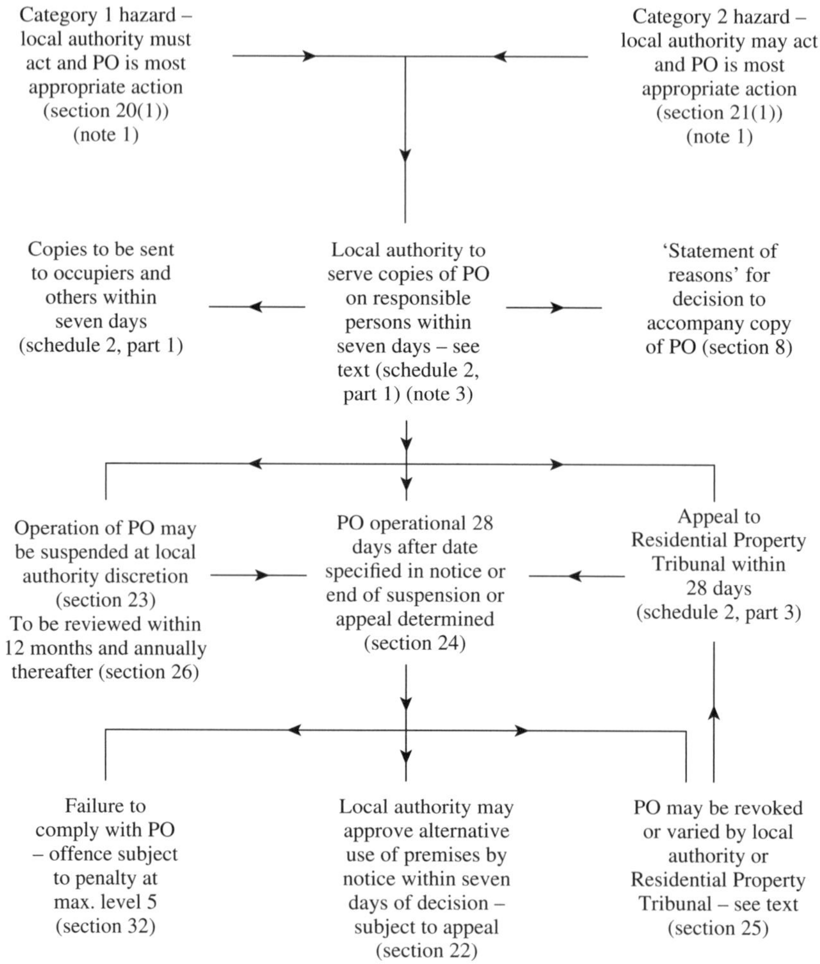

Notes

1. Where there are multiple hazards including both category 1 and category 2, the prohibition order may include both.
2. For emergency prohibition orders, see FC69.
3. The prohibition order is to be registered as a local land charge (section 37).

- Consider the availability of local accommodation for rehousing displaced occupants.
- Consider appropriate financial advice or assistance (Enforcement Guidance, paragraphs 5.15–5.23).

Content of prohibition orders

A prohibition order can relate to one or more hazards and may include both category 1 and category 2 hazards, either in a separate order or combined in the same prohibition order (sections 20(5) and 21(4)). Orders under both sections 20 and 21 must state for each hazard:

1. whether the hazard is considered to be category 1 or category 2;
2. the nature of the hazard and the premises where it exists;
3. the deficiency giving rise to the hazard;
4. the premises to which the prohibition relates;
5. any remedial action considered necessary for the prohibition order to be revoked;
6. the rights of appeal (section 22(2) and (6)).

The order may prohibit:

1. use for some or all purposes unless approved by the local authority;
2. occupation by a particular number of households or occupants or particular descriptions of people, e.g. by age (section 22 (3)–(5)).

Service of prohibition orders

Copies of a prohibition order are to be served within seven days of the order being made, with the first day being the day on which it is made. They are to be served on:

(a) an owner or occupier of the whole or part of the premises;
(b) anyone authorised to permit persons to occupy;
(c) a mortgagee.

The requirement relating to occupiers will be met by fixing a copy to a conspicuous part of the premises (schedule 2, part 1).

Operation of prohibition order

Unless there is an appeal or suspension, prohibition orders come into effect 28 days after they are made (section 24).

Approvals of use by local authority

Where the prohibition relates to the use of the whole or part of the premises without the approval of the local authority, that approval must not be unreasonably withheld. Where use is not approved, the local authority must notify the person who made application within seven days and include the reasons for the decision together with their rights of appeal. There is an appeal against refusal to the Residential Property Tribunal within 28 days (section 22(7)–(9)).

Suspension of prohibition order and reviews

The local authority may suspend the operation of a prohibition order at its discretion and may:

- specify an event, such as the breach of an undertaking, as ending the suspension; or
- specify a time at which the suspension ends (section 23).

The local authority must review a suspended prohibition order within a year of its being made and at least annually thereafter (section 26).

Revocation and variation

Revocation and variation may take place either on application or on the initiative of the local authority.

The local authority must revoke the prohibition order if it is satisfied that the hazard that necessitated the order no longer exists. Where there are multiple hazards, the revocation applies individually to each and, where the revocation is for less than them all, the prohibition order will need to be varied. There is then a discretionary power to revoke in relation to a category 1 hazard if satisfied that there are special circumstances. Revocation orders made in agreement come into force on the day they are made, otherwise 28 days after the decision.

The local authority may vary a prohibition order either with the agreement of every person on whom a copy has been served or, for a suspended order, to alter the time or events to which the suspension is to end (section 25).

The service and content of notices are prescribed in schedule 2, part 2.

Appeals

There is provision for appeals against the making of a prohibition order, refusals to approve use and decisions relating to revocation and variation. Such appeals are to be made to the Residential Property Tribunal within 28 days of the making of the order or decision.

The grounds of appeal include that the prohibition order was not the most appropriate course of action, which would have been an improvement notice, a hazard awareness notice or a demolition order (schedule 2, part 3).

Enforcement of prohibition order

Failure to comply with a prohibition order is an offence punishable by a fine of up to level 5 (section 32).

Determinations to purchase

In circumstances where the local authority would be required to make a prohibition order and it appears to them that the building can be rendered capable of providing accommodation that is adequate for the time being, the local authority may purchase the premises instead of making a prohibition order (Housing Act 1985, section 300).

Recovery of possession

Where it is necessary for the owner to take possession of the premises or part of it in order to comply with the order, this may be done by a notice to quit. The restrictions on the recovery of possession in the Rent Act 1977, the Rent (Agriculture) Act 1976 and part 1 of the Housing Act 1988 will not apply (section 33).

Compensation

Compensation is payable by the local authority to every owner of premises subject to a prohibition order (Housing Act 1985, section 584A) but is repayable where the prohibition order is revoked under section 25 (Housing Act 1985, section 584B).

EMERGENCY PROHIBITION ORDERS

References

Housing Act 2004, sections 43–45 and schedule 1, part 1.
Enforcement Guidance, ODPM, February 2006, part 5, paragraphs 5.35–5.37.

Extent

These provisions apply to England and Wales.

Scope

Where the local authority is satisfied that:

1. a category 1 hazard exists on a residential premises;
2. the hazard involves an imminent risk of serious harm to the health or safety of any of the occupants of that or other residential premises;
3. no management order is in force,

they may make an emergency prohibition order as one of the options available to comply with their duty to act (section 43(1)–(3)).

An emergency prohibition order is an order imposing, with immediate effect, such prohibition/s on the use of any premises as are specified in the order (section 43(2)).

FC69 Emergency prohibition orders (EPOs) – Sections 43 to 45 Housing Act 2004

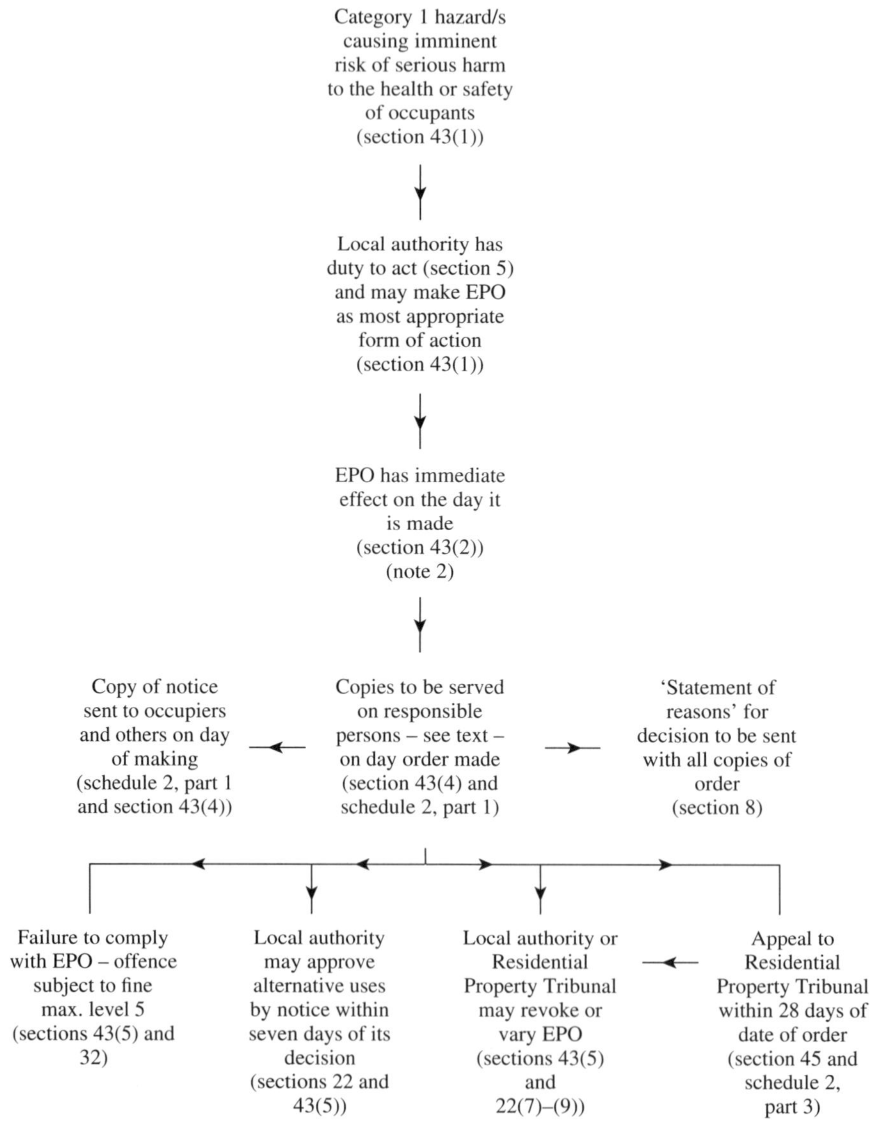

Notes

1. For prohibition notices, see FC68.
2. The emergency prohibition order is to be registered as a local land charge (section 37).

The order may prohibit:

- use for some or all purposes unless approved by the local authority;
- occupation by a particular number of households or occupants or particular descriptions of people, e.g. by age (section 43(3) and section 22(3)–(5)).

Contents of emergency prohibition order

The emergency prohibition order must specify:

- the hazard/s and the residential premises on which it exists;
- the deficiency giving rise to the hazard;
- the premises on which the prohibition is to operate;
- any remedial action which would cause the local authority to revoke the order.

The notice must also contain the rights of appeal and the period within which this may be done (section 44). The provisions of section 22(3)–(5) and (7)–(9) relating to the content of the notice also apply to emergency prohibition orders as they apply to prohibition orders – page 291 (section 43(3)).

Service of emergency prohibition orders

Copies of an emergency prohibition order are to be served on the day that the order is made. They are to be served on:

- an owner or occupier of the whole or part of the premises;
- anyone authorised to permit persons to occupy;
- a mortgagee.

The requirement relating to occupiers will be met by fixing a copy to a conspicuous part of the premises (section 43(4) and schedule 2, part 1).

Operation of emergency prohibition order

The emergency prohibition order has immediate effect on its making (section 43(2)).

Approvals of use

Where the emergency prohibition relates to the use of the whole or part of the premises without the approval of the local authority, that approval must not be unreasonably withheld. Where use is not approved, the local authority must notify the person who made application within seven days and include the reasons for the decision together with their rights of appeal. There is an appeal against refusal to the Residential Property Tribunal within 28 days (section 43(5) and section 22(7)–(9)).

Revocation and variation

Revocation and variation may take place either on application or on the initiative of the local authority. The local authority must revoke the emergency prohibition order if it is satisfied that the hazard that necessitated the order no longer exists. Where there are multiple hazards, the revocation applies individually to each and, where the revocation is for less than them all, the emergency prohibition order will need to be varied. There is then a discretionary power to revoke in relation to a category 1 hazard if satisfied that there are special circumstances. Revocation orders made in agreement come into force on the day they are made, otherwise 28 days after the decision.

The local authority may vary an emergency prohibition order either with the agreement of every person on whom a copy has been served or, for a suspended order, to alter the time or events to which the suspension is to end (section 25).

Appeals

There is an appeal to the Residential Property Tribunal against the making of an emergency prohibition order by those in receipt of a copy of it. This must be made within 28 days (the Residential Property Tribunal may agree a longer period) of the date on which the emergency prohibition order is made, but the order continues in effect until determination of the appeal. The Residential Property Tribunal may confirm, vary or revoke the emergency prohibition order. There is also an appeal against refusal to agree to alternative uses (section 45).

Enforcement

Failure to comply with an emergency prohibition order is an offence punishable by a fine of up to level 5 and daily fine of £20 (section 43(5)(b) and section 32).

Recovery of possession

Where it is necessary for the owner to take possession of the premises or part of it in order to comply with the order, this may be done by a notice to quit. The restrictions on the recovery of possession in the Rent Act 1977, the Rent (Agriculture) Act 1976 and part 1 of the Housing Act 1988 will not apply (section 43(5)(b) and section 33).

OVERCROWDING NOTICES FOR NON-HOUSES IN MULTIPLE OCCUPATIONS

References

Housing Act 1985, part 10.
The Housing (Prescribed Forms) Regulations 1990 as amended.

FC70 Overcrowding notices for non-HMOs – Sections 324 to 344 Housing Act 1985

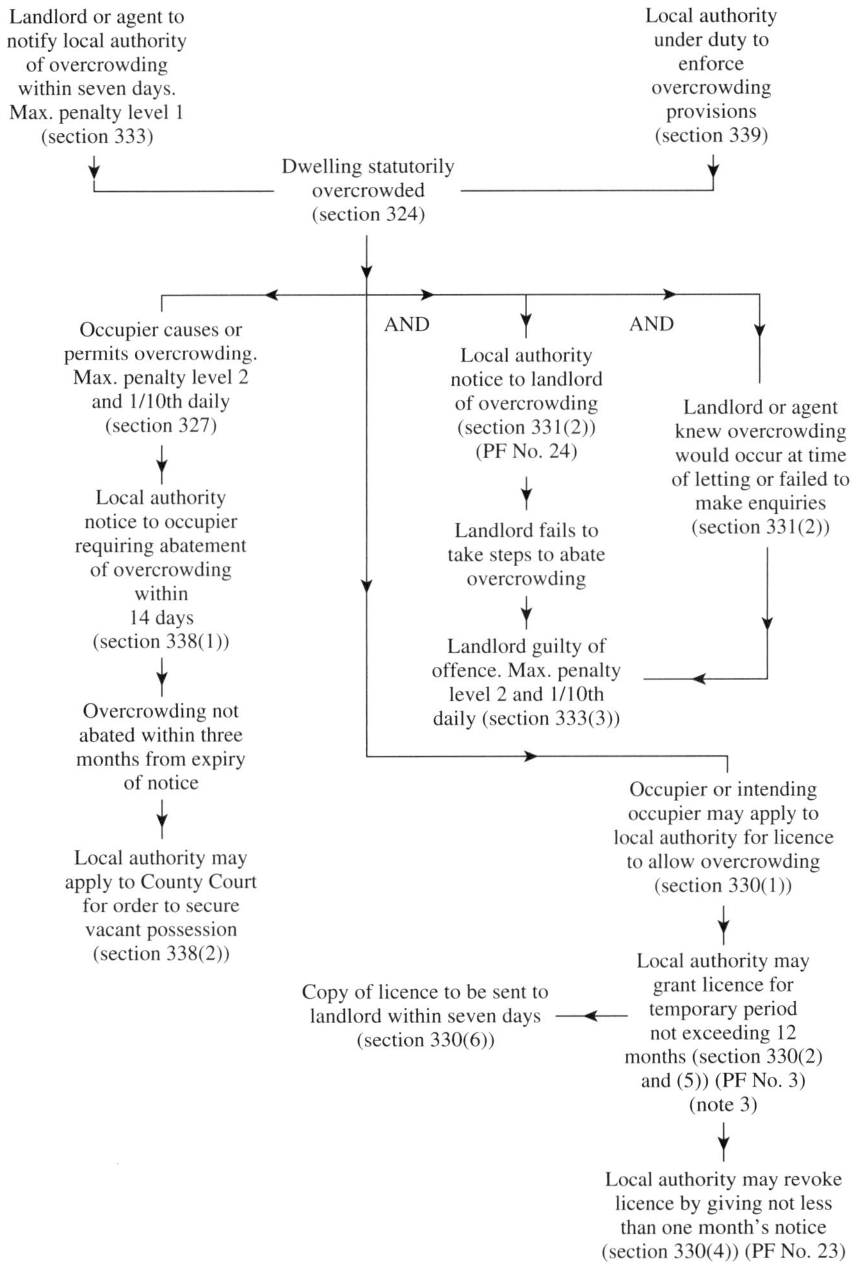

Notes

1. For overcrowding in registered HMOs, see FC76, and FC71 for non-registrable HMOs.
2. There are no appeal provisions relating to overcrowding notices.
3. The prescribed forms in this procedure are specified in the Housing (Prescribed Forms) (No. 2) Regulations 1990, except for the licence for temporary occupation, which is contained with the Housing (Prescribed Forms) Regulations 1990.
4. In this flowchart, PF = prescribed form.

Extent

These provisions apply to England and Wales.

Note

Section 216 of the Housing Act 2004 enables the Secretary of State by order to modify the operation of this procedure and the standards incorporated into it. No such order has yet been made, but a government strategy was published in December 2007 – 'Tackling overcrowding in England: An action plan'.

Scope

This procedure applies to any dwelling, which is defined as 'premises used or suitable for use as a separate dwelling'. It is a standard base on sleeping capacity and is not related to the provision of facilities etc.

Statutory overcrowding

A dwelling-house is deemed to be overcrowded when either:

(a) the number of persons sleeping there is such that any two persons ten years old or more of opposite sexes, not being man and wife, must sleep in the same room (i.e. a room normally used in the locality either as a bedroom or as a living room and is available as sleeping accommodation); or
(b) the permitted number as calculated below is exceeded.

No account is taken of any child under one year and a child of an age between one year and under ten years is counted as half a unit (sections 325 and 326).

Permitted number

This is the smaller of a number of persons calculated by the following two methods:

(a) where a house consists of:
    one room – two persons
    two rooms – three persons
    three rooms – five persons
    four rooms – seven and a half persons
    five rooms or more – ten persons, with additional two persons in respect of each room in excess of five (no regard is had to rooms of less than 50 sq. ft.); **or**
(b) where the aggregate for all rooms is determined by reference to the following:
    110 sq. ft. or more – two persons
    90–100 sq. ft. – one and a half persons
    70–90 sq. ft. – one person
    50–70 sq. ft. – half person
    Less than 50 sq. ft. – nil (section 326).

Offences by occupiers

The occupier of a dwelling-house who causes or permits it to be overcrowded is guilty of an offence unless:

(a) the overcrowding is caused only by virtue of a child attaining the age of one or ten years and the occupier has applied to the local authority for suitable alternative accommodation (defined in section 342) provided:
  (i) alternative accommodation is refused; or
  (ii) it subsequently becomes reasonably practicable for a person living in the house, but not a member of the family, to move; or
(b) the overcrowding is caused only because a member of the family is sleeping in the house temporarily; or
(c) the local authority has issued a licence authorising temporary overcrowding (below) (section 327).

Offences by landlords

The landlord is guilty of an offence if he has either:

(a) received a notice of overcrowding from the local authority but fails to take steps reasonably open to him to abate the overcrowding; or
(b) when letting the house, he had reasonable cause to believe it would become overcrowded or failed to make enquiries of the proposed occupier; or
(c) failed to notify the local authority of overcrowding within seven days of becoming aware of it (sections 331 and 333).

Licence for temporary overcrowding

An occupier, or intending occupier, may apply to the local authority for a licence authorising the sleeping occupation of a house by a number of persons in excess of the permitted number.

Such a licence may be granted by the local authority where it thinks it to be expedient, having had regard to any exceptional circumstances, including any seasonal increase in the general population of this district.

Licences may not operate for in excess of 12 months and may be revoked by the giving of at least one month's notice. There is no appeal against a refusal by a local authority to grant a licence or against revocation (section 330).

Definitions

In this procedure:

Agent in relation to the landlord of a dwelling:

(a) means a person who collects rent in respect of the dwelling on behalf of the landlord, or is authorised by him to do so; and
(b) in the case of a dwelling occupied under a contract of employment under which the provision of the dwelling for his occupation forms part of the

occupier's remuneration, includes a person who pays remuneration on behalf of the employer, or is authorised by him to do so.

Dwelling means premises used or suitable for use as a separate dwelling.
Landlord in relation to a dwelling:

(a) means the immediate landlord of an occupier of the dwelling; and
(b) in the case of a dwelling occupied under a contract of employment under which the provision of the dwelling for his occupation forms part of the occupier's remuneration, includes the occupier's employer.

Owner in relation to premises:

(a) means a person (other than a mortgagee not in possession) who is for the time being entitled to dispose of the fee simple, whether in possession or in reversion; and
(b) includes also a person holding or entitled to the rents and profits of the premises under a lease of which the unexpired term exceeds three years (section 343).

OVERCROWDING NOTICES FOR HOUSES IN MULTIPLE OCCUPATION

Reference

Housing Act 2004, sections 139–144.

Extent

These provisions apply to England and Wales.

Scope

This procedure deals with the control of overcrowding in houses in multiple occupation (HMOs) that are not required to be registered under part 2 (see FC76) and which are not the subject of operative interim management orders or final management orders (see FC78 and FC79) (section 139(1)). The notices relate to situations where, having regard to the rooms available, the local authority considers that an excessive number of persons is being, or is likely to be, accommodated in the house in multiple occupation (section 139(2)).

The level of occupation in licensed houses in multiple occupation is controlled through the licence in that it will only permit occupation by a specified number.

Service of overcrowding notices

At least seven days before any overcrowding notice is served, the local authority must inform the following persons of its intention:

(a) persons having an estate or interest in the house;

FC71 Overcrowding notices (ONs) for houses in multiple occupation

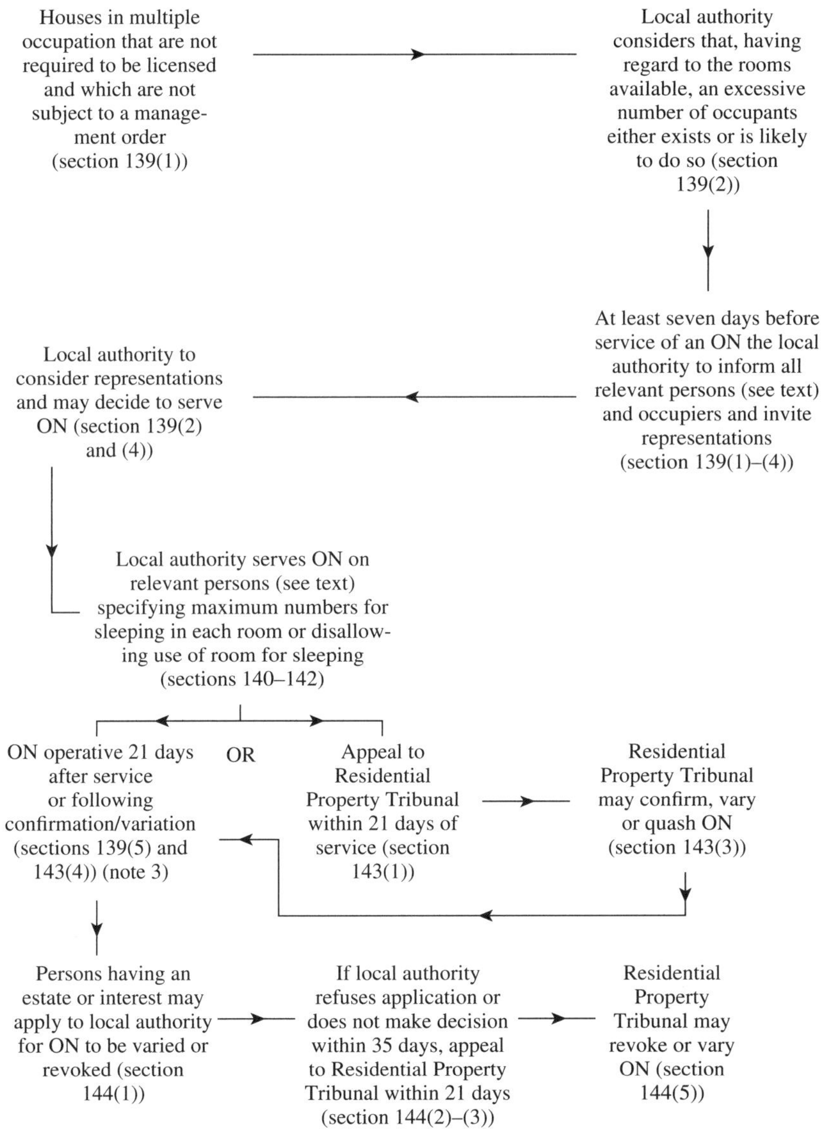

Notes

1. For the control of occupancy in licensed houses in multiple occupation, see FC76.
2. For the control of overcrowding in houses other than houses in multiple occupation, see FC70.
3. Offence to breach ON with penalty up to level 4 (section 139 (7)–(8)).

(b) the person managing or in control of it; and
(c) as far as is reasonably possible, all occupiers.

In the case of (a) and (b), the notification must be in writing.

Each of these must be given the opportunity to make representations for the local authority to consider. The overcrowding notice is then served on one or more of the relevant persons ((a) and (b)) above (section 139(1)–(4)).

Contents of overcrowding notice

The notice must state in relation to each room either:

(i) the maximum number of persons who may occupy it as sleeping accommodation at any one time; or
(ii) that the room is regarded as unsuitable for sleeping accommodation.

Special maxima may be applied relating to the age of occupants.

The notice must also *either*:

(a) require that the terms of the notice are not breached;
(b) disallow numbers of occupants of the house such as results in occupation of a sleeping room by opposite sexes who are not living together as man and wife (ignoring children under the age of ten); *or*

apply (a) and (b) but only in relation to new occupants not resident when the notice was served (sections 140–142).

Operation of an overcrowding notice

The notice becomes operative 21 days after service. However, where an appeal is lodged, it becomes effective after confirmation by the Residential Property Tribunal or subsequently by the Lands Tribunal (sections 139(5) and 143(4)).

Appeals

An appeal against the service of the overcrowding notice may be made to the Residential Property Tribunal within 21 days of service, although that period may be extended by the Residential Property Tribunal if it feels that there is good reason. The Residential Property Tribunal may confirm, vary or quash the notice with a further appeal to the Lands Tribunal (section 143).

Revocation and variation

On the application of a person having an estate or interest or a person managing or having control, the local authority may, at any time:

- revoke the overcrowding notice; or
- vary it to allow higher occupancy.

The local authority must make a decision and give written notice of it within 35 days. There is an appeal to the Residential Property Tribunal against a refusal or where the matter has not been dealt with within the 35 days. Appeals are to be made within 21 days but the Residential Property Tribunal has power to extend that period if there is good reason. The Residential Property Tribunal may revoke the notice or confirm it (section 144).

Offences

Any person contravening an overcrowding notice is liable, on conviction, to a fine up to level 4. There is a defence of 'reasonable excuse' (section 139(7)–(8)).

SELECTIVE LICENSING OF RESIDENTIAL ACCOMMODATION

Reference

Housing Act 2004, sections 79 to 100.

Extent

These provisions apply to England and Wales.

Scope

The Housing Act 2004 provides for the mandatory licensing of certain houses in multiple occupation (see page 316), and the discretionary licensing of other houses in multiple occupation and other privately rented property. Here the provisions relate to selective licensing of privately rented dwellings.

Designation of selective licensing areas

Local authorities can designate all or part of their district for selective licensing of rented dwellings (section 80(1)).

Before making this designation, it must consider:

(a) that the area is, or is likely to become, an area of low housing demand; and
(b) that making a designation will, when combined with other measures, contribute to the improvement of the social or economic conditions in the area (section 80(3)).

It must take into account (among other matters):

(a) the value of residential premises in the area, in comparison to the value of similar premises in other areas which the authority consider to be comparable (whether in terms of types of housing, local amenities, availability of transport or otherwise);
(b) the turnover of occupiers of residential premises;
(c) the number of residential premises which are available to buy or rent and the length of time for which they remain unoccupied (section 80(4));

(d) whether the area is experiencing a significant and persistent problem caused by anti-social behaviour;
(e) whether some or all of the private sector landlords who have let premises in the area are failing to take action to combat the problem that it would be appropriate for them to take; and
(f) whether making a designation will, when combined with other measures taken in the area by the local authority, lead to a reduction in, or the elimination of, the problem (section 80(6)).

Before making a designation, the local authority must:

(a) take reasonable steps to consult persons who are likely to be affected by the designation; and
(b) consider any representations made in accordance with the consultation and not withdrawn (section 80(9)).

The local authority must also ensure that any exercise of the power is consistent with the authority's overall housing strategy (section 81(2)) and must also seek to adopt a coordinated approach in dealing with homelessness, empty properties and anti-social behaviour.

It must consider whether there are any other courses of action available to them that might provide an effective method of achieving the objective or objectives that the designation would be intended to achieve, and they consider that making the designation will significantly assist them to achieve the objective or objectives (whether or not they take any other course of action as well).

The designation must be submitted to and be confirmed by the Secretary of State or comply with prior approved schemes (section 82).

If confirmed, a designation comes into force on a date specified by the local authority, no earlier than three months after the date on which the designation is confirmed (section 80(3) and 80(4)).

Notification

As soon as the designation is confirmed or made, the local authority must publish a notice stating:

(a) that the designation has been made;
(b) whether the designation was required to be confirmed or that a general approval applied to it (giving details of the approval in question);
(c) the date on which the designation is to come into force; and
(d) any other information which may be prescribed (section 83(2)).

Copies of the designation, and prescribed information must be made available (section 83(3)).

Duration, review and revocation of designations

Unless previously revoked, a designation ceases to have effect at the time that is specified in the designation, no later than five years after the date on which the

FC72 Designation of areas for selective licensing of rented dwellings – Sections 80 to 84 Housing Act 2004

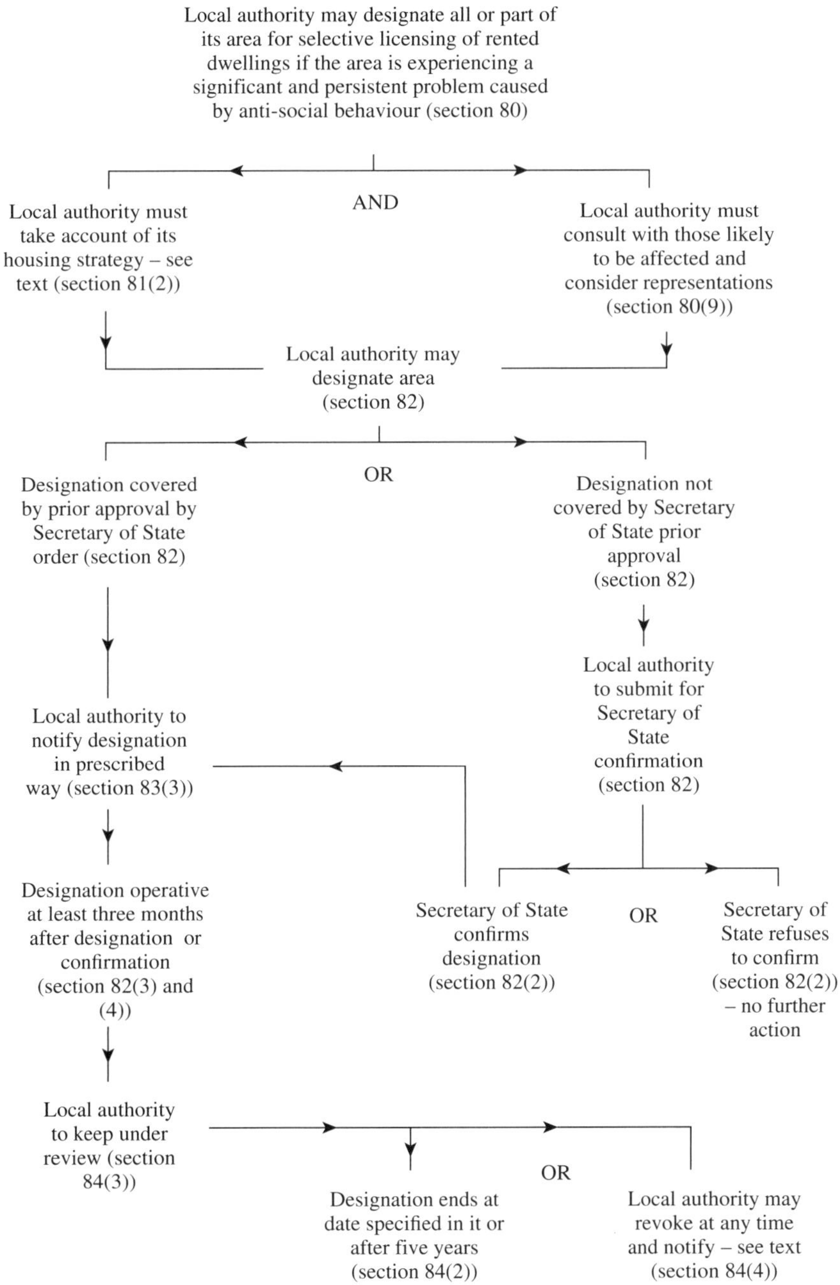

designation came into force. The local authority must from time to time review the operation of any designation made by them.

On review, if they consider it appropriate, the local authority may revoke the designation ceasing to have effect on the date specified.

The local authority must publish notice of the revocation (section 84).

The designation applies to all privately rented dwellings except houses in multiple occupation in the designated area. The local authority must take all reasonable steps to secure that applications for licences are made to them for houses in their area which are required to be licensed (section 85).

Temporary exemption

A person having control of or managing a house which is required to be licensed may notify the local authority of his intention to take steps to ensure the house is no longer required to be licensed. If the local authority thinks fit, it may serve a temporary exemption notice in respect of the house.

A temporary exemption notice is in force for three months beginning with the date on which it is served. If the local authority receives a further notification of intention to take steps to ensure licensing is not required, a second temporary exemption notice may be served for a further three months after the date when the first notice ceases to be in force (section 86(5)).

If the authority decides not to serve a temporary exemption notice, they must without delay serve on the person concerned a notice informing him of:

(a) the decision;
(b) the reasons for it and the date on which it was made;
(c) the right to appeal against the decision; and
(d) the period within which an appeal may be made.

The person concerned may appeal to the Residential Property Tribunal against the decision within the period of 28 days beginning with the date of the decision. The tribunal may confirm or reverse the decision of the authority, and if it reverses the decision, must direct the local authority to issue a temporary exemption notice with effect from such date as the tribunal directs (section 86).

Applications for licences

An application for a licence must be made to the local authority accompanied by a fee fixed by the authority. In fixing the fee for application, the local authority should consider all costs incurred in carrying out their functions under the licensing scheme (section 87).

Grant or refusal of licence

On receiving an application for licensing, the authority must either grant or refuse a licence.

FC73 Selective licensing of rented dwellings – temporary exemption notices – Section 86 Housing Act 2004

Person having control of or managing a house required to be licensed notifies local authority of intention to take steps to ensure the house is no longer required to be licensed (section 86(1))

Local authority satisfied and serves a temporary exemption notice (section 86(2))

↓

Local authority receives further notification of intention to take steps to ensure licensing is not required

↓

A second temporary exemption notice served for a further three months (section 86(5))

Local authority decides not to serve a temporary exemption notice and serves on the person concerned a notice informing him of the decision, the reasons for it and the date on which it was made, the right to appeal against the decision, and the period within which an appeal may be made

↓

Appeal to a Residential Property Tribunal against the decision within 28 days of the date of the decision

↓

Residential Property Tribunal confirms or reverses the decision of the authority, and if it reverses the decision, must direct the local authority to issue a temporary exemption notice (section 86)

FC74 Selective licensing of rented dwellings – Sections 87 to 93 Housing Act 2004

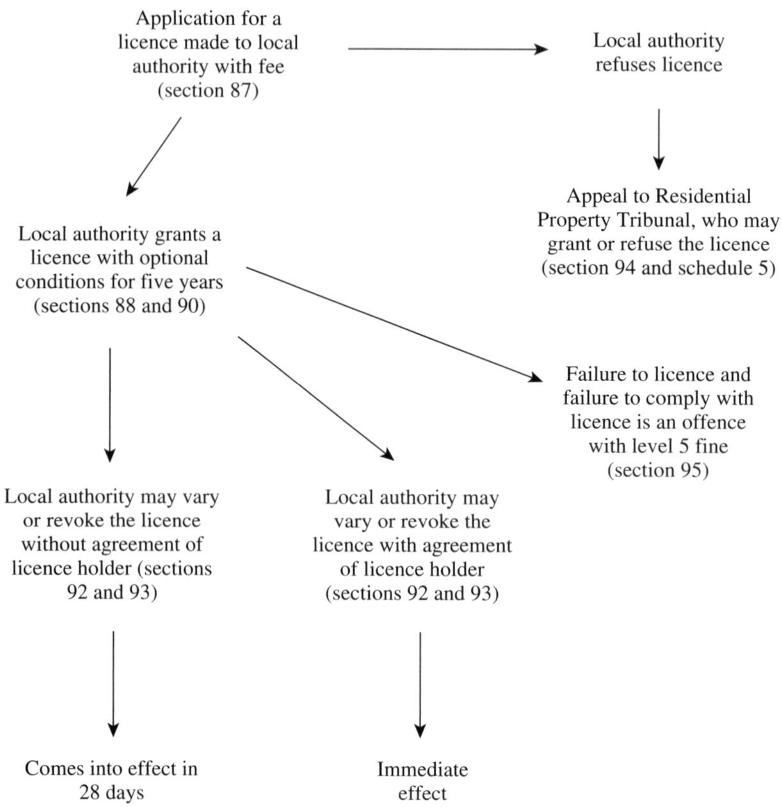

Notes

1. Local authorities must not use licensing to remove or reduce category 1 or category 2 hazards in the house (section 90).
2. The local authority can vary the licence on its own initiative or on application from the licence holder or a 'relevant person', or a person managing or having control of the house or a person who is subject to any restriction or obligation imposed by the licence (section 92).

If the local authority is satisfied:

(a) that the proposed licence holder is a fit and proper person and is the most appropriate person to be the licence holder;
(b) that the proposed manager of the house is either the person having control of the house, or a person who is an agent or employee of the person having control of the house;
(c) that the proposed manager of the house is a fit and proper person; and
(d) that the proposed management arrangements for the house are satisfactory,

they may grant a licence either to the applicant or to some other person, if both he and the applicant agree (section 88).

Tests for fitness etc. and satisfactory management arrangements

In deciding whether a person is a fit and proper person to be the licence holder or the manager of the house, the local authority must have regard to evidence that shows that the person has:

(a) committed offences involving fraud, other dishonesty, violence or drugs, or of a sexual nature;
(b) practised unlawful discrimination on grounds of sex, colour, race, ethnic or national origins or disability in, or in connection with, the carrying on of any business; or
(c) contravened any provision of the law relating to housing or of landlord and tenant law.

They must also consider evidence that shows that any person associated or formerly associated with the person (whether on a personal, work or other basis) has done any of the things above and it appears to the authority that the evidence is relevant to the question of whether the person is a fit and proper person to be the licence holder or the manager of the house.

In deciding whether the proposed management arrangements for the house are satisfactory, the local authority must have regard to:

(a) whether any person proposed to be involved in the management of the house has a sufficient level of competence to be involved;
(b) whether any person proposed to be involved in the management of the house (other than the manager) is a fit and proper person to be involved; and
(c) whether any proposed management structures and funding arrangements are suitable (section 89).

Licences

The local authority may attach conditions to a licence and they may include:

(a) restrictions or prohibitions on the use or occupation of particular parts of the house by persons occupying it;

(b) requiring the taking of reasonable and practicable steps to prevent or reduce anti-social behaviour by persons occupying or visiting the house;
(c) requiring facilities and equipment to be made available in the house;
(d) requiring such facilities and equipment to be kept in repair and proper working order;
(e) requiring works needed for any such facilities or equipment to be made available (section 90(1) to (3)).

Local authorities must not use licensing to remove or reduce category 1 or category 2 hazards in the house but this does not prevent the authority from imposing licence conditions relating to the installation or maintenance of facilities or equipment (section 90).

Licenses apply to only one house and come into force at the specified time and, unless previously terminated or revoked, continue in force for the specified period for at least five years. A licence cannot be transferred to another person. If the holder of the licence dies while the licence is in force, the licence ceases to be in force on his death. But for the three months after the licence holder's death, the house is to be treated as if on that date a temporary exemption notice had been served in respect of the house. If, at any time during that period, the personal representatives of the licence holder requests the local authority to do so, the authority may serve on them a notice which has the same effect as a temporary exemption notice (section 91).

Variation of licences

The local authority may vary a licence with the agreement of the licence holder, or if they consider that there has been a change of circumstances since the time when the licence was granted, including any discovery of new information. A variation made with the agreement of the licence holder takes effect at the time when it is made. Otherwise, it comes into force 28 days later, unless there is an appeal against the variation.

The local authority can vary the licence on its own initiative or on application from the licence holder or a 'relevant person' (who has an estate or interest in the house concerned), or a person managing or having control of the house or a person who is subject to any restriction or obligation imposed by the licence (section 92).

Revocation of licences

The local authority may revoke a licence:

(a) with the agreement of the licence holder;
(b) where the licence holder or any other person has committed a serious breach of a condition of the licence;
(c) where the licence holder is no longer considered a fit and proper person to be the licence holder; and
(d) where the management of the house is no longer being carried on by persons who are fit and proper persons to be involved in its management.

A revocation made with the agreement of the licence holder takes effect at the time when it is made. Otherwise, it takes effect 28 days from the notification of revocation, unless there is an appeal against the revocation to the Residential Property Tribunal.

The local authority can revoke a licence on its own initiative or on application made by the licence holder or a 'relevant person' (see above).

The procedural requirements and appeals against licence decisions and contained in schedule 5 of the Housing Act 2004, which deals with procedural requirements relating to the grant, refusal, variation or revocation of licences and appeals against licence decisions.

Offences

It is an offence not to license a house which should be licensed and to fail to comply with any condition of the licence.

It is a defence that a temporary exemption notice is in force or an application for a licence had been made. It is also a defence to have a reasonable excuse for having control of or managing the house or for failing to comply with the condition.

On summary conviction, offences are subject to a fine either not exceeding £20,000 or not exceeding level 5 on the standard scale (section 95).

Definitions

Dwelling means a building or part of a building occupied or intended to be occupied as a separate dwelling.
House means a building or part of a building consisting of one or more dwellings (section 99).
Private sector landlord does not include a non-profit registered provider of social housing or a registered social landlord (section 80).

HOUSES IN MULTIPLE OCCUPATION

This section deals exclusively with houses in multiple occupation, but the Housing Health and Safety Rating System and its enforcement procedures in the previous sections also apply to houses in multiple occupation, except in those cases where the existence of a management order removes it from that procedure. These situations are noted in the text relating to that procedure.

Definitions

Section 254 of the Housing Act 2004 defines three categories of privately rented buildings that are deemed to be houses in multiple occupation. These are as follows:

1. Converted blocks of flats as defined in section 257, which are buildings or parts of buildings which have been converted into and consist of self-contained flats and:

(a) the works of conversion did not, and still do not, comply with appropriate building standards of the Building Regulations; and
(b) less than two-thirds of the self-contained flats are owner-occupied.

These are known as 'poorly converted buildings' or PCBs. Their inclusion as houses in multiple occupation will be made by regulations to be made by the Secretary of State (note – Houses in Multiple Occupation (Certain Blocks of Flats) (Modifications to the Housing Act 2004 and Transitional Provisions for section 257 HMOs) (England) Regulations 2007/1904).

2. Buildings or parts thereof which are the subject of a house in multiple occupation declaration under section 255 – see FC75 below.
3. Buildings or parts thereof that meet one of the tests set out in section 254. These are as follows:
   (a) The 'standard test' where persons who do not form a single household (this term is defined in section 258), and occupy as their sole or main residence, share basic amenities in living accommodation that is not a self-contained flat or flats, or that the living accommodation lacks one or more of those amenities. In order to be a house in multiple occupation, the property must be used as the tenant's only or main residence and it should be used solely or mainly to house tenants. Properties let to students and migrant workers will be treated as their only or main residence and the same applies to properties that are used as domestic refuges (see section 259 and regulation 5 of the Licensing and Management of HMOs and other Houses (Miscellaneous Provisions) (England) Regulations 2006).
   (b) The 'self-contained flat test' where, in part of a building, persons of different households share basic amenities or the flat lacks one or more of those amenities.
   (c) The 'converted building test' where privately rented converted building unrelated occupiers share basic facilities, including flats where the basic amenities for the exclusive use of the occupant are situated outside of the main living accommodation.

The basic amenities referred to are:

- a toilet;
- personal washing facilities;
- cooking facilities (Housing Act 2004, section 254).

Persons do not form a single household unless they are members of the same family (defined in detail in section 258(3)–(5)) or they form a prescribed relationship under the Licensing and Management of Houses in Multiple Occupation and Other Houses (Miscellaneous Provisions) (England) Regulations 2006. These 'prescribed relationships' include persons carrying out domestic work, carers and fostering families (section 258).

In order to satisfy any of the three tests above, the property must be used as the tenant's only or main residence and it should be used solely or mainly to house tenants. (But see houses in multiple occupation declarations below.)

A reasonable precis of all of this is provided in 'Houses in multiple occupation (HMO): Licensing scope and schemes – May 2006' (DCLG).

A house in multiple occupation, subject to exemptions (see below) is either:

(a) an entire house which is let to three or more tenants who form two or more households and who share basic facilities;
(b) a house that has been converted entirely into bedsits or other non-self-contained accommodation and which is let to three or more tenants and who form two or more households and who share basic facilities;
(c) a converted house which contains one or more flats which are not wholly self-contained and which is occupied by three or more tenants who form two or more households;
(d) a building which is converted entirely into self-contained flats if the conversion did not meet the building regulation standards and more than one-third of the flats are let on short-term tenancies.

Exemptions

In relation to the licensing, management and overcrowding provisions of part 2, but not the enforcement provisions of part 1, the following properties are not to be regarded as houses in multiple occupation:

(a) those managed by local authorities, registered social landlords and other specified bodies (see the Houses in Multiple Occupation (Specified Educational Establishments) (England) Regulations 2012);
(b) those subject to other regulatory schemes specified in schedule 1 of the Licensing and Management of Houses in Multiple Occupation and other Houses (Miscellaneous Provisions) (England) Regulations 2006;
(c) student halls of residence;
(d) buildings occupied by religious communities;
(e) those occupied predominantly by owner-occupiers at a level to be specified in regulations;
(f) those occupied only by persons who form two households (section 254(5) and schedule 14).

Section 254(6) empowers the Secretary of State by regulations to alter any of these definitions of a house in multiple occupation if it is considered appropriate.

HOUSES IN MULTIPLE OCCUPATION DECLARATIONS

Reference

Housing Act 2004, sections 255–256.

Extent

These provisions apply to England and Wales.

Scope

To satisfy any of the three tests which define a house in multiple occupation (see above), all occupants need to live there as their sole or main residence. Because of the rapidly changing occupancy of many such properties, this may lead to difficulties in the enforcement of standards. This procedure allows a local authority to designate a house in multiple occupation where the building, or part of it, is occupied by persons who do not form a single household, and occupancy as their sole or main residence forms only a 'significant use' of that accommodation. 'Significant use' is not defined. Such designation then removes doubt about its status, and the provisions of the housing authorities relating to houses in multiple occupation can then be applied (section 255(1)–(3)).

Service of house in multiple occupation declaration

The declaration is made by service of a notice on all 'relevant persons' (see below) within seven days of the decision which:

1. states the date of the decision to make the declaration;
2. states the day on which it will come into force if no appeal is made being not less than 28 days after the decision;
3. sets out the rights of appeal (section 255(4)).

Relevant persons

For the purposes of this procedure, these are:

(a) those having an estate or interest but not a tenant under a lease with three years or less unexpired;
(b) those managing or having control (section 255(12)).

Appeals

Any 'relevant person' may appeal to the Residential Property Tribunal within 28 days of the date of decision. And in this event the effect of the declaration is delayed until the appeal is determined. The Residential Property Tribunal may confirm or revoke the declaration. If confirmed, the declaration comes into effect 28 days after the Residential Property Tribunal decision (section 255(7) and (9)).

Revocation

The local authority may revoke a declaration at any time, either on its own initiative or following representation from a 'relevant person', if satisfied that the 'significant use' test is no longer fulfilled. There is no requirement for notification of a decision to revoke. There is, however, a requirement for the local authority to serve notice on any person who made an application for revocation where that request is refused. In this case, a notice must be served giving:

(a) the decision and the date on which it was made;

FC75 House in multiple occupation (HMO) declarations – Sections 255 and 256 Housing Act 2004

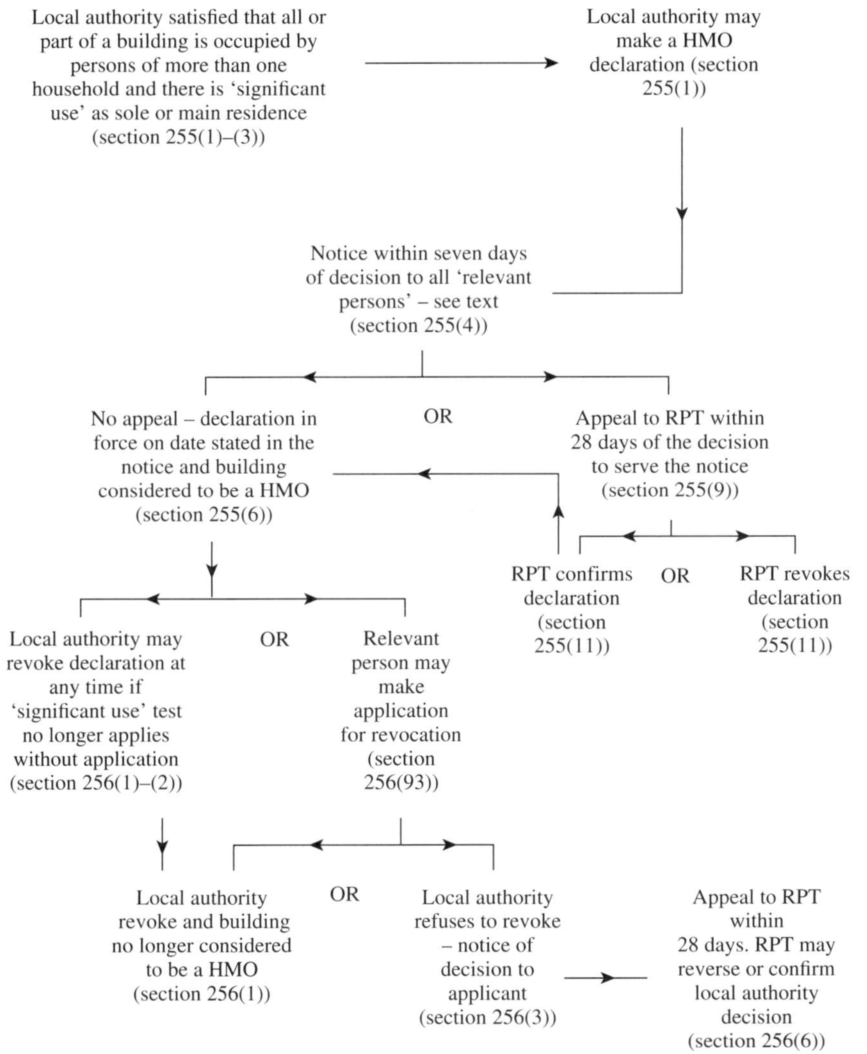

Note

1. RPT in this flow chart = residential property tribunal.

(b) the reasons for that decision;
(c) the rights of appeal.

A person served with a refusal notice may, within 28 days, appeal to the Residential Property Tribunal, which may confirm or reverse the decision by revoking the declaration (section 256).

LICENSING OF HOUSES IN MULTIPLE OCCUPATION

References

Housing Act 2004, part 2.
Management of Houses in Multiple Occupation (England) Regulations 2006.
The Licensing of Houses in Multiple Occupation (Prescribed Descriptions) (England) Order 2006 (Prescribed Description Order).
The Licensing and Management of Houses in Multiple Occupation and other Houses (Miscellaneous Provisions) (England) Regulations 2006 (Miscellaneous Provisions Regulations) as amended.
Enforcement Guidance, ODPM February 2006, part 6, paragraphs 6.1–6.5.

Extent

These provisions apply to England only but there are similar provisions for Wales.

Scope

Houses in multiple occupation on three or more storeys with five or more tenants in two or more households must be licensed with the local authority, except those:

(i) where an exemption notice is in force (FC80);
(ii) subject to an interim or final management order (FC78 and FC79) (section 55(2), 61(1) and Prescribed Description Order 2006).

The intention behind the licensing requirements for this group of houses in multiple occupation is that they present the highest risk to occupiers and therefore need to have additional controls to ensure adequate standards.

In addition, local authorities may designate areas of their district in which there is to be licensing for other types of houses in multiple occupation. The procedure for designation is at FC77 but the licensing procedure is the same as for other houses in multiple occupation (section 56(1)).

A licence for a house in multiple occupation is defined as ‘authorising occupation of the house concerned by not more than a maximum number of households or persons specified in the licence’ (section 61(2)).

FC76 Licensing of houses in multiple occupation (HMOs) – Sections 55 to 71 Housing Act 2004

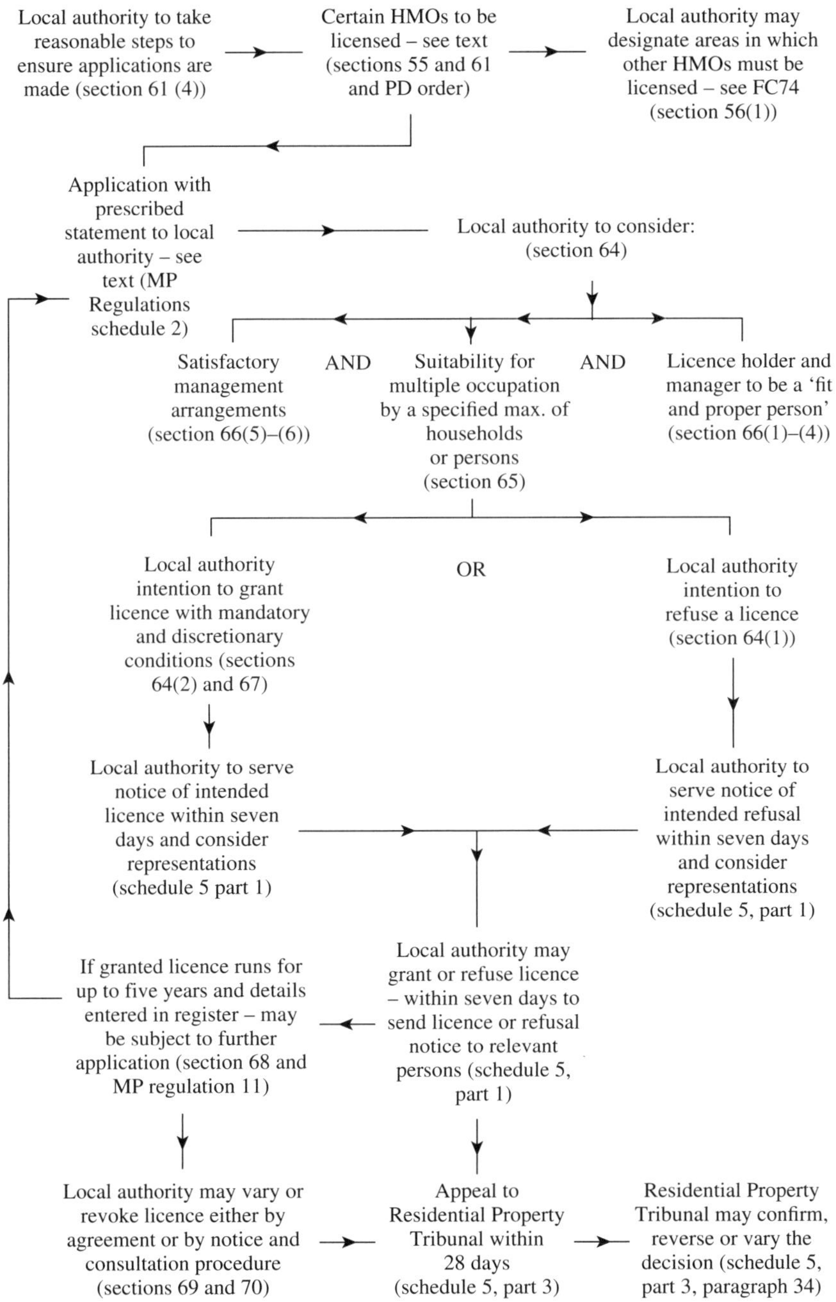

Notes

1. PD order = The Licensing of Houses in Multiple Occupation (Prescribed Descriptions) (England) Order 2006.
2. MP regulations = The Licensing and Management of Houses in Multiple Occupation and other Houses (Miscellaneous Provisions) (England) Regulations 2006.

Local authority duties connected with licensing

The local authority is required to:

1. promote the implementation of licensing – this includes a duty to take reasonable steps to ensure that applications are made where they appear to be required;
2. ensure that it deals with applications promptly;
3. satisfy itself that there are no part 1 enforcement functions remaining to be discharged in respect of a licensed house in multiple occupation as soon as possible but in any event not longer than five years from the first application (section 55(5) and 61(4)).

Applications for licences

The requirements are prescribed in section 63 and the Miscellaneous Provisions Regulations 2006 and these include:

- a prescribed statement from the local authority requiring the applicant to inform all persons with an interest etc. of the application and details relating to it (schedule 2, paragraph 1);
- details to be included in the application (regulation 7(2) and schedule 2);
- payment of a fee at the discretion of the local authority but subject to a maximum specified in the regulations (section 63(3)).

Grounds for consideration

A licence must be granted if the following criteria are satisfied:

(a) The house is suitable (see below) for occupation by a certain number of persons or households as specified in the application or by the local authority, or can be rendered suitable by the imposition of conditions.
(b) The proposed licence holder is a fit and proper person (see below) as well as being the most appropriate person to be granted a licence.
(c) The proposed manager is the person having control of the house or an agent or employee of that person and is also a fit and proper person (see below).
(d) The proposed management arrangements are satisfactory (see below).

If these criteria are not met, the application must be refused (section 64).

Suitability for multiple occupation

A house in multiple occupation is deemed reasonably suitable for occupation by a particular maximum of households or persons if it meets the standards contained in schedule 3 of the Miscellaneous Provisions Regulations 2006. These standards relate primarily to:

(i) heating;
(ii) washing facilities (bathrooms, toilets, washbasins and showers);

(iii) kitchens (areas for food storage, preparation and cooking);
(iv) units of living accommodation without shared basic facilities; and
(v) fire precautions.

If it does not meet these standards, a licence cannot be granted. A local authority has the discretion to decide that the house is not suitable for occupation even if it complies with the prescribed standards and may require higher standards in relation to a maximum number of persons (section 65).

Fit and proper person

In determining whether a person is a fit and proper person to be a licence holder or manager, the local authority must have regard to whether that person, or a relevant associate, e.g. a spouse or business partner, has committed offences involving fraud, dishonesty, violence, drugs or sexual offences, unlawful discrimination in business, contravention of housing law or breach of an approved code of management, e.g. Housing (Codes of Management Practice) (Student Accommodation) (England) Order 2010 (section 66(1)–(4)).

Satisfactory management arrangements

In deciding if the proposed management arrangements are satisfactory, the local authority must have regard, amongst other matters, to:

(a) the competence of the manager;
(b) fitness of other persons involved in the management structure;
(c) management structure; and
(d) funding (section 66(5) and (6)).

Grant and refusal of licences

Before granting a licence, the local authority must:

(a) serve notice on the applicant and other relevant persons of its intention to grant the licence, together with a copy of it;
(b) consider any representations made within the period specified.

If, following this consultation, the local authority intends to modify the proposed licence, they must again serve notice and consider representations.

Before refusing to grant a licence, the local authority must:

(a) serve notice on the applicant and other relevant persons of its intention; and
(b) state the reasons for refusal and indicate the length of the consultation period;
(c) consider representations made.

Following a grant or refusal, the local authority must serve notice within seven days on the applicant and other relevant persons setting out:

(a) the local authority's decision;

(b) the reasons for the decision and the date on which it was made; and
(c) rights of appeal and the appeal period (section 65 and schedule 5, part 1).

Licence conditions

Mandatory conditions

These are set out in schedule 4 and deal mainly with the safety of occupiers, including furniture and electrical, gas and fire safety appliances. These must be included in each licence (section 67(3) and schedule 4).

Discretionary conditions

The local authority may include any conditions relating to the management, use and occupation of the house and its contents and condition. These might include:

(a) restrictions on the use of particular parts;
(b) the reduction of anti-social behaviour;
(c) the installation and maintenance of facilities and equipment to meet the standards required under section 65 above;
(d) carrying out works to facilities and equipment within specified periods.

Local authorities are not to use licence conditions to deal with matters of health and safety but should use the enforcement procedures of part 1 (section 67).

General requirements and duration

A licence can only relate to a single house in multiple occupation and is valid for the period specified in it, which cannot exceed five years. They are not transferable but, on death of a licence holder, a three months' temporary exemption from the licensing requirement is given (section 68) (see page 338). If the house is to continue in a licensable status after the expiration of the licence, there must be a further application for a new licence, which will be subject to the same procedure.

Variation

A local authority may vary a licence either by agreement with the owner or if there is a change of circumstances, e.g. the discovery of new information about issues present before the licence was granted, or there is a need to vary the appropriate number of households or persons. Variation may be instigated by the local authority or on application from the owner or a relevant person. The procedure to be used in varying a licence is contained in part 2 of schedule 5 and provisions for appeal against refusals to vary in part 3.

Variations with the agreement of the owner come into effect immediately, otherwise when the period for appeal has elapsed or any appeal has disposed of (section 69).

Revocation

A local authority may revoke a licence:

- with the agreement of the licence holder;
- where the licence holder has committed a serious breach of a licence condition or there are repeated breaches of conditions;
- where the local authority considers that the holder is no longer a fit and proper person; and
- where the local authority believes that the property no longer meets the standards required for a licence.

The procedure to be adopted is set out in part 2 of schedule 5, with appeals dealt with in part 3 of schedule 5. Revocations with the agreement of the owner come into effect immediately, otherwise when the period for appeal has elapsed or any appeal has disposed of (section 70).

Appeals to the Residential Property Tribunal

Appeals against licensing decisions are dealt with in part 3 of schedule 5 and deal with appeals:

(a) against the refusal or grant of a licence, including licence conditions; and
(b) decisions to vary or revoke, or refusals to do so.

Generally appeals are to be made within 28 days of the date specified in the local authority notice.

Register of licences

A local authority must keep a register of the licences granted containing the details laid out in regulation 11 of the Miscellaneous Provisions Regulations. The register must be available to the public at the head office of the local authority and copies of it, or extracts from it, must be provided by the local authority on payment of a fee (section 232).

Offences

It is an offence, punishable by a fine of up to £20,000, to own or manage a house in multiple occupation without a licence or knowingly allow occupation by more persons or households than permitted. Breaching the licence conditions carries a fine of up to level 5. There is a defence of 'reasonable excuse' in each case (section 72).

There is also a provision for the making of rent repayment orders (See FC 82) by the Residential Property Tribunal in relation to unlicensed houses in multiple occupation, whereby there is a penalty equivalent to the rent received during unlicensed operation of up to 12 months (sections 73–74).

Management regulations

The Management of Houses in Multiple Occupation (England) Regulations 2006, as amended, apply to all houses in multiple occupation, except those converted blocks of flats under section 257. They impose duties on managers concerning:

- provision of information to occupiers;
- safety measures, including fire safety;
- maintenance of water, drainage, gas and electricity;
- maintaining common parts and living accommodation; and
- providing waste disposal facilities.

They also place a duty on occupiers to ensure that the manager can effectively carry out these duties.

There are no provisions for the service of notices by a local authority to secure compliance but a person in breach of these requirements is liable to a fine of up to level 5.

It should be noted that there are duties placed on the manager of a house in multiple occupation through the Licensing and Management of Houses in Multiple Occupation (Additional Provisions) (England) Regulations 2007.

DESIGNATION OF AREAS BY LOCAL AUTHORITIES FOR THE ADDITIONAL LICENSING OF HMOs

References

Housing Act 2004, sections 55–60.
The Licensing and Management of Houses in Multiple Occupation and other Houses (Miscellaneous Provisions) (England) Regulations 2006.

Extent

These provisions apply to England and Wales.

Scope

This procedure allows a local authority to designate the whole or part of its district as an area in which all or specified types of houses in multiple occupation will require to be registered in addition to those already designated by the Licensing of Houses in Multiple Occupation (Prescribed Descriptions) (England) Order 2006 (section 56(1)). Following designation, the procedure for their licensing is the same as in FC76 above.

Before making a designation, the local authority must be satisfied that a significant proportion of the houses in multiple occupation to be included are being managed sufficiently ineffectively as to give rise, or be likely to give rise, to one or more particular problems, either for their occupants or for members of

FC77 Designation of areas for additional licensing of houses in multiple occupation – Sections 55 to 60 Housing Act 2004

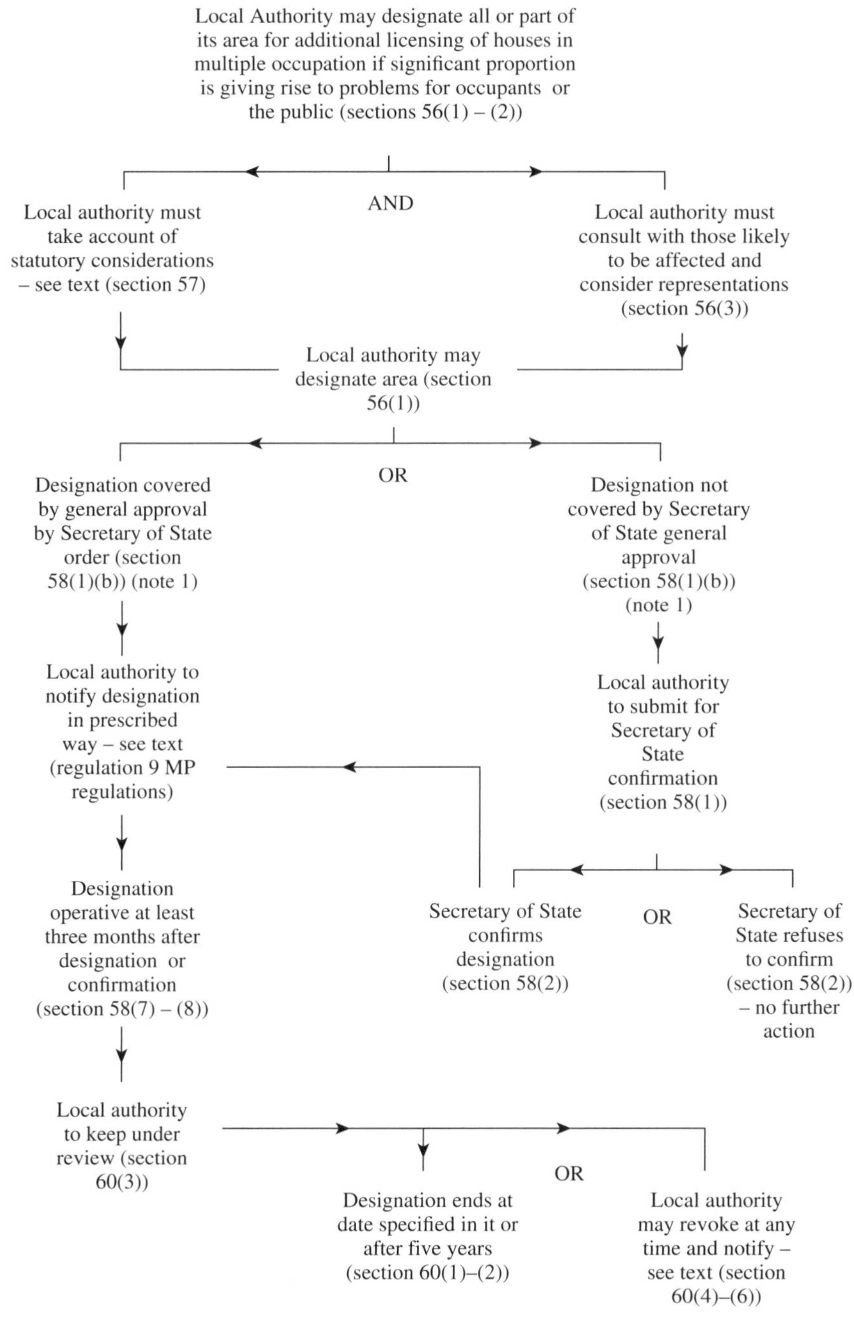

Notes

1. No order has yet been made.
2. MP regulations in this flow chart – The Licensing and Management of Houses in Multiple Occupation and other Houses (Miscellaneous Provisions) (England) Regulations 2006.

the public, e.g. anti-social behaviour. In forming this opinion, the local authority must have regard to any Codes of Practice issued under section 233 (section 56(2)–(6)).

Designation

Before making a designation, the local authority must consult with those most likely to be affected and take account of any representations made (section 56(3)).

There are further matters for the local authority to consider:

- The designation must be consistent with the local authority's overall housing strategy.
- There must be a coordinated approach to dealing with homelessness, empty properties and anti-social behaviour in the private rented sector, which considers both combining additional licensing with other possible courses of action and measures to be taken by others.
- The local authority must also consider alternative actions.
- The local authority must be satisfied that designation will significantly assist in dealing with the problems of the area (section 57).

Confirmation by Secretary of State and operation

Unless covered by a general approval (none have yet been issued), the designation requires confirmation by the Secretary of State (section 58(1)). There is no specified procedure for this part of the process. The Secretary of State may confirm or refuse to confirm a designation. The designation comes into effect no earlier than three months after it has been confirmed or, in the case of general approvals, three months after the designation was made (section 58).

Notification of designation

As soon as designation is confirmed or made, the local authority must publicise it by:

1. within seven days:
   (a) placing notices on municipal buildings in or closest to the area;
   (b) publishing the notice on the internet site;
   (c) publishing it in the next edition of two local newspapers and five times more at intervals of between two and three weeks;
2. within two weeks, sending a copy of the notice to:
   (a) those who responded to the consultation;
   (b) organisations representing landlords, tenants, agents;
   (c) organisations providing advice on landlord and tenant matters, e.g. housing advice centres.

The notice must contain:

(a) the date of designation;

(b) details of whether the designation requires confirmation;
(c) a brief description of the area;
(d) contact details for the local authority, the place where the designation may be inspected and where applications for licences may be obtained;
(e) a statement advising persons affected to seek advice from the local authority;
(f) a warning of the consequences of not applying for a licence (section 59(2) and the Licensing and Management of Houses in Multiple Occupation and other Houses (Miscellaneous Provisions) (England) Regulations 2006, regulation 9).

Duration

The designation must be reviewed from time to time and may be revoked at any time at the discretion of the local authority but ends no later than five years after it has been made (section 60(1)–(3)). Revocation does not require a confirmation by the Secretary of State.

Revocation

When a scheme is revoked, the local authority must within seven days:

- place notices on municipal buildings in or closest to the area;
- publish the notice on the internet site;
- publish it in the next edition of two local newspapers.

The notice must give:

- a brief description of the area;
- a summary of the reasons for revocation;
- the date revocation takes effect;
- contact details for the local authority (section 60(6) and the Licensing and Management of Houses in Multiple Occupation and other Houses (Miscellaneous Provisions) (England) Regulations 2006, regulation 10).

INTERIM MANAGEMENT ORDERS

References

Housing Act 2004, chapter 1, part 4, sections 102 to 112.
Housing (Interim Management Orders) (Prescribed Circumstances) (England) Order 2006.
Housing (Management Orders and Empty Dwelling Management Orders) (Supplemental Provisions) (England) Regulations 2006.

Extent

These provisions apply to England only. There are similar provisions for Wales.

Scope

An interim management order is an order made by a local authority in relation to a house (see below) which cannot be licensed or where there is a management problem that requires intervention by the local authority. It runs for up to 12 months and is intended to secure that:

- immediate steps are taken to protect the health, safety or welfare of occupants or persons having an estate or interest in the vicinity; and
- appropriate steps are taken to secure the proper management of the house pending the grant of a house in multiple occupation licence, or one under part 3, or a final management order (see FC79) (section 101(3)).

It is therefore an interim measure designed to provide time for a longer-term management solution to be found. Through an interim management order, the local authority takes over most of the rights and duties of the landlord and right to the possession of the property – see below.

In this procedure, a 'house' is defined as a house in multiple occupation or a part 3 house (selective licensing of other residential accommodation – see page 303) (section 101(5)).

That part of a house occupied by a resident landlord may be excluded from the terms of the order (section 102(8)).

Making of an interim management order

An interim management order *must* be made when:

(a) a house in multiple occupation or part 3 house ought to be licensed but is not and either there is no reasonable prospect of the house becoming licensed in the near future or the health and safety condition is met (see below); or
(b) the local authority intends to revoke an existing licence and either there is no prospect of a new licence being issued in the near future or the health and safety condition will be satisfied (section 102(1)–(3)).

The local authority *may* make an interim management order for:

(a) a house in multiple occupation not licensable under part 2 where the Residential Property Tribunal considers that the health and safety condition is satisfied (see below) and takes account of any relevant, approved Code of Practice (section 101(4)–(6)); and
(b) a house which satisfies the conditions of section 103 relating to special management orders and the Residential Property Tribunal authorises the local authority to make the interim management order (section 102(7)).

These discretionary interim management orders therefore require the submission to the Residential Property Tribunal, who may authorise them, either on the

FC78 Interim management orders (IMOs) – Sections 102 to 112 Housing Act 2004

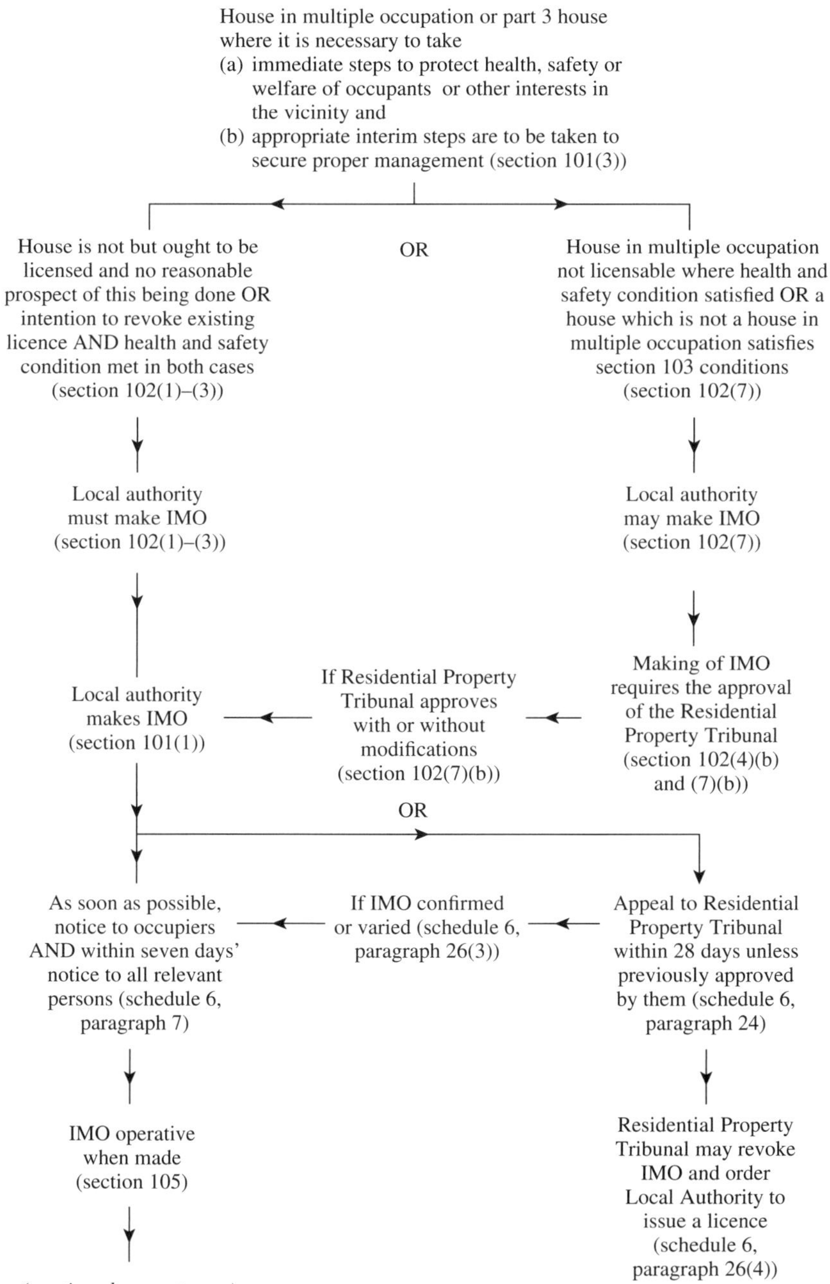

(continued on next page)

FC78 Interim management orders (IMOs) – continued

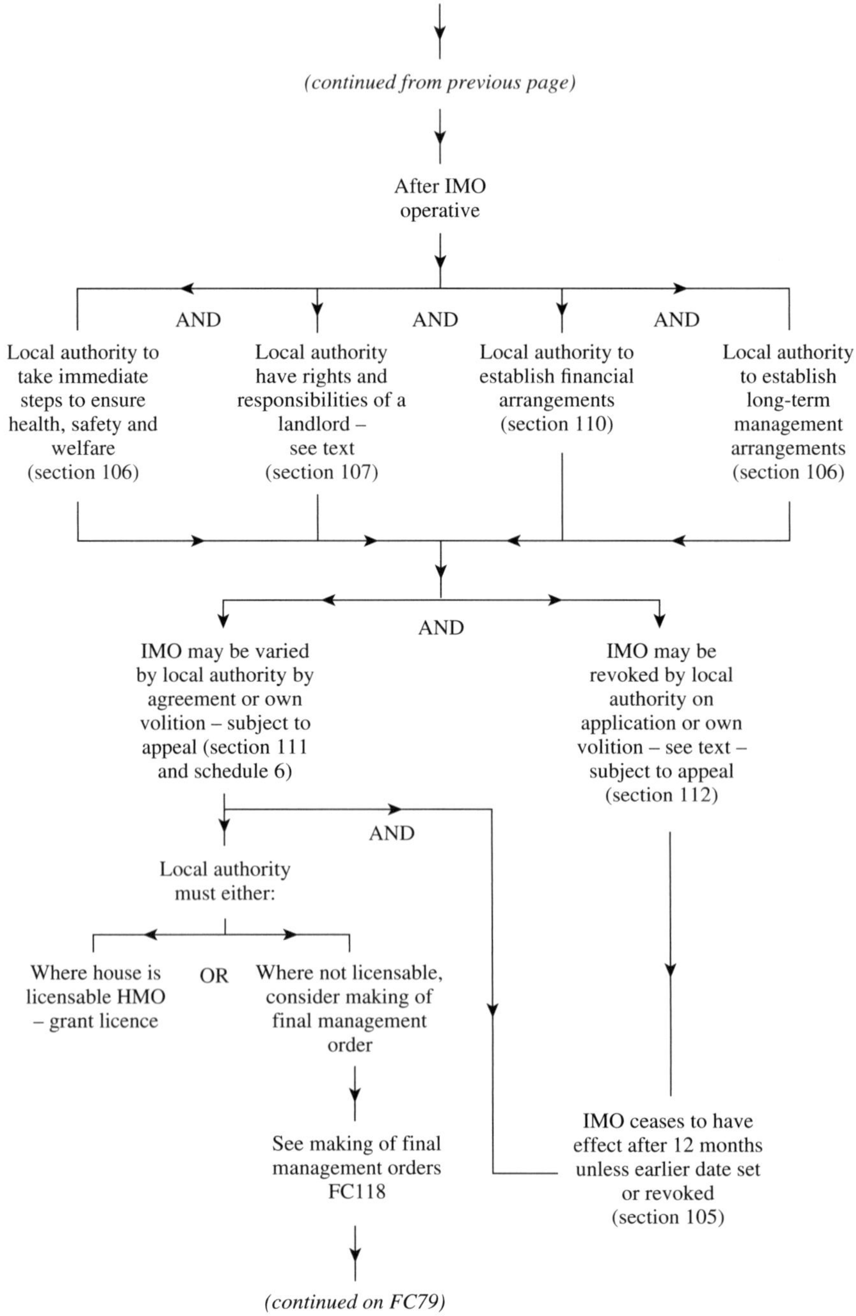

terms contained in the draft order or on such other terms as they consider appropriate (section 102 (4(b)) and 7(b)).

The procedure following the making of an interim management order is contained in schedule 6, part 1, paragraph 7. As soon as possible for occupiers and within seven days for relevant persons, the local authority should serve notices with copies of order stating:

(a) reasons for making the order;
(b) its general effect;
(c) date of operation;
(d) details of appeals to the Residential Property Tribunal to be made within 28 days.

The health and safety condition

This condition is satisfied when it is necessary to make an interim management order in order to protect the health, safety or welfare of the occupants of the house or persons occupying or owning property in its vicinity. This condition can be satisfied if there is a threat to evict occupiers in order to avoid licensing under part 2 but cannot be met where the threat relates to a category 1 or category 2 hazard and the appropriate course of enforcement action can be taken (section 104).

Operation and duration of interim management order

The interim management order comes into effect when it is made, except when it is made to follow a revocation of a licence, in which case it takes effect upon that revocation being made. The interim management order ceases to have effect after 12 months, unless an earlier date had been set or is continued in force pending the disposal of an appeal against the making of a final management order (section 105).

Local authority duties under an interim management order

These are to:

(a) take immediate steps to ensure the health, safety or welfare of occupants and those involved with property in the vicinity;
(b) put in place the long-term management arrangements;
(c) where the house is licensable, grant a licence or make a final management order;
(d) where the house is not licensable, consider if it should make a final management order or revoke the interim management order and take no further action (section 106).

General effect of an interim management order

Whilst the interim management order is operative, the local authority takes over most of the rights and responsibilities of the landlord, including the right to possession. With the consent of the landlord, the local authority may grant occupation rights (section 107). Tenancies and leases granted by the local authority in this period are to be treated as if they were legal grants by the landlord (section 108). There are special provisions in the Housing (Management Orders and Empty Dwelling Management Orders) (Supplemental Provisions) (England) Regulations 2006 which apply where the local authority is to be treated as the lessee.

Effect of interim management order on landlords and mortgagees

The landlord is no longer able to receive rents and cannot exercise any management functions or grant tenancies. His right to sell the house remains, as do the rights of any mortgagee, except where they are being used to grant tenancies or licences under section 107 (section 109).

Compensation

Compensation is payable to third parties (a person with an estate or interest in the house other than the immediate landlord and any tenants under leases granted by the local authority) for any interference or loss with their rights due to the interim management order. Compensation disputes are dealt with by the Residential Property Tribunal (section 128).

Financial arrangements

The local authority is allowed to spend the rent receipts on 'relevant expenditure', which is that reasonably incurred in performing their duties under section 106(1)–(3) (see above), with any balance going to the landlord. The local authority must keep accounts and make them available for inspection. Persons entitled to review the accounts may seek a declaration from the Residential Property Tribunal that the amount of expenditure by the local authority is unreasonable and seek an order making financial adjustments (section 110).

At the end of an interim management order, any surpluses must be paid to the landlord unless a following final management order disallows this. Deficits on balance may be recovered from the landlord (section 129).

Variation of interim management order

Interim management orders may be varied at the request of a relevant person or by the local authority without application. The procedure for dealing with this is set out in schedule 6, part 2 and includes:

- notices to relevant persons;
- consideration of representations;
- notice of decision within seven days;
- details of appeals to the Residential Property Tribunal to be within 28 days (section 111 and schedule 6).

Revocation of interim management order

The local authority may revoke an interim management order where:

(a) the house ceases to need a licence;
(b) the local authority grants a licence;
(c) the local authority makes a final management order; or
(d) other appropriate circumstances apply.

The revocation comes into effect after 28 days of being made or the resolution of any appeal. The procedure for dealing with revocations is the same as for variations above (section 112).

Appeals

A relevant person may appeal to the Residential Property Tribunal against:

(i) the making of an interim management order;
(ii) the terms of it;
(iii) decision or refusal to vary it;
(iv) decision or refusal to revoke it;
(v) compensation payable.

In the case of (i) and (ii), there is no appeal if the interim management order had previously been authorised by the Residential Property Tribunal. Generally the appeal must be made within 28 days of the making or decision involved and the Residential Property Tribunal has powers to confirm, vary or revoke the local authority's order or other decisions (schedule 6, part 3).

Powers of entry

The local authority has a power of entry at reasonable times to carry out works. If prevented from doing so, the local authority may apply to a magistrates' court for an order which carries a penalty of up to level 5 if not followed (section 131).

Special interim management orders

These may be applied to properties that are not houses in multiple occupation but which could be licensed under part 3 (see page 303) and have circumstances as defined in the Housing (Interim Management Orders) (Prescribed Circumstances) (England) Order 2006. They must be confirmed by the Residential Property Tribunal. This procedure is not covered here (section 103).

FINAL MANAGEMENT ORDERS

References

Housing Act 2004, part 4, chapter 1, sections 113 to 122.
Housing (Management Orders and Empty Dwelling Management Orders) (Supplemental Provisions) (England) Regulations 2006.

Extent

These provisions apply in England only but there are similar provisions in Wales.

Scope

The making of a final management order is one of the options a local authority can take after the ending of an interim management order; the others being revocation or the granting of a licence (section106(4)).

A final management order is an order made by a local authority in relation to a house (see below) which cannot be licensed or at which there is some management problem which requires intervention by the local authority. A final management order runs for up to five years and is intended to secure the proper management of the house on a long-term basis in accordance with a management scheme contained in the order (section 101(4)).

In this procedure, a 'house' is defined as a house in management order or a part 3 house (for selective licensing of other residential accommodation, see page 303) (section 101(5)).

The final management order may be made to exclude part of a property occupied by a person who is the owner or long leaseholder of the entire house (section 113(7)).

Making of a final management order

The local authority:

(a) *must* make a final management order to replace an interim management order that is expiring if the house is licensable under part 2 or 3 and the local authority cannot grant a licence;
(b) *may* make a final management order for a non-licensable house where an interim management order is ending and the local authority considers it necessary to protect on a long-term basis the health, safety or welfare of the occupants or of those in the vicinity (section 113(1)–(5)).

The procedural requirements for the making of a final management order are as follows.

Before making the order

(a) Serve notice containing details of and copy of order on relevant persons, together with end date of consultation period.
(b) Consider any representations.

FC79 Final management orders (FMOs) – Sections 113 to 122 Housing Act 2004

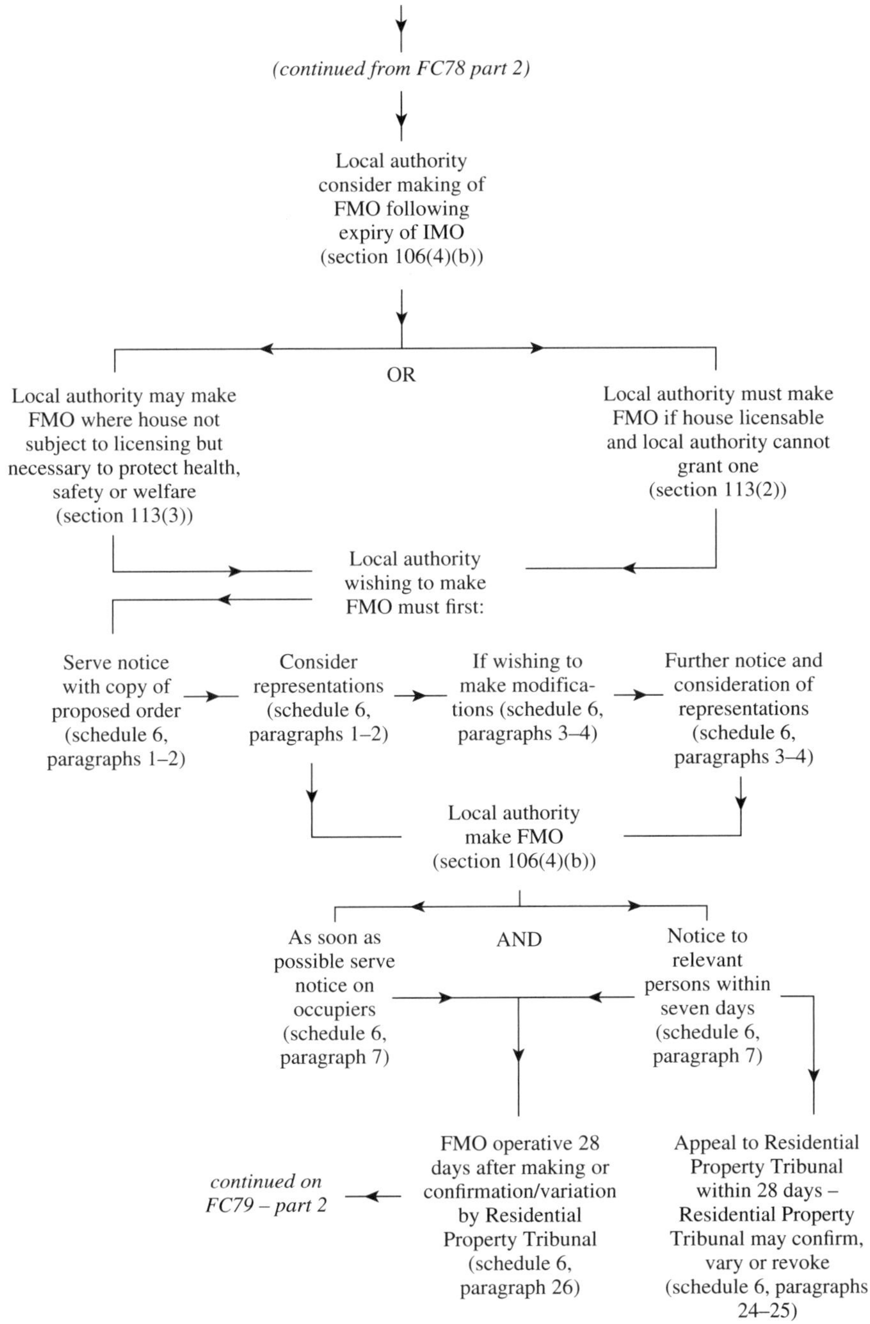

Notes

1. Relevant landlord or third party may appeal to Residential Property Tribunal if management by local authority is not in accordance with scheme (section 120).
2. There is an appeal where the local authority refuses an application to vary or revoke (schedule 6, paragraph 28).
3. In this procedure, IMO = interim management order.

FC79 Final management orders (FMOs) – continued

Continued from FC79 part 1

After FMO operative:

Local authority must take steps to secure the proper management of the house (section 115)

AND

Local authority has rights and responsibilities of landlord (section 116)

AND

Local authority must operate management scheme – text (section 119) (note 1)

AND

Local authority must keep FMO under review (section 115(3))

Local authority may vary FMO on application or at own volition subject to procedure and appeal – see text (section 121 and schedule 6, part 2) (note 2)

AND

Local authority may revoke FMO on application or at own volition subject to procedure and appeal – see text (section 122 and schedule 6, part 2) (note 2)

FMO ceases to have effect after five years or earlier if specified in the order (section 114)

Local authority may or must consider – see text – new FMO to follow this one. This will be through the same procedure (section 113(4)–(6))

Before making the final management order with modifications

(a) Serve notice of proposals and reasons and indicate length of consultation period.
(b) Consider representations.

After making final management order

Local authority must serve notices with copies of order:

(a) within seven days on all relevant persons, and
(b) as soon as possible on occupiers, stating:
    (i) reasons for making the order;
    (ii) its general effect;
    (iii) a description of the way the house is to be managed in accordance with the management scheme;
    (iv) date of operation;
    (v) details of appeals to the Residential Property Tribunal to be made within 28 days (schedule 6 part 1).

Operation of final management order

A final management order will come into force 28 days after being made or, on appeal, 28 days after confirmation by the Residential Property Tribunal. If there is a further appeal to the Lands Tribunal, it will become operative on the date of confirmation by the Land Tribunal.

It ceases to have effect after five years or on an earlier date if specified in the order. If its cessation is to be followed by a new final management order (see below), which is the subject of an appeal, the original final management order will continue until the appeal has been determined (section 114).

New final management orders

On the expiry of a final management order, the local authority must or may, depending on the particular situation (see ‘Making of a final management order’ above), make a further final management order if the circumstances causing the existing final management order to be made are still present (section 113(4)–(6)). The procedure to be followed will be the same as for the making of the existing final management order.

Local authority duties when final management order becomes operative

The local authority must take steps to secure the proper management of the house through the management scheme (see below). They must review the order and the scheme from time to time and consider whether the final management order should remain in force (section 115).

General effect of a final management order

The local authority takes over the duties and responsibilities of the landlord and, unlike the position with an interim management order, does not need the consent of the landlord before granting occupation rights, unless that right is for a fixed term to expire after the final management order is to end. Tenancies and leases granted by the local authority in this period are to be treated as if they were legal grants by the landlord (section 116). There are special provisions in the Housing (Management Orders and Empty Dwelling Management Orders) (Supplemental Provisions) (England) Regulations 2006 which apply where the local authority is to be treated as the lessee.

Sections 126–127 apply detailed provisions to deal with the status of the local authority as landlord in legal proceedings and the supply of furniture.

The landlord is no longer able to receive rents and cannot exercise any management functions or grant tenancies. His right to sell the house remains, as do the rights of any mortgagee, except where they are being used to grant tenancies or licences (sections 116–118). The occupiers retain the same legal status as before the final management order was made and nothing prevents the local authority from granting a private sector-type tenancy, although the tenant does not become a secure tenant (section 124).

Management schemes and accounts

The final management order must contain a management scheme which only becomes operative after the landlord has had the opportunity to agree its terms or after an appeal to the Residential Property Tribunal is settled. The scheme is to be divided into two parts. The first will deal with financial matters, including:

- carrying out of works;
- estimates of expenditure to be incurred;
- amounts of rents to be sought;
- amounts and provision of compensation;
- any payments to the landlord;
- financial arrangements with the landlord when the final management order ends.

The second part of the scheme will cover those issues that caused the local authority to make the order, including:

- steps to be taken to require occupiers to comply with their obligations under lease or general law;
- description of any repairs required (section 119).

Any relevant landlord or third party may apply to the Residential Property Tribunal if it is considered that management of the house by the local authority is not in accordance with the management scheme. The Residential Property Tribunal may make an order requiring the local authority to act in accordance with the scheme and award damages. It may also revoke the final management order (section 120).

At the end of a final management order, any surpluses must be paid to the landlord unless any following final management order disallows this. Deficits on balance may be recovered from the landlord (section 129).

Variation of final management order

Final management orders may be varied at the request of a relevant person or by the local authority without application (section 121). The procedure for dealing with this includes:

- notices to relevant persons with reasons for the decision and indicating the length of the consultation period;
- consideration of representations;
- notice of decision to vary within seven days;
- appeals to the Residential Property Tribunal within 28 days.

Before refusing to vary a final management order, the local authority must:

(a) give notice to each relevant person with reasons and length of consultation period;
(b) consider representations;
(c) if still intending to revoke, give notice to same persons within seven days giving the decision, the reasons for it and the rights of appeal to be made within 28 days (schedule 6, part 2).

Revocation of final management order

The local authority is able to revoke a final management order either on its own volition or following a request to do so:

(a) if the house ceases to be licensable;
(b) if the local authority grants a licence;
(c) if the local authority makes a further final management order; or
(d) in other appropriate circumstances (section 122).

The procedure for dealing with the revocation of a final management order is contained in schedule 6, part 2 and mirrors that for the variation of them above.

Compensation

Compensation is payable to third parties (a person with an estate or interest in the house other than the immediate landlord and any tenants under leases granted by the local authority) for any interference or loss with their rights due to the final management order. Compensation disputes are dealt with by the Residential Property Tribunal (section 128).

Power of entry

The local authority has a power of entry at reasonable times to carry out works. If prevented from doing so, the local authority may apply to a magistrates' court for an order which carries a penalty of up to level 5 if not followed (section 131).

Appeals

A relevant person may appeal to the Residential Property Tribunal against:

- the making of a final management order;
- the terms of it and of the management scheme;
- decision or refusal to vary it;
- decision or refusal to revoke it;
- compensation payable.

Generally the appeal must be made within 28 days of the making or decision involved and the Residential Property Tribunal has powers to confirm, vary or revoke the local authority's order or other decisions (schedule 6, part 3).

TEMPORARY EXEMPTION NOTICES FOR HMOs

Reference

Housing Act 2004, section 62.

Extent

These provisions apply in England and Wales.

Scope

This procedure allows a local authority to grant a temporary exemption from the need to licence a house in multiple occupation in circumstances where the person having control or management notifies the local authority of his intention to take steps to create a situation where the house will no longer require a licence, e.g. by reducing occupants or households (section 62(1)).

Notification

Having notified the local authority of his intent, the person does not commit the offence of using a house as a house in multiple occupation without a licence until a temporary exemption notice expires. This applies both to licensing requirements under the Licensing of Houses in Multiple Occupation (Prescribed Descriptions) (England) Order 2006 and those houses in multiple occupation included in designated areas (section 72(4)(a)).

Temporary exemption notices (TENs)

The local authority upon receipt of such a notification may, at its discretion, serve a temporary exemption notice, which will be in force for three months from the date of service (section 62(4)(a)).

FC80 Temporary exemption notices (TENs) – Section 62 Housing Act 2004

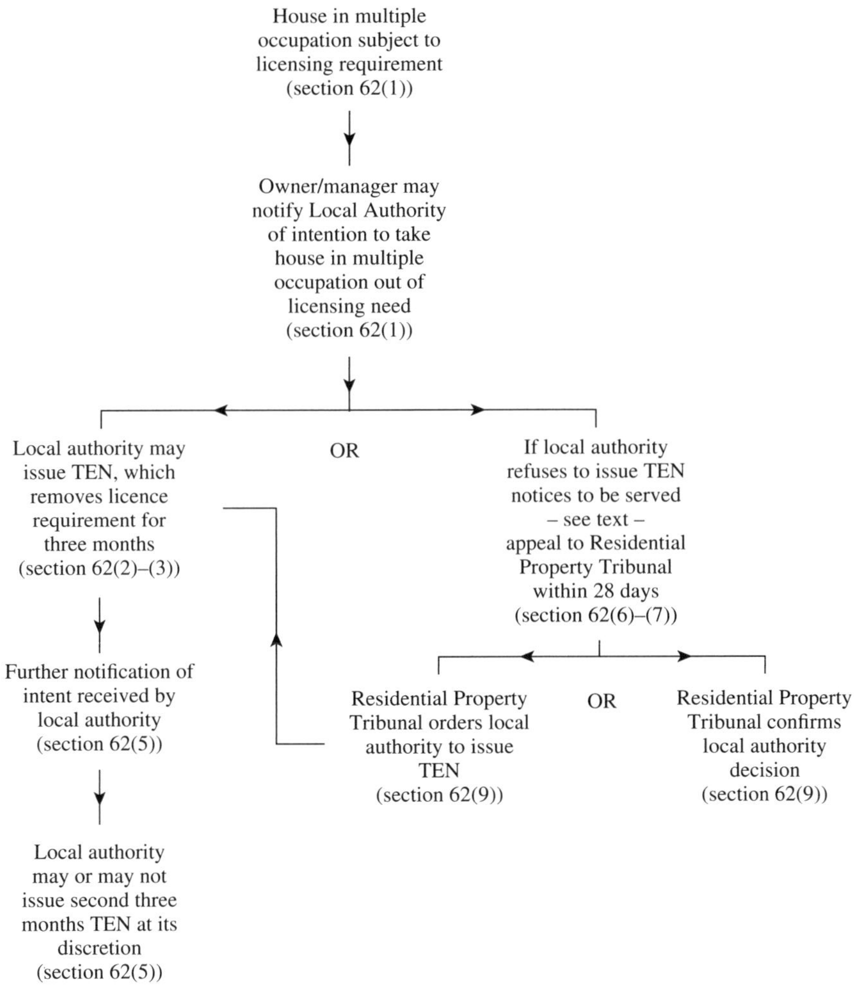

If the local authority receives a further notification, and considers that there are exceptional circumstances which justify it, they may serve a second temporary exemption notice, which will operate for three months from the date of expiration of the first notice. No further temporary exemption notices may be issued for that property (section 62(5)).

Appeals

If the local authority refuses to serve a temporary exemption notice following a first notification, they must, without delay, serve notice on the person informing him of:

(a) the decision;
(b) the reasons for it and the date on which it was made;
(c) the rights of appeal; and
(d) the period in which an appeal can be made (section 62(6)).

The person concerned may appeal to the Residential Property Tribunal within 28 days of the date of the decision and the Residential Property Tribunal may confirm or reverse the local authority's decision. If the latter, the Residential Property Tribunal will direct the local authority to issue a temporary exemption notice, which will come into force on a date which the Residential Property Tribunal directs (section 62(7)–(9)).

Temporary exemption notices where there is a death of a holder of a house in multiple occupation licence

Upon the death of a house in multiple occupation licence holder, the licence ceases to have effect but, for a period of three months from the date of death, the house is treated as if a temporary exemption notice has been served. This period may be extended for a further three months by the service of a temporary exemption notice at the request of the person's representatives made during that initial period. There is an appeal against a local authority decision not to serve a temporary exemption notice for that additional period and the same notice procedure as for temporary exemption notices generally, above, is then applicable (section 68(7)–(10)).

RENT REPAYMENT ORDERS

References

Housing Act 2004, sections 79 to 100.
Rent Repayment Orders (Supplementary Provisions) (England) Regulations 2007.

Extent

These provisions apply to England but there are similar provisions in Wales.

FC81 Rent repayment orders – Sections 96 to 98 Housing Act 2004

Local authority applies to Residential Property Tribunal for rent repayment order (section 96)

Rent repayment order issued

Local authority to send copy of the notice to the department of the authority responsible for administering the housing benefit (section 96)

Keep department informed of any matters in connection with the administration of housing benefit (section 96)

Housing benefit department pays rent to the local authority

Local authority subsequently grants a licence which may include a condition requiring the licence holder to pay to the authority any amount payable to them under the rent repayment order not recovered by them (section 97)

Scope

Rent repayment orders can be used for tackling unlicensed houses in multiple occupation.

A Residential Property Tribunal can make a rent repayment order on application of the local authority or an occupier of the whole or part of the house. The tribunal must be satisfied:

(a) that within that previous 12 months, the appropriate person has committed an offence in relation to the house (whether or not he has been charged or convicted);
(b) that housing benefit has been paid in respect of periodical payments payable for the occupation of the whole or any part or parts of the house (rent) during any period during that the offence was being committed;
(c) that the authority has served on the appropriate person a notice (a 'notice of intended proceedings'):
  (i) informing him that the authority are proposing to make an application;
  (ii) setting out the reasons why;
  (iii) stating the amount that they will seek to recover and how that amount is calculated; and
  (iv) inviting him to make representations to them within a period specified in the notice of not less than 28 days;
(d) that such period must have expired; and
(e) that the authority has considered any representations made to them within that period by the appropriate person.

Where a local authority serves a notice of intended proceedings, they must ensure that a copy of the notice is received by the department of the authority responsible for administering the housing benefit and that that department is subsequently kept informed of any matters relating to the proceedings that are likely to be of interest to it in connection with the administration of housing benefit (section 96).

If the tribunal is satisfied:

(a) that a person has been convicted of an offence; and
(b) that housing benefit was paid in respect of rent,

the tribunal must make a rent repayment order requiring the appropriate person to pay to the authority an amount equal to the total amount of housing benefit paid or an amount reasonable in the circumstances.

If the total of rent is less than the total amount of housing benefit paid, the amount required to be paid through a rent repayment order is limited to the rent total.

A rent repayment order made may not require the payment of any amount which the tribunal is satisfied that it would be unreasonable for that person to be required to pay.

Any amount payable to a local authority under a rent repayment order does not constitute an amount of housing benefit recovered by them, and is, until recovered by them, a legal charge on the house.

If the authority subsequently grants a licence in respect of the house to the appropriate person or any person acting on his behalf, the conditions contained in the licence may include a condition requiring the licence holder to pay (possibly by instalments) to the authority any amount payable to them under the rent repayment order not recovered by them (section 97).

A local authority that has made an application for a rent repayment order can ask the Residential Property Tribunal to amend its application where it believes that there has been an overpayment of housing benefit so that the application is in respect of the amount of housing benefit that the local authority believes is properly payable (regulation 2).

The money received by a local authority under a rent repayment order may be used for the reimbursement of the authority's costs and expenses (whether administrative or legal) incurred in, or associated with:

(a) the making of the application for a rent repayment order;
(b) the registration and enforcement of any legal charge in relation to a rent repayment order;
(c) dealing with any application for the grant of a licence in respect to a house in multiple occupation or subject to selective licensing of other residential accommodation;
(d) the prosecution of the appropriate person for an offence in relation to licensing of houses in multiple occupation;
(e) the making of an interim or final management order in relation to a house in multiple occupation;
(f) the management of the relevant property while an interim or final management order is in force;
(g) the execution of works undertaken while an interim management order is in force;
(h) the preparation of, or execution of, works under a management scheme while a final management order is in force (regulation 3).

Any money left after having been used for the purposes above must be paid into the Consolidated Fund (regulation 4).

PRIVATE LANDLORDS' ACCREDITATION SCHEMES

A private landlords' accreditation scheme gives recognition to private landlords who offer good-quality, private rented accommodation. It usually covers all types of private rented accommodation – from purpose-built flats and houses in multiple occupation (HMOs), through to larger single houses and uses standards based on existing legislation, with additional facilities normally considered to constitute a satisfactory decent standard.

The objectives of a scheme are to recognise and encourage landlords who are prepared to provide good-quality accommodation at an appropriate rent and to improve and promote the public image of the private rented sector and private rented sector housing conditions. Schemes also aim to improve liaison and communication between landlords and local authorities and provide tenants with confidence in the quality and management of the accommodation they are renting.

Landlords in the scheme can gain a market advantage over other providers when attracting new tenants; can generate a better image with potential lenders; gain recognition that the landlord provides a high standard of accommodation, with fewer transient tenants, which means regular on-time rent payments; and have lower management costs and fewer void periods.

Tenants get clearer information on properties available to let and it is easier for them to identify good-quality accommodation. Schemes give confidence that landlords are professional and reputable, and that properties are safe, secure and free from hazards, as identified under the Housing Health and Safety Rating System, with improved energy efficiency measures.

Accreditation

Accreditation provides official recognition that landlord accommodation is of good quality, free from hazards and well managed. Accredited properties are those which have been inspected by an accreditation officer, who has certified that they meet the requirements stated in the scheme's Code of Standards, that the landlord displays good management standards and practices and that the relevant physical standards relating to the property have been verified.

An accreditation certificate is issued to the landlord confirming that the property meets the required standards. In addition, a property information folder is also presented to the landlord – this is to be kept at the property and contains useful information, contact details and copies of all relevant safety certificates.

Schemes are usually established by a local authority in partnership with landlords and others. Officers from the council oversee the promotion and day-to-day running of the scheme.

Schemes are usually voluntary, some with no cost involved in applying for accreditation. If all of the landlords' properties do not currently comply with the relevant standards, they can still participate in the scheme because it is property-specific. It is quite possible to have both accredited and non-accredited properties in a landlord's portfolio – though landlords will be actively encouraged to work towards accredited status for all of their properties.

Local authorities will inspect non-accredited properties, identify any work which is needed to bring them up to the accredited standard and, if necessary, prepare and issue schedules of work.

Standards can be chosen for each scheme and are usually drawn up by the local authority.

Property standards are defined and detailed guidance is given.

Accreditation status will normally last for three years. At the end of this period, further checks and a re-inspection of the property will be undertaken before the accredited status is renewed.

Relevant certificates for each property will also be required annually and spot-check inspections of properties may be carried out within the three-year accreditation period.

Accredited status may be suspended where:

- the property has fallen below the required standard and the landlord does not undertake the necessary work;
- relevant certification is not submitted;
- the landlord contravenes the qualifying criteria for the scheme;
- any information provided supporting an application was found to be fraudulent or inaccurate.

Landlords of accredited properties are expected to:

- at all times comply with the management Code of Practice agreed with the accreditation officer;
- comply with the accreditation scheme's Code of Standards;
- once accreditation is achieved, maintain the property to those standards;
- ensure that any staff or persons acting on their behalf comply with the requirements of the scheme and any Codes of Practice.

All breaches of the requirements of the scheme in relation to the property standards will be reviewed and, if the landlord does not take satisfactory remedial action, then the property concerned will be removed from the accreditation scheme.

If the breach relates to the management standards and Code of Practice, then the incident will be referred to the accreditation officer, who will consider the nature of the breaches, the representations of the landlord and any other relevant parties, and decide on any appropriate sanctions.

Some schemes operate a complaint service. An accreditation officer will act as a mediator for complaints between landlords and tenants. The complainant may be either the landlord or the tenant.

If a complaint is from the tenant:

- a visit will be made to the tenant to determine the nature of the complaint and decide whether it is justified;
- discussions will take place with the landlord to inform them of the nature of the complaint;
- the Accreditation Officer will agree with the landlord any remedial works to be carried out and a timescale for those works.

If this fails, then the complaint will be passed to the relevant local authority department for statutory action.

If the complaint concerns management standards (and not the physical state of the property) then the complaint will be referred to the accreditation officer.

If a complaint is from the landlord, the complaint will be referred to the accreditation officer. The accreditation officer will discuss the nature of the problem with the landlord and provide the appropriate advice, support and assistance as necessary to attempt to resolve the problem.

MEANS OF ESCAPE FROM CERTAIN HIGH BUILDINGS

Reference

Building Act 1984, section 72.

Extent

These provisions apply to England and Wales

Scope

The section applies to existing or proposed buildings which are, or are to be let as, flats or tenement dwellings and which exceed two storeys in height and in which the floor of any upper storey is more than 20 ft above the surface of the street or ground on any side of the building (section 72(1) and (6)).

The section is, however, now only a residual section and does not cover premises subject to the Fire Precautions Act 1971 or houses in multiple occupation, which are dealt with in the Housing Act 2004 (see FCs 65–69, FC78 and FC79).

Notices

The provisions of sections 99 and 102 apply to appeals against, and the enforcement of, notices under this procedure where works are required (pages 447–449).

Appeals

No appeal is provided against notices which do not require works, but in any subsequent proceedings it is open to the defendant to question the reasonableness of the requirements (section 72(4)).

LICENSING OF CAMPING SITES

Reference

Public Health Act 1936, section 269 (as amended by Caravan Sites and Control of Development Act 1960).

Extent

These provisions apply to England and Wales.

Scope

The procedure applies to all moveable dwellings (other than caravans) and licences are required for:

FC82 Means of escape from certain high buildings – Section 72 Building Act 1984

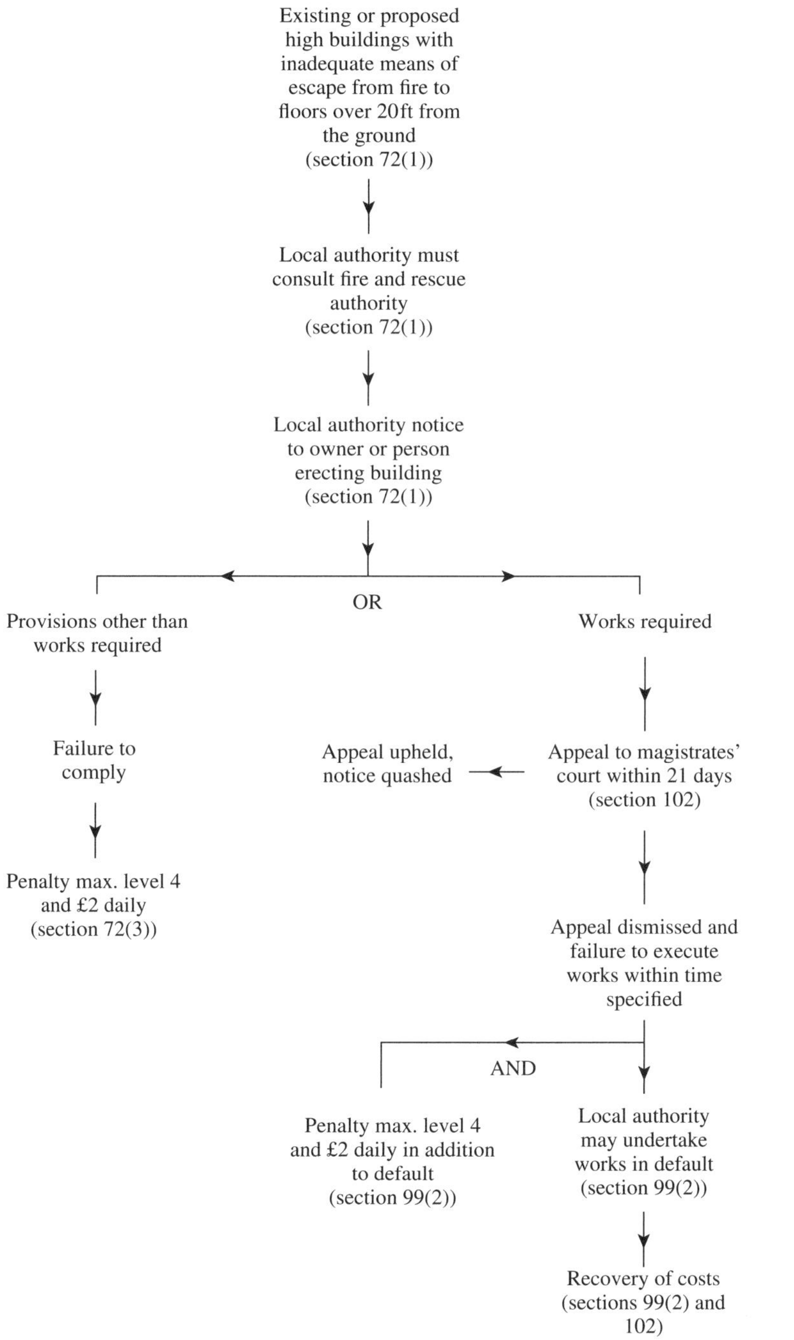

Note

1. For means of escape from fire in houses in multiple occupation, see FCs 65–68, FC76 and FC77.

(a) the use of land for camping purposes on more than 42 consecutive days or more than 60 days in any 12 consecutive months; or
(b) the keeping of a moveable dwelling on any one site, or two or more sites in succession if any of those sites is within 100 yards of another of them, for more than the same periods as in (a).

In respect of (a), land which is in the occupation of the same person, and within 100 yards of a site on which a moveable dwelling is stationed during any part of a day, is regarded as being used for camping on that day (section 269(1)–(3)). The removal of a moveable dwelling for not more than 48 hours does not constitute an interruption of the 42 days' period (section 269(8)).

Exemptions

Apart from the use of land or moveable dwellings for periods less than those in (a) and (b) above, the following situations do not require licensing by the local authority:

(a) moveable dwellings kept by the owner on land occupied by him in connection with his dwelling if used only by him or members of his household;
(b) moveable dwellings kept by the owner on agricultural land occupied by him and used for habitation only at certain times of the year and only by persons employed in farming operations on that land;
(c) moveable dwellings not in use for human habitation kept on premises by an occupier who does not permit moveable dwellings to be kept there for habitation;
(d) organisations granted a certificate of exemption by the minister, being satisfied that:
  (i) the camping sites belonging to, provided by or used by the organisation are properly managed and kept in good sanitary condition; and
  (ii) the moveable dwellings are used so as not to give rise to any nuisance (section 269(5) and (6)).

Applicants

Applications for licences may be made by:

(a) occupiers in respect of either the use of land or the stationing of moveable dwellings on that land; or
(b) a person intending to station a moveable dwelling on land (section 269(2) and (3)).

Conditions

Local authorities may attach to licences such conditions as they think fit with respect to:

(a) for licences authorising the use of land:
  (i) number and classes of moveable dwellings;

FC83 Licensing of camping sites – Section 269 Public Health Act 1936

Local authority may grant licences for:
(a) land to be used for camping for exceeding 42 consecutive days or 60 days in any consecutive 12 months, or
(b) particular moveable periods (section 269(2))

← Exemptions (see text) (section 269(5) and (6))

→ Offence to use land or moveable dwelling without a licence, max. penalty level 1 and £2 daily (section 269(7))

↓

Application to local authority for licence from either:
(a) occupier of land or
(b) for licence for particular moveable dwelling, occupier of land or person using the site (section 269(2))

↓

Application approved, licence issued with or without conditions (section 269(1))

OR

Applicant not notified of refusal within four weeks deemed granted unconditionally (section 269(4))

OR

Application refused

↓

Appeal against refusal or conditions to be attached to magistrates' court within 21 days of notice of decision being served (section 269(4))

↓

Appeal succeeds and local authority directed to issue licence

OR

Appeal succeeds and court amends conditions

OR

Appeal fails and licence refused or conditions unaltered

Notes

1. This procedure does not apply to caravans. For licensing of caravan sites, see FC84.
2. For removal of unauthorised campers under the Criminal Justice and Public Order Act 1994, see FC105.

(ii) the space between them;
(iii) water supply; and
(iv) for securing sanitary conditions;

(b) for licences authorising the use of a moveable dwelling for:
(i) the use of that dwelling, including space to be kept free between dwellings;
(ii) securing its removal at the end of a specified period; and
(iii) for securing sanitary conditions (section 269(1)).

Definitions

Caravan – see page 353.
Moveable dwelling includes any tent, any van or other conveyance (other than a caravan) whether on wheels or not, and … any shed or similar structure, being a tent, conveyance or structure which is used either regularly or at seasons only, or intermittently, for human habitation but does not include a structure to which the building regulations apply (section 269(8)).

LICENSING OF CARAVAN SITES (PARK HOMES)

References

Caravan Sites and Control of Development Act 1960 Part I (as amended by Local Government (Miscellaneous Provisions) Act 1982).
Model Standards 2008 for Caravan Sites in England: Caravan Sites and Control of Development Act 1960 – section 5, April 2008.

NB. This licensing procedure and the Model Standards for Park Homes were under review by the Office of the Deputy Prime Minister in early 2006 – see Park Homes Licensing – Proposals for Reform and Revising the Model Standards for Park Homes – Consultation Paper on Revised Standards and Guidance – both ODPM 2006. There has been little movement by the government to implement the recommendations of the review.

Extent

The procedure applies in England, Wales and with modifications in Scotland. It does not apply in Northern Ireland (section 50).

Scope

The occupier of any land must not cause or permit the land to be used as a caravan site unless he holds a site licence (section 1(1)).

FC84 Licensing of caravan sites – Caravan Sites and Control of Development Act 1960

Offence to cause or permit land to be used as caravan site without a licence (section 1(1))

← Max. penalty level 4 (section 1(2))

→ Exemptions (see text) (section 2 and schedule 1)

↓

Written application by occupier to local authority (section 3(1) and (2))

↓

If planning permission has been granted etc., local authority must issue site licence within two months of final information or application (six weeks if planning permission received after application made) (section 3(4) and (5)) (note 3)

→ Register of licences to be kept (section 25(2)) and transfers to future occupiers to be notified to and approved by local authority (section 10)

↓

Local authority may attach conditions to licence subject to consultation with the Fire and Rescue Authority (section 5) (note 4)

→ Local authority may alter, cancel or make new conditions at any time after giving opportunity for representations and consulting the Fire and Rescue Authority where appropriate (section 8(1))

↓

If person convicted of failing to comply with conditions on at least three occasions, court may revoke licence (section 9(2))

↓

Local authority or person convicted may apply to magistrates' court for extension of period allowed before revocation (section 9(2))

AND

Where non-compliance with conditions requiring works

↓

Local authority may undertake works in default and recover costs (section 9(3))

↓

Appeal against conditions to magistrates' court within 28 days of date of issue of licence (section 7(1))

↓

Court alters or cancels conditions if found to be unduly burdensome (section 7(1))

OR

If:

(a) no planning permission has been granted; or
(b) planning permission will expire within six months of site licence application; or
(c) applicant has had a site licence revoked within previous three years

↓

Licence must be refused (sections 3(2), 9(2) and 18) (note 2)

Notes

1. For licensing of camping sites (other than for caravans), see FC83.
2. There is no power for the local authority to refuse the licence for reasons other than these.
3. The licence is not to be issued for a limited period unless this is the case with the planning consent, in which case the two must be brought together (section 4).
4. Maximum penalty for failure to comply with conditions – level 4 (section 9(1)).

Exemptions

The long list of exemptions from licensing requirements is set out in schedule 1 of the Act in the following categories:

(a) incidental use within the curtilage of a dwelling-house;
(b) a single caravan for not more than two nights and 28 days in 12 months;
(c) holdings of five acres or more if not more than 28 days in 12 months and maximum three caravans at a time;
(d) sites occupied and supervised by organisations exempted by the minister;
(e) sites approved by exempted organisations for up to five caravans;
(f) meetings organised by exempted organisations;
(g) agriculture and forestry workers;
(h) building and engineering sites;
(i) travelling showmen;
(j) sites occupied by a local authority; and
(k) gypsy sites occupied by county councils or regional councils (section 2 and schedule 1).

Applications

The occupier of the land may apply to the local authority in writing and must specify the land in question and, either at the time of making the application or later, give the local authority such information as it may reasonably require (section 3). No fee is payable.

Issue of licence

The local authority cannot issue the licence if:

(a) there has not been a formal grant of planning permission (section 3(3)); or
(b) the applicant has had a site licence revoked within the previous three years (section 3(6)).

Licences must be issued within two months of the giving to the local authority of any information which it requires, or such longer period as may be agreed in writing between them (section 3(4)). If it does not occur, no offence is committed in relation to not having a site licence (section 6).

Licences may not be time limited unless this is the case with the planning permission, in which case the licence must expire on the same date (section 4).

Conditions

The local authority may attach to the licence any conditions which it sees necessary or desirable to impose in the interest of persons living in the caravans, or of other classes of persons, or of the public at large. No conditions may however control the material to be used in the construction of the caravans. In forming conditions, the local authority must have regard to model standards issued by the minister from time to time.

In respect of conditions which require the carrying out of works, the works may be required within a specified period and caravans may be prohibited or restricted until the works have been completed. The local authority has power to undertake these works in default at the end of the period allowed and recover costs (sections 5 and 9(3)).

The local authority is required to consult with the Fire and Rescue Authority before granting a site licence or making or varying conditions relating to fire precautions (section 5(3A) and (3B)).

Where the Regulatory Reform (Fire Safety) Order 2005 applies to the land concerned, no condition may be attached to the site licence in relation to any requirements or prohibitions which are or could be imposed under that Order (section 5 (2A)).

Power of entry

An authorised officer of the local authority upon:

(a) producing, if required, a duly authenticated document giving his authority; and
(b) giving 24 hours' notice to the occupier

may demand entry to any land used as a caravan site or in respect of which a site licence application has been made for the purpose of:

(a) determining the conditions to be attached to licences;
(b) ascertaining if there have been or are any contraventions;
(c) ascertaining if the local authority should take any action or carry out any work; or
(d) to take action or undertake any work authorised by the Act.

Where a Justice of the Peace is satisfied that:

(a) admission has been refused or is apprehended; or
(b) the occupier is temporarily absent and the case is one of urgency; or
(c) that an application for admission would defeat the object of entry; and
(d) there is reasonable ground for entering the land,

he may authorise the local authority by warrant to enter, if need be by force. The officer exercising the warrant may take with him any other person as may be necessary. The maximum penalty for obstruction is level 1 (section 26).

Definitions

Caravan means any structure designed or adapted for human habitation which is capable of being moved from one place to another (whether by being towed, or by being transported on a motor vehicle or trailer) and any motor vehicle so designed or adapted, but does not include:

(a) any railway rolling stock which is for the time being on rails forming part of a railway system; or
(b) any tent (section 29(1)),

but a structure designed or adapted for human habitation which:

(a) is composed of not more than two sections separately constructed and designed to be assembled on a site by means of bolts, clamps or other devices; and
(b) is, when assembled, physically capable of being moved by road from one place to another (whether by being towed, or by being transported on a motor vehicle or trailer)

shall not be treated as not being (or as not having been) a caravan within the meaning of part 1 of the Caravan Sites and Control of Development Act 1960, by reason only that it cannot be so moved on a highway when assembled.

In addition, the expression 'caravan' shall not include a structure designed or adapted for human habitation which falls within paragraphs (a) and (b) of the foregoing subsection if its dimensions when assembled exceed any of the following limits, namely:

(a) length (exclusive of any drawbar): 60 ft (18.288 m);
(b) width: 20 ft (6.096 m);
(c) overall height of living accommodation (measured internally from the floor at the lowest level to the ceiling at the highest level): 10 ft (3.048 m) (Caravan Sites Act 1968, section 13).

Caravan site means land upon which a caravan is stationed for the purposes of human habitation and land which is used in conjunction with land on which a caravan is so stationed (section 1(4)).
Occupier means, in relation to any land, the person who, by virtue of an estate or interest therein held by him, is entitled to possession thereof or would be so entitled but for the rights of any other person under any licence granted in respect of the land (section 1(3)).
Works – any reference … to the carrying out of works shall include a reference to the planting of trees and shrubs and the carrying out of other operations for preserving or enhancing the amenity of the land (section 29(2)).

Chapter 6

PUBLIC HEALTH

ANIMAL BOARDING ESTABLISHMENTS

Reference

Animal Boarding Establishments Act 1963 as amended by the Animal Welfare Act 2006.

Extent

This licensing procedure applies in England, Wales and Scotland but not in Northern Ireland (section 7(3)).

Scope

A person keeping an animal boarding establishment for cats and/or dogs is required to hold a licence issued by the local authorities (section 1(1)).

The Act defines the keeping of an animal boarding establishment as:

> to the carrying on by him at premises of any nature (including a private dwelling) of a business of providing accommodation for other people's animals:
>
> Provided that:
>
> (a) a person shall not be deemed to keep a boarding establishment for animals by reason only of his providing accommodation for other people's animals in connection with a business of which the provision of such accommodation is not the main activity; and
> (b) nothing in this Act shall apply to the keeping of an animal at any premises in pursuance of a requirement imposed under, or having effect by virtue of, the Animal Health Act 1981.
>
> (section 5(1))

Applications

Applications made to the local authority must specify the premises concerned and a fee as determined by the local authority must be paid before any licence is

FC85 Animal boarding establishments – Animal Boarding Establishments Act 1963

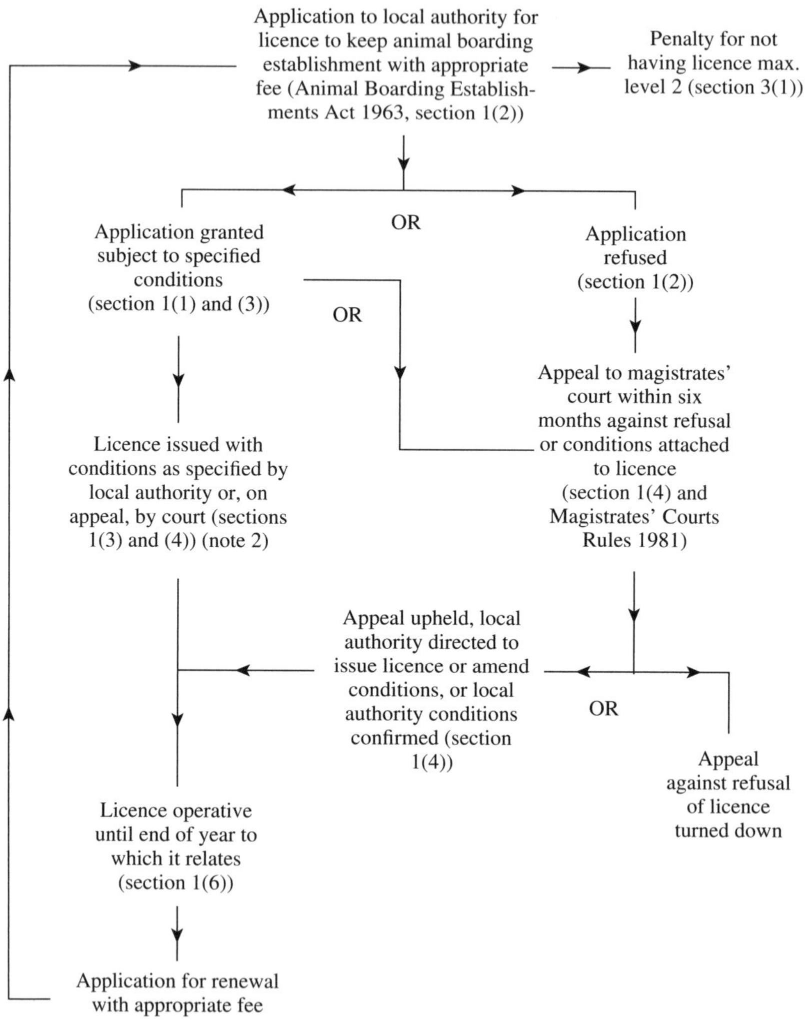

Notes

1. This procedure applies in Scotland but not in Northern Ireland (section 7(3)).
2. Penalty for contravening licence conditions is max. level 2 fine and up to three months' imprisonment, or both (section 3(1)).
3. Conditions may only be varied on the renewal of licence.

granted (section 1(2)). The factors which may be taken into account in deciding whether or not to issue a licence are listed under 'Conditions' below but these are without prejudice to the local authority's discretion to refuse a licence on any other grounds (section 1(3)).

Licences

A licence normally remains in force until the end of the year in which it is granted. However, the applicant determines whether it comes into force on the date it is granted or the first day of the following calendar year (section 1(5) and (6)).

In the event of the death of a licence holder, the licence passes to his personal representative and operates for a period of three months from the date of death before expiring. The local authority may, on application, agree to extension of that period (section 1(7)).

Conditions

The local authority is required to specify such conditions in the licence as appears to it necessary or expedient for securing all or any of the following objectives:

(a) that animals will at all times be kept in accommodation suitable as respects construction, size of quarters, number of occupants, exercising facilities, temperature, lighting, ventilation and cleanliness;
(b) that animals will be adequately supplied with suitable food, drink and bedding material, adequately exercised, and (so far as necessary) visited at suitable intervals;
(c) that all reasonable precautions will be taken to prevent and control the spread among animals of infectious or contagious diseases, including the provision of adequate isolation facilities;
(d) that appropriate steps will be taken for the protection of the animals in case of fire or other emergency;
(e) that a register will be kept containing a description of any animals received into the establishment, date of arrival and departure, and the name and address of the owner, such register to be available for inspection at all times by an officer of the local authority, veterinary surgeon or veterinary practitioner authorised under section 2(1) of this Act (section 1(3)).

Guidance on conditions relating to these licences is given in two booklets published by the Chartered Institute of Environmental Health in November 1995, 'Model Licence Conditions and Guidance for Dog Boarding Establishments', (www.cieh.org/policy/dog_guidance.html), and 'Model Licence Conditions and Guidance for Cat Boarding Establishments', (www.cieh.org/policy/cat_guidance.html). The Institute encourages local authorities to discuss with establishment owners appropriate standards and timescales based on the documents.

Power of entry

local authority officers, veterinary surgeons or practitioners authorised in writing by the local authority for this purpose, may, upon producing their authority, if required, inspect a licensed animal boarding establishment and any animals found there at all reasonable times. Persons wilfully obstructing or delaying authorised officers are subject on conviction to a maximum penalty at level 2 on the standard scale (sections 2 and 3(2)).

There is no power of entry to unlicensed premises.

Disqualifications and cancellations

In making a conviction under the Animal Boarding Establishment Act 1963 or under the Protection of Animals Act 1911, the Protection of Animals (Scotland) Act 1912, the Pet Animals Act 1951 and the Animal Welfare Act 2006, the court may cancel any animal boarding establishment licence held by the person and may disqualify him from holding such a licence, whether or not he currently holds one, for any specified period. The cancellation or disqualification may be suspended by the court pending an appeal (sections 3(3) and (4)).

Local authorities must refuse applications for animal boarding establishments licences from persons disqualified under:

(a) the Animal Boarding Establishment Act 1963;
(b) the Pet Animals Act 1951 from keeping a pet shop;
(c) the Protection of Animals (Amendment) Act 1954 from having the custody of animals (section 1(2));
(d) the Animal Welfare Act 2006.

Definition

Animal means any cat or dog (section 5(2)).

DANGEROUS WILD ANIMALS

References

Dangerous Wild Animals Act 1976 as amended by the Animal Welfare Act 2006.
Dangerous Wild Animals Act 1976 (Modification) (No. 2) Order 2007.
DEFRA Circular 1/2002 The Keeping of Wild Animals: An Introductory Guide to: The Protection of Animals Act 1911, The Performing Animals (Regulation) Act 1925, The Pet Animals Act 1951, The Dangerous Wild Animals Act 1976, The Zoo Licensing Act 1981.

Extent

This licensing procedure applies in England, Wales and Scotland but not in Northern Ireland (section 10(3)).

Scope

Any person keeping any dangerous wild animal is required to hold a licence from a local authority (section 1(1)).

The animals covered by this provision are listed in the schedule to the Act and this has been amended from time to time. The latest modification was in 2008 (Scotland) and the animals requiring their keeper to be licensed are given in paragraph 1 of schedule 1 of the Act and paragraph 1 of the schedule in the Dangerous Wild Animals Act 1976 (Modification) (No. 2) Order 2007. The common names of the animals in the schedules are reproduced in 'Animals requiring a licence' below.

This licensing procedure does not apply to animals kept in:

(a) a zoo within the meaning of the Zoo Licensing Act 1981;
(b) a circus;
(c) pet shops; and
(d) places which are designed establishments under the Animals (Scientific Procedures) Act 1986 (section 5).

A person is held to be the keeper of the animal if he has it in his possession and the assumption of possession continues even if the animal escapes or it is being transported, etc. This removes the need for carriers or veterinary surgeons to be licensed (section 7(1)).

Animals requiring a licence

1. The Tasmanian devil
2. Grey kangaroos, the euro, the wallaroo and the red kangaroo
3. New-world monkeys (including capuchin, howler, saki, spider, uakari and woolly monkeys)
4. Old-world monkeys (including baboons, the drill, colobus monkeys, the gelada, guenons, langurs, leaf monkeys, macaques, the mandrill, mangabeys, the patas and proboscis monkeys and the talapoin)
5. Leaping lemurs (including the indri and sifakas)
6. Large lemurs (the broad-nosed gentle lemur and the grey gentle lemur are excepted)
7. Anthropoid apes (including chimpanzees, gibbons, the gorilla and the orang-utan)
8. The giant armadillo
9. The giant anteater
10. Crested porcupines
11. The giant panda and the red panda
12. Jackals, wild dogs, wolves and the coyote (foxes, the raccoon-dog and the domestic dog are excepted)
13. The bobcat, caracal, cheetah, jaguar, lion, lynx, ocelot, puma, serval, tiger and all other cats (the domestic cat is excepted)
14. Hyaenas (except the aardwolf)
15. Badgers (except the Eurasian badger), otters (except the European otter) and the tayra, wolverine, fisher and ratel (otherwise known as the honey badger)

16. The African, large-spotted, Malay and large Indian civets, and the fossa
17. The walrus, eared seals, sealions and earless seals (the common and grey seals are excepted)
18. Elephants
19. Asses, horses and zebras (the donkey, domestic horse and domestic hybrids are excepted)
20. Rhinoceroses
21. Tapirs
22. The aardvark
23. The pronghorn
24. Antelopes, bison, buffalo, cattle, gazelles, goats and sheep (domestic cattle, goats and sheep are excepted)
25. Camels
26. The moose or elk and the caribou or reindeer (the domestic reindeer is excepted)
27. The giraffe and the okapi
28. The hippopotamus and the pygmy hippopotamus
29. Old-world pigs (including the wild boar and the wart hog) (the domestic pig is excepted)
30. New-world pigs (otherwise known as peccaries)
31. Mammalian hybrids with a parent (or parents) of a specified kind
32. Cassowaries
33. The ostrich
34. Alligators and caimans
35. Crocodiles and the false gharial (otherwise known as the gavial)
36. Mole vipers and certain rear-fanged venomous snakes (including the moila and montpellier snakes, twig snakes, the boomslang, the red-necked keel-back and the yamakagashi, otherwise known as the Japanese tiger-snake)
37. Certain front-fanged venomous snakes (including cobras, coral snakes, the desert black snake, kraits, mambas, sea snakes, adders, the barba amarilla, the bushmaster, the copperhead, the fer-de-lance, moccasins, rattlesnakes and vipers and all Australian poisonous snakes, including the death adders)
38. The gila monster and the (Mexican) beaded lizard
39. Wandering spiders
40. The Sydney funnel-web spider and its close relatives
41. Brown recluse spiders (otherwise known as violin spiders)
42. The black widow spider (otherwise known as red-back spider) and its close relatives
43. Buthid scorpions

Added in 2007:

44. The Argentine black-headed snake, the Peruvian racer, the South American green racer and the Amazon false viper
45. The Middle-Eastern thin-tailed scorpion
46. The dingo

FC86 Dangerous wild animals – Dangerous Wild Animals Act 1976

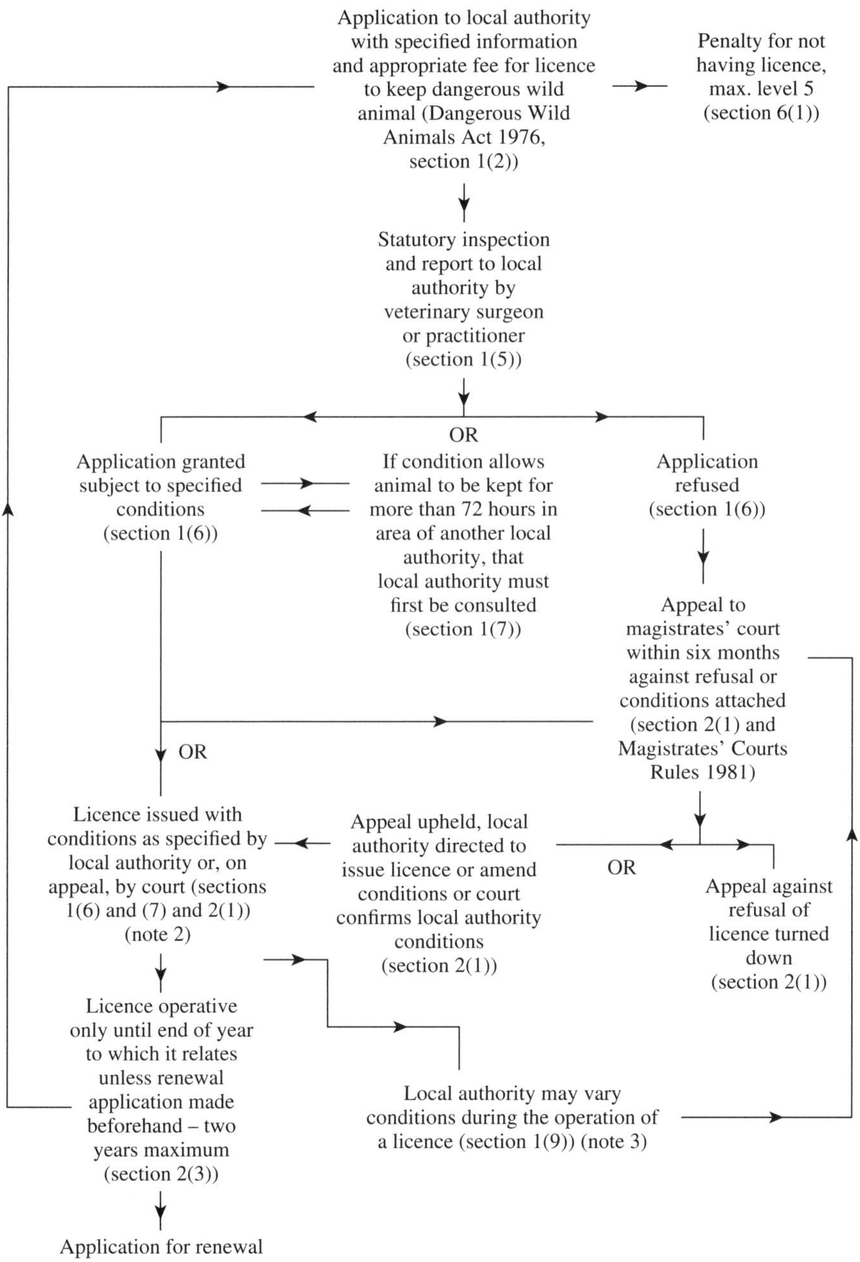

Notes

1. This procedure applies in England, Wales and Scotland but not in Northern Ireland (section 10(3)).
2. Penalty for contravening licence condition is maximum level 5 (section 6(1)).
3. Unless the variation was requested by the licence holder, the local authority must notify him of the variation and allow a reasonable time for compliance.

Applications for licence

Any application made to a local authority for a licence must be made (unless in exceptional circumstances) by the person who proposes to own and possess the animal and must:

- (a) specify the species and number of animals to be kept;
- (b) specify the premises where the animals will normally be kept;
- (c) be made to the local authority for those premises;
- (d) be made by a person 18 years of age or over and not disqualified from holding a licence under the Act; and
- (e) be accompanied by a fee stipulated by the local authority at a level sufficient to meet the direct and indirect costs involved.

Applications not complying with these requirements may not be granted (section 1(2) and (4)).

Reports

Before granting any licence, the local authority is required to consider a report of an inspection of the premises by a veterinary surgeon or practitioner authorised by it (section 1(5)).

Matters for consideration

The local authority may not grant a licence unless:

- (a) it will not be contrary to the public interest on grounds of safety, nuisance or otherwise to issue a licence;
- (b) the applicant is suitable;
- (c) animals will:
    - (i) be held in secure accommodation suitable in size for the animals kept and which is suitable as regards construction, temperature, lighting, ventilation, drainage and cleanliness; and
    - (ii) have adequate and suitable food, drink and bedding and be visited at regular intervals;
- (d) be appropriately protected in case of fire or other emergency;
- (e) be subject to precautions to control infectious diseases;
- (f) be provided with adequate exercise facilities (section 1(3)).

Licences

According to the wishes of the applicant, a licence either comes into force on the day on which it is granted, in which case it expires on 31 December of that same year, or it comes into force on 1 January of the next year, in which case it expires on 31 December of that next year. If an application for renewal is made before the date of expiration, the licence continues until the application is determined.

On the death of a licence holder, the licence continues in the name of the personal representative for 28 days only and then expires unless application is made for a new licence within that time, in which case it continues until the new application is determined (sections 2(2), (3) and (4)).

Conditions

The local authority is required to specify conditions which:

(a) require the animals to be kept only by persons specified in the licence;
(b) require the animals to be normally held at the premises specified in the licence;
(c) require the animals not to be moved from those premises unless in circumstances allowed for in the licence;
(d) require the licence holder and person keeping the animals to be insured against liability for damage caused by the animals to the satisfaction of the local authority;
(e) restrict the species and numbers of animals;
(f) require a copy of the licence to be made available by the licence holder to persons entitled to keep the animals; and
(g) are necessary or desirable to secure the objectives specified in paragraphs (c)–(f) listed under 'Matters for consideration' above (section 1(6) and (7)).

The local authority may attach any other conditions which it thinks fit but if it is to permit the animal to be taken into another local authority area for more than 72 hours, it must consult that local authority (section 1(8)).

Conditions not required by the Act to be attached to the licence may be revoked or modified by the local authority or new conditions may be added. These variations come into effect immediately if they were requested by the licence holder but otherwise the local authority must notify him and allow a reasonable time for compliance (section 1(10)).

Disqualifications and cancellations

Where a person is convicted of an offence under the Dangerous Wild Animals Act 1976 or under the:

(a) Protection of Animals Act 1911;
(b) Protection of Animals (Scotland) Act 1912–64;
(c) Pet Animals Act 1951;
(d) Performing Animals (Regulation) Act 1925;
(e) Animals (Cruel Poisons) Act 1962;
(f) Animal Boarding Establishments Act 1963;
(g) Riding Establishments Act 1964 and 1970;
(h) Breeding of Dogs Act 1973;
(i) Animals Welfare Act 2006, sections 4–9 and 11,

the court may cancel any licence he may hold to keep a dangerous wild animal and disqualify him, whether or not he is a current holder, from holding such a licence for such period as the court thinks fit. The cancellation or disqualification may be suspended by the court in the event of an appeal (section 6(2) and (3)).

Power of entry

Local authorities may authorise competent persons to enter premises either licensed under the Act or specified in an application for a licence, at all reasonable times, producing, if required, their authority, and the authorised officers may inspect these premises and an animal in them. The local authority may charge the person making the application, for the inspection.

The penalty for wilfully obstructing or delaying an authorised officer is a maximum fine at level 5 (sections 3 and 6(1)).

Seizure of animals

If a dangerous wild animal is being kept without the authority of a licence or in contravention of a licence condition, the local authority may seize the animal and retain it, destroy it or otherwise dispose of it. The local authority is not liable to compensation and may recover costs from the keeper of the animal at the time of this seizure (section 4).

Definitions

Circus includes any place where animals are kept or introduced wholly or mainly for the purpose of performing tricks or manoeuvres;
Premises includes any place (section 7(4)).

DOG BREEDING ESTABLISHMENTS

References

Breeding of Dogs Acts 1973 (as amended by the Animal Welfare Act 2006) and 1991.
Breeding and Sale of Dogs (Welfare) Act 1999.
The Breeding of Dogs (Licensing Records) Regulations 1999.
Home Office Circular 53/1999.

Extent

This licensing procedure applies in England, Wales and, with modifications, Scotland but not in Northern Ireland (section 7(2)).

Scope

A licence is required to carry out a *business* of the breeding of dogs for sale. The Act indicates that the keeping of bitches who give birth to five or more litters in any period of 12 months is presumed to constitute such a business and required to be licensed, although local authorities may conclude in the circumstances of a particular case that licensing is required for a lesser number. Where there is no

FC87 Dog breeding establishments – Breeding of Dogs Act 1973 as amended

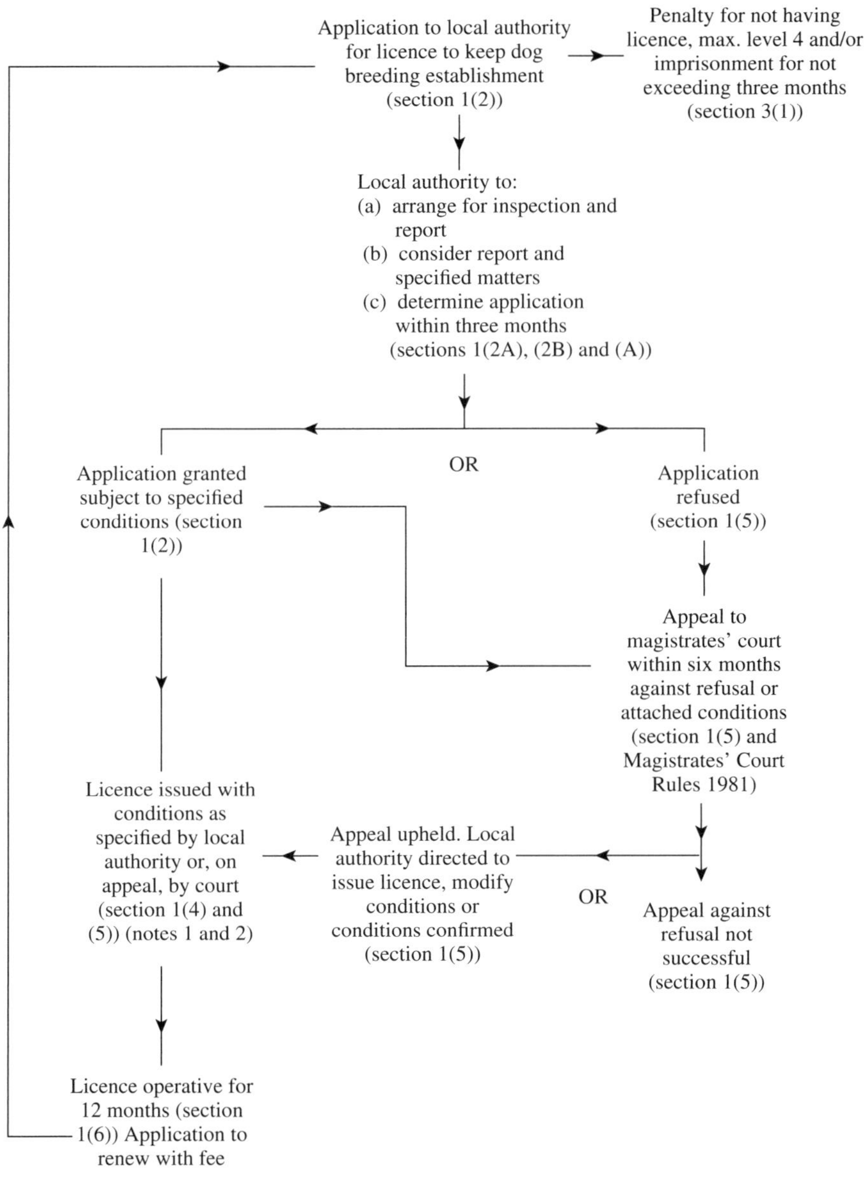

Notes

1. Penalty for non-compliance with conditions is max. fine of level 4 and/or imprisonment for not exceeding three months (section 3(1)).
2. Conditions may only be varied on the renewal of a licence.

sale of any puppies during the 12 months being considered, licensing is not required.

Bitches count towards the qualifying total if they are kept at any time during the 12-month period by the applicant/licence holder or by their relatives at the premises, by the applicant/licence holder elsewhere or by someone else as part of a breeding agreement. This is intended to prevent evasion of licensing by distributing bitches amongst different people and/or premises (section 4A).

Inspection

Upon receipt of an application for a licence for the first time, the local authority must arrange for an inspection by a veterinary surgeon/practitioner *and* by an officer of the local authority. Either or both may carry out inspections relating to subsequent applications (section 1(2A)).

Report

The local authority must arrange for a report of inspection to be prepared and considered before making its decision (section 1(2B)).

Matters for consideration

In deciding whether or not to grant a licence, the local authority must have regard to the following matters (but without prejudice to its right to refuse a licence on other grounds):

(a) that the dogs will at all times be kept in accommodation suitable as respects construction, size of quarters, number of occupants, exercising facilities, temperature, lighting, ventilation and cleanliness;
(b) that the dogs will be adequately supplied with suitable food, drink and bedding material, adequately exercised, and visited at suitable intervals;
(c) that all reasonable precautions will be taken to prevent and control the spread among dogs of infectious or contagious diseases;
(d) that appropriate steps will be taken for the protection of the dogs in case of fire or other emergency;
(e) that all appropriate steps will be taken to secure that the dogs will be provided with suitable food, drink and bedding material and adequately exercised when being transported to or from the breeding establishment;
(f) that bitches are not mated if they are less than one year old;
(g) that bitches do not give birth to more than six litters each;
(h) that bitches do not give birth to puppies before the end of the period of 12 months beginning with the day on which they last gave birth to puppies; and
(i) that accurate records are kept at the premises and made available for inspection for any authorised officer of the local authority to examine. The particular records to be kept are listed in the Breeding of Dogs (Licensing Records) Regulations 1999 (section 1(4)).

Persons disqualified under the following provisions may not be granted a dog breeding licence:

(a) from keeping a dog breeding establishment;
(b) from keeping a pet shop under the Pet Animals Act 1951;
(c) from keeping an animal boarding establishment under the Animal Boarding Establishments Act 1963;
(d) from having the custody of animals under the Protection of Animals (Amendment) Act 1954;
(e) under section 34(2), (3) or (4) of the Animal Welfare Act 2006 (section 1(2) and (4));
(f) under subsection (1) of section 40 of the Animal Health and Welfare (Scotland) Act 2006.

Licences

The local authority must determine the application within three months (section 1(5A)). Licences run for 12 months from the start date, being either the date requested by the applicant or the date of issue, whichever is the later (section 1(6) and (7)). On the death of a licence holder, licences pass to his personal representative for a period of three months from the day of his death and then expire. The local authority may extend that three-month period at its discretion (section 1(8)).

Conditions

On granting a licence, the local authority must specify such conditions as appear necessary or expedient to achieve the objectives set out in (a)–(i) in the paragraph headed 'Matters for consideration' above (section 1(4)).

Disqualifications and cancellations

In making a conviction under this Act, the court may cancel any dog breeding licence held by the convicted person, disqualify him from holding a licence for such period as the court thinks fit and disqualify him from having custody of any dog for such period as specified. The cancellation or disqualification may be suspended by the court pending appeal (section 3(3) and (4)).

Fees

Local authorities may charge fees for both the applications for licences and for the related inspections. The level of fees may reflect the reasonable administrative and enforcement costs incurred and may be varied to take account of different circumstances (section 3A).

Appeals

Any aggrieved person may appeal to a magistrates' court (sheriff court in Scotland) against a refusal to grant a licence and against the imposition of any condition. The court may direct in either case, as it thinks proper (section 1(5)).

Offences/penalties

Persons operating without a licence or breaching licence conditions are guilty of an offence and on conviction liable to a penalty of either up to three months' imprisonment or a fine not exceeding level 4, or both (sections 1(9) and 3(1)).

Power of entry

Local authority officers, veterinary surgeons or practitioners authorised in writing by the local authority may, upon producing if required their authority, enter any premises licensed as a dog breeding establishment at any reasonable time and inspect it and any animals found there (section 2(1)). Persons wilfully obstructing or delaying authorised officers are liable on conviction to a maximum fine of level 3 (section 2(2) and 3(2)).

Authorised officers are also given powers of entry into premises other than private dwellings, although entry into garages, outhouses and other structures forming part of the private dwelling is provided for where it is suspected that a person is operating without a licence. Before entry is requested, a warrant must be obtained from a Justice of the Peace and must be effected at reasonable times. The penalty for obstruction against these powers is also a maximum fine of level 3 (1991 Act, section 1).

PET SHOPS

References

Pet Animals Act 1951 as amended by the Pet Animals Act 1951 (Amendment) Act 1983 and the Animal Welfare Act 2006.
DEFRA Circular 1/2002.

Extent

This licensing procedure applies in England, Wales and Scotland but not in Northern Ireland (section 8).

Scope

A person keeping a pet shop requires a licence from the local authority (section 1(1)). The Act defines 'the keeping of a pet shop' as:

> the carrying on at premises of any nature (including a private dwelling) of a business of selling animals as pets, and as including references to the keeping of animals in any such premises … with a view to their being sold in the course of such a business, whether by the keeper thereof or by any other person.
>
> Provided that:
>
> (a) a person shall not be deemed to keep a pet shop by reason only of his keeping or selling pedigree animals bred by him, or the offspring of an animal kept by him as a pet;

(b) where a person carries on a business of selling animals as pets in conjunction with a business of breeding pedigree animals, and the local authority are satisfied that the animals so sold by him (in so far as they are not pedigree animals bred by him) are animals which were acquired by him with a view to being used, if suitable, for breeding or show purposes but have subsequently been found by him not to be suitable or required for such use, the local authority may if they think fit direct that the said person shall not be deemed to keep a pet shop by reason only of his carrying on the first-mentioned business.

References in this Act to the selling or keeping of animals as pets shall be construed in accordance with the following provisions, that is to say:

(a) as respects cats and dogs, such references shall be construed as including references to selling or keeping, as the case may be, wholly or mainly for domestic purposes; and
(b) as respects any animal, such references shall be construed as including references to selling or keeping, as the case may be, for ornamental purposes.

(section 7(1) and (2))

Applications

Applications made to the local authority must specify the premises concerned and a fee as determined by the local authority must be paid before a licence is granted (section 1(2)).

The factors which may be taken into account in deciding whether or not to issue a licence are listed under 'Conditions' below but these are without prejudice to the local authority's discretion to refuse a licence on any other grounds (section 1(3)).

Licences

Licences remain in force until the end of the year to which they relate and the latter is determined by the applicant as being either the year in which it is granted or the following year. In the first case, i.e. the year in which it is granted, the licence comes into force on the day it is granted and expires on 31 December of that year; in the second case, i.e. the year following that in which it is granted, it comes into force on 1 January of that year and expires on 31 December of that year (section 1(5) and (6)).

Conditions

The local authority must attach any conditions which it considers to be necessary or expedient for securing all or any of the following objectives:

(a) that animals will at all times be kept in accommodation suitable as respects size, temperature, lighting, ventilation and cleanliness;

FC88 Pet shops – Pet Animals Act 1951

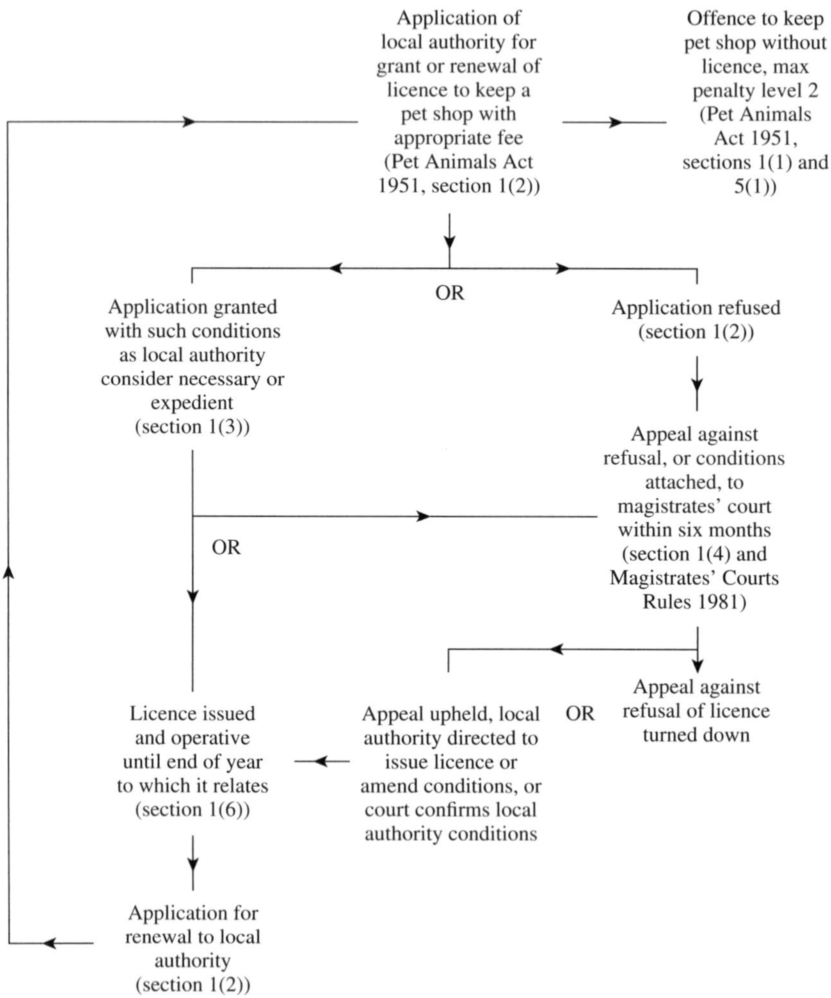

Notes

1. Penalty for contravening licence conditions, maximum level 2 (section 5(1)).
2. This procedure applies in Scotland but not in Northern Ireland (section 8).
3. Conditions may only be varied on the renewal of a licence.

(b) that animals will be adequately supplied with suitable food and drink and (so far as necessary) visited at suitable intervals;
(c) that animals, being mammals, will not be sold at too early an age;
(d) that all reasonable precautions will be taken to prevent the spread among animals of infectious diseases;
(e) that appropriate steps will be taken in case of fire or other emergency (section 1(3)).

No conditions may be specified which relate to a matter dealt with by the Regulatory Reform (Fire Safety) Order 2005.

Power of entry

Local authority officers, veterinary surgeons or practitioners authorised in writing by the local authority for this purpose, may, upon producing their authority if required, inspect a licensed pet shop and any animals there at all reasonable times. Persons wilfully obstructing authorised officers are subject on conviction to a maximum fine of level 2 (sections 4 and 5(1)).

There does not appear to be a power of entry for premises which do not hold a pet shop licence, e.g. which may be suspected of operating an unlicensed pet shop or in respect of which an application has been made. In the latter case, refusal of the licence application would be the obvious remedy.

Disqualifications and cancellations

In making a conviction under the Pet Animals Act 1951, the Protection of Animals Act 1911, the Protection of Animals (Scotland) Act 1912 or the Animal Welfare Act 2006, sections 4–9 and 11, the court may cancel any pet shop licence held by the person and may disqualify him from holding such a licence, whether or not he currently holds one, for any specified period. The cancellation or disqualification may be suspended by the court pending an appeal (section 5(3) and (4)).

Where proceedings for cruelty or neglect are brought under the Protection of Animals Act 1911, the Protection of Animals (Amendment) Act 2000 enables courts to make orders regarding the care, disposal or slaughter of animals kept for sale in pet shops.

Local authorities must refuse licence applications from persons currently disqualified by a court from holding a pet shop licence (section 1(2)). This provision differs from that relating to animal boarding establishments, where disqualifications under other Acts are also relevant.

Definitions

Animal includes any description of vertebrate.
Pedigree animal means an animal of any description which is, by its breeding, eligible for registration with a recognised club or society keeping a register of animals of that description (section 7(3)).

There is no definition of the word 'premises', this having been removed by the Amendment Act of 1983, but the sale of animals as pets as a business is prohibited in any part of a street or public place or at a stall or barrow in a market (section 2 as amended).

RIDING ESTABLISHMENTS

Reference

Riding Establishments Acts 1964 and 1970 as amended by the Animal Welfare Act 2006.

Extent

This licensing procedure applies in England, Wales and Scotland but not in Northern Ireland (section 9(2)).

Scope

A person keeping a riding establishment is required to hold a licence issued by the local authority (section 1(1)).

The 1964 Act defines the keeping of a riding establishment as:

> the carrying on of a business of keeping horses for either or both of the following purposes, that is to say, the purpose of their being let out on hire for riding or the purpose of their being used in providing, in return for payment, instruction in riding, but as not including a reference to the carrying on of such a business:
>
> (a) in a case where the premises where the horses employed for the purposes of the business are kept are occupied by or under the management of the Secretary of Defence; or
> (b) solely for police purposes; or
> (c) by the Zoological Society of London; or
> (d) by the Royal Zoological Society of Scotland.
>
> (section 6(1))

Horses kept by a university providing veterinary courses are also exempt and the place at which a riding establishment is run is to be taken as the place at which the horses are kept (section 6(2) and (3)).

Applications

Applications made to the local authority must specify the premises concerned and a fee as determined by the local authority must be paid before any licence is granted (section 1(2)).

FC89 Riding establishments – Riding Establishments Act 1964 as amended

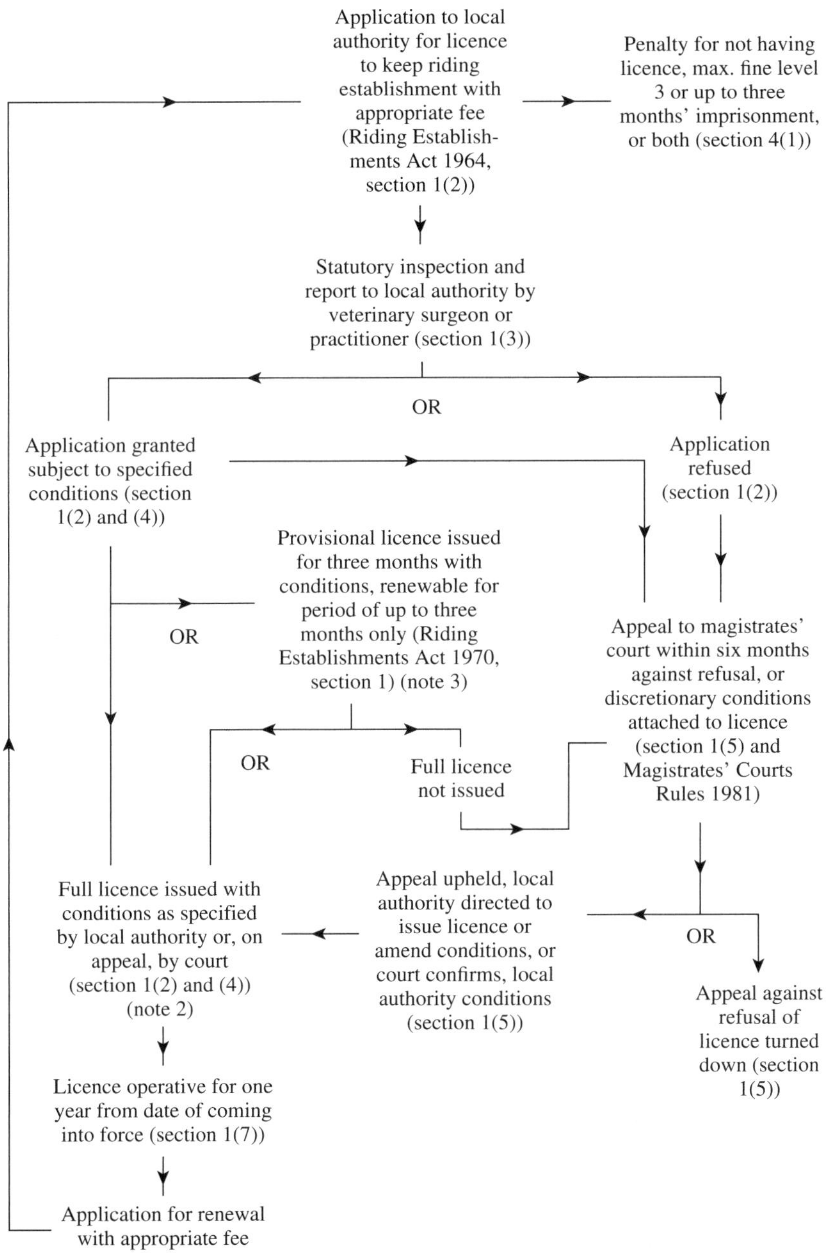

Notes

1. This procedure applies in England, Wales and Scotland but not in Northern Ireland (section 9(2)).
2. Penalty for contravening licence condition is maximum level 3 and up to three months' imprisonment, or both (section 4(1) as amended).
3. There is no appeal against the decision to issue a provisional licence only or against the conditions attached to it.
4. Conditions may only be varied at the renewal of the licence.

Reports

The local authority is required to receive a report by a listed veterinary surgeon or practitioner (see 'Power of entry' below) before making a decision, and the report must be based on an inspection made not more than 12 months before the application was received (section 1(3)).

Matters for consideration

In deciding whether or not to grant a licence, or provisional licence, the local authority must, without prejudice to its right to refuse a licence on other grounds, have regard to the following matters:

(a) whether that person appears to them to be suitable and qualified, either by experience in the management of horses or by being the holder of an approved certificate or by employing in the management of the riding establishment a person so qualified, to be the holder of such a licence; and

(b) the need for securing:

   (i) that paramount consideration will be given to the condition of the horses and that they will be maintained in good health, and in all respects physically fit and that, in the case of a horse kept for the purpose of its being let out on hire for riding or a horse kept for the purpose of its being used in providing instruction in riding, the horse will be suitable for the purpose for which it is kept;

   (ii) that the feet of all animals are properly trimmed and that, if shod, their shoes are properly fitted and in good condition;

   (iii) that there will be available at all times, accommodation for horses suitable as respects construction, size, number of occupants, lighting, ventilation, drainage and cleanliness and that these requirements be complied with not only in the case of new buildings but also in the case of buildings converted for use as stabling;

   (iv) that in the case of horses maintained at grass there will be available for them at all times during which they are so maintained adequate pasture and shelter and water and that supplementary feeds will be provided as and when required;

   (v) that horses will be adequately supplied with suitable food, drink and (except in the case of horses maintained at grass, so long as they are so maintained) bedding material, and will be adequately exercised, groomed and rested and visited at suitable intervals;

   (vi) that all reasonable precautions will be taken to prevent and control the spread among horses of infectious or contagious disease and that veterinary first-aid equipment and medicines shall be provided and maintained in the premises;

   (vii) that appropriate steps will be taken for the protection and extrication of horses in case of fire and, in particular, that the name, address and telephone number of the licence holder or some other responsible person will be kept displayed in a prominent position on the outside of the premises and that instructions as to action to be taken in the event of fire, with particular regard to the extrication of horses, will be kept displayed in a prominent position on the outside of the premises;

(viii) that adequate accommodation will be provided for forage, bedding, stable equipment and saddlery (section 1(4)).

Persons under 18 years old or persons or bodies corporate disqualified under the following provisions may not be given a licence:

(a) from keeping a riding establishment under the Riding Establishment Act 1964;
(b) from keeping a pet shop under the Pet Animals Act 1951;
(c) from having custody of animals under the Protection of Animals (Amendment) Act 1954;
(d) from keeping an animal boarding establishment under the Animal Boarding Establishment Act 1963;
(e) under section 30(2), (3) or (4) of the Animal Welfare Act 2006;
(f) under subsection (1) of section 40 of the Animal Health and Welfare (Scotland) Act 2006 (asp 11) (section 1(2)).

Licences

Full licences continue for one year beginning on the day on which they came into force and then expire. The date of operation, depending upon the wishes of the applicant, is either the day on which it is granted or the 1 January next (section 1(6) and (7)).

Provisional licences operate for three months from the day on which they are granted and are used where the local authority is satisfied that it would not be justified in issuing a full licence. The three months' period of operation may, on application before the expiration of the three months, be extended for a further period of not exceeding three months so long as this would not exceed a six-month period in one year (Riding Establishments Act 1970, section 1).

On the death of a licence holder, licences pass to his personal representative for a period of three months and then expire. The three-month period may be extended at the local authority's discretion (section 1(8)).

Any person guilty of an offence under any provision of this Act other than section 2(4) is liable to a fine not exceeding level 3 or three months' imprisonment, or both (section 4(1)).

Conditions

On granting a licence, the local authority is required to specify conditions as appear necessary or expedient to achieve all the objectives set out in (b)(i)–(viii) in the paragraph headed 'Matters for consideration' above. In addition, the following conditions are required by the Act, whether specified in the licence or not:

(a) a horse found on inspection of the premises by an authorised officer to be in need of veterinary attention shall not be returned to work until the holder of the licence has obtained at his own expense and has lodged with the local authority a veterinary certificate that the horse is fit for work;
(b) no horse will be let out on hire for riding or used for providing instruction in riding without supervision by a responsible person of the age of 16 years or over unless (in the case of a horse let out for hire for riding) the holder of the licence is satisfied that the hirer of the horse is competent to ride without supervision;
(c) the carrying on of the business of a riding establishment shall at no time be left in the charge of any person under 16 years of age;
(d) the licence holder shall hold a current insurance policy which insures him against liability for any injury sustained by those who hire a horse from him for riding and those who use a horse in the course of receiving from him in return for payment, instruction in riding and arising out of the hire or use of a horse and which also insures such persons in respect of any liability which may be incurred by them in respect of injury to any person caused by, or arising out of, the hire or use of a horse;
(e) a register shall be kept by the licence holder of all horses in his possession aged three years and under and usually kept on the premises, which shall be available for inspection by an authorised officer at all reasonable times (section 1(4A)).

Disqualifications and cancellations

In making a conviction under this Act, under the Protection of Animals Act 1911, the Protection of Animals (Scotland) Act 1912, the Pet Animals Act 1951, the Animal Boarding Establishment Act 1963 or sections 4–9 and 11 of the Animal Welfare Act 2006, the court may cancel any riding establishment licence held by the convicted person and may disqualify him from holding such a licence, whether or not he is a current holder, for such period as the court thinks fit. The cancellation or disqualification may be suspended by the court pending an appeal (section 4(3) and (4)).

Power of entry

Local authority officers and veterinary surgeons and practitioners, authorised in writing by the local authority, may, upon producing their authority if required, enter and inspect the following premises at all reasonable times:

(a) licensed riding establishments;
(b) premises subject to an application for a licence to run a riding establishment; and
(c) unlicensed premises suspected of being used as a riding establishment.

This power extends to the inspection of any horses found on the premises (section 2(1) to (3)).

Persons wilfully obstructing or delaying authorised officers are subject on conviction to a maximum fine not exceeding level 2 (sections 2(4) and 4(1)).

Definitions

Approved certificate means:

(a) any one of the following certificates issued by the British Horse Society, namely, Assistant Instructor's Certificate, Instructor's Certificate and Fellowship;
(b) Fellowship of the Institute of Horse; or
(c) any other Certificate for the time being prescribed by order of the Secretary of State.

Horse includes any mare, gelding, pony, foal, colt, filly or stallion and also any ass, mule or jennet.
Premises includes land (section 6(4)).

PUBLIC HEALTH (CONTROL OF DISEASE) ACT 1984
GENERAL PROCEDURAL PROVISIONS

The following provisions on power of entry are applicable to procedures under the Public Health (Control of Disease) Act 1984.

Power of entry

An authorised officer of a council, producing if required a duly authenticated document showing his authority, has a right to enter any premises at all reasonable hours for the following purposes:

(a) ascertaining if there has been a contravention of the Act or bye-laws made under it;
(b) seeing if circumstances exist which would authorise or require the council to take any action or execute work;
(c) to take any action or execute any work authorised or required;
(d) generally for the council's performance of its functions.

For premises other than a factory, workshop or workplace, at least 24 hours' notice must be given before entry may be demanded.

Where there is reasonable ground for entry under (a)–(d) above and:

(a) admission has been refused; or
(b) refusal is apprehended; or
(c) the premises are unoccupied; or
(d) the occupier is temporarily absent; or
(e) the case is one of urgency; or
(f) application for admission would defeat the object of entry,

the local authority may make application to a Justice of the Peace by way of sworn information in writing, and the Justice of the Peace may authorise entry by an authorised officer, if necessary by force. Unless the case falls as one of (c)–(f) above, notice to apply for the warrant must be given to the occupier (section 61). The authorised officer may take with him any other persons as may be necessary but must leave any unoccupied premises as secure as he found them (section 62).

The penalty for obstructing any person acting in the execution of these Acts, regulations and orders is a maximum of level 1 on the standard scale (section 63).

INFECTIOUS DISEASES CONTROL

References

Public Health (Control of Disease) Act 1984 as amended by the Health and Social Care Act 2008.
Health protection legislation guidance 2010 – Department of Health.
Health Protection Regulations 2010 Toolkit – Chartered Institute of Environmental Health, Department of Health and Lewes District Council.
The following provisions are applicable to procedures under the Act and regulations made under the Act.

Extent

This Act does not apply in Scotland or Northern Ireland (section 79(3)).

Power of entry

An authorised officer of a council, producing if required a duly authenticated document showing his authority, has a right to enter any premises at all reasonable hours for the following purposes:

(a) ascertaining if there has been a contravention of the Act or bye-laws made under it;
(b) seeing if circumstances exist which would authorise or require the council to take any action or execute work;
(c) to take any action or execute any work authorised or required;
(d) generally for the council's performance of its functions.

For premises other than a factory, workshop or workplace, at least 24 hours' notice must be given before entry may be demanded.

Where there is reasonable ground for entry under (a)–(d) above and:

(a) admission has been refused; or
(b) refusal is apprehended; or
(c) the premises are unoccupied; or
(d) the occupier is temporarily absent; or
(e) the case is one of urgency; or
(f) application for admission would defeat the object of entry,

the local authority may make application to a Justice of the Peace by way of sworn information in writing, and the Justice of the Peace may authorise entry by an authorised officer, if necessary by force. Unless the case falls as one of (c)–(f) above, notice to apply for the warrant must be given to the occupier (section 61). The authorised officer may take with him any other persons as may be necessary but must leave any unoccupied premises as secure as he found them.

The officer may for the purpose for which entry is authorised:

(a) search the premises;
(b) carry out measurements and tests of the premises or of anything found on them;
(c) take and retain samples of the premises or of anything found on them;
(d) inspect and take copies or extracts of any documents or records found on the premises;
(e) require information stored in an electronic form and accessible from the premises to be produced in a form in which it can be taken away and in which it is visible and legible or from which it can readily be produced in a visible and legible form; and
(f) seize and detain or remove anything which the officer reasonably believes to be evidence of any contravention relevant to the purpose for which entry is authorised (section 62).

The penalty for obstructing any person acting in the execution of these Acts, regulations and orders is a maximum of level 1 on the standard scale (section 63).

KEY TO POWERS AVAILABLE FOR THE CONTROL OF DISEASE

The following powers are available to local authorities for the control of communicable disease in the Health Protection (Local Authority Powers) Regulations 2010:

- power to order health measures in relation to persons;
- power to order health measures in relation to things;
- power to order health measures in relation to premises;;
- orders in respect of groups;
- procedure for making, varying and revoking Part 2A orders;
- enforcement of Part 2A orders.

Note

Special arrangements exist for the control of infectious diseases in ships and aircraft in the Public Health (Aircraft) Regulations 1979 and the Public Health (Ships) Regulations 1979 and, in respect of the Channel Tunnel, the Public Health (International Trains) Regulations 1994 (see 'Port Health').

PROPER OFFICER

All of the powers and procedures for the notification and control of communicative diseases identified in this chapter are enforced by the local authority (for definition, see page 8). Local authorities are, however, advised that the majority of such powers should be delegated to the Consultant in Communicative Disease Control (CCDC) appointed by each Health Protection Agency (Annex B to NHS Management Executive Guidance HSG (93)56). This will require a formal resolution of the local authority detailing each of the powers being delegated and is to be done in the context of a local communicable disease plan which identifies the role and responsibilities of the various agencies and individuals.

Local authorities need to provide appropriate professional and other staff support to the Consultant in Communicative Disease Control and ensure that proper arrangements have been made for the receipt of notifications (FC90).

NOTIFICATION OF DISEASES

References

Health Protection (Notification) Regulations 2010.
Public Health (Control of Disease) Act 1984, section 11.

Extent

These regulations apply to England only but there are similar provisions in Wales.

Diseases to be notified

The diseases (confirmed or suspected) which are required to be notified by this procedure are:

Acute encephalitis
Acute infectious hepatitis
Acute meningitis
Acute poliomyelitis
Anthrax
Botulism
Brucellosis
Cholera
Diphtheria
Enteric fever (typhoid or paratyphoid fever)
Food poisoning
Haemolytic uraemic syndrome (HUS)
Infectious bloody diarrhoea
Invasive group A streptococcal disease and scarlet fever

Legionnaires' disease
Leprosy
Malaria
Measles
Meningococcal septicaemia
Mumps
Plague
Rabies
Rubella
SARS
Smallpox
Tetanus
Tuberculosis
Typhus
Viral haemorrhagic fever (VHF)
Whooping cough
Yellow fever

(Schedule 1 Health Protection (Notification) Regulations 2010)

Note – for the notification of occupation-related diseases, see FC54.

Duty to notify

A registered medical practitioner who becomes aware or suspects that a patient whom he is attending:

- has a notifiable disease; or
- has an infection or is contaminated in a manner which presents or could present significant harm to human health

must notify the proper officer of the local authority within three days of the patient being attended to by the practitioner.

The duty does not apply where the registered medical practitioner believes the proper officer has already been notified (regulation 2).

Details to be notified

The form of certificate to be used (or similar) is set out as shown here: www.hpa.org.uk/Topics/InfectiousDiseases/InfectionsAZ/NotificationsOfInfectious-Diseases/ReportingProcedures/

Notification must include the following details of the patient:

(a) name, date of birth and sex;
(b) home address, including postcode;
(c) current residence (if not home address);
(d) telephone number;
(e) NHS number;
(f) occupation (if considered relevant);

(g) the name, address and postcode of his/her place of work or education (if considered relevant);
(h) relevant overseas travel history;
(i) ethnicity;
(j) contact details for a parent of the patient (where a child) (where the child is living);
(k) the disease or infection which the patient has or is suspected of having, or the nature of the patient's contamination or suspected contamination;
(l) the date of onset of symptoms;
(m) the date of diagnosis; and
(n) name, address and telephone number of the registered medical practitioner (regulation 2(2))

Notification etc. by a proper officer

The proper officer of the local authority must disclose the notification and its contents to:

(a) the Health Protection Agency;
(b) the proper officer of the local authority in whose area the patient usually resides (if different); and
(c) the proper officer of the port health authority or local authority in whose district or area a ship, hovercraft, aircraft or international train is or was situated from which the patient has disembarked (if known to the disclosing proper officer and if that officer considers disclosure appropriate).

The disclosure must be made in writing within three days, beginning with the day that the proper officer receives the notification.

If the proper officer considers that the case is urgent, disclosure must be made orally as soon as reasonably practicable.

In determining whether a case is urgent, the proper officer must have regard to:

(a) the nature of the disease, infection or contamination or the suspected disease, infection or contamination notified;
(b) the ease of spread of the disease, infection or contamination;
(c) the ways in which the spread of the disease, infection or contamination can be prevented or controlled; and
(d) where known, the patient's circumstances (including age, sex and occupation) (regulation 6).

In addition, proper officers should inform the Department of Health of any case or outbreak of infectious disease which may have wider than local significance (Annex B HSG (93) 56) – this is to allow consideration of action at national level, e.g. withdrawal of a food for sale. Proper officers also have a responsibility to consult as appropriate with the Communicable Disease Surveillance Centre, specifically with cases of outbreaks which present unusual features (HSG (93)56 Annex B).

FC90 Notification of diseases – Health Protection (Notification) Regulations 2010

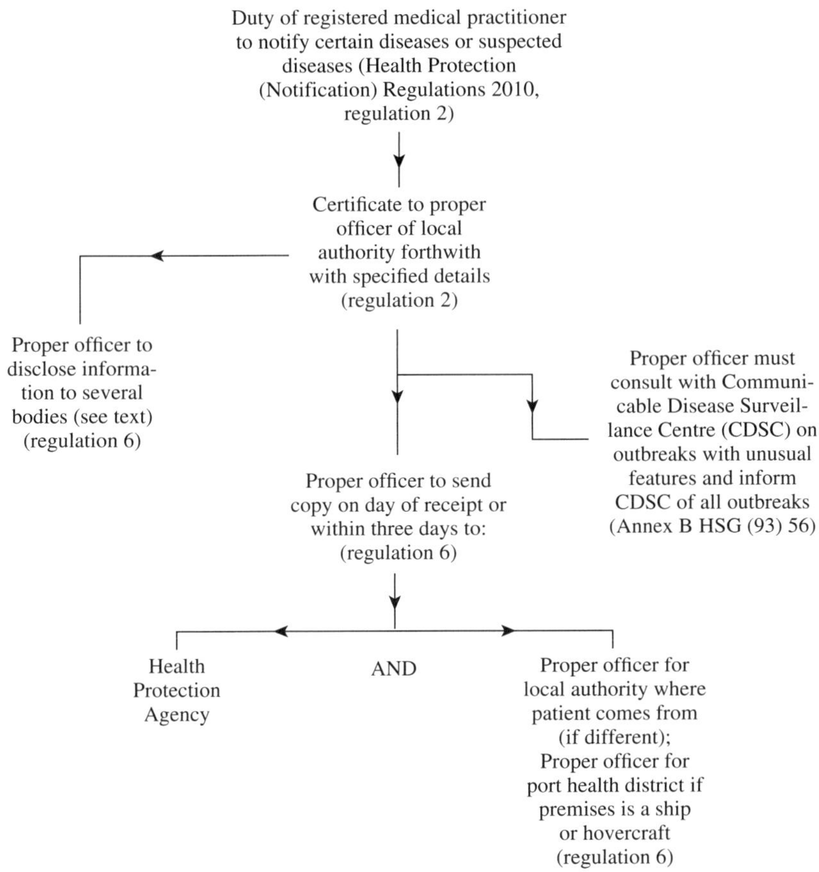

Notes

1. This procedure does not apply to Scotland or Northern Ireland.
2. For special procedures relating to the notification of infectious diseases on board ships or aircraft, see the Public Health (Ships) Regulations 1979 and the Public Health (Aircraft) Regulations 1979.

The proper officer must make weekly and quarterly returns of notifications to the Registrar General and these are collated by the Communicable Disease Surveillance Centre for weekly publication in the Communicable Disease Report.

Notifications, information, disclosures, lists and reports, which are required to be in writing, may be communicated electronically if:

(a) the recipient has consented in writing to receiving the notification, information, disclosure list or report (as the case may be) by an electronic communication; and
(b) the communication is sent to the number or address specified by the recipient when giving that consent (regulation 7).

Causative agents

The operator of a diagnostic laboratory must notify the Health Protection Agency in accordance with this regulation where the diagnostic laboratory identifies a causative agent in a human sample.

The notification must include the following information insofar as it is known to the operator of the diagnostic laboratory:

(a) name and address of the diagnostic laboratory;
(b) details of the causative agent identified;
(c) date of the sample;
(d) nature of the sample;
(e) name of person from whom the sample was taken;
(f) the person's date of birth and sex, current home address, including postcode, current residence (if not home address), ethnicity and NHS number;
(g) and the name, address and organisation of the person who solicited the test which identified the causative agent.

The notification must be provided in writing within seven days, beginning with the day on which the causative agent is identified.

If the operator of the diagnostic laboratory considers that the case is urgent, the notification must be provided orally as soon as reasonably practicable.

In determining whether the case is urgent, the operator of the diagnostic laboratory must have regard to:

(a) the nature of the causative agent;
(b) the nature of the disease which the causative agent causes;
(c) the ease of spread of the causative agent;
(d) the ways in which the spread of the causative agent can be prevented or controlled; and
(e) where known, the person's circumstances (including age, sex and occupation).

This regulation does not apply where the operator of the diagnostic laboratory reasonably believes that the Health Protection Agency has already been notified in accordance with this regulation by the operator of another diagnostic laboratory in relation to the same causative agent being found in a sample from the same person.

It is an offence for the operator of a diagnostic laboratory to fail without reasonable excuse to comply with this regulation.

Any person who commits an offence under this regulation is liable on summary conviction to a fine not exceeding level 5 on the standard scale (regulation 4).

Duty to provide information to the Health Protection Agency

The Health Protection Agency may request that the person who solicited the laboratory test which identified the causative agent to which the notification relates provides to it the information included in the notification in writing within three days, beginning with the day on which the request is made.

If the Health Protection Agency considers the case to be urgent, the information must be provided orally as soon as reasonably practicable.

In determining whether the case is urgent, the Health Protection Agency must have regard to:

(a) the nature of the causative agent to which the notification relates;
(b) the nature of the disease which the causative agent causes;
(c) the ease of spread of the causative agent;
(d) the ways in which the spread of the causative agent can be prevented or controlled; and
(e) where known, the circumstances of the person from whom the sample was taken (including age, sex and occupation).

List of causative agents

Bacillus anthracis
Bacillus cereus (only if associated with food poisoning)
Bordetella pertussis
Borrelia spp
Brucella spp
Burkholderia mallei
Burkholderia pseudomallei
Campylobacter spp
Chikungunya virus
Chlamydophila psittaci
Clostridium botulinum
Clostridium perfringens (only if associated with food poisoning)
Clostridium tetani
Corynebacterium diphtheriae
Corynebacterium ulcerans
Coxiella burnetii
Crimean-Congo haemorrhagic fever virus
Cryptosporidium spp

Dengue virus
Ebola virus
Entamoeba histolytica
Francisella tularensis
Giardia lamblia
Guanarito virus
Haemophilus influenzae (invasive)
Hanta virus
Hepatitis A, B, C, delta, and E viruses
Influenza virus
Junin virus
Kyasanur Forest disease virus
Lassa virus
Legionella spp
Leptospira interrogans
Listeria monocytogenes
Machupo virus
Marburg virus
Measles virus
Mumps virus
Mycobacterium tuberculosis complex
Neisseria meningitidis
Omsk haemorrhagic fever virus
Plasmodium falciparum, vivax, ovale, malariae, knowlesi
Polio virus (wild or vaccine types)
Rabies virus (classical rabies and rabies-related lyssa viruses)
Rickettsia spp
Rift Valley fever virus
Rubella virus
Sabia virus
Salmonella spp
SARS coronavirus
Shigella spp
Streptococcus pneumoniae (invasive)
Streptococcus pyogenes (invasive)
Varicella zoster virus
Variola virus
Verocytotoxigenic Escherichia coli (including E.coli O157)
Vibrio cholerae
West Nile virus
Yellow fever virus
Yersinia pestis

(Schedule 2 Health Protection (Notification) Regulations 2010)

Definitions

Causative agent means:

(a) a causative agent listed in schedule 2; or
(b) evidence of an infection caused by such an agent.

Child means a person under the age of 18 years.
Parent has the meaning given to it by section 576 of the Education Act 1996.

LOCAL AUTHORITY POWERS

The following procedures are available to local authorities as the regulations confer discretionary powers on local authorities (including powers to impose restrictions and requirements) for the purposes of preventing, protecting against, controlling or providing a public health response to the incidence or spread of infection or contamination which presents or could present significant harm to human health.

Reference

Health Protection (Local Authority Powers) Regulations 2010.

Extent

These regulations apply to England only.

Requirement to keep a child away from school

Where a local authority is satisfied that:

(a) a child is or may be infected or contaminated;
(b) the infection or contamination is one which presents or could present significant harm to human health;
(c) there is a risk that the child might infect or contaminate others;
(d) it is necessary to keep the child away from school in order to remove or reduce that risk; and
(e) keeping the child away from school is a proportionate response to the risk to others presented by the child,

the local authority may, by serving notice on the child's parent, require that he or she keep the child away from school.

The notice must include the following information:

(a) the date from which the requirement commences;
(b) the duration of the requirement (up to a maximum of 28 days);
(c) why the requirement is believed to be a necessary and proportionate measure;
(d) the penalty for failing to comply with the notice; and

FC91 Requirement to keep a child away from a school – Regulation 2 Health Protection (Local Authority Powers) Regulations 2010

Local authority satisfied child infected or contaminated and could present significant harm to human health etc.

Local authority serves notice/informs headteacher of school notice served and its contacts (regulation 2)

Local authority serves notice on child's parents to keep child away from school (regulation 2)

Parent may request review at any time before notice elapses (regulation 2)

Local authority must review the notice within five working days beginning with the day on which the request is made where the parent is requesting a review in respect of that notice for the first time (regulation 2)

Local authority may vary, revoke or confirm the review (regulation 2)

Offence for parent to fail to comply (level 2) (regulation 2)

Note

1. There seems to be no appeal against this notice.

(e) contact details for an officer of the local authority who is able to discuss the notice.

The local authority must, as soon as reasonably practicable after serving, inform the headteacher of the child's school:

(a) that it has served such a notice in relation to the child; and
(b) of the contents of that notice.

The parent may request that the local authority review the notice at any time before the requirement lapses.

The local authority:

(a) must review the notice within five working days, beginning with the day on which the request is made where the parent is requesting a review in respect of that notice for the first time; or
(b) may review the notice in the case of all other requests.

The local authority must inform the parent and the headteacher of the child's school of the outcome of any review it conducts as soon as reasonably practicable after the review is concluded.

The local authority may vary or revoke a notice and must, as soon as reasonably practicable, inform the parent and the headteacher of the child's school that the notice has been varied or revoked and, if varied, the nature of the variation.

The local authority may serve consecutive notices and must inform the parent and the headteacher of the child's school as soon as reasonably practicable where a notice has expired and no further notice is to be served.

It is an offence for the parent to fail without reasonable excuse to comply with a notice.

Any person who commits an offence under this regulation is liable on summary conviction to one or both of:

(a) a fine not exceeding level 2 on the standard scale;
(b) a further fine not exceeding an amount equal to 50 per cent of level 1 on the standard scale for each day on which the default continues after conviction (regulation 2).

Requirement to provide details of children attending school

A local authority may serve notice to require a headteacher of a school in its area to provide it with a list of the names, addresses and contact telephone numbers for all the pupils of that school, or such group of pupils attending that school as it may specify, where

(a) a person who is or has recently been on the school's premises is or may be infected or contaminated;
(b) the infection or contamination is one which presents or could present significant harm to human health;
(c) there is a risk that the person may have infected or contaminated pupils at the school;

(d) it is necessary for the local authority to have the list in order to contact those pupils with a view to ascertaining whether they are or may be infected or contaminated; and
(e) requiring the list (and contacting those pupils who may be infected or contaminated) is a proportionate response to the risk presented by the person.

The notice must:

(a) specify a time limit for meeting the requirement;
(b) specify an address where the list is to be sent;
(c) provide contact details for an officer of the local authority who is able to discuss the notice.

It is an offence for a headteacher to fail without reasonable excuse to comply with a notice.

Any person who commits an offence under this regulation shall be liable on summary conviction to a fine not exceeding level 1 on the standard scale (regulation 3).

Requests for cooperation for health protection purpose

A local authority may serve notice on any person or group of persons requesting that the person or group of persons do, or refrain from doing, anything for the purpose of preventing, protecting against, controlling or providing a public health response to the incidence or spread of infection or contamination which presents or could present significant harm to human health. The notice must provide contact details for an officer of the local authority who is able to discuss the notice.

The local authority may offer compensation or expenses in connection with its request (regulation 8).

Restriction of contact with and access to dead bodies

Where a local authority is satisfied that:

(a) a dead body is or may be infected or contaminated;
(b) the infection or contamination is one which presents or could present significant harm to human health;
(c) there is a risk that the dead body might infect or contaminate people;
(d) it is necessary to restrict contact with or access to the dead body in order to remove or reduce that risk; and
(e) prohibiting any person from having contact with or access to the dead body is a proportionate response to the risk presented by that dead body,

the local authority may serve on the person having charge or control of the premises in which the dead body is located a notice prohibiting any person from having contact with the dead body or entering the room where it is located.

FC92 Request for cooperation for health protection purpose – Regulation 8 Health Protection (Local Authority Powers) Regulations 2010

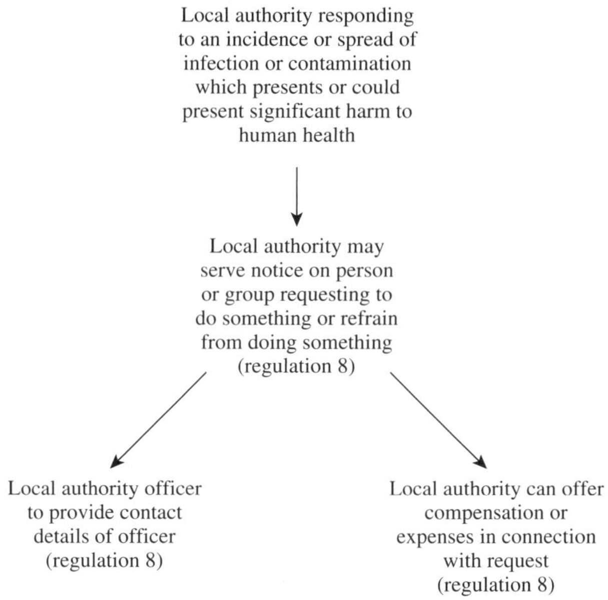

Note

1. There seems to be no appeal against this notice.

On receipt of a notice served on the person having charge or control of the premises, he/she must arrange for a copy of the notice to be conspicuously displayed at each of the entry points to the room and/or near the dead body without delay. The notice must include:

(a) a statement to the effect that entry into the room and contact with the body near which the notice has been displayed is prohibited;
(b) a statement to the effect that breach of the prohibition is a criminal offence;
(c) contact details for an officer of the local authority who is able to discuss the notice; and
(d) the legal authority for the prohibition.

It is an offence for:

(a) any person having charge or control of the premises to fail to arrange for a copy of the notice to be displayed in the appropriate places;
(b) any person to remove or deface a notice displayed; or
(c) any person to fail to comply with a notice displayed,

but no offence is committed if

(a) the person has the local authority's consent to enter the room and/or have contact with the dead body; or
(b) the person is exercising the functions of a coroner or is acting under the authority of a coroner (regulations 9 and 10).

Relocation of dead bodies

Where a local authority is satisfied that (a) to (e) above apply and it is necessary to relocate the dead body in order to remove or reduce that risk and relocating the body is a proportionate response to the risk to people presented by the dead body in its current location, it can relocate, or cause to be relocated, the dead body to a place where it considers that the risk of the dead body infecting or contaminating people is reduced or removed.

The local authority cannot relocate, or cause to be relocated, the dead body if:

(a) a coroner has jurisdiction over the dead body; or
(b) it has failed to take reasonable steps to inform the person with charge or control of the premises where the dead body is located of its intention to relocate the body.

Any person having charge or control of the premises in which a dead body is located must cooperate with a local authority that intends to take action and it is an offence for any person to fail without reasonable excuse to comply (regulation 11).

Any person who commits an offence under regulations 9, 10 and 11 is liable on summary conviction to a fine not exceeding level 3 on the standard scale.

FC93 Restriction of contact with and access to dead bodies – Regulations 9 and 10 Health Protection (Local Authority Powers) Regulations 2010

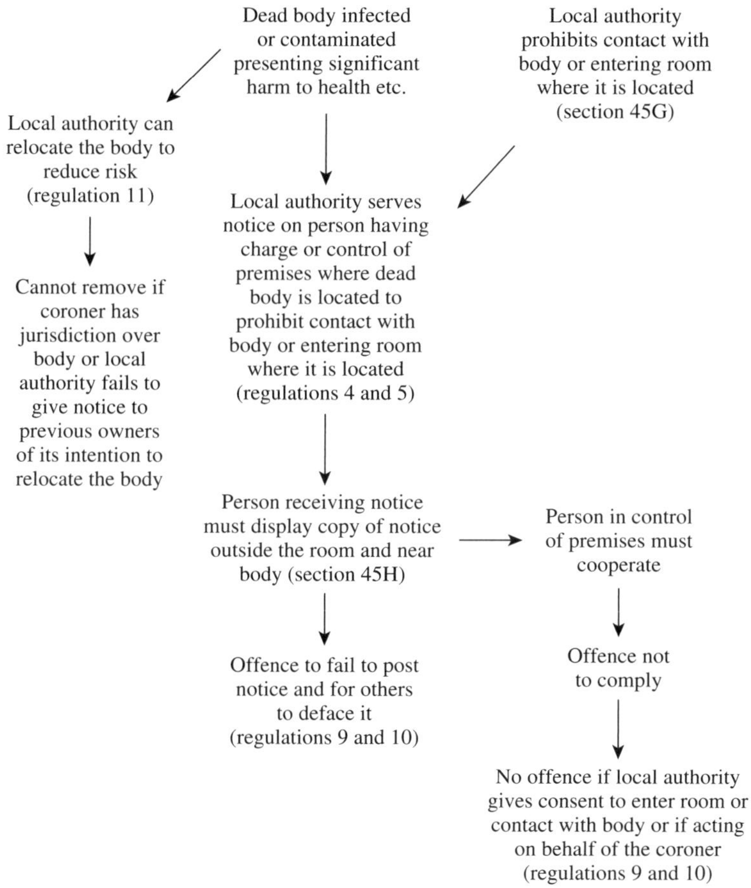

Definitions

Child means a person under 18 years of age.
Headteacher means the headteacher or, if the headteacher is absent, the person deputising for the headteacher.
Parent has the meaning given to it by section 576 of the Education Act 1996.
School has the meaning given to it by section 4 of the Education Act 1996.
Working day means any day which is not Saturday, Sunday, Christmas Day, Good Friday or a day which is a bank holiday in England under the Banking and Financial Dealings Act 1971.

POWER TO ORDER HEALTH MEASURES IN RELATION TO PERSONS, THINGS AND PREMISES (PART 2A ORDERS)

References

Public Health (Control of Disease) Act 1984.
Health Protection (Part 2A orders) Regulations 2010.
(Note – where section is mentioned, it refers to the above Act and, where regulation is mentioned, it refers to the above regulations.)

Extent

The Public Health (Control of Disease) Act 1984 applies in England and Wales. Health Protection (Part 2A orders) Regulations 2010 apply in England only, but there are similar provisions in Wales.

Part 2A orders and their content

A local authority can apply to a Justice of the Peace to make an order for health measures in relation to persons, things and premises (sections 45G, 45H and 45I).

If the Justice is satisfied that:

(a) a person, thing or premises is or may be infected or contaminated;
(b) the infection or contamination is one which presents or could present significant harm to human health;
(c) there is a risk that the person, thing or premises might infect or contaminate others; and
(d) it is necessary to make the order in order to remove or reduce that risk,

he may make an order to impose one or more of the following restrictions or requirements – that the *person*:

(a) submit to medical examination;
(b) be removed to a hospital or other suitable establishment;
(c) be detained in a hospital or other suitable establishment;
(d) be kept in isolation or quarantine;
(e) be disinfected or decontaminated;
(f) wear protective clothing;

(g) provide information or answer questions about the person's health or other circumstances (including, in particular, information or questions about the identity of a related party);
(h) be monitored in their health and the results reported;
(i) attend training or advice sessions on how to reduce the risk of infecting or contaminating others;
(j) be subject to restrictions on where the person goes or with whom the person has contact;
(k) abstain from working or trading.

The justice may also make an order to require that the person provide information or answer questions about the person's health or other circumstances (including, in particular, information or questions about the identity of a related party). This also applies to a person with parental responsibility (within the meaning of the Children Act 1989) for the person to secure that the person submits to or complies with the restrictions or requirements imposed by the order.

In relation to *things*, the order may impose one or more of the following restrictions or requirements – that

(a) the thing be seized or retained;
(b) the thing be kept in isolation or quarantine;
(c) the thing be disinfected or decontaminated;
(d) in the case of a dead body, the body be buried or cremated;
(e) in any other case, the thing be destroyed or disposed of.

The Justice may also make an order to require that the owner or person having control of the thing provide information or answer questions about the thing (including, in particular, information or questions about where the thing has been or about the identity of any related person or the whereabouts of any related thing) (section 45H).

In relation to *premises*, the order may impose one or more of the following restrictions or requirements – that

(a) the premises be closed;
(b) in the case of a conveyance or movable structure, the conveyance or structure be detained;
(c) the premises be disinfected or decontaminated;
(d) in the case of a building, conveyance or structure, the premises be destroyed.

The Justice may also make an order to require the owner or any occupier of the premises to provide information or answer questions about the premises (including, in particular, information about the identity of any related person or the whereabouts of any related thing) (section 45I)

The Justice has the power to make an order in relation to a group of persons, things or premises (section 45J).

Orders for health measures in relation to persons, things and premises may include such other restrictions or requirements as the Justice considers necessary for the purpose of reducing or removing the risk in question, including conditions and directions. It may also order the payment of compensation or expenses in connection with the taking of measures pursuant to the order and authorise those persons to whom it is addressed to do such things as may be necessary to give effect to it (section 45K).

Making, varying and revoking Part 2A orders

Local authorities may apply to a Justice of the Peace to make a Part 2A order and they must cooperate with each other in deciding which of them should apply for a Part 2A order in any particular case.

A Part 2A order may be varied or revoked by a Justice of the Peace on the application of:

(a) an affected person;
(b) a local authority; or
(c) any other authority with the function of executing or enforcing the order in question.

For an order involving a person, the following persons are affected persons:

(a) the person;
(b) a person with parental responsibility (within the meaning of the Children Act 1989) for the person;
(c) the person's husband, wife or civil partner;
(d) a person living with the person as the person's husband, wife or civil partner; and
(e) such other persons as may be prescribed by regulations.

For an order involving a thing, the following persons are affected persons:

(a) the owner of the thing;
(b) any person with custody or control of the thing; and
(c) such other persons as may be prescribed by regulations.

For an order involving premises, the following persons are affected persons:

(a) the owner of the premises;
(b) any occupier of the premises; and
(c) such other persons as may be prescribed by regulations.

Variation or revocation of a Part 2A order does not invalidate anything done under the order prior to the variation or revocation (section 45M).

Notice of Part 2A applications

A local authority must make reasonable enquiries as to the existence and location of persons subject to a Part 2A application.

Having made reasonable enquiries, the local authority must give notice of the application to the persons specified in the application, but if a Justice of the Peace considers it necessary to do so, the Justice may make a Part 2A order without a person having been given such notice.

The persons specified are:

(a) the person subject to the application;
(b) a person with parental responsibility for that person, if a child; and
(c) that person's decision-maker (if any).

If the subject of the application is a thing, the persons specified are:

(a) the owner of the thing; and
(b) the person with custody or control of the thing.

Where a dead body or human remains is concerned, the person specified is the deceased's next of kin.

Where premises are the subject of an application, the persons specified are:

(a) the owner of the premises; and
(b) the occupier of the premises, if any.

If it is likely the person will abscond or otherwise take steps to undermine the order applied for, or where exceptional circumstances exist which mean that notifying such a person would not be in the person's best interests, the local authority is not required to give notice (regulation 3).

Evidence required for a Part 2A application in relation to persons

Before an order can be granted, the local authority should provide the Justice of the Peace with the following evidence:

(a) a report which gives known and relevant details, or reasons for omission of details, of:
    (i) the signs and symptoms of the infection or contamination in the person who is the subject of the application;
    (ii) the person's diagnosis;
    (iii) the outcome of clinical or laboratory tests; and
    (iv) the person's recent contacts with, or proximity to, a source or sources of infection or contamination;
(b) a summary of the characteristics and effects of the infection or contamination which the person has or may have, which includes an explanation of:
    (i) the mechanism by which the infection or contamination spreads;
    (ii) how easily the infection or contamination spreads amongst humans; and
    (iii) the impact of the infection or contamination on human health (by reference to pain, disability and the likelihood of death);
(c) where relevant an assessment of the risk to human health that the person or related party presents, including a description of any acts or omissions, or anticipated acts or omissions, of the person which affect that risk, and an assessment of the options available to deal with the risk that the person or related party presents

The evidence must be given by persons who are suitably qualified to give the evidence and may be given orally or in writing (regulation 4).

FC94 Part 2 orders – Public Health (Control of Disease) Act 1984

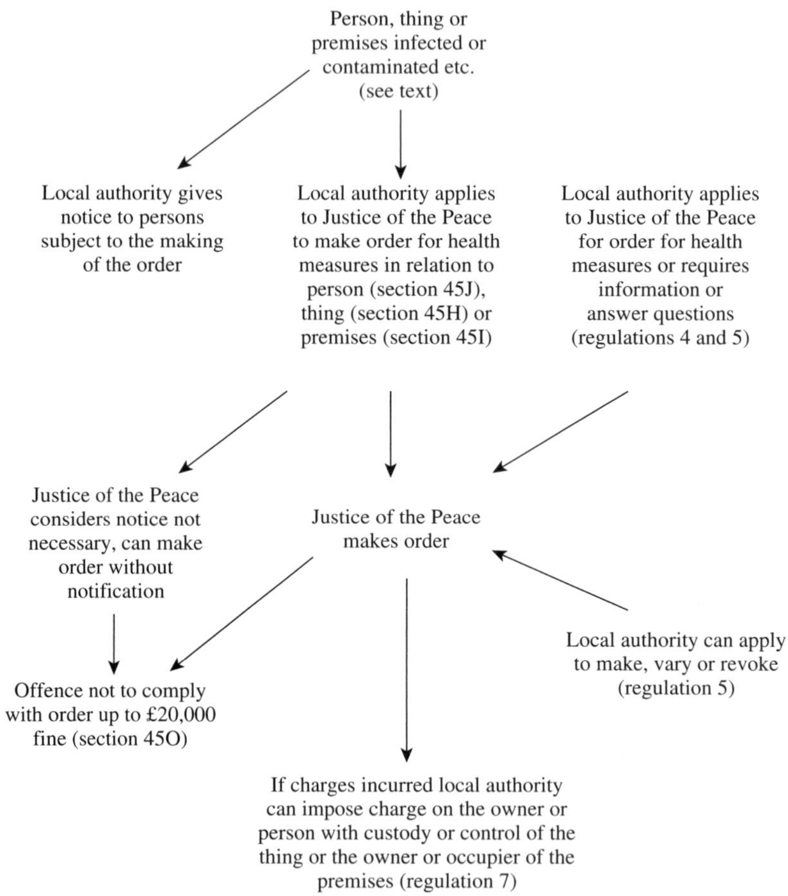

Notes

1. For groups of persons, things or premises, order can be made by Justice of the Peace (section 45J).

Period for which Part 2A order may be in force

The period for which the restriction or requirement imposed by or under the order may be in force must not exceed 28 days beginning with the day on which the order was made.

The period of any extension or variation of a restriction or requirement must not exceed 28 days (section 45I and regulation 5).

Charges

Where a local authority has incurred costs taking measures under a Part 2A order in relation to things and premises, the local authority may impose a charge on the owner or person with custody or control of the thing or the owner or occupier of the premises.

The amount of the charge imposed must not exceed the actual costs (including staff costs) incurred by the local authority in taking measures in relation to the thing or premises and must be reasonable in the circumstances.

A local authority cannot impose a charge in connection with orders which relate to a dead body or human remains (regulation 7).

Understanding of the person subject to the order

The local authority must take all reasonable steps to ensure that the person who is the subject of the order understands:

(a) the effect of the order, the reason it has been made, the power under which it has been made and the person's right to apply for a variation or revocation of the order; and
(b) the relevant support services available to the person (and how to access them).

Where the person is under the age of 18 years, the person with parental responsibility for the person must understand the relevant matters (regulation 8).

Welfare of dependants

Where a person is detained in a hospital or other suitable establishment or is kept in isolation or quarantine under a Part 2A order, the local authority must have regard to the impact of the order on the welfare of the person and his/her dependants, if any, for the duration of the order (regulation 9).

Report to the Health Protection Agency

A local authority must provide a written report to the Chief Executive of the Health Protection Agency each time it makes a Part 2A application and where it is varied or evoked.

The report must include:

(a) the name of the local authority;
(b) contact details for the officer of the local authority responsible for the report;

(c) a copy of the Part 2A application (with information that would enable the identification of the person who is the subject of the application removed);
(d) if an order is made, a copy of that order (with information that would enable the identification of the person who is the subject of the order removed); and
(e) if a Part 2A order is not made, the reason for it not being made.

The report must be provided as soon as practicable after the application is determined and no later than ten days beginning with the day on which the application is determined (regulations 10 and 11)

Offences

A person commits an offence if the person:

(a) fails without reasonable excuse to comply with a restriction or requirement imposed by or under a Part 2A order; or
(b) wilfully obstructs anyone acting in the execution of a Part 2A order.

A person guilty of an offence is liable on summary conviction to a fine not exceeding £20,000.

If a person is convicted of an offence above, and the court is satisfied that the failure or wilful obstruction constituting the offence has caused premises or things to become infected or contaminated or otherwise damaged them in a material way, it may order the person to take or pay for such remedial action as may be specified in the order.

If a Part 2A order imposes a requirement that a person be detained or kept in isolation or quarantine in a place, and the person leaves that place contrary to the requirement, a constable may take the person into custody and return the person to that place.

However, a person may not be taken into custody after expiry of the period for which the requirement is in force (section 45O).

The provisions of this part have effect in relation to the territorial sea adjacent to England or Wales (section 45S).

Definitions

Affected persons – in section 45G, the person's decision-maker (if any). In section 45H(2), in relation to a dead body or human remains, the deceased's next of kin.

Medical examination includes microbiological, radiological and toxicological tests.

Next of kin means the person accorded highest rank in the following list (but if two or more persons are accorded equal ranking, then each of those persons is to be treated as next of kin):

(a) a person with parental responsibility for the deceased person (P);
(b) P's husband, wife or civil partner;
(c) a person who had been living with P up to the time of P's death as P's husband, wife or civil partner;

(d) P's child where aged 18 years or over;
(e) P's parent;
(f) P's brother or sister where aged 18 years or over.

Person's decision-maker means the person's donee of enduring power of attorney or lasting power of attorncy under the Mental Capacity Act 2005 or a deputy appointed by the Court of Protection in relation to the person, where decisions in connection with Part 2A applications or orders are within the scope of that person's authority (regulation 6).
Related party means:

(a) a person who has or may have infected or contaminated the person; or
(b) a person whom the person has or may have infected or contaminated (section 45G).

Thing includes:

(a) human tissue;
(b) a dead body or human remains;
(c) animals; and
(d) plant material.

DISINFECTION OR DECONTAMINATION OF THINGS AND PREMISES

Reference

Health Protection (Local Authority Powers) Regulations 2010, regulations 4 to 7.

Disinfection or decontamination of things and premises on request of the owner or person with custody or control

A local authority may disinfect or decontaminate, or cause to be disinfected or decontaminated, a thing where requested to do so by the owner of the thing or the person with custody or control, and may charge the owner or person with custody or control for the disinfection or decontamination of the thing if the owner is made aware of the charge prior to disinfection or decontamination being carried out and agrees to pay it.

The local authority's charge must not exceed the cost incurred by the local authority in carrying out the disinfection or decontamination (regulations 4 and 5).

Where the thing is in the custody or control of a person, the local authority should only disinfect or decontaminate the thing if it is reasonably satisfied that the financial value of the thing will not be reduced as a consequence of the disinfection or decontamination.

This procedure also applies to premises. For premises with a tenant, he/she can request the action (regulations 6 and 7). See FC94 and FC95.

FC95 Disinfection or decontamination of premises and things on request of owner – Regulations 4 to 7 Health Protection (Local Authority Powers) Regulations 2010

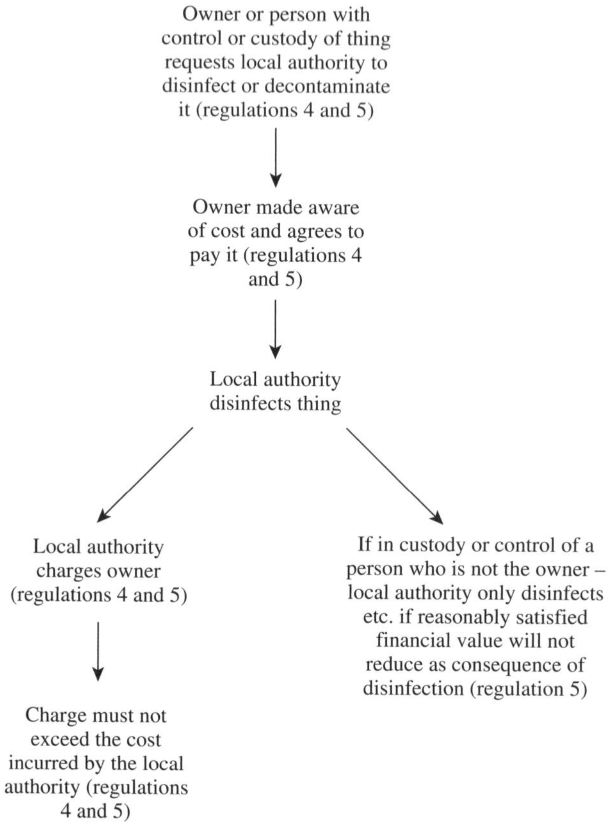

Notes

1. This procedure applies to things. The same procedure applies to premises by substituting 'premises' for 'things' and 'regulations 4 and 5' with 'regulations 6 and 7'.
2. Also substitute 'tenant' for 'person with control or custody of thing'.

CLEANSING OF FILTHY OR VERMINOUS PREMISES, ARTICLES AND PERSONS

References

Public Health Act 1936, sections 83 to 85 (as amended by Public Health Act 1961, section 35).
Public Health Act 1936, section 8 (as amended by Public Health Act 1961, section 35).

Extent

These provisions apply to England and Wales only.

Scope

These provisions apply to any premises except those forming part of a mine or quarry but including ships, boats, tents, vans and sheds (sections 83(4), 267(4) and 268(1)), and cover situations which are either:

(a) filthy or unwholesome so as to be prejudicial to health; or
(b) verminous.

Section 84 Public Health Act 1936 deals with articles that are filthy and verminous.

Section 85 Public Health Act 1936 deals with persons that are verminous.

Works required

For *premises*, the steps which are required to be taken must be specified in the notice and may include:

(a) removal of wallpaper or other wall covering;
(b) destruction or removal of vermin;
(c) interior surfaces of houses, shops or offices to be papered, painted or distempered and in all other cases painted, distempered or whitewashed (section 83(1)).

A special procedure is set out for situations where the local authority wishes to use gas to destroy vermin, and this includes the power to require vacation of both the affected premises and ones adjoining at a date specified in the notice, provided that temporary accommodation has been provided, whether it be dwelling accommodation or otherwise. The period of vacation must be specified in the notice (Public Health Act 1961, section 36).

Where a certificate from the proper officer of the authority that any *article* in any premises:

(a) is in so filthy a condition as to render its cleansing, purification or destruction necessary in order to prevent injury, or danger of injury, to the health of any person in the premises; or
(b) is verminous, or likely to be verminous,

FC96 Cleansing of filthy or verminous premises – Section 83 Public Health Act 1936

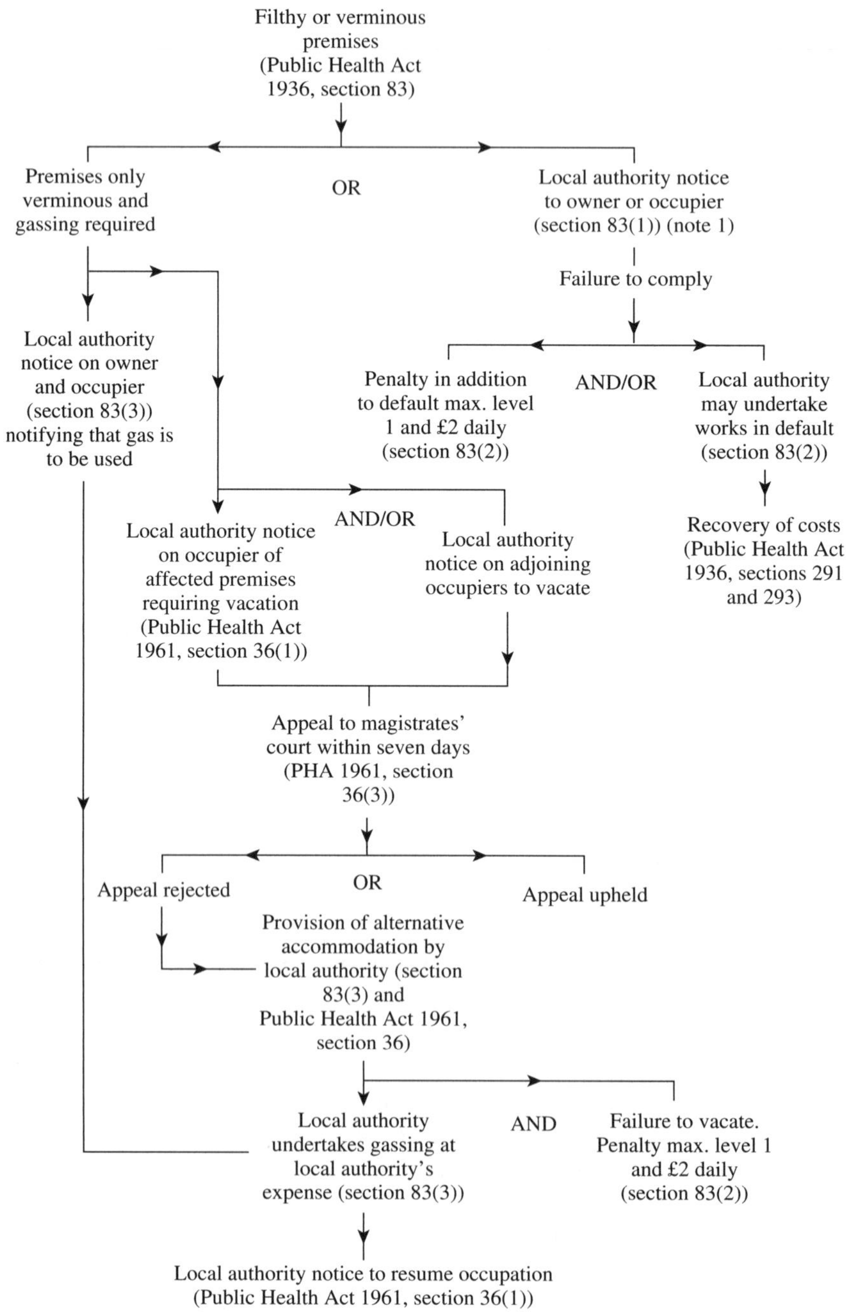

Notes

1. There is no appeal against this notice.
2. For cleansing of verminous articles, see Public Health Act 1936, section 84, verminous persons and clothing, section 85, and sale of verminous articles, Public Health Act 1961, section 37.

FC97 Cleansing of verminous persons and their clothing – Section 85 Public Health Act 1936

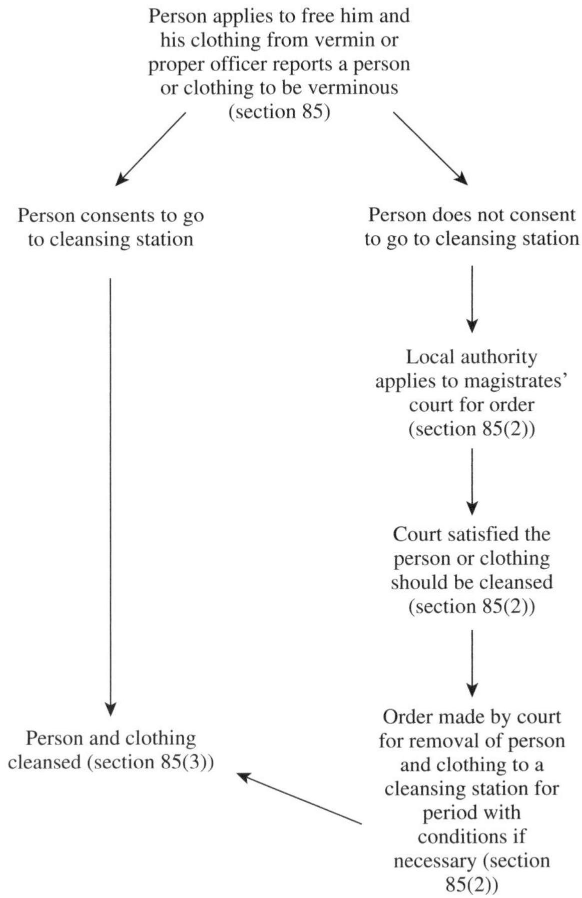

Notes

1. Cleansing of females done by a registered medical practitioner or woman authorised by the proper officer (section 85(4)).
2. No charge for cleansing of person or clothing (section 85(6)).

the local authority shall cleanse, purify, disinfect or destroy at their expense and, if necessary, remove it from the premises.

Notices

Notices under these provisions are not subject to section 290, Part 12 requirements (nature of works), but must be in writing and the time allowed for compliance must be reasonable. In the case of notices indicating the local authority's intention to gas, the period of notice should be not less than the appeal period, i.e. seven days.

Appeals

There is no appeal against the service of notices under section 83 (except those relating to gassing), but in any subsequent proceedings by the local authority for recovery of costs following work in default, the defendant can question the reasonableness of the requirements and the fact that the notice was served on him and not the owner or occupier as the case may be (section 83(2)).

Cleansing of verminous persons and their clothing

Where a verminous person applies to a county council or a local authority, they may take measures necessary to free him and his clothing from vermin. However, where the proper officer reports that any person, or his clothing, is verminous, and the person consents, he can be taken to a cleansing station. Yet if he does not consent, they may apply to the magistrates' court, and if satisfied that it is necessary that he or his clothing should be cleansed, the court may make an order for his removal to a cleansing station to detain him for a period with conditions specified in the order. The local authority must take such measures necessary to free him and his clothing from vermin.

No charge can be made for cleansing of a person or his clothing or removal to, or maintenance in, a cleansing station (section 85).

Definition

Vermin, in its application to insects and parasites, includes their eggs, larvae and pupae, and the expression 'verminous' shall be constructed accordingly (section 90(1)).

REMOVAL OF PERSONS IN NEED OF CARE

References

National Assistance Act 1948, section 47.
National Assistance (Amendment) Act 1951.

Extent

This procedure applies to England, Wales and Scotland.

Scope

The procedure is aimed at securing the necessary care and attention of persons who:

(a) are suffering from grave chronic illness or, being aged, infirm or physically incapacitated, are living in insanitary conditions; and
(b) are unable to devote themselves, and are not receiving from other persons, proper care and attention (section 47(1)).

The procedure is usually used together with filthy and verminous persons and premises procedures.

Certificate of proper officer

If the proper officer designated by the local authority for this procedure, after thorough enquiry and consideration of:

(a) the interests of the person; or
(b) the need to prevent injury to the health of, or serious nuisance to, other persons

is satisfied that the situation as in (a) and (b) in 'Scope' above exists, he certifies such in writing to the local authority (section 47(2)).

If the proper officer, joined in the certificate by another registered medical practitioner, considers that it is necessary in the interests of the person to remove him without delay, the accelerated procedure may be invoked (section 1(1) National Assistance (Amendment) Act 1951).

Removal orders

The application for an order is made by the local authority (or its proper officer if authorised to do so) to a magistrates' court or, if the accelerated procedure is being used, a single Justice of the Peace. If the court or Justice of the Peace is satisfied that it is expedient to do so, they may order the removal of the person by a specified officer of the local authority to a suitable hospital or other place and order his detention and maintenance there.

The court will not order the removal of a person to any premises, unless either the person managing the premises has been heard in the proceedings or seven clear days' notice has been given to him of the intended application and of the time and place at which it is proposed to be made (section 47(7)).

Once the order is made, seven days' notice must also be given to the person or person in control of him of the removal order (section 47(3)).

FC98 Removal of persons in need of care: normal procedure – Section 47 National Assistance Act 1948

Certificate by proper officer of necessity to remove person to care (National Assistance Act 1948, section 47(2))

Local authority to give seven days' clear notice before application to the person to be removed or some person in charge (section 47(7))

AND

Local authority to give seven days' clear notice to manager of premises to which person is to be removed and manager must be heard at hearing of application (section 47(3))

Magistrates' court may order removal (section 47(3)) for up to three months (section 47(4))

If person is maintained at premises outside NHS, social services authority or Part 3 accommodation, cost to be borne by local authority (section 47(8))

Costs recoverable from the person removed or person liable to maintain that person (section 47(9))

AND

Court may subsequently extend order for further periods of up to three months (section 47(4))

Order may later be modified by court in respect of the place to which the person is removed providing that the manager of the premises to receive that person has been given seven days' notice or has been heard (section 47(5))

AND

At any time after six weeks from the making of the order the person or someone on his behalf may apply to the court for revocation (section 47(6))

Seven days' notice of application to be given to proper officer (section 47(7))

Court may revoke the order (section 47(6))

Notes

1. For accelerated procedure, see FC99.
2. Maximum penalty for wilfully disobeying or obstructing the execution of an order is £50 (section 47(11)).
3. All notices may be served by post (section 47(14)).

FC99 Removal of persons in need of care: accelerated procedure – Section 47 National Assistance Act 1948

Certificate by proper officer and another registered medical practitioner of necessity to remove person to care without delay (National Assistance Act 1948, section 47(2) and National Assistance (Amendment) Act 1951, section 1(1))

↓

Local authority or proper officer (if authorised) may apply to magistrates' court or single Justice of the Peace for removal order (National Assistance (Amendment) Act 1951, section 1)

↓

Court or Justice of the Peace may order removal but manager of premises to which person is to be removed must be heard (National Assistance Act 1948, section 47(3) and National Assistance (Amendment) Act 1951, section 1(4))

↓

Order may be made for a period not exceeding three weeks (National Assistance (Amendment) Act 1951, section 1(4))

↓

Orders may be extended by adoption of the normal procedure in FC98

If person is maintained at premises outside NHS, social services department or Part 3 accommodation, cost to be borne by LA (National Assistance Act 1948, section 47(8))

↓

Costs recoverable from the person removed or person liable to maintain that person (section 47(a))

Notes

1. For normal procedure, see FC98.
2. There is no provision to allow application for the revocation of an order made under this accelerated procedure.
3. The maximum penalty for wilfully disobeying or obstructing the execution of a removal order is £50 (section 47(11)).

Period of order

An order may be made for not exceeding three months (three weeks for accelerated procedure) but may be extended by further periods of up to three months. If the first order was made by use of the accelerated procedure, subsequent applications need to be made by the normal procedure, with appropriate notice of application etc. being made (section 47(4)).

Applications for revocation of order (other than one made under the accelerated procedure) may be made to the magistrates' court at any time after six weeks from the making of the order (section 47(6)).

SKIN PIERCING AND OTHER SPECIAL TREATMENTS

References

Local Government (Miscellaneous Provisions) Act 1982, sections 13–17 as amended by the Local Government Act 2003 (section 120).
Body Art – Skin piercing, cosmetic therapies and other special treatments – CIEH Good Practice Guidelines 2001.
Local Government Act 2003: Regulation of cosmetic piercing and skin colouring businesses – guidance on section 120 and schedule 6, Department of Health, 26 February 2004.

Extent

The provisions apply only in England and Wales.

In London, similar powers also exist through the London Local Authorities Act 1991.

Scope

The activities which may be controlled by registration and subsequent application of bye-laws are:

(a) acupuncture;
(b) tattooing;
(c) semi-permanent skin colouring;
(d) cosmetic piercing;
(e) electrolysis (sections 14–15).

Businesses carried on under the supervision of a registered medical practitioner do not require registration under these provisions (sections 14(8) and 15(8)) and they do not apply to skin piercing on animals (section 16(12)).

The only one of these activities which is defined in the Act is 'semi-permanent skin-colouring', which is defined as 'the insertion of semi-permanent colouring into a person's skin' (section 15(9)).

FC100 Skin piercing and other special treatments – Sections 13 to 17 Local Government (Miscellaneous Provisions) Act 1982

Resolution of local authority to adopt all or some of these provisions (Local Government (Miscellaneous Provisions) Act 1982, section 13(2) and (3))

↓

Local authority to publish notice for two consecutive weeks in local newspaper, the first at least 28 days before operative date (section 13(6) and (7))

↓

Resolution operative on date(s) specified not less than one month of date of resolution (section 13(2))

↓

Offence to practise any/all activities in area covered by resolution without registration of person and premises (sections 14(1) and 15(1). Max. penalty level 3 (section 16(1))

AND

Local authority may make bye-laws (sections 14(7) and 15(7))

↓

AND

Application for registration of person and premises to local authority accompanied by particulars specified by local authority and fee (sections 14(3) to (6) and 15(3) to (6))

↓

Local authority must register (sections 14(3) and 15(3)) and send certificate of registration (note 1)

↓

(continued on next page)

Note

1. Where a previous registration has been cancelled by that local authority, future registrations require the consent of the magistrates' court which convicted the applicant (section 16(8)(b)).

FC100 Skin piercing and other special treatments – continued

(Continued from previous page)

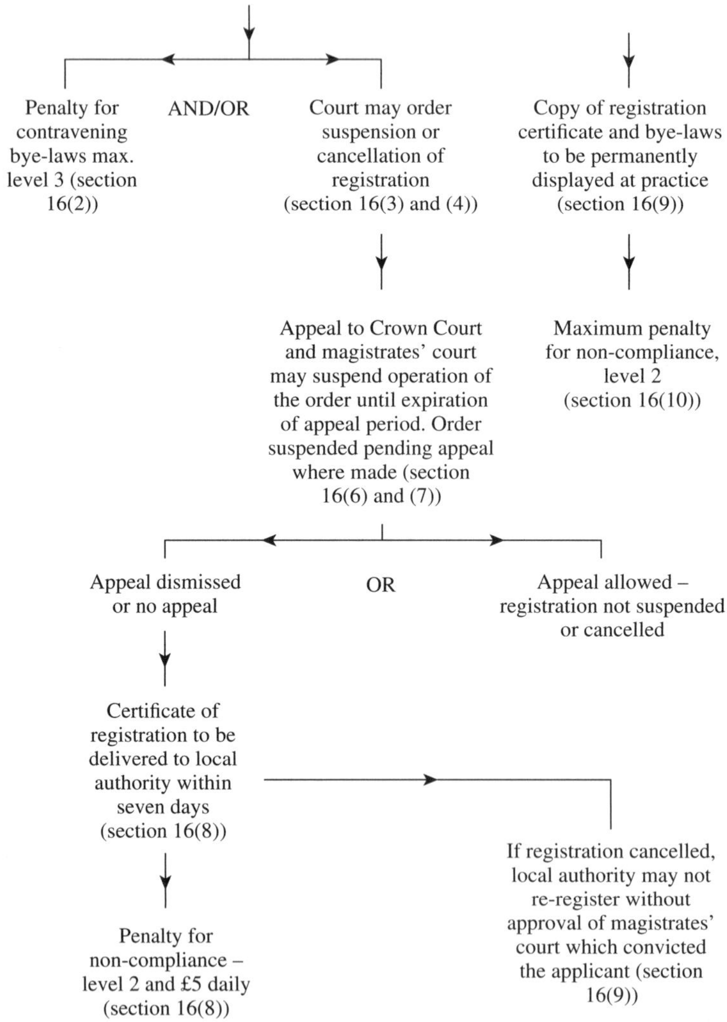

Adoption

These provisions do not operate unless they are adopted by a local authority in respect of its own area by the procedure shown in FC100. The adoption may relate to all or any of the activities mentioned below and different dates of operation may be applied to different activities (section 13).

Registration

Both the practitioner and the premises at which the activity is to be carried out are required to be registered (sections 14(1) and (2) and 15(1) and (2)). Where a registered practitioner sometimes visits people to give treatment, the premises where the treatment takes place is not required to be registered (sections 14(2) and 15(2)).

Applications are to be accompanied by such particulars as the local authority may reasonably require and these include details of:

(a) the premises where the applicant desires to practise; and
(b) any convictions under section 16 of this Act,

but the local authority cannot require information about people to whom treatment services have been given (sections 14(5) and 15(5)).

Registration is required to be effected by the local authority unless the applicant has had a previous registration cancelled by a court, in which case the consent of that court is required (sections 14(3), 15(3) and 16(8)(b)).

Any person who fails to register themselves and the premises they operate from is guilty of an offence and liable on summary conviction to a fine not exceeding level 3 and if they contravene bye-laws the court can suspend or cancel their registration.

Fees

The local authority may charge reasonable fees for registration at their discretion (sections 14(6) and 15(6)).

Bye-laws

The local authority operating these procedures may (but does not need to) make bye-laws which may cover:

(a) cleanliness of premises and fittings;
(b) cleanliness of persons;
(c) cleansing and sterilisation of instruments, materials and equipment (sections 14(7) and 15(7)).

Power of entry

This is only available through a warrant by a Justice of the Peace and may be granted if the Justice of the Peace is satisfied that there is reasonable ground for entry on suspicion that an offence is being committed and that:

(a) admission has been refused; or
(b) admission is apprehended; or
(c) the case is one of urgency; or
(d) application for admission would defeat the object of entry.

Unless the situation falls under (c) or (d) above, notice of intention to apply for the warrant must be given to the occupier.

The warrant is operative for seven days or until entry is secured, whichever is the shorter period. There is no mention in these provisions of entry by force and authorised officers effecting entry may be required to show their authority.

The maximum penalty for refusing to permit entry to an authorised officer acting under warrant is a fine not exceeding level 3 (section 17).

DOG CONTROL ORDERS (DCOs)

References

The Clean Neighbourhoods and Environment Act 2005, sections 55 to 67.
The Dog Control Orders (Prescribed Offences and Penalties etc.) Regulations 2006.
The Dog Control Orders (Procedures) Regulations 2006.
The Environmental Offences (Fixed Penalties) (Miscellaneous Provisions) Regulations 2007.
The Environmental Offences (Use of Fixed Penalty Receipts) Regulations 2007.
The Controls on Dogs (Non-application to Designated Land) Order 2009.
DEFRA Guidance on the Clean Neighbourhoods and Environment Act 2005 – Dog Control Orders 2006 and Fixed Penalty Notices 2006.

Extent

This procedure extends to England and Wales only.

Scope

This procedure allows local authorities and parish and town councils (in this procedure all are referred to as local authorities) to make dog control orders that apply offences aimed at the control of dogs to designated land in their areas (section 55). It replaces the earlier dog controls through bye-laws (although existing bye-laws remain in force until the land is made subject to a dog control order) and the Dogs (Fouling of Land) Act 1996 (although orders made under it continue to have effect if designated before 6 April 2006 unless the land is superseded by a dog control order or is revoked).

Designation of land

Dog control orders may be made in respect of any land in the open air to which the public has access with or without payment. Land is considered to be in the

FC101 Dog control orders (DCOs) – The Clean Neighbourhoods and Environment Act 2005 Sections 55 to 67 and the Dog Control Orders (Procedures) Regulations 2006

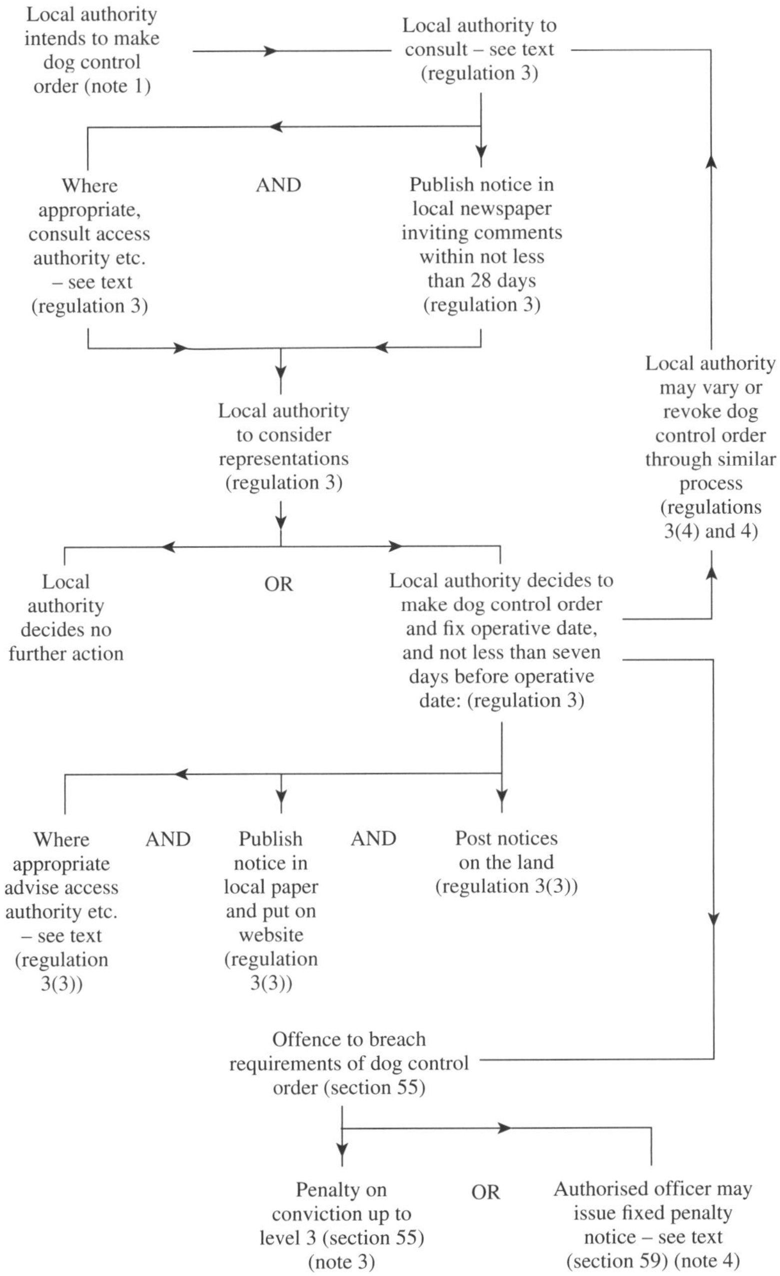

Notes

1. Local authority is local authority, town or parish council.
2. Regulation numbers refer to the Dog Control Order (Procedures) Regulations 2006.
3. Specified in Dog Control Orders (Prescribed Offences and Penalties etc.) Regulations 2006.
4. Specified in Environmental Offences (Fixed Penalties) (Miscellaneous Provisions) Regulations 2007.

open air if it is open on at least one side. The Secretary of State is able to designate types of land which are not to be the subject of dog control orders and these are listed in the Controls on Dogs (Non-application to Designated Land) Order 2006 as:

(a) forestry commission land for all dog control orders;
(b) roads (including highways) in respect of those orders that exclude dogs from land specified in the order (section 57).

Making of a dog control order

The process for designation is set out in the Dog Control Orders (Procedures) Regulations 2006 and is as follows:

- Consultation by publishing a notice in a local newspaper, with specified information contained in regulation 3(2), and inviting representations by a date not less than 28 days from the publication.
- Consult all other authorities with power to designate a dog control order for the land.
- Where access land is involved (see the Countryside and Rights of Way Act 2000), consult with the access authority and the local access forum or, outside of a national park, the Countryside Agency.
- Consider representations.
- Decision to make the order (or not) and determine the operative date.
- Then, and not less than seven days before operation of the dog control order:

(a) publish a notice in a local newspaper;
(b) where practicable, also publish the notice on the local authority website;
(c) where the order affects access land, copy to be sent to the access authority, the local access forum and the Countryside Agency;
(d) where practicable, signs to be placed summarising the order on land to which it applies;
(e) send the information to all other authorities having power to make a dog control order for that land.

The notice referred to in paragraph (a) shall:

(a) identify the land in respect of which the order is to apply, and, if any of the land is access land, state that this is the case;
(b) summarise the order;
(c) where the order refers to a map, state where the map may be inspected at an address within the authority's area, and that any inspection shall be free of charge at all reasonable hours during the period mentioned in sub-paragraph (d);
(d) state the period within which representations may be made in writing or by e-mail, such period being not less than 28 days after the publication of the notice; and
(e) state the address and e-mail address to which representations may be sent.

Offences

There are five types of 'model' offences described in the Dog Control Orders (Prescribed Offences and Penalties etc.) Regulations 2006 that may be included in dog control orders. These are:

(i) failing to remove dog faeces;
(ii) not keeping a dog on a lead;
(iii) not putting, and keeping, a dog on a lead when directed to do so by an authorised officer;
(iv) permitting a dog to enter land from which dogs are excluded;
(v) taking more than a specified number of dogs onto land.

All are subject to the tests of 'without reasonable excuse' and 'without the consent of the owner, occupier or other person or authority who has control of the land'. Also (i) and (ii) do not apply to a person who is registered as blind or to a person with a disability in respect of which he relies on dog assistance.

Offence (iv) does not apply to a person who is deaf in respect of a dog trained by a specified charity.

The offences to be included will be listed by the local authority in the schedule of the order and may deviate in some respects from the 'models' included in the regulations (section 55).

Penalties and fixed-penalty notices

The penalty for committing an offence is a fine of up to level 3 but there is the opportunity for an authorised officer to serve a fixed-penalty notice. The details for fixed penalties are in the Environmental Offences (Fixed Penalties) (Miscellaneous Provisions) Regulations 2007 and the level of fine is set at the discretion of the local authority between £50 and £80, with discounts for early payment (sections 59 to 60). There is a power for an authorised officer to require the name and address of an offender and it is an offence not to do so, with a level 3 fine (section 61). Chief constables may authorise community support officers and other 'accredited' people to issue fixed-penalty notices (section 62).

Amendments and revocation

Amendments, other than minor, must be drawn up in a specified way and advertised and brought into force in the same way as a new order.

Where the revocation of a dog control order is intended, again the process is similar to a new order with publication, consideration of representations and further publicity being required (2006 Procedure Regulations.).

SEIZURE OF STRAY DOGS

References

Environmental Protection Act 1990, sections 149 and 150 as amended by the Clean Neighbourhoods and Environment Act 2005 from a date to be decided.
The Environmental Protection (Stray Dogs) Regulations 1992.
The Control of Dogs Order 1992.

Extent

This procedure applies in England, Wales and Scotland but not in Northern Ireland, where there is separate legislation (section 164(4)).

Duty on local authorities

Local authorities must appoint an officer to deal with stray dogs (section 149(1)).

There are no particular powers of entry provided for any authorised local authority officers.

Scope

Where the authorised officer finds a stray dog in a public place or on any other land or premises, he must seize it and detain it provided, in the case of other than a public place, the owner or occupier agrees (section 149(3)). Dogs not wearing collars with the name and address of the owner may be seized and treated as a stray dog under this procedure (Control of Dogs Order 1992).

A member of the public may take possession of a stray dog and must then take it to the local authority. Unless the finder wishes to keep the dog, the local authority is required to deal with it by the procedure in FC102.

Notification to owner

Where the authorised officer knows the dog's owner or this is shown on its collar, the authorised officer must give that person notice in writing that the dog has been seized, inform them of where it is being kept and that he will dispose of the dog unless it is claimed within seven days and the local authority's expenses are paid (section 149(4)).

Identification of dog owners

The Control of Dogs Order 1992, which is enforced by local authorities, requires dogs, with certain exceptions, e.g. pack hounds, whilst on a highway or in a public place, to wear a collar with the name and address of the owner inscribed on it or on a place attached to it. If not so worn, the owner commits an offence under the Animal Health Act 1981. In these circumstances, local authorities have

FC102 Seizure of stray dogs by the public – Section 150 Environmental Protection Act 1990

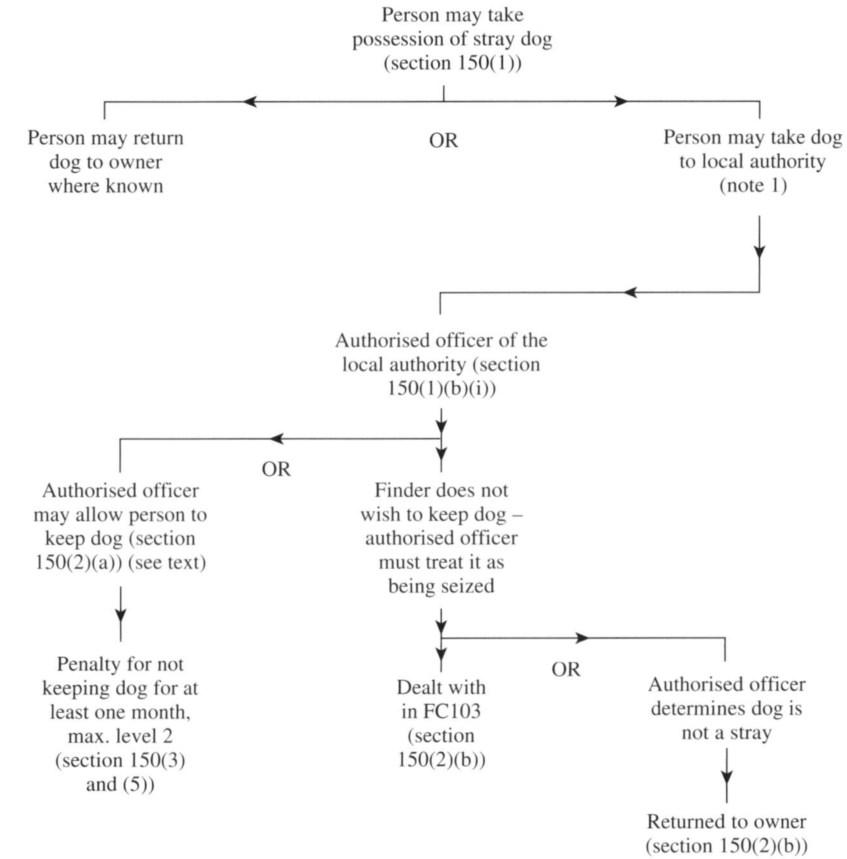

Notes

1. Having taken possession of a stray dog, the member of the public must bring it either to the owner or to the local authority – maximum penalty for not doing so, level 2 (section 150(1) and (5)).
2. In Scotland, a person who keeps a dog by virtue of this section for a period of two months without its being claimed by the person who has right to it shall at the end of that period become the owner of the dog (section 150(4)).

FC103 Seizure of stray dogs by local authorities

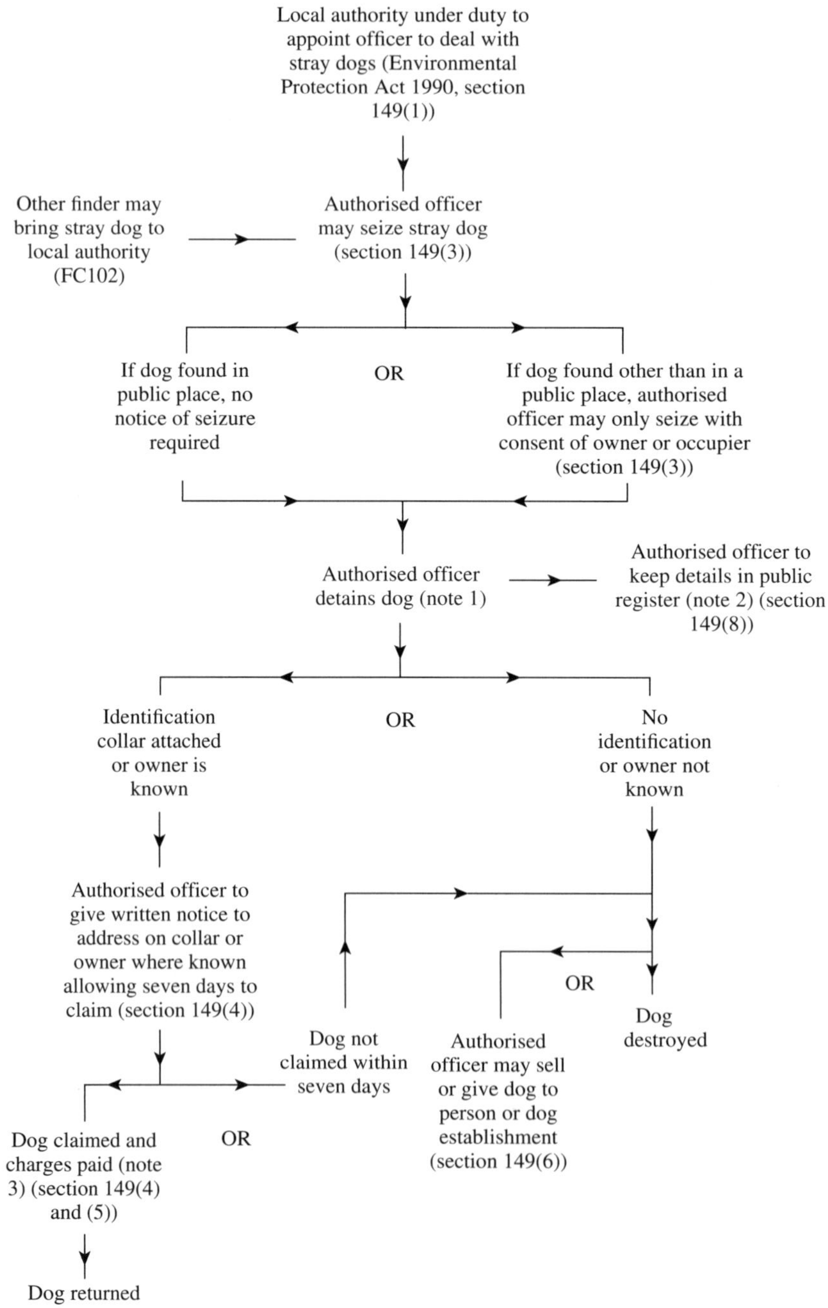

Notes

1. The authorised officer may arrange for the dog to be destroyed immediately if this is necessary to avoid suffering (section 149(10)).
2. The details to be kept in the register are prescribed in the Environmental Protection (Stray Dogs) Regulations 1992.
3. The charges to be made are controlled by the Environmental Protection (Stray Dogs) Regulations 1992.

been given the power to prosecute under the Environmental Protection Act (Environmental Protection Act 1990, section 151).

Detention of dogs

The local authority must provide or arrange for the detention of dogs and for them to be properly fed and confined. Costs involved in the kennelling of dogs are recoverable from owners who require return of the animal (section 149(9)) and additionally the owner is required to pay a sum prescribed by regulations (regulation 2).

Register of seized dogs

The authorised officer must keep a register of seized dogs, which is kept by the officer, including:

- (a) a brief description of each dog, including its breed (if known), and any distinctive physical characteristics or markings, tattoos or scars;
- (b) any information which is recorded on a tag or collar worn by, or which is otherwise carried by, the dog;
- (c) the date, time and place of the seizure;
- (d) where a notice has been served under section 149(4), the date of service of the notice, and the name and address of the person on whom it has been served;
- (e) where the officer disposes of the dog under section 149(6):
  - (i) the date of disposal;
  - (ii) whether disposal was by destruction, gift or sale, and if by sale, the price obtained;
  - (iii) the name and address of the purchaser, donee or person effecting the destruction; and
- (f) where the dog was returned to a person claiming to be its owner, the name and address of that person, and the date of return.

The register shall be available, at all reasonable times, for inspection by the public free of charge (section 149(8) and Environmental Protection (Stray Dogs) Regulations 1992, regulation 3).

Dog to be kept by finder

The finder of a stray dog may indicate to the authorised officer of the local authority a wish to keep the dog. The authorised officer must record specific details and contact the dog's owner where this proves to be possible. The owner may elect to collect it. However, where the owner is not known or does not wish to collect the dog, the authorised officer can allow the finder to keep the dog if he is satisfied that the person is fit and proper and is able to feed and care for it. The authorised officer must inform the finder both verbally and in writing that he must keep the dog for at least one month and that failure to do so is a criminal offence (section 150(2) and Environmental Protection (Stray Dogs) Regulations 1992, regulation 4).

Disposal of dogs

The authorised officer may dispose of the dog by either:

(a) giving it or selling it to:
  (i) any person who will care for it;
  (ii) an establishment for the reception of stray dogs; or
(b) destroying it so as to cause as little pain as possible.

No dog may be sold or given away for vivisection (section 149(6)).

Definitions

Public place means:

(a) as respects England and Wales, any highway and any other place to which the public are entitled or permitted to have access;
(b) as respects Scotland, any road (within the meaning of the Roads (Scotland) Act 1984) and any other place to which the public are entitled or permitted to have access (section 149(11)).

EXHUMATIONS

Bodies may need to be exhumed for many reasons: lack of burial space; the redevelopment of old cemeteries or crypts; archaeological reasons; individual requests for reburial, repatriation or cremation; criminal investigations; or even to correct errors made at the time – for instance, the person was buried in the wrong plot.

Reference

Burial Act 1857, section 25.

Extent

These provisions apply in England, Scotland and Wales.

Scope

A body cannot be removed from a consecrated place without a licence to do so. Licences are available from the Department of Constitutional Affairs and the licence will include conditions or precautions to be taken. It is an offence to remove a body or body remains without a licence or to neglect to observe the precautions prescribed under the conditions of the licence. A person guilty of such an offence is liable, on summary conviction, to a level 1 fine (section 25).

FC104 Exhumations – Burial Act 1857

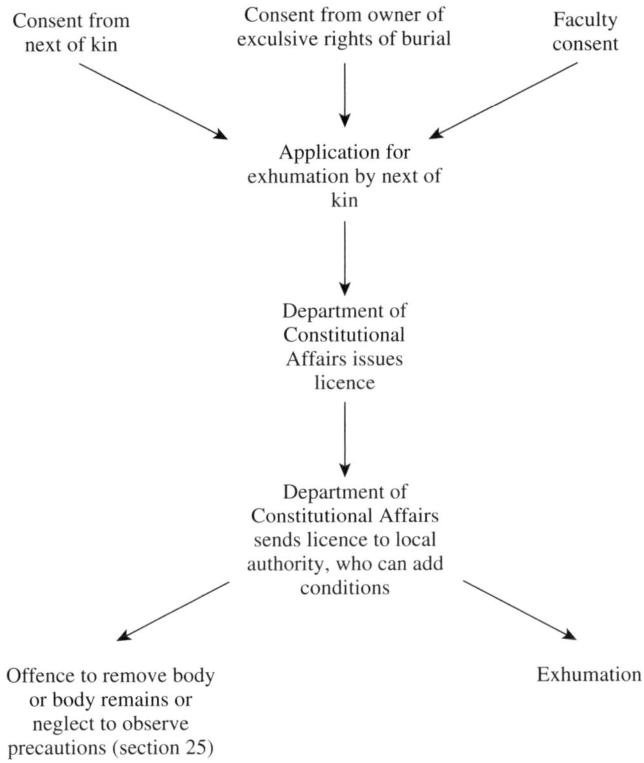

Practice

There are three categories of legitimate exhumation – under Department of Constitutional Affairs licence; ecclesiastical faculty, where remains are exhumed from consecrated ground; or a coroner's order.

Of these, environmental health practitioners are only required for those exhumations sanctioned under the Department of Constitutional Affairs licence. The Department of Constitutional Affairs licence will stipulate that exhumation should be carried out early in the morning. In practice, the licence is sent to the local authority, who may devise and attach other conditions.

It is the responsibility of the licence holder to ensure the conditions are adhered to. This may be a relative of the deceased or a company if a licence has been applied for on behalf of an individual.

Consents

Any person who wishes to exhume buried remains has to apply to the Department for Constitutional Affairs for a licence to permit it. There are a number of factors which the Department of Constitutional Affairs considers before issuing a licence to exhume human remains. The Department of Constitutional Affairs will normally grant exhumation licences to the next of kin, subject to any other necessary consents, where the application is made for private family reasons. The consents of all the next of kin of the deceased are required. The priority given is in accordance with that set out in the Administration of Estates Act 1925 or the Civil Partnerships Act 2004.

If contact has been lost with any of the surviving relatives, it is the applicant's responsibility to undertake a search for that relative. Licences are unlikely to be issued without all of the required consents. Applications will also be considered from any person, but in such cases it will be important to explain why the application is not being made by a relative.

Where a non-family member, e.g. the coroner, applies for a licence, they will need the consent of the next of kin/relatives.

Consent will also be needed from the owner of the exclusive rights of burial relating to the grave. This is because a Department of Constitutional Affairs licence is permissive only: it does not require the remains to be removed, and the grave owner must therefore be prepared to grant access. If the grave is a public or common one, which could mean that unrelated remains may be buried there, and if so may have to be disturbed, then the permission of any surviving relatives would be required.

If the remains are buried within consecrated ground, then consent from the Churches of England, Scotland and Wales and the Roman Catholic Church will also be required. If the present burial plot is within consecrated ground, you may need to apply for faculty consent. It is the applicant's responsibility to obtain these permissions.

A licence from a straightforward application would be given a three-month timescale from the date of issue unless there are other extraneous circumstances.

Objections

It is the applicant's responsibility, even if they are the nearest surviving relative, to state whether there could be objections to the exhumation from other relatives, as it may be necessary to take such objections into account. However, this does not necessarily mean that a licence would be refused.

If the consent of the grave owner is not given, then the Department for Constitutional Affairs would not normally be prepared to issue a licence.

The Department for Constitutional Affairs is unable to become involved in any family dispute and will not normally issue a licence until objections or disagreements between next of kin with the same degree of kinship have been resolved.

Definitions

Consecrated means a burial ground that has been consecrated in accordance with the rites of the Church of England and not merely blessed by the vicar during the funeral service, e.g. consecrated by the Roman Catholic Church or the Church in Wales.
Exhumation means 'any disturbance' of buried human remains.
Faculty means consent from the ecclesiastical courts of the Church of England to exhume.

Note

The responsibility for protecting the health and ensuring the safety of the EHP in attendance lies with the company carrying out the exhumation. See 'Controlling the risks of infection at work from human remains: A guide for those involved in funeral services (including embalmers) and those involved in exhumation', Health and Safety Executive.

REMOVAL OF UNAUTHORISED CAMPERS

References

Criminal Justice and Public Order Act 1994, sections 77 to 80.
Department of the Environment Circular 18/94 – Gypsy Sites Policy and Unauthorised Camping (amended 2000).
Guidance on Managing Unauthorised Camping, Office of the Deputy Prime Minister, 27 February 2004.

Extent

This procedure is applicable in England and Wales but not in Scotland or Northern Ireland.

Scope

Local authorities are able to direct persons residing unlawfully in vehicles:

(a) on highway land;
(b) on unoccupied land; or
(c) on occupied land without the consent of the occupier

to leave the land and remove their vehicles and any other property they have with them (section 77(1)).

The procedure may be applied to all persons residing in vehicles (as defined below); however, in Department of the Environment Circular 18/94, local authorities are asked not to use these powers needlessly in relation to gypsy encampments, particularly where they are causing no nuisance (paragraphs 6 to 9). Complementary provisions in section 61 give powers to the police to deal with collective trespass or nuisance on land in circumstances where the trespassers have refused requests from the occupier to leave and either:

(a) they have caused damage to land/property; or
(b) threatening, abusive or insulting behaviour has been used; or
(c) there are at least six vehicles (section 61).

The operation of both provisions therefore requires close liaison between the police and the local authorities.

In implementing these procedures, local authorities are asked to take account of the government's objective of ensuring that the settled communities and gypsies and travellers live together peacefully. Stress is also placed on the importance of choosing the most appropriate power to deal with any situation and that this should be done within the umbrella of a protocol between a local authority, the police and other involved agencies.

Under part 7 of the Anti-Social Behaviour Act 2003, the police have powers to move on unauthorised camping provided that there is adequate site provision elsewhere.

Vehicles

Both provisions deal with the presence of vehicles, defined as including:

(a) any vehicle, whether or not it is in a fit state for use on roads, and includes any body, with or without wheels, appearing to have formed part of such a vehicle, and any load carried by, and anything attached to, such a vehicle; and
(b) a caravan as defined in section 29(1) of the Caravan Sites and Control of Development Act 1960;

and a person may be regarded for the purposes of this section as residing on any land notwithstanding that he has a home elsewhere (section 61(9) and 77(6)).

FC105 Removal of unauthorised campers – Section 77 Criminal Justice and Public Order Act 1994

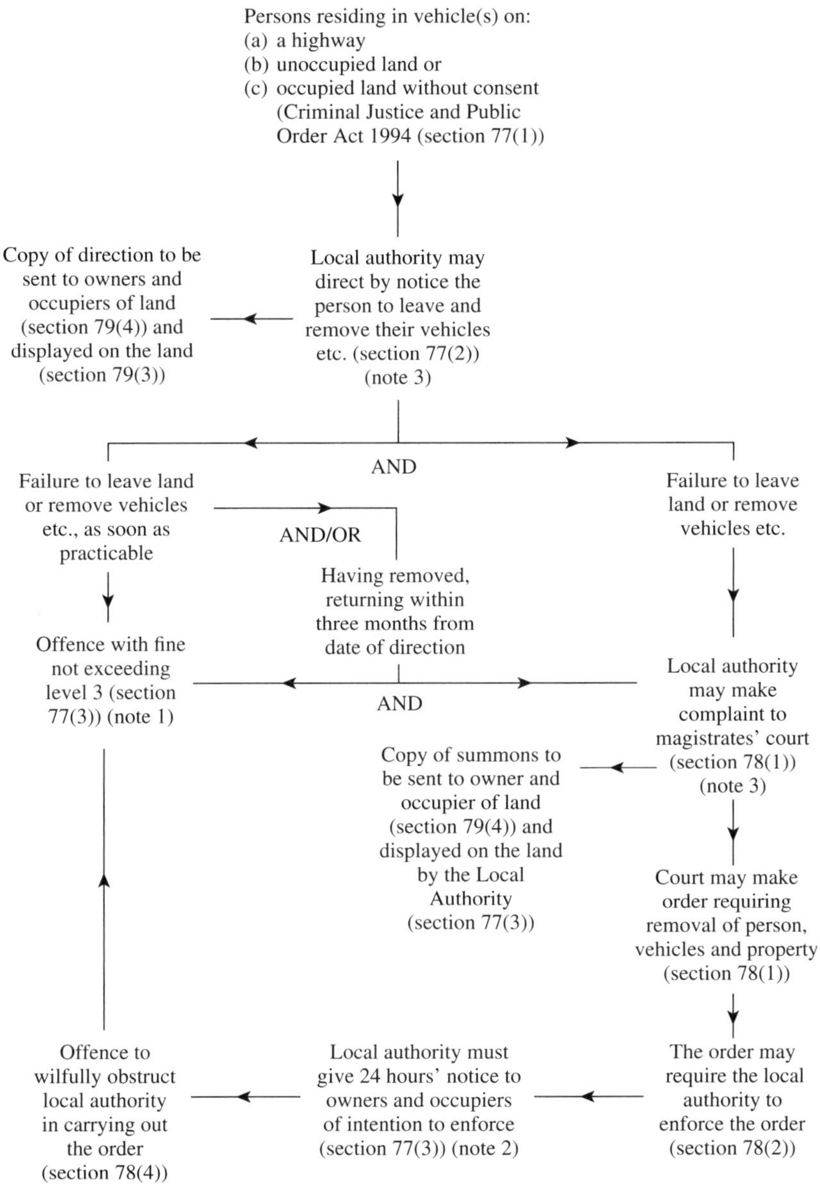

Notes

1. It is a defence to show that failure to leave or remove vehicles etc. was due to illness, mechanical breakdown or other immediate emergencies (section 77(5)).
2. Notice is not required where, after reasonable enquiry, the local authority is unable to obtain the names and addresses of the person(s) subject to the court direction (section 78(3)).
3. For service of notice of direction and summons of complaint, see text.
4. For police powers to remove unauthorised camping where there is adequate provision elsewhere, see part 7 of the Anti-Social Behaviour Act 2003.

Gypsies

In Circular 18/94, local authorities are asked in enforcing this procedure to take account of the definition of 'gypsies' given by the Court of Appeal in *R* v *South Hams DC, ex parte Gibb* (*The Times*, 8 June 1994; i.e. 'gypsies' meant any persons who wandered or travelled for the purpose of making or seeking their livelihood, and did not include persons who moved from place to place without any connection between their movement and their means of livelihood).

Directions

Directions to be given by local authorities must be served on the persons to whom the direction is to be applied but where there is more than one person the direction can merely specify the land and be addressed to all occupants of the vehicles without naming them (section 77(2)).

Where it is impracticable to serve on the persons named in the direction, a copy may be fixed prominently on the vehicle concerned and, if directed to, unnamed persons may be served in the same manner on every vehicle (section 79(2)).

Copies of the direction must also be displayed on the land (other than on a vehicle) in such a way that it can easily be seen by persons camping on the land and copies sent to the owner and any occupier of the land, unless after reasonable enquiry they cannot be found (section 79(3) and (4)).

The direction requires the persons concerned to leave the land and remove the vehicles etc., as soon as practicable. It is an offence:

(a) not to comply; and
(b) having removed the vehicles etc., to again enter the land with a vehicle within three months of the day on which the direction was given (section 77(3)).

On summary conviction, level 3 penalties are available to the court.

Defences

In any proceedings taken by the local authority for non-compliance with a direction, it will be a defence to show that non-compliance was due to:

(a) illness; or
(b) mechanical breakdown; or
(c) other immediate emergency (section 77(5)).

Complaint to magistrates' court

In addition to seeking a fine for non-compliance with the directive, the local authority may apply to the magistrates' court for an order requiring the removal of the vehicles and other property (section 78(1)).

Any order made may authorise the local authority to enforce it by entering the land and removing the vehicles etc., but, in relation to occupied land, at least 24 hours' notice must be given to the owner and lawful occupiers (section 78(3)).

There is no provision for the recovery of the local authority costs in enforcing these orders.

Provisions for the service of summons in relation to a local authority complaint are similar to the service of directions above.

To obtain an order for removal from the magistrates' court, it is necessary to obtain reports on the following:

- availability of sites in local gypsy sites;
- availability of school places in the locality for children in the group subject to the order;
- social services about the needs of the group subject to the order.

Definition

Land means land in the open air.

CONTROL OF RATS AND MICE

Reference

Prevention of Damage by Pests Act 1949, sections 2 to 7.

Extent

This procedure applies in England, Wales and, with amendments, to Scotland but not in Northern Ireland (sections 1 and 29).

Duty of local authority

Each local authority is under a statutory duty to take such steps as may be necessary to secure so far as practicable that its district is free from rats and mice including:

(a) carrying out inspections;
(b) destroying rats and mice on premises which the local authority occupies and keeping the premises free from infestation;
(c) enforcing duties on owners and occupiers under Prevention of Damage by Pests Act 1949 (section 2).

Occupier's duty to notify local authority

When an occupier of any land becomes aware of an infestation by rats or mice in substantial numbers, he must notify the local authority in writing but this does not apply to agricultural land, or where food in certain categories of premises is infested when notification is then given under part 2 to the Minister of the

Environment, Food and Rural Affairs or, in Scotland, the Secretary of State (section 3).

Notices

Notices, which must be in writing, may be served on the owner or the occupier or, where the owner is not the occupier, on both. The notice may require either or both of the following:

(a) treatment at times which may be prescribed;
(b) the carrying out of structural or other works.

The time allowed for compliance must be reasonable and should be not less than the period allowed for appeal, i.e. 21 days (section 4 and sections 290, 300 and 302, Public Health Act 1936). Works in default are allowed under section 5.

Block treatment

If rats or mice are found in substantial numbers in a group of premises in different occupation, the local authority may take such steps as it considers necessary (other than carrying out structural works), after giving at least seven days' notice to each occupier specifying the steps which it intends to take, and may subsequently recover its costs (section 6).

Power of entry

Authorised officers of local authorities, producing if required evidence of authority, may enter any land after, in the case of occupied land, giving at least 24 hours' notice, for the following purposes:

(a) inspection;
(b) compliance with notices or other requirements placed on owners or occupiers;
(c) enforcing notices served under sections 5 and 6 including the carrying out of treatment of works (section 22).

Definitions

Land includes land covered with water and any building or part of a building (section 28).

INFECTIOUS DISEASES CONTROL – AIRCRAFT

Reference

Public Health (Aircraft) Regulations 1979 as amended.

Extent

These provisions apply in England and Wales only.

FC106 Control of rats and mice – Sections 2 to 7 Prevention of Damage by Pests Act 1949

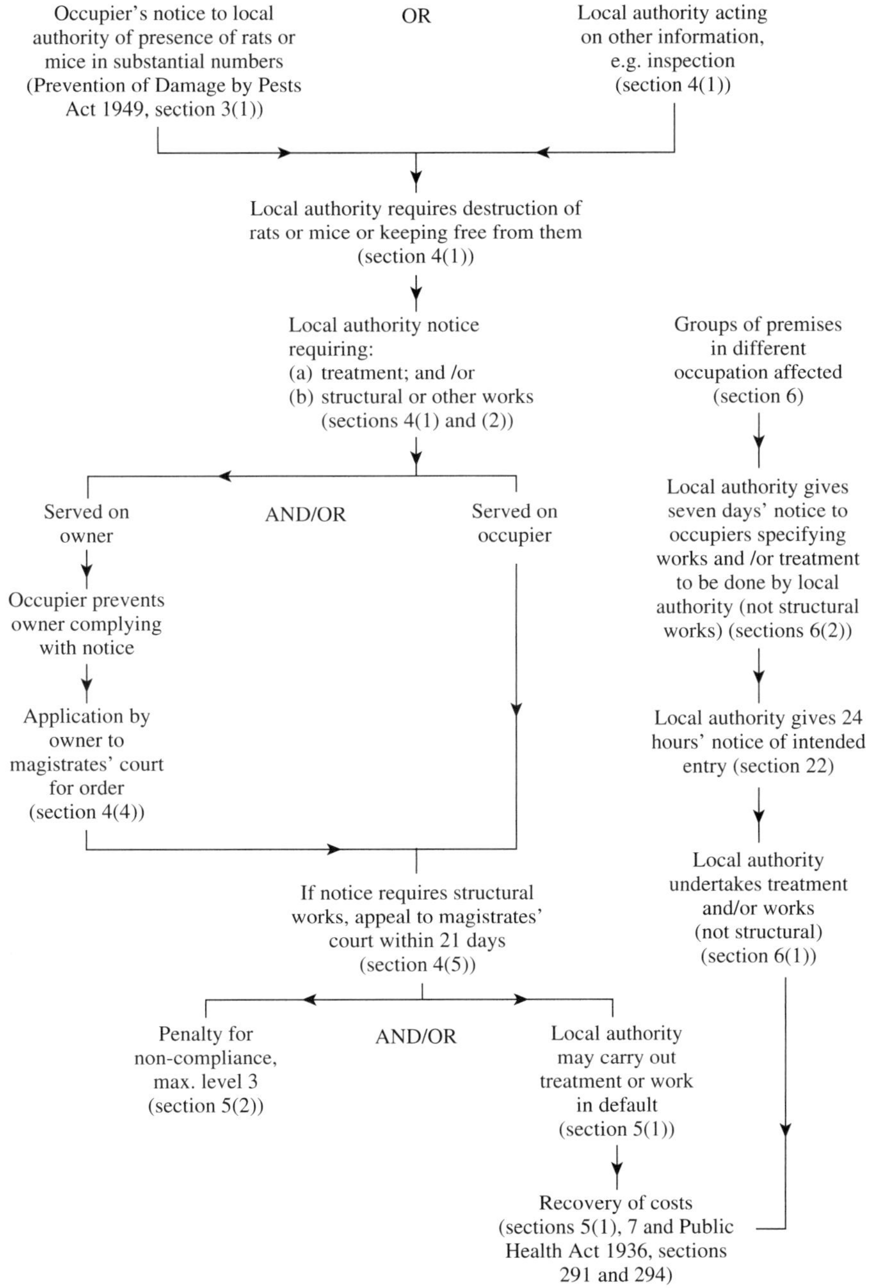

Notes

1. The following general provisions of the Public Health Act 1936 are applicable to this procedure:
   (a) authentication and service of notices (sections 284–286);
   (b) appeals against notices (only where structural works are required) (section 290(3)–(5) inclusive and 300–302 inclusive);
   (c) recovery of expenses (sections 291 and 294). These provisions are set out on page 447.
2. This procedure applies with certain modifications to non-seagoing ships and to aircraft (section 23).

Scope

The purpose of the legislation is the prevention of spread of disease of international concern and to control chemical, biological and radiological risk. The national legislation implements the World Health Organisation International Health Regulations 2005. The Department of Health intends to introduce international health regulations, which may well replace the Public Health (Aircraft) Regulations.

Inspection

An authorised officer may inspect any aircraft at a customs airport.

If an authorised officer or medical officer has reasonable grounds to suspect there is case of infectious disease on board or if the commander of an aircraft notifies of a suspected infectious disease, the authorised officer must inspect the aircraft on arrival. The authorised officer can take samples of food or water for analysis or examination for the purpose of determining treatment of persons affected with any epidemic, endemic or infectious disease and for preventing the spread of such diseases or for preventing other danger to public health (regulation 7).

Examination of persons on aircraft

An authorised officer may detain people for examination and require their clothing and other articles belonging to them to be disinfected and, where necessary, disinsected, and any person found to be verminous to be disinsected. He may also prohibit any person so examined from leaving the aircraft or airport, or permit him to leave it on certain conditions. He can also require the commander to take steps reasonably necessary for preventing the spread of infection, for disinfection and the destruction of vermin, and for the removal of conditions on the aircraft likely to convey infection, including conditions the existence of which might facilitate the harbouring of insects or vermin (regulation 8).

Notice to customs officers by authorised officer

The authorised officer must inform the customs officer of any measure applied by him or at his direction, to an aircraft, any person on it, or its stores, equipment or cargo (regulation 10).

Supply of information by commanders

Commanders of aircraft must answer questions on health conditions on the aircraft put to him by an authorised officer. Commanders must also notify the authorised officer of any deaths and any cases of, or conditions likely to cause the spread of, infectious disease, including tuberculosis (regulation 11).

FC107 International travel – infectious disease control – Public Health (Aircraft) Regulations 1979

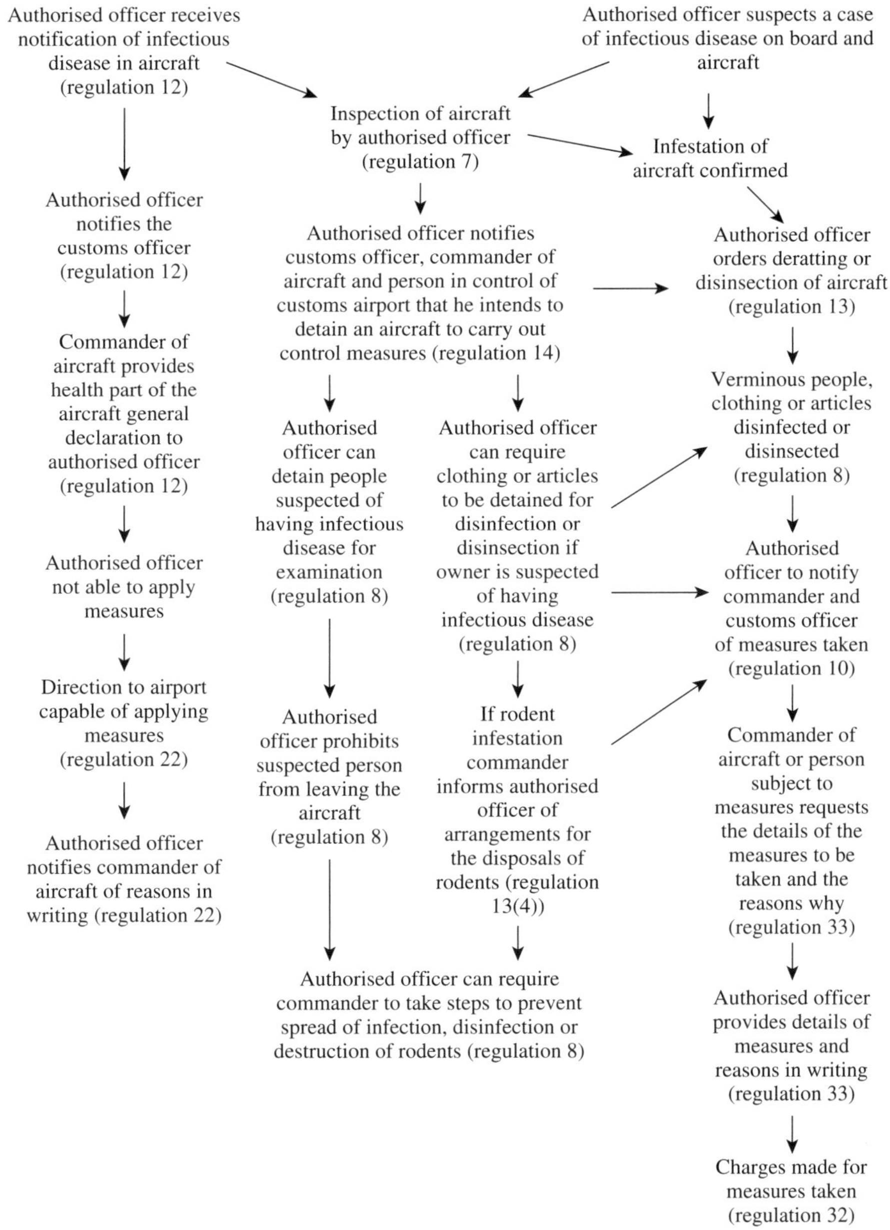

Notes

1. Samples can be taken by authorised officer (regulation 7).
2. Authorised officer must have regard to the need to free the aircraft as quickly as possible (regulation 24).

Notification of infectious disease on board

Where a radio message or other communication is sent to the authorised officer about the potential or existence of infectious disease on board an aircraft, he shall immediately notify the customs officer of its contents. This also applies to the person in control of the airport and the customs officer. The health part of the Aircraft General Declaration must be delivered to the authorised officer by the commander of the aircraft, or by a member of the crew acting on his behalf (regulation 12).

(In practice, if the local authority is contacted by an airline, they will liaise with the port medical officer (PMO). The PMO does a risk assessment on the information provided and will direct the environmental health officer as to what type of response is required, very much in the same way as infectious diseases are dealt with inland.)

Deratting and disinfection of aircraft

An authorised officer can require the aircraft to be deratted or disinfected if the presence of rodents is suspected on board or an infected animal, or suspected of being infected, is found on board on arrival. The methods used must be approved by the World Health Organisation, unless the authorised officer decides that other measures are as safe and reliable. Additional health measures can be applied for preventing danger to public health or the spread of infection, including isolation of the aircraft. The commander of the aircraft must inform the authorised officer of the arrangements made for the disposal of the rodents (regulation 13).

Detention of aircraft

The medical officer can detain an aircraft for medical inspection if he suspects there is infectious disease on board, and he must inform the person in charge of the customs airport and the customs officer (regulation 14). Customs officers and medical officers have similar duties of notification and powers of detention (regulations 15 to 18). Only three hours is allowed for this inspection before the aircraft must be released (regulation 17).

Release of aircraft

When the authorised officer releases an aircraft from detention, he must give notice in writing to the customs officer, to the commander of the aircraft and to the person in charge of the customs airport that the aircraft is free to proceed at or after a date and time stated in the notice (regulation 19).

Removal of infected persons from aircraft where required by commander

A commander of an aircraft may require the medical officer to remove any infected person or person suffering from tuberculosis from the aircraft (regulation 21).

Removal to airport able to apply measures

If an authorised officer considers that the airport cannot apply the necessary measures, he may direct the aircraft or the person to proceed to a customs airport that is able to apply the measures. He must give the commander the reasons for the direction in writing (regulation 22). In applying any measures, the authorised officer must have regard to the need for freeing aircraft from control as quickly as possible (regulation 24).

Outgoing aircraft

The medical officer has powers of examination, prohibition from embarking, surveillance and a duty to notify medical personnel in the destination country (regulation 27).

Examination of person proposing to embark

Regulation 28 provides for the declaration of infected areas where certain diseases are present or suspected, giving the medical officer and authorised officer powers to control infection spread.

Responsibilities

It is the duty of every person subject to direction, requirement or conditions given, made or imposed by an authorised officer or customs officer under these regulations, to comply and they must furnish required information that officer may reasonably require (including information about his name and intended destination and address to which he is going on leaving an aerodrome) (regulation 29). Currently there is no penalty for non-compliance with this duty.

Persons placed under surveillance must also comply with requests for medical examination and questions about personal health.

Charges

A responsible authority may charge the commander of an aircraft for a service to apply measures for preventing danger to public health from an aircraft arriving or the spread of infection from an aircraft leaving an airport where international flights arrive or depart. Charges must not exceed the actual cost of the service rendered and be published at least ten days in advance of being levied.

The charges may be required to be paid or deposited with them before the service is performed. If the commander or person to whom the measures have been taken request, the responsible authority must provide the commander of the aircraft with particulars of the measures taken and reasons why in writing free of charge (regulation 32). Charges are recoverable either summarily as a civil debt or as a simple contract debt in any court of competent jurisdiction (regulation 33).

Commanders can refuse to comply with an authorised officer's instructions provided the commander informs the authorised officer (regulation 36).

Note: A Port Health Plan is usually produced, which identifies the relevant organisations involved, along with their roles and responsibilities.

Definitions

Authorised officer – the medical officer, the proper officer, or any other officer authorised by the responsible authority.
Commander – the person for the time being in command of an aircraft.
Disinsecting – the operation in which measures are taken to kill the insect vectors of human disease.
IHR – the International Health Regulations (2005) of the WHO adopted by the fifty-eighth World Health Assembly on 23 May 2005.
Infected aircraft:

(a) an aircraft which has on board on arrival a case of plague, cholera, yellow fever, smallpox, rabies or viral haemorrhagic fever; or
(b) an aircraft on which a plague-infected rodent is found on arrival; or
(c) an aircraft which has had a case of smallpox on board during its voyage and which has not before arrival been subjected in respect of such case to appropriate measures equivalent to those provided in these regulations.

Infected person – a person who is suffering from plague, cholera, yellow fever or smallpox or who is considered by the medical officer to be infected with one of the diseases or with some other infectious or contagious disease other than venereal disease or tuberculosis.
Infectious disease – any infectious or contagious disease other than venereal disease or tuberculosis.
Responsible authority – in relation to an aerodrome or other place, the authority charged with the duty of enforcing and executing these regulations.

INFECTIOUS DISEASES CONTROL – TRAINS

Reference

Public Health (International Trains) Regulations 1994.

Extent

These provisions apply to England and Wales.

Stowaway animals on incoming trains

Where a train manager reports to the local authority the presence or suspected presence of a stowaway animal on board his train and the animal is or was capable of carrying rabies, plague or viral haemorrhagic fever, or is or may be contaminated, the enforcement authority may require that the train, together with such of its contents as they may reasonably specify, be deratted, disinfested or decontaminated, or be subject to other methods of control (regulation 6(1) to (3) and (5)).

The enforcement authority may require any or all of the measures to be taken at a designated customs-approved area, or a designated shuttle control area if the enforcement authority for that area agrees.

When control measures have been taken, the train operator must inform the enforcement authority for each area at which the control measures were undertaken of the arrangements made for the disposal of the stowaway animal(s) (regulations 6(6) and 6(7)).

Sick travellers on incoming trains

Train managers of an international train must notify the enforcement authority if they are aware of a sick traveller who is still on board the train, including details of the sick traveller's whereabouts and, if the sick traveller is no longer on board the train, details of the circumstances in which the traveller was identified as sick and where the traveller alighted from or was removed from the train.

The enforcement authority may require the disinfestation or decontamination of the train (regulation 8).

The enforcement authority has the power to ask travellers questions where there are reasonable grounds for suspecting that there is a significant danger to public health (regulation 9).

INFECTIOUS DISEASES CONTROL – SHIPS

Reference

Public Health (Ships) Regulations 1979 (as amended).

Extent

These provisions apply in England and Wales only.

Scope

The purpose of the legislation is the prevention of spread of disease of international concern and to control chemical, biological and radiological risk. The national legislation implements the World Health Organisation International Health Regulations 2005. The Department of Health intends to introduce international health regulations, which may well replace the Public Health (Ships) Regulations.

FC108 Infectious disease control – Public Health (Ships) Regulations 1979 (as amended)

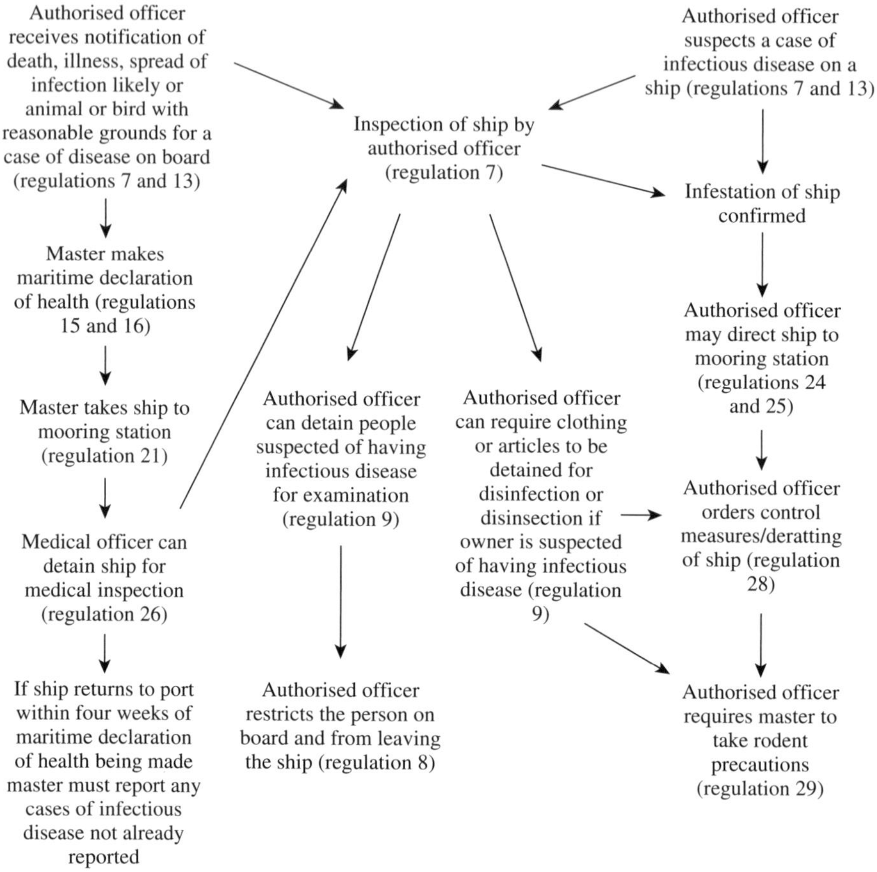

Notes

1. Local authorities can charge for measures taken (regulation 38).
2. Authorised officer must have regard to the need to free the ship as quickly as possible (regulation 27).

Incoming ships – inspection and notification of infectious disease

An authorised officer may inspect any ship on arrival or already in the district. The authorised officer must also inspect the ship if he receives a message of notification from the master of the ship that there has been:

(a) (i) the death of a person not being the result of an accident; or
(ii) illness where the person who is ill has or had a temperature of 38°C or greater accompanied by a rash, glandular swelling or jaundice, or where such temperature persisted for more than 48 hours; or
(iii) illness where the person has or had diarrhoea severe enough to interfere with work or normal activities;
(b) the presence on board of a person who is suffering from or who has symptoms of an infectious disease or tuberculosis;
(c) any other circumstances on board which are likely to cause the spread of infectious disease or other danger to public health; and
(d) the presence of animals or captive birds, and the occurrence of mortality or sickness amongst such animals or birds (regulations 7 and 13).

The authorised officer must also inspect the ship if he has reasonable grounds for believing that there is on board a case or suspected case of infectious disease (regulation 7(1)(b)).

The authorised officer can also take samples of food or water for analysis or examination (regulation 7(3)).

This analysis or examination must be:

(a) with a view to the treatment of persons affected with any epidemic, endemic or infectious disease and for preventing the spread of such diseases; or
(b) for preventing other danger to public health (regulation 7(4)).

An authorised officer may require a ship on arrival or already in the district to be brought to, and if necessary moored or anchored at, some safe and convenient place for the purpose of medical inspection (regulation 8).

The medical officer of health has the power to examine people suspected of suffering from an infectious disease etc. (regulation 9(1)).

Examination of persons on ships

An authorised officer may:

(a) detain persons suspected of suffering from an infectious disease for examination by the medical officer of health, either on the ship or at some place on shore appointed for the purpose;
(b) require the clothing and other articles belonging to any such person to be disinfected and, where necessary, disinsected, and any person found to be verminous to be disinsected;
(c) prohibit any such person from leaving the ship, or permit him to leave it subject to conditions the authorised officer considers reasonably necessary for preventing the spread of infection; and

(d) require the master to take or assist in taking such steps as in the opinion of the authorised officer are reasonably necessary for preventing the spread of infection, for disinfection and the destruction of vermin, and for the removal of conditions on the ship likely to convey infection, including conditions the existence of which might facilitate the harbouring of insects or vermin (regulation 9).

The medical officer of health has the power to remove a suspected person to hospital, place them under surveillance or prevent them from leaving the ship (regulation 10).

Maritime declaration of health

On the arrival of a ship, the master has a report to make about infectious disease on board or is directed by the medical officer to complete a maritime declaration of health. He must complete the declaration and deliver it to the authorised officer.

If, within four weeks after the master of a ship has delivered a maritime declaration of health, the ship arrives in a district or calls at another district, the master must report to the authorised officer any case or suspected case of infectious disease or tuberculosis which has occurred on board since the declaration was delivered and which has not already been reported (regulations 15 and 16).

Restriction on boarding or leaving of ships

An authorised officer can restrict the movement of persons on board or prevent them from leaving a ship until free pratique has been granted, and the master shall take all reasonable steps to secure compliance with this restriction etc.

Before granting permission to a person to leave the ship, the authorised officer may require him to state his name and his intended destination and address, and to give any other information which the authorised officer may think necessary for transmission to the medical officer for the area in which the intended destination of the person is situated.

If such a person cannot state his intended destination and address or arrives, within a period not exceeding 14 days after landing, to be specified to him by the authorised officer, at an address other than that which he has so stated, he shall immediately after his arrival at that address send particulars thereof to the authorised officer of the port where he left the ship (regulation 17).

Application for a ship sanitation control exemption certificate or ship sanitation control certificate

Where an owner or master of a ship applies in writing for a sanitation certificate, the authorised officer must inspect the ship to prevent danger to public health or spread of infection. The authorised officer may also take control measures with a view to issuing the ship sanitation certificate. If the ship is not in the area of an authorised port, an authorised officer must consult with a customs officer and

FC109 Ship sanitation control certificates – Public Health (Ships) Regulations 1979 (as amended) Regulations 18A to 18D

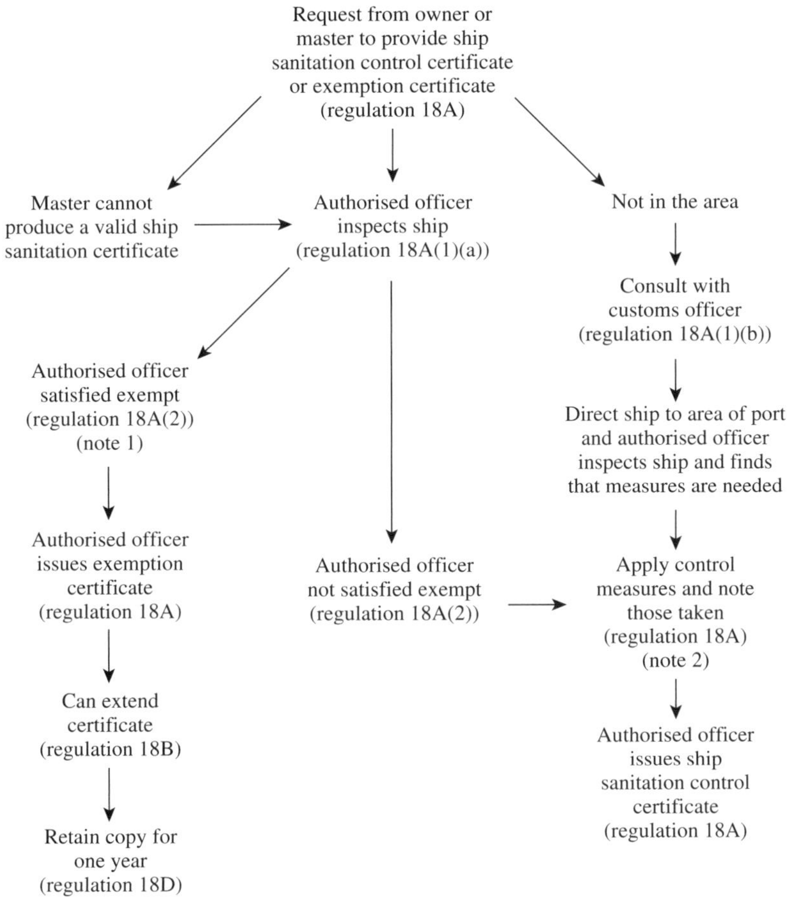

Note
1. Holds should be empty for ship sanitation exemption certificate to be issued (regulation 18A(3)).
2. Charges can be made for inspection and control measures taken (regulation 39).

direct the ship to an area of an authorised port convenient to the ship and the customs officer. If the authorised officer is satisfied that the ship is exempt from control measures, he must issue a ship sanitation control exemption certificate. If control measures have been completed to his satisfaction, he must issue a ship sanitation control certificate and note on the certificate the evidence found and the control measures taken.

A ship sanitation control exemption certificate can usually only be issued if the inspection of the ship was carried out when the ship and holds were empty, or a thorough inspection of the holds is possible (regulation 18A).

Ship sanitation certificate

If the master of a ship cannot produce a valid ship sanitation certificate, the authorised officer can inspect the ship for evidence of danger to public health or infection with a view to issuing a ship sanitation certificate or, if not in the area of the port, consult with a customs officer and direct the ship to proceed to an area of an authorised port.

The authorised officer must issue a ship sanitation control exemption certificate if he is satisfied that the ship is exempt from control measures.

If the authorised officer is not satisfied that the ship is exempt from control measures, he must order control measures necessary for the control of danger to public health or the spread of infection. If the ship is not within the area of an authorised port, the authorised officer must consult with a customs officer and direct the ship to proceed to a specified area of an authorised port.

When control measures have been completed to the satisfaction of an authorised officer, the authorised officer must:

(a) issue a ship sanitation control certificate; and
(b) note on the certificate the evidence found and the control measures taken (regulation 18B).

Control measures must be those advised by the World Health Organisation. The authorised officer has the power to take other equally effective measures but must report them to the national IHR (International Health Regulation) Focal Point (regulation 18C).

Ship sanitation certificates: form, period of validity, extension and retention

A ship sanitation control exemption certificate or ship sanitation control certificate must conform to the model schedule 3 of the regulations and is valid for six months beginning with the date of issue.

An authorised officer may extend the period of validity of a ship sanitation certificate by one month if:

(a) any inspection or control measures required cannot be carried out at the port;
(b) there is no evidence of danger to public health or infection; and

(c) the port is authorised to extend the validity of a ship sanitation certificate.

The local authority must retain a copy of any ship sanitation certificate for a period of one year beginning with the date of issue (regulation 18D).

Detention of ships and ships to be taken to mooring stations

Where an infected ship or a ship suspected of being infected (within last four weeks) has not had control measures, the master must take the ship to a mooring station (regulation 21).

The authorised officer can require the ship to be subject to a medical inspection, in writing to the master (regulation 22).

If there is a case or suspected case of, or animal infected with, plague, cholera, yellow fever, smallpox, rabies or viral haemorrhagic fever on board, the authorised officer may direct the master to take the ship to a mooring station and it must stay there until inspected by a medical officer (regulations 24 and 25).

The authorised officer may detain any ship for medical inspection at its place of mooring or at its place of discharge or loading (regulation 26).

The medical officer must inspect the ship and persons on board as soon as possible after it has been taken or directed to or detained at a mooring station. The authorised officer may apply control measures when the ship is at the mooring station (regulation 28).

The authorised officer may require the master of a ship to take all practicable measures to prevent the escape of rodents from the ship (regulation 29).

Removal of infected persons from ships where required by master

There are special control measures for specific diseases in schedule 4 of the regulations.

Outgoing ships – examination etc. of persons proposing to embark

There are similar powers available to the medical officer for ships departing from a port (regulation 33).

Infected places in England and Wales

Regulation 34 provides for the declaration of infected areas by the Secretary of State where certain diseases are present or suspected, giving the medical officer and authorised officer powers to control infection spread.

Charges

Local authorities can charge for services where there is a danger to public health from a ship arriving in its district, or the spread of infection from a ship leaving its district. Services include inspection and applying control measures in connection with issuing a ship sanitation certificate but must not exceed the actual cost of the service rendered.

At the request of the master or the person in relation to whom the measures have been taken, a local authority must provide him with particulars, free of charge, of measures taken for which a charge is made and the reasons why the measures were taken (regulation 38).

Every charge is recoverable either summarily as a civil debt or as a simple contract debt in any court of competent jurisdiction (regulation 39).

Note: A Port Health Plan is usually produced, which identifies the relevant organisations involved, along with their roles and responsibilities.

Definitions

Authorised officer – the medical officer, the proper officer or any other officer authorised by the local authority under regulation 4 to enforce and execute any of these regulations.

Authorised port – a port authorised to offer:

- (a) the issuance of a ship sanitation control certificate and the provision of the services referred to:
  - (i) in Annex 1 to the IHR; and
  - (ii) the form reproduced at schedule 3 to these regulations;
- (b) the issuance of a ship sanitation control exemption certificate following inspection of the ship, including a thorough inspection of the hold; or
- (c) the extension of a ship sanitation certificate for a period of one month.

Disinsecting – the operation in which measures are taken to kill the insect vectors of human disease.

Free pratique – permission for a ship to disembark and commence operation.

IHR – the International Health Regulations (2005) of the WHO adopted by the fifty-eighth World Health Assembly on 23 May 2005.

Infected person – a person who is suffering from plague, cholera, yellow fever or smallpox or who is considered by the medical officer to be infected with such a disease or with some other infectious or contagious disease other than venereal disease or tuberculosis.

Infected ship:

- (a) a ship which has on board on arrival a case of plague, cholera, yellow fever, smallpox, rabies or viral haemorrhagic fever; or
- (b) a ship on which a plague-infected rodent is found on arrival; or
- (c) a ship which has had on board during its voyage:
  - (i) a case of human plague which developed more than six days after the embarkation of the person affected; or
  - (ii) a case of cholera within five days before arrival; or
  - (iii) a case of yellow fever or smallpox;

  and which has not before arrival been subjected in respect of such case to appropriate measures equivalent to those provided for in these regulations.

Maritime declaration of health – means a declaration in the form set out in schedule 2 of the regulations.
Master – the person for the time being in charge of or in command of a ship.
Mooring station – a place, situated within the waters of a district, which is specified by the local authority, with the consent of the customs officer for the area in which the district is situated and the harbour master, or in such other district as the Secretary of State may allow, for the mooring of ships for medical inspection so that they do not come into contact with other ships or the shore.
Ship sanitation certificate – a ship sanitation control certificate or a ship sanitation control exemption certificate.
Ship sanitation control certificate – a certificate that conforms to the model in Annex 3 to the IHR reproduced at schedule 3 to these regulations that is issued in accordance with article 39 of the IHR (ship sanitation certificates).
Ship sanitation control exemption certificate – a certificate that conforms to the model in Annex 3 to the IHR reproduced at schedule 3 to these regulations that is issued in accordance with article 39 of the IHR.

PUBLIC HEALTH ACTS
GENERAL PROCEDURAL PROVISIONS

The following general provisions are those which are normally applicable to actions under the Public Health Acts 1936 and 1961 and are also generally applicable to procedures being taken under the Clean Air Act 1993 and the Prevention of Damage by Pests Act 1949. Unless otherwise indicated, they will apply to the procedures in this chapter.

Power of entry

An authorised officer of a council, producing if required a duly authenticated document showing his authority, has a right to enter any premises at all reasonable hours for the following purposes:

(a) ascertaining if there has been a contravention of this Act;
(b) seeing if circumstances exist which would authorise or require the council to take any action or execute work;
(c) to take any action or execute any work authorised or required;
(d) generally for the council's performance of its functions.

For premises other than a factory, workshop or workplace, at least 24 hours' notice must be given before entry may be demanded.

Where:

(a) admission has been refused; or
(b) refusal is apprehended; or
(c) the premises are unoccupied; or
(d) the occupier is temporarily absent; or
(e) the case is one of urgency; or
(f) application for admission would defeat the object of entry,

the local authority, following notice of intention to any occupier, may make application to a Justice of the Peace by way of sworn information in writing, and the Justice of the Peace may authorise entry by an authorised officer, if necessary by force. The authorised officer may take with him any other persons as may be necessary but must leave any unoccupied premises as secure as he found them (section 287).

The penalty for obstructing any person acting in the execution of these Acts, regulations and orders is a maximum level 1 on the standard scale and £20 daily for continuing offences (Public Health Act 1936, section 288).

Where an owner is prevented from executing any work which he has been required to undertake, he may apply to a magistrates' court for an order to the occupier to permit execution of the works (section 289).

Notices requiring the execution of works

These provisions (known as the Part 12 provisions) apply to any procedures where the section concerned specifically indicates this to be the case and this is shown in the procedures where appropriate.

(a) **Content**. The notice must indicate the nature of the works required and state the time within which they are to be executed. The time allowed should not be less than the period for appeal, i.e. 21 days.

(b) **Appeals**. Appeal against these notices may be made to a magistrates' court within 21 days of service on any of the following grounds:
  (i) the notice is not justified;
  (ii) there has been some material informality, defect or error in, or in connection with, the notice;
  (iii) the local authority has refused unreasonably to approve alternative works;
  (iv) the works required are unreasonable in character or extent or are unnecessary;
  (v) the time allowed for compliance is not reasonable;
  (vi) it would have been equitable for the notice to have been served on the occupier instead of the owner or vice versa;
  (vii) that some other person ought to contribute towards the cost of compliance.

(c) **Enforcement**. Following failure to comply with a notice to execute works within the time specified, subject to appeal, the local Authority may, at its discretion, carry out the work itself and recover expenses reasonably incurred from the person on whom the notice was served. In addition, the person concerned is liable to a fine not exceeding level 4 on the standard scale and a maximum daily penalty of £2 for a continuing offence (sections 290 and 300).

Appeals

Unless a longer period is specified, any appeals to magistrates' courts are to be made by way of complaint for an order within 21 days from the date on which the council's requirement, notice, etc. was served. Notices and other documents must state the rights of appeal and the time within which appeal may be made (section 300).

Persons aggrieved by decisions of the magistrates' court may appeal to the Crown Court, except where there is a provision for arbitration (Public Health Act 1936, section 301).

Recovery of costs

Costs incurred by local authorities in carrying out works in default of an owner are recoverable both by proceeding in the County Court and by the debt becoming a charge on the property. They may be recovered from an owner by instalments and with interest (section 291).

Power of a local authority to execute work on behalf of owners or occupiers at their request

A local authority, by agreement with an owner or occupier of any premises, may undertake on his behalf works which it has required him to carry out under the Act or undertake sewer or drain construction, alteration or repair which the owner or occupier is entitled to execute. The costs are chargeable to the owner or occupier (section 275).

BUILDING ACT 1984
GENERAL PROCEDURAL PROVISIONS

Notices requiring the execution of works

Where specifically indicated by the particular provisions relating to a procedure, the following apply:

(a) **Content.** The notice must indicate the nature of the works required and state the time within which they are to be executed. The time allowed should not be less than the period for appeal, i.e. 21 days (section 99(1)).

(b) **Appeals.** Appeal against these notices may be made to a magistrates' court within 21 days on any of the following grounds:

    (i) the notice is not justified;
    (ii) there has been some material informality, defect or error in, or in connection with, the notice;
    (iii) the local authority has refused unreasonably to approve alternative works;
    (iv) the works required are unreasonable in character or extent or are unnecessary;
    (v) the time allowed for compliance is not reasonable;
    (vi) it would have been equitable for the notice to have been served on the occupier instead of the owner or vice versa;
    (vii) that some other person ought to contribute towards the cost of compliance (sections 102 and 103).

(c) **Enforcement.** Following failure to comply with a notice to execute works within the time specified, subject to appeal, the local authority may, at its discretion, carry out the work itself and recover expenses reasonably incurred from the person on whom the notice was served. In addition, the person concerned is liable to a fine not exceeding level 4 and a maximum daily penalty of £2 for a continuing offence (section 99(2)).

Recovery of costs

Costs incurred by local authorities in carrying out works in default of an owner are recoverable either by proceedings in the County Court or by becoming a charge on the property. They may be recovered from an owner by instalments and with interest (sections 107(1) and 108).

Power of entry

An authorised officer of a council, producing, if required, a duly authenticated document showing his authority, has a right to enter any premises at all reasonable hours for the following purposes:

(a) ascertaining if there has been a contravention of building regulations;
(b) seeing if circumstances exist which would authorise or require the council to take any action or execute work;
(c) to take any action or execute any work authorised or required;
(d) generally for the council's performance of its functions.

For premises other than a factory, workshop or workplace, at least 24 hours' notice must be given before entry may be demanded.

Where:

(a) admission has been refused; or
(b) refusal is apprehended; or
(c) the premises are unoccupied; or
(d) the occupier is temporarily absent; or
(e) the case is one of urgency; or
(f) application for admission would defeat the object of entry,

the local authority may make application to a Justice of the Peace by way of sworn information in writing and the Justice of the Peace may authorise entry by an authorised officer, if necessary by force. The authorised officer may take with him any other person as may be necessary but must leave any unoccupied premises as secure as he found them (sections 95 and 96).

The penalty for obstructing a person acting in the execution of this Act is a fine not exceeding level 1 on the standard scale (section 112).

Where an owner is prevented from executing any work which he has been required to undertake, he may apply to a magistrates' court for an order to the occupier to permit execution of those works (section 98).

Appeals

Any appeals to magistrates' courts are to be made by way of complaint for an order within 21 days from the date on which the council's requirement, notice, etc., was served (unless a longer period is specified). Notices and other documents must give the rights of appeal and state the time within which appeal may be made (section 103).

Persons aggrieved by decisions of the magistrates' court may appeal to the Crown Court, except where there is a provision for arbitration (section 86).

Power of local authority to execute work on behalf of owners or occupiers

A local authority, by agreement with an owner or occupier of any premises, may undertake on his behalf works which it has required him to carry out under the Act or undertake sewer construction, alteration or repair which the owner or occupier is entitled to execute. The costs are chargeable to the owner or occupier (section 97).

DEFECTIVE SANITARY CONVENIENCES

Reference

Public Health Act 1936, section 45 and part 12.

Extent

This procedure applies to England and Wales only.

Scope

This procedure deals with closets provided in a building which are in such a state as to be prejudicial to health or a nuisance and can, without reconstruction, be put into a satisfactory condition. This includes both repair and cleansing (Public Health Act 1936, section 45(1)).

It does not apply to closets in a factory, workshop or workplace (section 45(4)). In such cases, the provisions of the Health and Safety at Work etc. Act 1974 can be used (FC52).

Notices etc.

A local authority can serve a notice on the owner or occupier requiring defective sanitary conveniences to be repaired. The provisions of Part 12 of the Public Health Act 1936 apply to appeals against and the enforcement of notices requiring repairs but not to those requiring cleansing only (above page 449).

FC110 Defective sanitary conveniences – Section 45 Public Health Act 1936

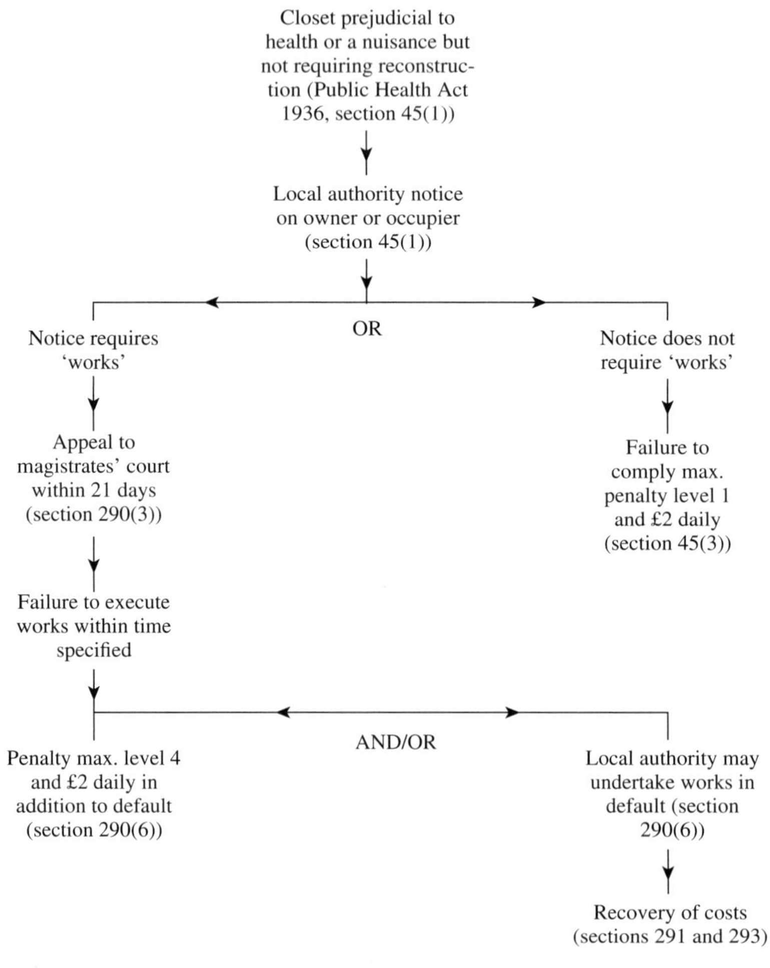

Notes

1. For closets prejudicial to health or a nuisance but requiring reconstruction, see Building Act 1984, section 64 (FC113).
2. At any time after service of the local authority notice, the local authority may undertake the required works at the request of the owner or occupier and charge the costs (section 275).

Appeals

In relation to a notice not requiring works, i.e. requiring cleansing etc., the defendant in subsequent proceedings for penalty may question the reasonableness of the requirements and the decision to address the notice to him and not the owner or occupicr as the case may be (section 45(3)).

Definitions

Closet includes privy (section 90(1)). It is suggested that the word includes the structure in which the appliance is housed as well as the appliance itself.
Prejudicial to health means injurious, or likely to be injurious to health.

SANITARY CONVENIENCES: PROVISION/REPLACEMENT

Reference

Building Act 1984, sections 64 and 65.

Extent

This procedure applies to England and Wales only.

Scope

The procedure applies to:

(a) insufficient or replacement accommodation in buildings (other than workplaces) where the existing closet accommodation is insufficient or is prejudicial to health or a nuisance and requires reconstruction. The notices may require the provision of additional or replacement closets, but water closets may not be required unless a sufficient water supply and sewer are available (section 64):
  (i) sewer available – must be within 100 ft of the building at a level which makes connection reasonably practicable, is a sewer which the owner of the building is entitled to use and the intervening land is land through which he is entitled to construct a drain;
  (ii) water supply available – water is laid on or can be laid on from a point within 100 ft of the building and the intervening land is land through which the owner of the building is entitled to lay a pipe (section 125);

(b) sanitary conveniences in workplaces. In deciding whether or not sanitary conveniences for buildings used as workshops are sufficient and satisfactory, regard must be paid to the number of persons employed in or in attendance at the building and the need to provide separate accommodation for the sexes, although the local authority may waive this latter provision if it so wishes (section 65(1)).

FC111 Sanitary conveniences: provision/replacement – Sections 64 and 65 Building Act 1984

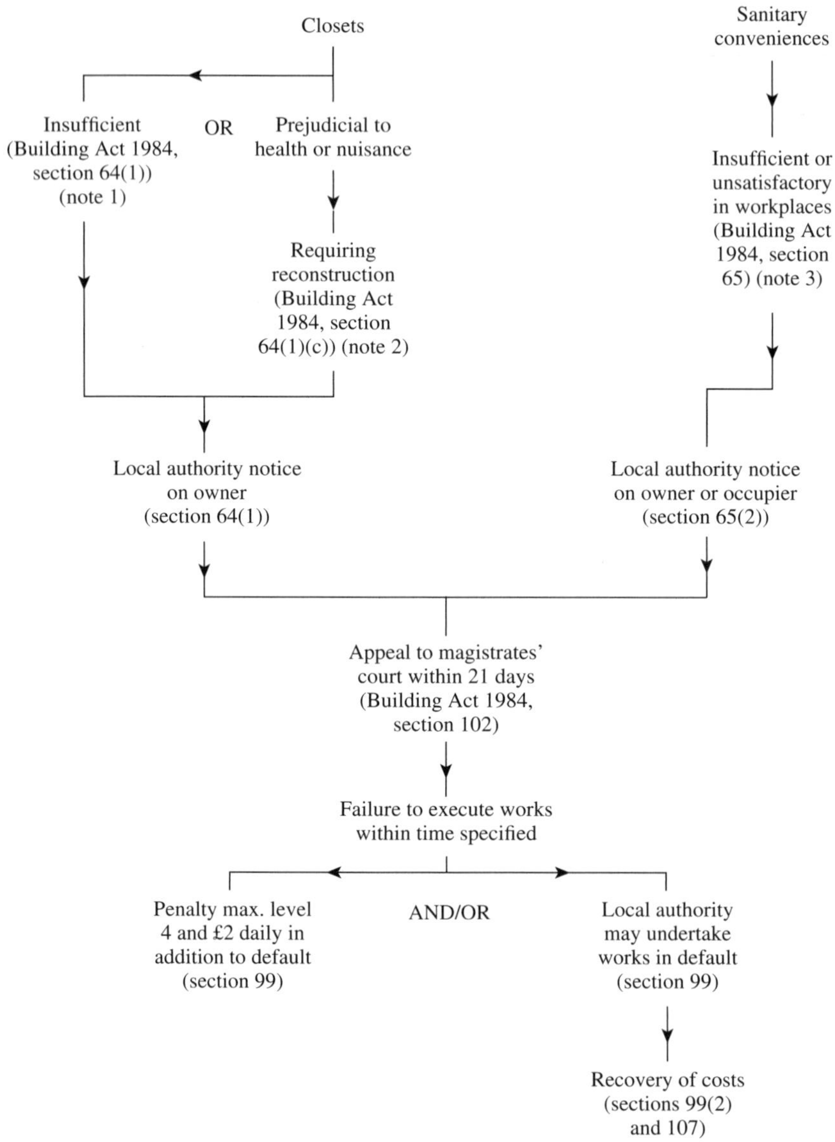

Notes

1. This relates to buildings, or parts of buildings occupied as a separate dwelling, which are without sufficient closet accommodation but does not include workplaces.
2. The procedure for dealing with existing closets of buildings which are prejudicial to health or a nuisance is split between those which can be put into satisfactory condition without construction (section 45, Public Health Act 1936, FC110) and those which cannot (section 64, Building Act 1984).
3. This section covers all workplaces (does not include factories). Alternatively the situation could be dealt with by using the enforcement procedures of the Health and Safety at Work etc. Act 1974, see FC52.
4. For sanitary conveniences in a place of public entertainment etc., see FC112.
5. For the power of a local authority to require replacement of earth closets by water closets at the joint expense of the owner and local authority, see FC113.

Notices, appeals, etc.

The provisions of sections 99 and 102 apply to notices used in this procedure (pages 447–449).

Definition (see also page 458)

Closet includes privy (section 126).

SANITARY CONVENIENCES AT PLACES OF ENTERTAINMENT ETC.

References

Local Government (Miscellaneous Provisions) Act 1976, sections 20 and 21.
Chronically Sick and Disabled Persons Act 1970, section 6.

Extent

This procedure does not apply in Scotland or Northern Ireland (section 83).

Scope

The premises covered by these provisions are termed 'relevant places' and are premises used, or proposed to be used, occasionally or permanently, for the following purposes:

(a) the holding of any entertainment, exhibition or sporting event to which the public is admitted as spectators or otherwise;
(b) the sale of food and drink to the public for consumption on the premises;
(c) a betting office (section 20(9)).

In respect of (a), a licence for the holding of entertainment may also be required under the Licensing Act 2003 – see FC142.

Notices

Notices must be in writing on either the owner or the occupier of the premises and can be served in anticipation of the use of a premises as a 'relevant place'. Notices may require:

(a) the provision of a specified number and type of sanitary appliances at specified positions within a defined period, which should be not less than six weeks;
(b) the maintenance and cleaning of the appliances;
(c) the provision and maintenance of washing facilities, including hot and cold water and other facilities for use in connection with the sanitary appliances, e.g. hand drying facilities etc.;

FC112 Sanitary conveniences at places of entertainment etc. – Section 20 Local Government (Miscellaneous Provisions) Act 1976

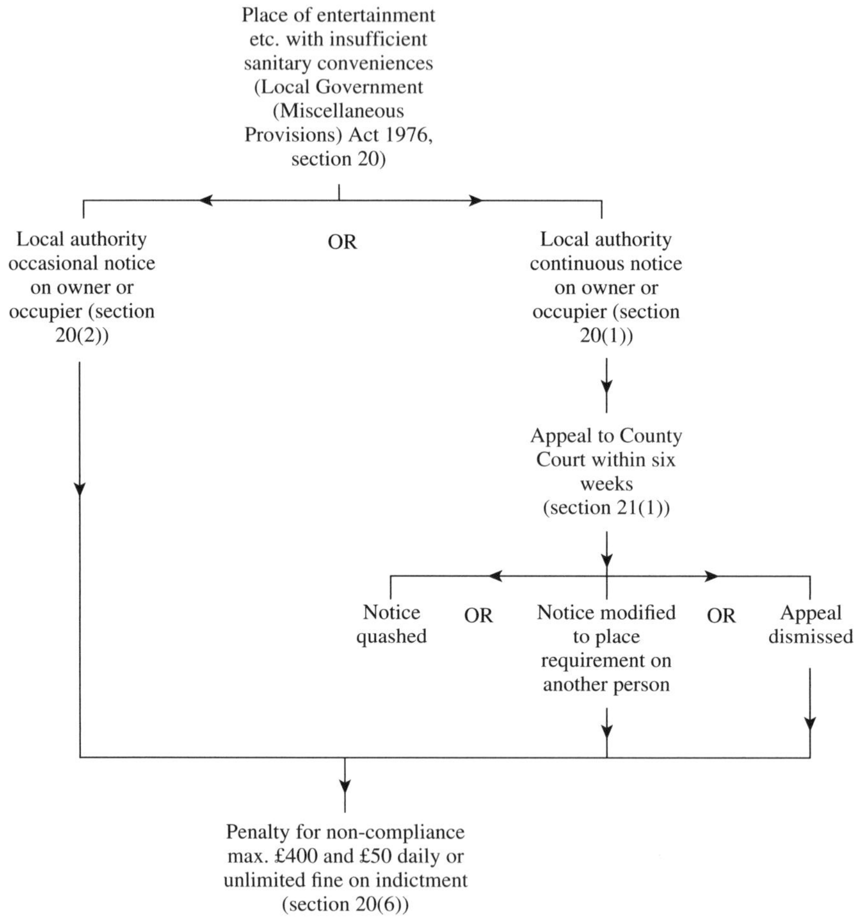

Notes

1. Notices under this section may be served by a district council, London borough council, the Common Council and the Council of the Isles of Scilly (Local Government (Miscellaneous Provisions) Act 1976, sections 20(1) and 44(1)).
2. There is no provision for appeal against 'occasional' notices.
3. The notice must be in writing (Local Government (Miscellaneous Provisions) Act 1976, section 44(1)).
4. For the licensing of public entertainment etc. under the Licensing Act 2003, see FCs 140–145.

(d) that the appliance and associated facilities should be made available to the public resorting to the premises, if required by the notice, free of charge.

The notice may require that existing sanitary appliances must be made available but cannot rcquire either moveable sanitary appliances at a betting office or, in the case of new buildings, provisions in excess of requirements of the building regulations (section 20(1), (2) and (3)).

Occasional notices

As an alternative to permanent provision, the notice may require provision of the specified appliances and associated facilities only on occasions which are specified in the notice (section 20(2)).

Disabled persons

A person upon whom a section 20 notice is served must make provision for the needs of the disabled so far as is practicable and reasonable (Chronically Sick and Disabled Persons Act 1970, section 6). The notice served under section 20 must draw attention to the requirements of the Chronically Sick and Disabled Persons Act 1970 and to BS 5810/1979, which contains a code at practice relating to Access for the Disabled to Buildings (sections 20(11) and 20(12)).

Appeals

So far as notices requiring permanent provision of facilities are concerned, appeal may be made within six weeks of service to the County Court on the following grounds:

(a) a requirement is unreasonable;
(b) it would have been fairer to have served the notice on the owner or occupier as the case may be (section 21).

There is no appeal against an occasional notice but in any subsequent proceedings it will be a defence to prove situations as in (a) and (b) above.

Defences

In proceedings for non-compliance with notices served under section 20, it will be a defence to prove:

(a) that at the time of the failure the person on whom the notice was served was neither the owner nor occupier and that he did not cease to be the owner or occupier with a view to avoiding compliance with the notice; or
(b) where the contravention relates to a particular day, that the relevant place was closed to members of the public or not used as a relevant place; or
(c) in respect of occasional notices only that:
    (i) the requirement was unreasonable; or

(ii) that it would have been fairer to have served the notice on another person (section 20(7) and (8)).

Power of entry

Authorised officers producing evidence of such may enter a 'relevant place' at any reasonable time to determine if a notice should be served or is being complied with. Penalty for obstruction is max. level 3 (section 20(5)).

Definitions

Betting office means premises, other than a track, for which a betting office licence within the meaning of the Gambling Act 2005, is in force (section 20(9)).
Owner in relation to any land, place or premises, means a person who, either on his own account or as agent or trustee for another person, is receiving the rackrent of the land, place or premises or would be entitled to receive it if the land, place or premises were let at a rackrent and 'owned' shall be construed accordingly (section 44(1)).
Sanitary appliances means water closets, other closets, urinals and wash basins (section 20(9)).

CONVERSION OF EARTH CLOSETS ETC. TO WATER CLOSETS

Reference

Building Act 1984, section 66.

Extent

This procedure applies to England and Wales only.

Scope

The procedure allows a local authority to secure the replacement of any closet, other than a water closet, by a water closet, even though the existing closet may be sufficient and not prejudicial to health or a nuisance, provided that the costs are equally shared between the local authority and the owner of the building and that a water supply and sewer are available.

Water supply and sewer available

Sewer available – this must be within 100 ft of the building at a level which makes connection reasonably practicable and be a sewer which the owner of the building is entitled to use, and the intervening land should be land through which he is entitled to construct a drain.

FC113 Conversion of earth closets etc. to water closets – Section 66 Building Act 1984

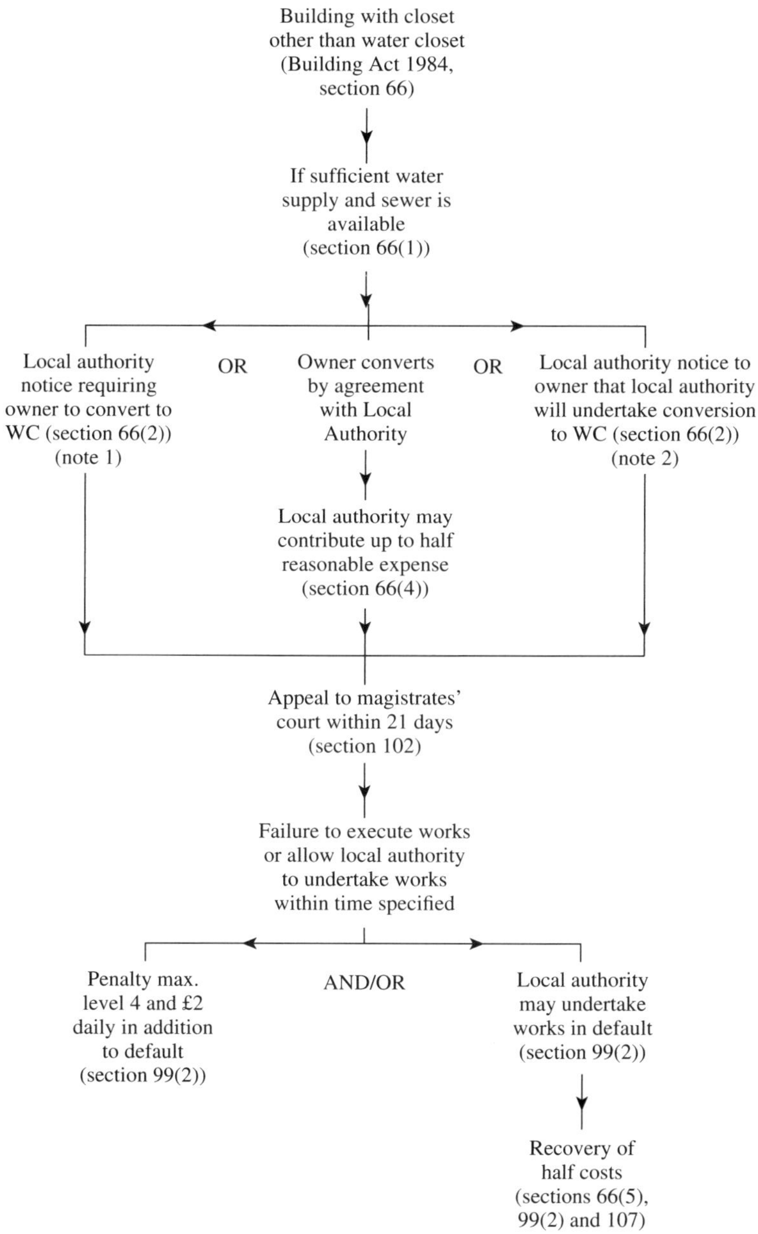

Notes

1. In this case, the owner may recover half of the reasonable cost from the local authority.
2. In this case, the local authority may recover half of the reasonable cost from the owner.

Water supply available – water is laid on or can be laid on from a point within 100 ft of the building and the intervening land should be land through which the owner of the building is entitled to lay a pipe (section 125).

The limit of 100 ft does not apply if the local authority undertakes to bear so much of the expenses reasonably incurred in constructing, and in maintaining and repairing, a drain to communicate with a sewer or, as the case may be, in laying and in maintaining and repairing a pipe for the purpose of obtaining a supply of water, even though the distance of the sewer or water supply exceeds 100 ft (section 125(3)).

The availability of a sewer or water supply in these terms is not necessary for the local authority to contribute towards the costs of an owner converting by agreement.

Notices etc.

The provisions of sections 99, 102 and 66(5) (pages 447–449) apply to appeals against, and the enforcement of notices served under, this procedure. Note that there is no appeal that the works are unnecessary (section 66(5)). Notices to be served on owner. Section 66(3) allows half of the expenses to be shared between the owner and the local authority.

Definitions (also page 453)

Closet includes privy (section 126).
Water closet means a closet which has a separate fixed receptacle connected to a drainage system and separate provision for flushing from a supply of clean water, either by the operation of mechanism or by automatic action (section 126).

BLOCKED PRIVATE SEWERS

Reference

Local Government (Miscellaneous Provisions) Act 1976, section 35.

Extent

This procedure applies to England and Wales only.

Scope

This procedure allows a notice requiring the unblocking of a private sewer to be served on owners or occupiers of all premises draining into it (section 35(1)).

Notices

Notice may be served on each of the persons who is an owner or occupier of premises served by the sewer or on any of these as the local authority sees fit.

FC114 Blocked private sewers – Section 35 Local Government (Miscellaneous Provisions) Act 1976

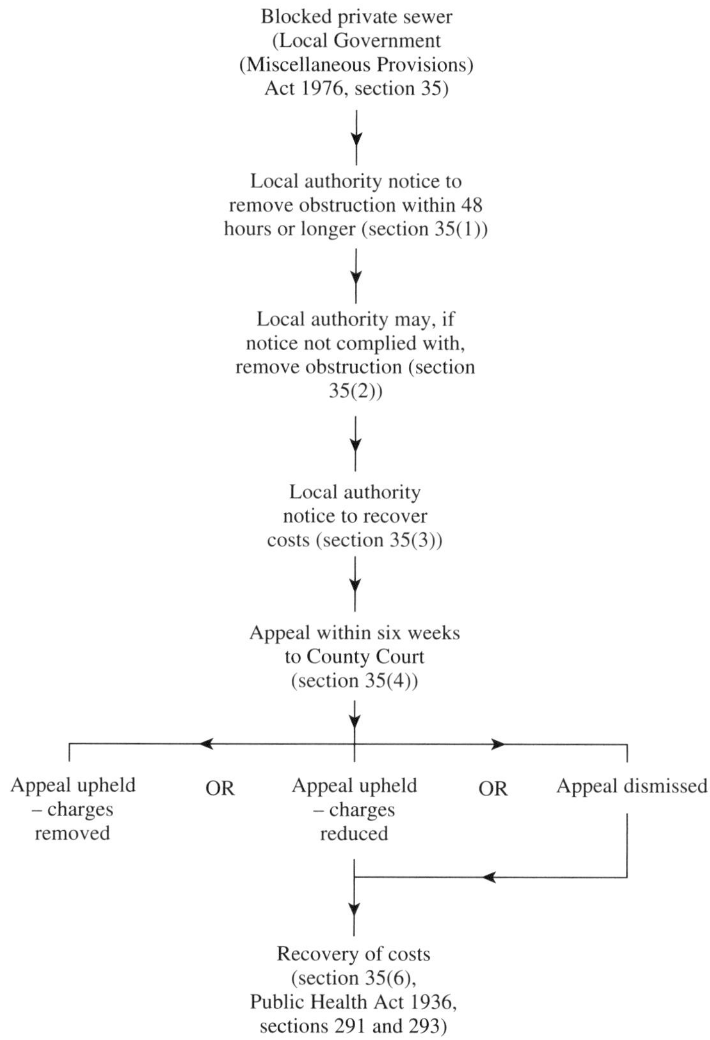

Notes

1. There is no appeal against the notice to remove the blockage under section 35(1) and neither is there a penalty for non-compliance.
2. For defective private sewers, see FC115 and FC117.

The time specified within which the works are required must not be less than 48 hours but can be longer at the discretion of the local authority (section 35(1)). Notices must be in writing (Public Health Act 1936, section 283).

Notices apportioning costs

Following works in default (section 35(2)), the local authority must specify by notice to those persons in receipt of the notice to clear the blockage, the amount to be recovered from each and which other persons are to be charged and the amount. In apportioning cost, the local authority may have regard to any matters relating to the cause of the blockage and to any agreements relating to the cleansing of that private sewer (section 35(3)). Particular regard should be paid to any determinations under section 22 of the Building Act 1984, dealing with the drainage of buildings in combination.

Appeals against notices to recover costs

In considering an appeal that whole or part should be paid by some other person, the County Court may either dismiss the appeal or order the whole or part to be paid by other owners or occupiers of premises served by the sewer, provided that such owners or occupiers have been given eight days' notice of the appeal (section 35(5)).

Definitions (see also pages 115, 456, 462)

Notice means a notice in writing (Local Government (Miscellaneous Provisions) Act 1976, section 44).
Owner in relation to any land, place or premises means a person who, either on his own account or as agent or trustee for another person, is receiving the rackrent of the land, place or premises or would be entitled to receive it if the land, place or premises were let at a rackrent (Local Government (Miscellaneous Provisions) Act 1976, section 44).

STOPPED-UP DRAINS, PRIVATE SEWERS, ETC.

Reference

Public Health Act 1961, section 17 (as substituted by section 27 of the Local Government (Miscellaneous Provisions) Act 1982).

Extent

This procedure applies to England and Wales only.

Scope

This procedure may be applied to any drain, private sewer, water closet, waste pipe or soil pipe on any premises which is stopped-up. There is no exclusion

FC115 Stopped-up drains, private sewers, etc. – Section 17 Public Health Act 1961

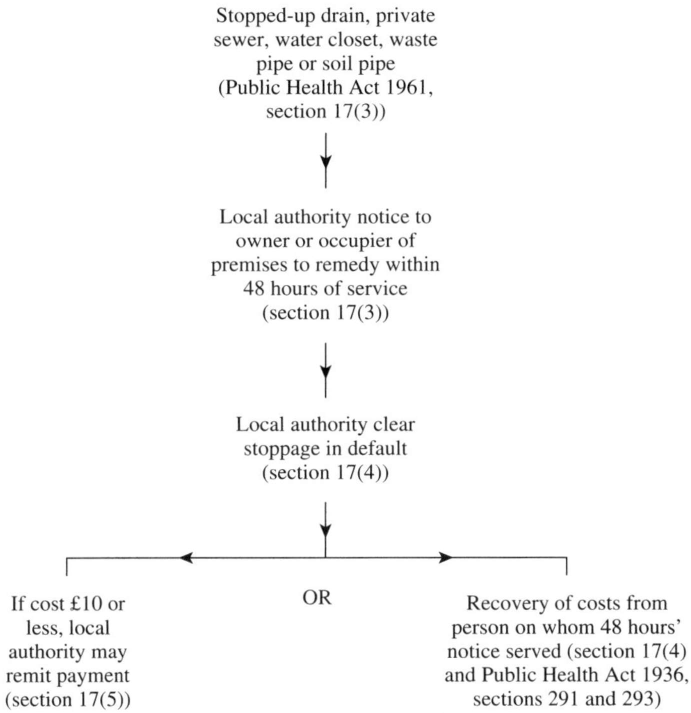

Notes

1. At the request of the owner or occupier, a local authority may cleanse or repair drains, water closets, sinks or gullies and recover costs. This could be done before the expiration of the 48 hours' notice by agreement (Public Health Act 1961, section 22).
2. There is no appeal against a notice served under section 17, neither is there a penalty for non-compliance.
3. For defective drainage, see FC116 and FC117.
4. For alternative procedure for blocked private sewers, see Local Government (Miscellaneous Provisions) Act 1976, section 35, at FC114.

here for premises owned by statutory undertakers as there is in relation to other parts of section 17 (see FC117).

Because section 17 appears to limit the service of the notice to the owner or occupier of the premises at which the drain etc. is stopped-up, blocked private sewers are best dealt with under section 35 (Local Government (Miscellaneous Provisions) Act 1976) since owners of other premises draining into the sewer may be involved (FC114).

Notices

Notices requiring remedy of the defect within 48 hours (this period is specified by the section and should not be shorter or longer) must be in writing, but are not subject to section 290 of the Public Health Act 1936 provisions and are to be served on either the owner or the occupier of the premises on which the drain, private sewer, etc. is stopped-up (section 17(3)).

Recovery of costs

If the cost to the local authority of undertaking the works in default does not exceed £10, recovery may be waived, otherwise costs are recoverable from persons on whom the notices were served (section 17(5)). In any proceedings to recover expenses, the magistrates' court may inquire:

(a) whether any requirement in the notice was reasonable; and
(b) whether the expenses ought to be borne wholly or in part by someone other than the defendant, but due notice of the proceedings and an opportunity to be heard must be given to these other persons (section 17(6)).

Definitions

Drain means a drain used for the drainage of one building or any buildings or yards appurtenant to buildings within the same curtilage.
Owner means the person for the time being receiving the rackrent of the premises ... whether on his own account or as agent or trustee for any other person, or who would so receive the same if those premises were let at a rackrent.
Premises includes messuages, buildings, lands, easements and hereditaments of any tenure.
Private sewer means a sewer which is not a public sewer.
Sewer does not include a drain (see definition above) ... but ... includes all sewers and drains used for the drainage of buildings and yards appurtenant to buildings (Public Health Act 1936, section 343).

DEFECTIVE DRAINAGE TO EXISTING BUILDINGS

Reference

Building Act 1984, sections 59 and 60.

Extent

This procedure applies to England and Wales only.

Scope

The procedure is applied to buildings which have:

(a) unsatisfactory provision for drainage (section 59(1)(a));
(b) cesspools, private sewers, drains, soil pipes, rainwater pipes, spouts, sinks or other necessary appliances which are insufficient or, in the case of a private sewer or drain communicating with a public sewer, is so defective as to admit subsoil water (section 59(1)(b));
(c) cesspools etc. as detailed in (b) above, in such a condition as to be prejudicial to health or a nuisance, and this also covers cesspools, private sewers and drains no longer in use (section 59(1)(c) and (d));
(d) rainwater pipes being used for foul waste, soil pipes from water closets not properly ventilated and surface water pipes acting as vents to foul drains or sewers (section 60).

In relation to (a), unsatisfactory drainage is not defined.

Notices etc.

The provisions on pages 447–449 apply to appeals against and the enforcement of notices served under this procedure. Notices are to be served on the owner of the building concerned. Appeal provisions are included in section 102 of the Building Act 1984. Works and default powers, penalties and recovery of costs are included in sections 99 and 107 of the Building Act 1984.

Testing of drains etc.

Local authorities have the power to examine sanitary conveniences, drains, private sewers or cesspools and to apply tests, other than tests by water under pressure, where they suspect deficiencies under section 59. This includes exposing the pipes by opening the ground (section 48, Public Health Act 1936).

Notice of intention to repair drains etc.

Before carrying out repairs to drains or private sewers, the person undertaking the work must give at least 24 hours' notice to the local authority, except if the work is required in an emergency, in which case he must not cover over the drain or sewer without such notice (Building Act, section 61).

FC116 Defective drainage to existing buildings – Sections 59 and 60 Building Act 1984

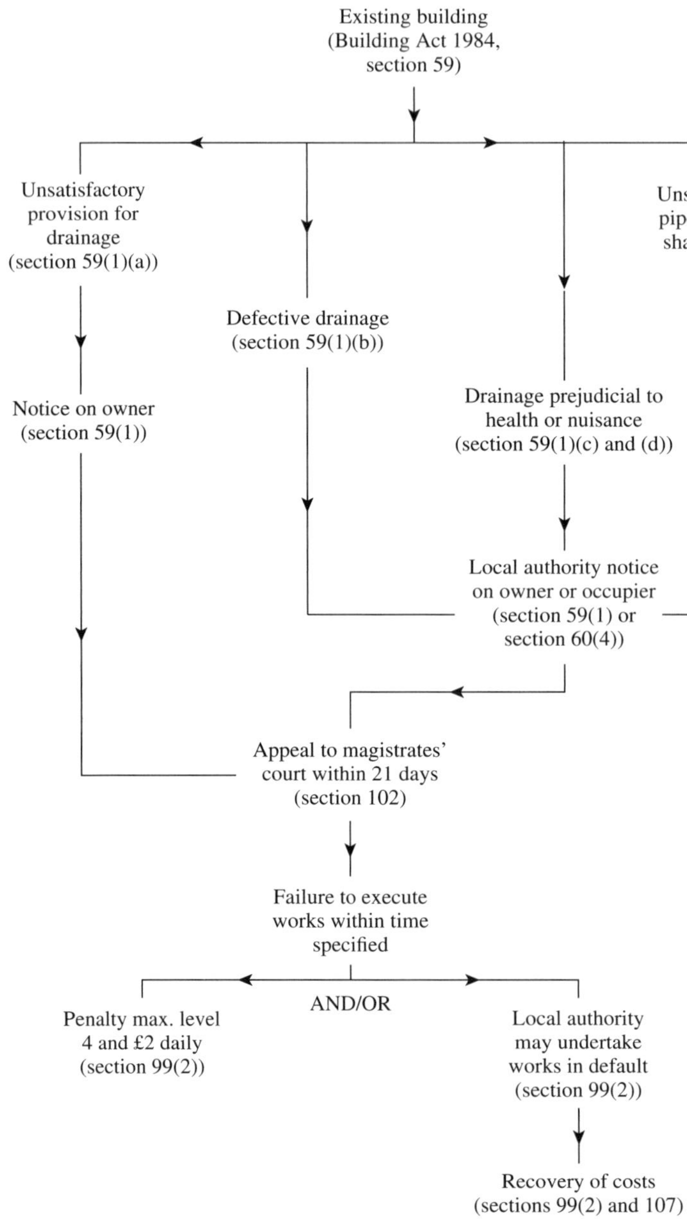

Notes

1. At the request of an owner or occupier, a local authority may cleanse or repair drains, water closets, sinks or gullies and recover costs. This could be done by agreement before the expiration of any notice served under section 59 and whether or not a notice has been served (Public Health Act 1961, section 22).
2. See also Public Health Act 1961, section 17 as amended (FC117).

Definitions (see also page 458)

Surface water includes water from roofs.
Water closet means a closet that has a separate fixed receptacle connected to a drainage system and separate provision for flushing from a supply of clean water, either by the operation of a mechanism or by automatic action (section 126).

REPAIR OF DRAINS, PRIVATE SEWERS, ETC. BY LOCAL AUTHORITIES

Reference

Public Health Act 1961, section 17 (as substituted by section 27 of the Local Government (Miscellaneous Provisions) Act 1982).

Extent

This procedure applies to England and Wales only.

Scope

The procedure applies to any drain, private sewer, water closet, waste pipe or soil pipe which is:

(a) not sufficiently maintained and kept in good repair; and
(b) can be sufficiently repaired at a cost not exceeding £250 (section 17(1)).

Works by local authorities on land belonging to statutory undertakers and held or used by them for those purposes is not authorised unless it affects houses or buildings used as offices or showrooms, other than those forming part of a railway station (section 17(10) and (11)).

The procedure under this section to deal with blocked or stopped-up (as opposed to defective) drains, private sewers etc. is covered separately in FC115.

Persons concerned

Notices are to be served on the person or persons concerned, which means:

(a) in relation to a water closet, waste pipe or soil pipe, the owner or occupier of the premises; and
(b) in relation to a drain or private sewer, any person owning any premises drained by means of it and also, in the case of a sewer, the owner of the sewer (section 17(2)).

Notices

Notices must be in writing (Public Health Act 1936, section 283) but are not subject to the provisions of section 290.

FC117 Repair of drains, private sewers, etc. by local authorities – Section 17 Public Health Act 1936

Drain, private sewer, water closet, waste pipe or soil pipe not sufficiently maintained and kept in good repair (Public Health Act 1961, section 17)

↓

Cost of repair not exceeding £250 (section 17(1))

↓

Minimum seven days' notice to persons concerned by local authority (section 17(1))

↓

Local authority repair (section 18(1))

↓

If cost £10 or less, local authority may remit payment (section 17(5))

OR

Recovery of costs up to £250 from persons concerned (section 17(1) of the Public Health Act 1936, sections 291 and 293)

Notes

1. For alternative procedure, see Building Act 1984, section 59, FC116.
2. For blocked drains and private sewers, see FC115.
3. There is no appeal against a notice served under section 17, neither is there a penalty for non-compliance.
4. Local authorities may clear obstructions in drains or private sewers by agreement before the expiration of the notice (Public Health Act 1961, section 22).

Recovery of costs

Costs not exceeding £250 incurred by the local authority are recoverable from the persons concerned (as defined above) in such proportions, if there is more than one person concerned, as the local authority may determine. Costs of £10 or less may be remitted (section 17(4) and (5)).

In any proceedings to recover the local authority's expenses of up to £250, the court must consider whether or not the local authority was justified in concluding that the drain, private sewer, etc. was not sufficiently maintained and kept in good repair and decide whether any apportionment between different persons is fair (section 17 (6)).

PAVING OF YARDS AND PASSAGES

Reference

Building Act 1984, section 84.

Extent

This procedure applies to England and Wales only.

Scope

The section applies to any court or yard appurtenant to, or any passage giving access to, buildings as defined below, including those used in common by two or more houses provided it is not repairable by the inhabitants at large as a highway (section 84). It does not apply to garden paths.

The procedure applies to buildings which are houses, industrial or commercial buildings.

Paving and drainage

Courts, yards and passages must be so formed, flagged, asphalted or paved, or provided with such works on, above or below its surface, as to allow satisfactory drainage of its surface or subsoil to a proper outfall (section 84(1)). Regard should be made to the need to remove water from the court, yard or passage and also to the need to dispose of it satisfactorily in the course of or after its removal.

Notices

The provision relating to enforcement of notices and appeals on pages 447–449 applies to this procedure.

Notices, which are to be served on the owner, must indicate the nature of the works required and state the time within which they must be executed. The time allowed must be reasonable and not less than the period for appeal, i.e. 21 days (sections 99 and 102).

FC118 Paving of yards and passages – Section 84 Building Act 1984

Courts, yards or passages of certain buildings (Building Act 1984, section 84(1))

↓

Unsatisfactory paving etc. or drainage (section 84(1))

↓

Local authority notice to owner (section 84(1))

↓

Appeal to magistrates' court within 21 days (section 102)

↓

Failure to execute works within time specified

↓

Penalty max. level 4 and £2 daily in addition to default (section 99)

AND/OR

Local authority may undertake works in default (section 99(2))

↓

Recovery of costs (sections 99(2) and 107)

Grounds of appeal

These are as in section 102 set out on pages 447–449.

OVERFLOWING AND LEAKING CESSPOOLS

Reference

Public Health Act 1936, section 50 and part 12.

Extent

This procedure applies to England and Wales only.

Scope

The section does not apply in relation to the effluent from a properly constructed tank for the reception and treatment of sewage provided that the effluent is not prejudicial to health or a nuisance. Otherwise it applies to any leaking or overflowing cesspool even though the situation may not be prejudicial to health or a nuisance (section 50(1)).

Person responsible

Notices are served by the local authority on the person by whose act, default or sufferance the soakage or overflow occurs or continues (section 50(1)).

Notices etc.

Part 12 of the Public Health Act 1936 applies to appeals against the enforcement of notices served under this procedure (pages 447–449).

Appeals against notices

The grounds of appeal against notices requiring works are set out on pages 447–449. Notices which require periodic emptying are not subject to the provisions of section 290 and no appeal is provided for. In defending any proceedings for penalty because of non-compliance with this form of notice, it is open to the defendant to question the reasonableness of the requirements (section 50(3)).

Definition

Cesspool includes a settlement tank or other tank for the reception or disposal of foul matter from buildings (section 90(1)).

FC119 Overflowing and leaking cesspools – Section 50 Public Health Act 1936

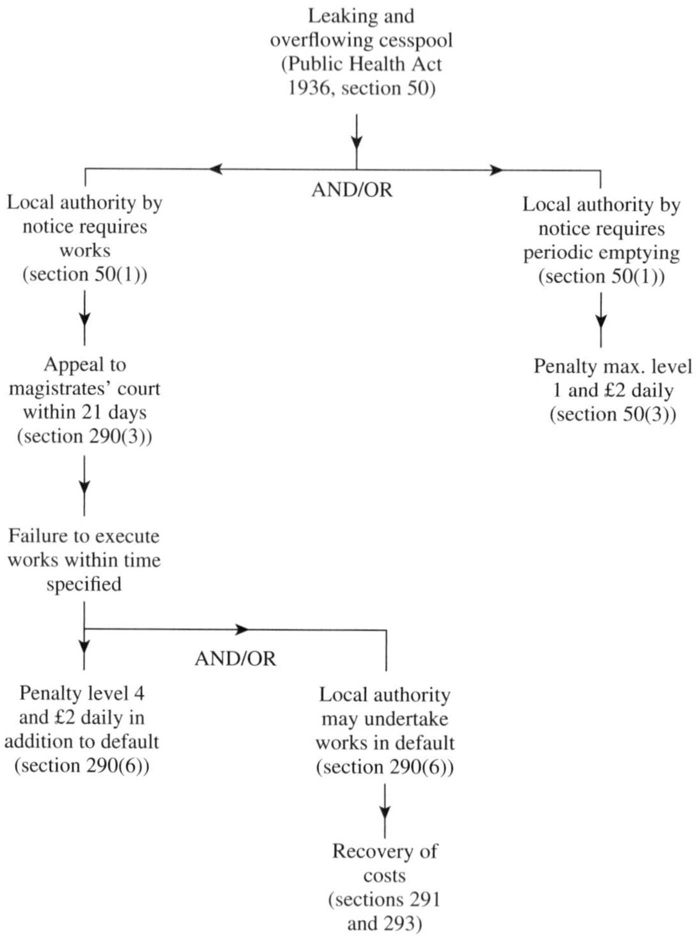

Notes

1. There is no provision for appeal against notices not requiring works under this section (e.g. periodic emptying), but it is open to the defendant to question the reasonableness of the notice in subsequent proceedings.
2. The local authority may at any time undertake any works which it has requested if asked to do so by the owner or occupier (Public Health Act 1936, section 275).

STATUTORY NUISANCES

References

Environmental Protection Act 1990, sections 79–82 and schedule 3 (as amended by the Noise and Statutory Nuisance Act 1993, the Environment Act 1995, the Pollution Prevention and Control Act 1999 and the Clean Neighbourhoods and Environment Act 2005).
The Statutory Nuisance (Appeals) Regulations 1995 (as amended 2006).
The Statutory Nuisance (Insects) Regulations 2006.
The Statutory Nuisance (Artificial Lighting) (Designation of Relevant Sports) (England) Order 2006.
DEFRA Guidance on the CNEA 2005 – Statutory Nuisances 2006, Noise 2006 and Fixed Penalty Notices 2006.

Extent

The 1990 Act applies in England and Wales and, with certain amendments, in Scotland. It does not apply in Northern Ireland.

Statutory nuisances

These are nuisances to which the abatement procedures of Part 3 of the Environmental Protection Act 1990 have been applied and are as follows:

1. Any premises in such a state as to be prejudicial to health or a nuisance (section 79(1)(a)). (For an expedited procedure to deal with the defective premises, see Building Act 1984, section 76, FC57.)
2. Smoke emitted from premises so as to be prejudicial to health or a nuisance; but this does not apply to:
   (a) premises occupied by the Crown for military or Ministry of Defence purposes;
   (b) smoke emitted from the chimney of a house within a smoke control area;
   (c) dark smoke from the chimney of a building or of a furnace attached to a building or installed on any land;
   (d) smoke from a railway locomotive steam engine;
   (e) dark smoke from any industrial or trade premises (section 79(1)(b), (2) and (3)).
3. Fumes or gases emitted from private dwellings so as to be prejudicial to health or a nuisance (section 79(1)(c) and (4)).
4. Any dust, steam (other than from a railway locomotive engine), smell or other effluvia arising on industrial, trade or business premises and being prejudicial to health or a nuisance (section 79(1)(d) and (5)).
5. Any accumulation or deposit which is prejudicial to health or a nuisance (section 79(1)(e)).
6. Any animal kept in such a place or manner as to be prejudicial to health or a nuisance (section 79(1)(f)).

7. Noise (except that from aircraft other than model aircraft) emitted from premises so as to be prejudicial to health or a nuisance; but this does not apply to Crown premises used for military or Ministry of Defence purposes (section 79(1)(ga), (2) and (6) as amended).
8. Noise that is prejudicial to health or a nuisance and is emitted from or caused by a vehicle, machinery or equipment in a street (other than noise made by traffic, by any military force or by political demonstration or a demonstration supporting or opposing a cause or campaign) (section 79(1)(ga) and (6A)).
9. Insects (other than those listed in schedule 45 of the Wildlife and Countryside Act 1981) emanating from relevant industrial, trade or business premises (certain types of farmland and land within sites of special scientific importance and certain waters are excluded – see the Statutory Nuisances (Insects) Regulations 2006) and being prejudicial to health or a nuisance (section 79 (1)(fa)).
10. Artificial light emitted from industrial or trade premises or from lights used only to illuminate outdoor relevant sports facilities so as to be prejudicial to health or a nuisance. Certain types of premises are excluded, e.g. those used for defence, some transport-related premises, airports and prisons (section 79 (1)(fb) and 80 (aza)).
11. Any other matter declared by any enactment to be a statutory nuisance (section 79(1)(h)) and these include:
    (a) any well, tank, cistern or water butt used for the supply of water for domestic purposes which is so placed, constructed or kept as to render the water therein liable to contamination prejudicial to health (Public Health Act 1936, section 141);
    (b) any pond, pool, ditch, gutter or watercourse which is so foul or in such a state as to be prejudicial to health or a nuisance (Public Health Act 1936, section 259(1)(a));
    (c) any part of a watercourse, not being a part ordinarily navigated by vessels employed in the carriage of goods by water, which is so choked or silted up as to obstruct or impede the proper flow of water and thereby to cause a nuisance or give rise to conditions prejudicial to health (Public Health Act 1936, section 259(1)(b));
    (d) a tent, van, shed or similar structure used for human habitation:
        (i) which is in such a state, or so overcrowded, as to be prejudicial to the health of the inmates; or
        (ii) the use of which, by reason of the absence of proper sanitary accommodation, or otherwise, gives rise, whether on the site or on other land, to a nuisance or to conditions prejudicial to health (Public Health Act 1936, section 268(2));
    (e) a shaft or outlet of certain abandoned or disused mines where:
        (i) it is not provided with a properly maintained device designed and constructed to prevent persons from accidentally falling down the shaft or accidentally entering the outlet; or
        (ii) by reason of its accessibility from a highway or a place of public resort, it constitutes a danger to the public (Mines and Quarries Act 1954, section 151);

(f) a quarry which is not provided with an efficient and properly maintained barrier so designed and constructed as to prevent persons from accidentally falling into it and which, by reason of its accessibility from a highway or place or public resort, constitutes a danger to the public (Mines and Quarries Act 1954, section 151).

In implementing the statutory nuisance procedures, local authorities may not have regard to the radioactive state of any substance, article or premises. These are matters to be dealt with under the Environmental Permitting Regulations 2010 as amended.

Unless the Secretary of State has granted consent, a local authority may not bring summary proceedings for causing a statutory nuisance where proceedings may be brought under the Integrated Pollution Prevention Control regime under the Pollution Prevention and Control Act 1999 – see FC32 (section 79(10)).

However, activities that are not covered by Integrated Pollution Prevention Control, even though they are on the same site as an Integrated Pollution Prevention Control installation, may be regulated by the statutory nuisance procedures. The restriction also does not affect the right of members of the public to use private proceedings under section 82 of the Environmental Protection Act 1990 (see pages 479–480) or prevent a local authority from taking proceedings in the High Court (see page 479).

Abatement notices

Where the local authority (definition on page 8) is satisfied that a statutory nuisance:

(a) exists; or
(b) is likely to occur; or
(c) is likely to recur,

it is required to serve an abatement notice:

(a) requiring the abatement of the nuisance or prohibiting or restricting its occurrence;
(b) requiring the execution of such works or steps as necessary for those purposes;
(c) specifying the time or times within which the notice is to be complied with (section 80(1)).

The notice must also indicate the rights for and times of appeal (schedule 3, paragraph 6). The period allowed for compliance must be reasonable ((d) under 'Appeals' below) but may nevertheless be short, e.g. to deal with noise from a party. Where the period allowed is less than the time allowed for appeal, the notice is suspended in certain circumstances ('Suspension of notices' below). For the service of notices, see pages 24–26, 56, 140.

The notice may be served by a local authority to deal with a nuisance which exists or has occurred from outside its area (section 81(2)).

FC120 Statutory nuisances: local authority action – Section 80 Environmental Protection Act 1990

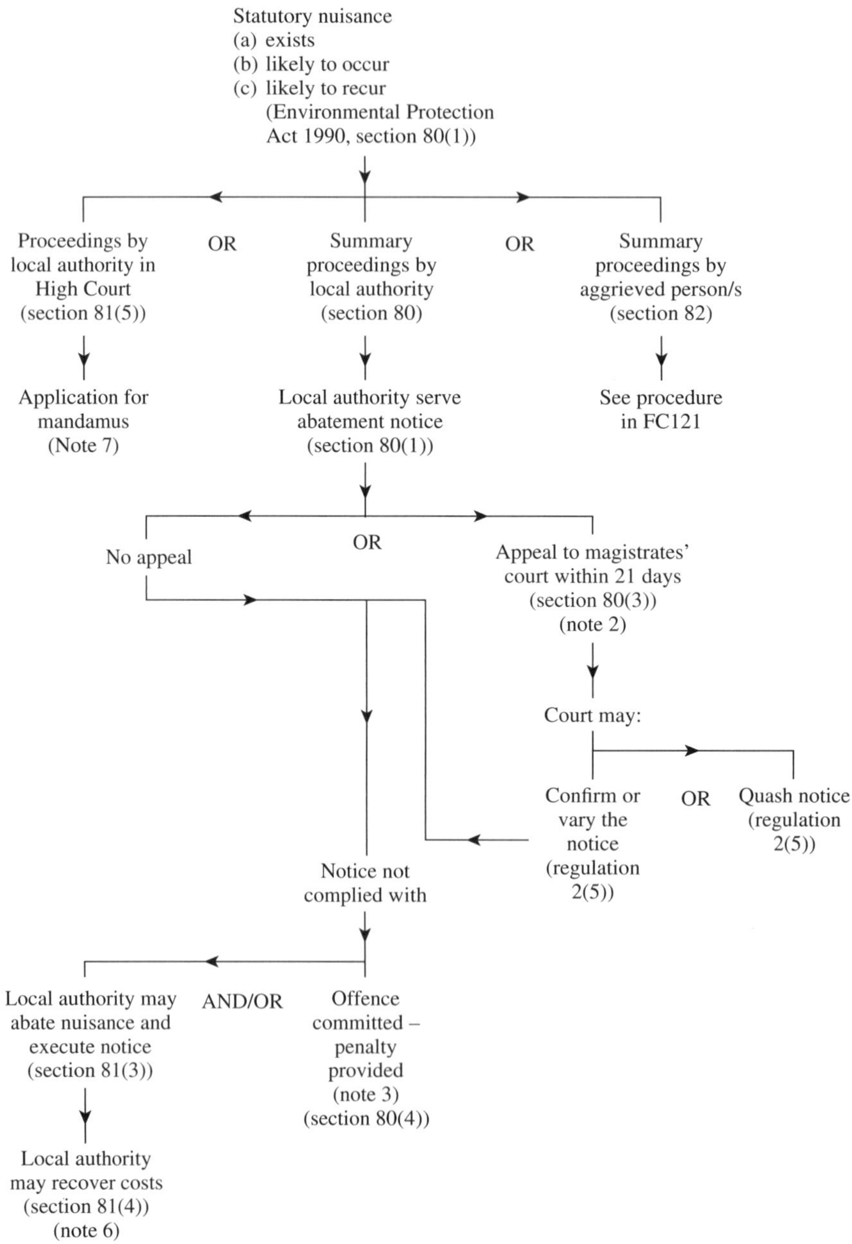

Notes

1. Regulation numbers refer to the Statutory Nuisance (Appeals) Regulations 1995.
2. For the suspension of some notices pending hearing of the appeal, see page 478.
3. Details of penalties are given on page 11.
4. For expedited procedure to deal with the defective premises, see FC57.
5. For service of notices under the Environmental Protection Act 1990, see page 24–26, 56, 140.
6. Or local authority may put a charge on the property (section 81A).
7. An order of mandamus is issued by a superior court to a lower court or public authority to do or refrain from doing some act which that body is obliged to do under the law.

FC121 Statutory nuisances: action by aggrieved persons – Section 82 Environmental Protection Act 1990

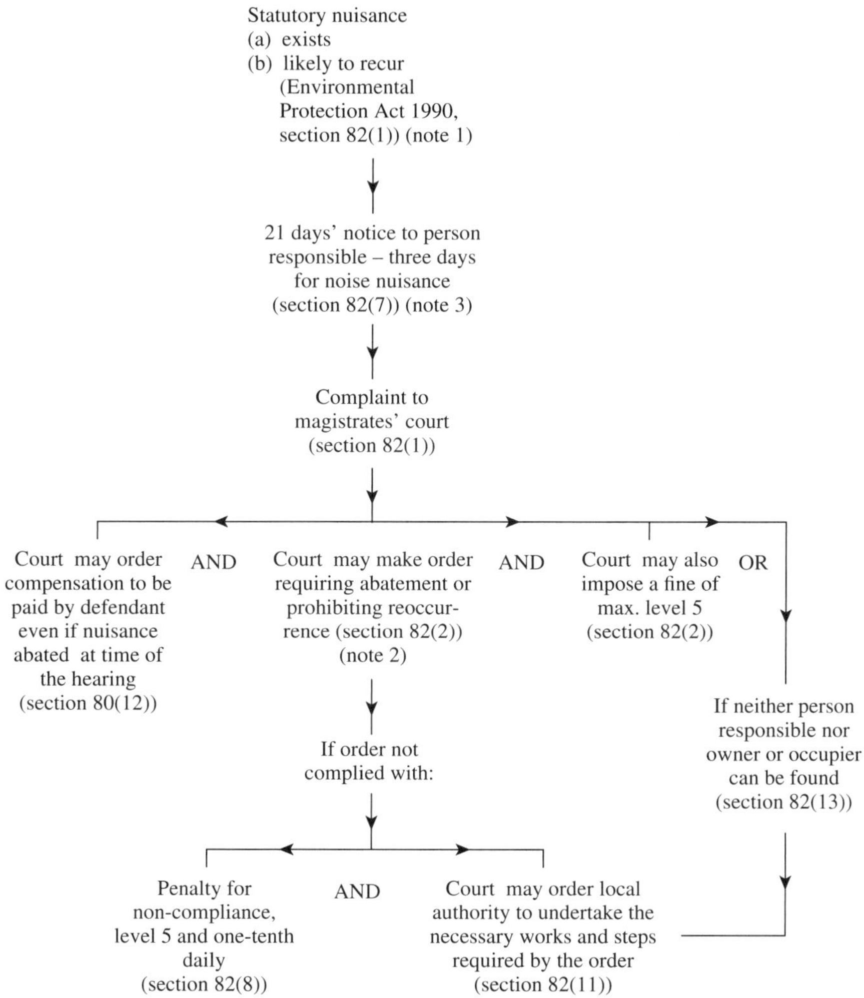

Notes

1. Use of this procedure is not possible to take action to prohibit an anticipated nuisance which does not exist or has not already occurred.
2. The court may also order the closure of a premises rendered unfit for human habitation by the nuisance until they are rendered fit (section 82(3)).
3. This includes noise nuisances for vehicles, machinery or plant in streets.

Deferral of duty to serve abatement notice

In dealing with a statutory nuisance caused by noise, the local authority may defer the service of the abatement notice for seven days while it takes other appropriate action to persuade the person who would receive the notice to abate the nuisance. After the seven days, or before if the local authority judges it right, it may proceed by way of the usual abatement notice (section 80(2A)).

Persons responsible

The abatement notice is to be served:

(a) on the person/s responsible for the nuisance except as in (b) and (c) below;
(b) on the owner where the nuisance arises from any defect of a structural character;
(c) where the person responsible cannot be found or the nuisance has not yet occurred, on the owner or occupier of the premises (section 80(2)).

Where more than one person is responsible, the notice must be served on each (section 81(1)).

For noise nuisances from vehicles, machinery or plant in streets, the notice is to be served:

(a) on the person responsible for that vehicle etc.; or
(b) where that person cannot be found, by fixing it to the vehicle etc. (section 80A(2)).

Appeals

Any person served with an abatement notice may appeal to a magistrates' court within 21 days of the date of service (section 80(3), schedule 3, paragraphs 1–4) on one or more of the following grounds:

(a) that the abatement notice is not justified by section 80 of the 1990 Act;
(b) that there has been some material informality, defect or error in, or in connection with, the abatement notice or in, or in connection with, any copy of the abatement notice served under section 80A(3) (which relates to notices in respect of vehicles, machinery or equipment);
(c) that the authority has refused unreasonably to accept compliance with alternative requirements, or that the requirements of the abatement notice are otherwise unreasonable in character or extent, or are unnecessary;
(d) that the time, or, where more than one time is specified, any of the times, within which the requirements of the abatement notice are to be complied with is not reasonably sufficient for the purpose;
(e) where the nuisance to which the notice relates:
    (i) is a nuisance falling within (1) premises, (4) dust etc., (5) accumulations etc., (6) animals, (7) noise, (8) noise in streets, (9) insects or (10) artificial lighting in 'Statutory nuisances' above, and arises on industrial, trade or business premises or, in the case of a vehicle,

equipment or machinery, is being used for industrial, trade or business purposes; or

(ii) is a nuisance falling within (2) smoke etc. of the same section above and the smoke is emitted from a chimney;

that the best practicable means were used to prevent, or to counteract the effects of, the nuisance;

(f) that, in the case of a nuisance under (7) noise or (8) noise in streets, the requirements imposed by the abatement notice are more onerous than the requirements for the time being in force, in relation to the noise to which the notice relates, of:
  (i) any notice served under section 60 or 66 of the Control of Pollution Act 1974; or
  (ii) any consent given under section 61 or 65 of the Control of Pollution Act 1974; or
  (iii) any determination made under section 67 of the Control of Pollution Act 1974; or
  (iv) in relation to noise in streets, of any condition of consent given under paragraph 1 of schedule 2 of the Environmental Protection Act 1990;

(g) that the abatement notice should have been served on some person instead of the appellant, being:
  (i) the person responsible for the nuisance; or
  (ii) the person responsible for the vehicle, machinery or equipment; or
  (iii) in the case of a nuisance arising from any defect of a structural character, the owner of the premises; or
  (iv) in the case where the person responsible for the nuisance cannot be found or the nuisance has not yet occurred, the owner or occupier of the premises;

(h) that the abatement notice might lawfully have been served on some person instead of the appellant being:
  (i) in the case where the appellant is the owner of the premises, the occupier of the premises; or
  (ii) in the case where the appellant is the occupier of the premises, the owner of the premises;

  and that it would have been equitable for it to have been so served (copy of the appeal notice to be served on the person/s to be implicated);

(i) that the abatement notice might lawfully have been served on some person in addition to the appellant, being:
  (i) a person responsible for the nuisance; or
  (ii) a person who is also an owner of the premises; or
  (iii) a person who is also an occupier of the premises; or
  (iv) a person who is also the person responsible for the vehicle, machinery or equipment (copy of appeal notice to be served on the person/s to be implicated);

  and that it would have been equitable for it to have been so served (copy of the appeal notice to be served on the person/s to be implicated) (regulation 2 of the Statutory Nuisance (Appeals) Regulations 1995).

Any party may appeal to the Crown Court against any decision of the magistrates' court (schedule 3, paragraph 1(3)). The court may either:

(a) quash the notice; or
(b) vary it; or
(c) dismiss the appeal and may make orders about the responsibility of persons for costs of workers.

Suspension of notices

Where the nuisance:

(a) is injurious to health; or
(b) is likely to be of limited duration; or
(c) any expenditure incurred in compliance with the notice would not be disproportionate to the public benefit expected; and
(d) the notice has specified the existence of these circumstances,

the abatement notice is *not* suspended pending the hearing of any appeal. In any other case where:

(a) expenditure is involved in attaining compliance; or
(b) in the case of noise nuisance, the noise is caused in the course of the performance of a duty imposed by the local authority,

the notice is suspended pending the determination of the appeal (regulation 3).

Offences

Upon conviction for non-compliance with the requirements of an abatement notice, a person is liable on summary conviction to a fine not exceeding level 5 and one-tenth of that level daily for a continuation. Where the offence is committed on industrial, trade or business premises, the maximum fine is £20,000 (section 80(5) and (6)).

Defences

It will be a defence to prove that the best practicable means have been used to prevent or counteract the effects of the nuisance except:

(a) where the nuisance arises on *other than* industrial, trade or business premises *and* involves nuisance categories (1) premises, (4) dust etc., (5) accumulations etc., (6) animals and insects, (7) noise or (8) noise in streets;
(b) in the case of a category (2) smoke nuisance, except where it arises from smoke emitted from a chimney; and
(c) in relation to category (3) fumes etc. and (8) statutory nuisances declared by Acts other than the Environmental Protection Act 1990 (sections 80(7) and (8)).

In relation to noise nuisances only, including noise in streets, it will be a defence to prove that:

(a) the situation was covered by a notice under section 60 or a consent under sections 61 or 65 of the Control of Pollution Act 1974 relating to construction sites; or
(b) where a section 66 Control of Pollution Act 1974 noise reduction notice was in force, the level of noise was below that specified in the notice; or
(c) although a section 66 notice was not in force, there was a section 67 notice (Control of Pollution Act 1974) relating to new buildings liable to an abatement order and the noise was less than the specified level (section 80(9)).

Defaults

When an abatement notice has not been complied with, the local authority may, in addition to prosecuting for non-compliance, do whatever is necessary in the terms of the notice to abate or prevent the nuisance (section 81(3)). This may include the confiscation of sound amplification equipment (FC29). The special procedures for dealing with noise from vehicles, machinery and plant in streets are also dealt with in the Noise section (pages 116–119).

The costs of abating or preventing the nuisance are recoverable from the person/s by whose act or default the nuisance was caused (section 80(4)). Where the owner of any premises is the person responsible for the nuisance, the local authority may, with prior notification to that person and subject to appeal, recover costs in executing notices by a charge on the property (section 81A). A local authority may also recover its costs by instalments (section 81B).

High Court proceedings

If a local authority considers that the taking of summary proceedings (service of abatement notice etc.) would afford an inadequate remedy in relation to the abatement, prohibition or restriction of any statutory nuisance, it may take proceedings in the High Court or, in Scotland, in any court of competent jurisdiction (section 81(5)).

In such cases respecting noise nuisance, it will be a defence to show that the noise was authorised by either a notice under section 60 or a consent under section 61 of the Control of Pollution Act 1974 dealing with construction sites (see FC23). Otherwise the defences identified above relating to summary proceedings are not applicable to the High Court procedure (section 81(6)).

Summary proceedings by persons aggrieved (FC121)

It is possible for an aggrieved person or persons to proceed directly to abate or prevent the recurrence of a statutory nuisance by making a complaint to the magistrates' court. For example, this would include the tenant of a house, local authority or privately owned, seeking a remedy to disrepair. At least 21 days' notice shall be given to the person responsible for the nuisance (three days for noise nuisances, including noise from vehicles, machinery or equipment in streets).

If the court is satisfied, it may make an order for either or both the abatement or prohibition of the nuisance by the carrying out of specified works within the time specified. The court may also fine the defendant up to a maximum of level 5. The defences available parallel those where the local authority serves the abatement notice, as identified above.

In the event of non-compliance with the order, the court may impose further fines of up to level 5 and one-tenth of that level daily for its continuation. The court may also, after giving the local authority an opportunity of being heard, order the local authority to undertake whatever works or steps are necessary to abate or prohibit the nuisance.

At the discretion of the court and upon the making of an order, compensation may be payable by the defendant to the aggrieved persons making the complaint (section 82).

Powers of entry

Specific powers relating to power of entry to deal with statutory nuisances are provided in schedule 3 to the Environmental Protection Act 1990.

Authorised officers of the local authority, upon producing if required their authority, may enter premises at any reasonable time to see whether or not a statutory nuisance exists or to execute works. Unless in an emergency, 24 hours' notice of entry is required to the occupier of residential premises. Where admission is refused or apprehended, where premises are unoccupied, in an emergency or where application for entry would defeat the object, application may be made to the magistrates' court for a warrant.

The maximum penalty for obstruction is level 3 (schedule 3, paragraphs 2 and 3).

Definitions

Best practicable means is to be interpreted by reference to the following provisions:

(a) practicable means reasonably practicable, having regard among other things to local conditions and circumstances, to the current state of technical knowledge and to the financial implications;
(b) the means to be employed include the design, installation, maintenance and manner and periods of operation of plant and machinery, and the design, construction and maintenance of buildings and structures;
(c) the test is to apply only so far as it is compatible with any duty imposed by local authority;
(d) the test is to apply only so far as it is compatible with safety and safe working conditions, and with the exigencies of any emergency or unforeseeable circumstances;

and, in circumstances where a Code of Practice under section 71 of the

Control of Pollution Act 1974 (noise minimisation) is applicable, regard shall also be had to guidance given in it (section 79(7), (8) and (9)).

Chimney includes structures and openings of any kind from or through which smoke may be emitted.
Dust does not include dust emitted from a chimney as an ingredient of smoke.
Equipment includes a musical instrument.
Fumes means any airborne solid matter smaller than dust.
Gas includes vapour and moisture precipitated from vapour.
Industrial, trade or business premises means premises used for any industrial, trade or business purposes or premises not so used on which matter is burnt in connection with any industrial, trade or business process, and premises are used for industrial purposes where they are used for the purpose of any treatment or process as well as where they are used for the purposes of manufacturing.
Noise includes vibration.
Person responsible:

(a) in relation to a statutory nuisance, means the person to whose act, default or sufferance the nuisance is attributable;
(b) in relation to a vehicle, includes a person in whose name the vehicle is for the time being registered under the Vehicles (Excise) Act 1994 and any other person who is for the time being the driver of the vehicle;
(c) in relation to machinery or equipment, includes any person who is for the time being the operator of the machinery or equipment.

Prejudicial to health means injurious, or likely to cause injury, to health.
Premises includes land and any vessel (other than one powered by steam reciprocating machinery).
Private dwelling means any building, or part of a building, used or intended to be used as a dwelling.
Smoke includes soot, ash, grit and gritty particles emitted in smoke.
Street means a highway and any other road, footway, square or court that is for the time being open to the public (section 79(8)).

WASTE ON LAND

Definitions: general

Environment consists of all, or any, of the following media, namely land, water and the air.
Pollution of the environment means pollution of the environment due to the release or escape (into any environmental medium) from:

(a) the land on which controlled waste is treated;
(b) the land on which controlled waste is kept;
(c) the land in or on which controlled waste is deposited;
(d) fixed plant by means of which controlled waste is treated, kept or disposed of,

of substances or articles constituting or resulting from the waste, and capable (by reason of the quantity or concentrations involved) of causing harm to man or any other living organisms supported by the environment.

This applies in relation to mobile plant by means of which controlled waste is treated or disposed of as it applies to plant on land by means of which controlled waste is treated or disposed of.

For the purposes of the above, **harm** means harm to the health of living organisms or other interference with the ecological systems of which they form part, and in the case of humans includes offence to any of his senses or harm to his property; and **harmless** has a corresponding meaning.

Disposal of waste includes its disposal by way of deposit in or on land and, subject to the paragraph below, waste is '**treated**' when it is subjected to any process, including making it re-usable or reclaiming substances from it, and **recycle** (and cognate expressions) shall be construed accordingly.

Land includes land covered by waters where the land is above the low water mark of ordinary spring tides and references to land on which controlled waste is treated, kept or deposited are references to the surface of the land (including any structure set into the surface).

Substance means any natural or artificial substance, whether in solid or liquid form or in the form of a gas or vapour (section 29).

The definition of waste

Waste means anything that is waste within the meaning of Article 3(1) of Directive 2008/98/EC of the European Parliament and of the council on waste. That is, 'waste' means any substance or object which the holder discards or intends or is required to discard.

Controlled waste means household, industrial and commercial waste or any such waste.

Household waste means waste from:

(a) domestic property, that is to say, a building or self-contained part of a building which is used wholly for the purposes of living accommodation;
(b) a caravan (as defined in section 29(1) of the Caravan Sites Control and Development Act 1960) which usually and for the time being is situated on a caravan site (within the meaning of that Act);
(c) a residential home;
(d) premises forming part of a university or school or other educational establishment;
(e) premises forming part of a hospital or nursing home.

Industrial waste means waste from any of the following premises:

(a) any factory (within the meaning of the Factories Act 1961);
(b) any premises used for the purposes of, or in connection with, the provision to the public of transport services by land, water or air;
(c) any premises used for the purposes of, or in connection with, the supply to the public of gas, water or electricity or the provision of sewerage services; or

(d) any premises used for the purposes of, or in connection with, the provision to the public of postal or telecommunications services.
(e) any mine or quarry or any premises used for agriculture.

Commercial waste means waste from premises used wholly or mainly for the purposes of a trade or business or the purposes of sport, recreation or entertainment excluding:

(a) household waste;
(b) industrial waste;
(c) waste of any other description prescribed by regulations made by the Secretary of State for the purposes of this paragraph (section 75).

The Controlled Waste Regulations 1992 (as amended), the Special Waste Regulations 1996 and DoE Circular 14/92 clarify the definitions of household, commercial and industrial waste in relation to the application of the duty of care, the obligations of collection authorities and the charging regime.

A detailed explanation of the definition of waste is given in Annex 2 of DoE Circular 11/94.

Local authorities

The waste collection authorities are:

(a) for any district in England not within Greater London, the council of the district;
(b) in Greater London, the following:
  (i) for any London borough, the council of the borough;
  (ii) for the City of London, the Common Council;
  (iii) for the Temples, the Sub-Treasurer of the Inner Temple and the Under-Treasurer of the Middle Temple, respectively;
(c) for any county or county borough in Wales, the council of the county or county borough;
(d) in Scotland, a council constituted under the Local Government (Scotland) Act 1994 (section 30(3)).

Waste regulation authority:

(a) in relation to England and Wales, is a reference to the Environment Agency; and
(b) in relation to Scotland, is a reference to the Scottish Environment Protection Agency (section 30(1)).

LITTER

Definitions

Principal litter authority (PLA) (also see page 499)

In England and Wales, the following are PLAs:

(a) a county council;
(b) a county borough council;
(c) a district council;
(d) a London borough council;
(e) the Common Council of the City of London; and
(f) the Council of the Isles of Scilly.

This definition includes the unitary authorities in England and Wales. The Secretary of State may, by order, designate other descriptions of local authorities as litter authorities for the purposes of this part; and any such authority shall also be a PLA (section 86(2)).

In Scotland, the PLAs are councils constituted under the Local Government etc. (Scotland) Act 1994, i.e. unitary councils, and joint boards.

Litter

Litter is defined as including:

(a) the discarded ends of cigarettes, cigars and like products; and
(b) discarded chewing-gum and the discarded remains of other products designed for chewing (section 98(5A)).

The Litter (Animal Droppings) Order 1991 includes as refuse for the whole of Part 4 of the Environmental Protection Act 1990 dog faeces (for other controls over the fouling of land by dogs, see FC101) on land of the following description which is not heath or woodland used for the grazing of animals:

(a) any public walk or pleasure ground;
(b) any land, whether enclosed or not, on which there are no buildings or of which no more than one-twentieth part is covered with buildings, and the whole or the remainder of which is laid out as a garden or is used for the purposes of recreation;
(c) any part of the seashore (that is to say, every cliff, bank, barrier, dune, beach, flat or other land adjacent to and above the place to which the tide flows at mean high water springs) which is:
    (i) frequently used by large numbers of people; and
    (ii) managed by the person having direct control of it as a tourist resort or recreational facility;
(d) any esplanade or promenade which is above the place to which the tide flows at mean high water springs;
(e) any land not forming part of the highway, or, in Scotland, a public road,

which is open to the air, which the public are permitted to use on foot only, and which provides access to retail premises;

(f) a trunk road picnic area provided by the minister under section 112 of the Highways Act 1980 or, in Scotland, by the Secretary of State under section 55 of the Roads (Scotland) Act 1984;

(g) a picnic site provided by a local planning authority under section 10(2) of the Countryside Act 1968 or, in Scotland, a picnic place provided by an islands or district council or a general or district planning authority under section 2(2)(a)(i) of the Local Government (Development and Finance) (Scotland) Act 1964;

(h) land (whether above or below ground and whether or not consisting of or including buildings) forming or used in connection with off-street parking places provided in accordance with section 32 of the Road Traffic Regulation Act 1984.

Relevant land

The duties placed on various public bodies, statutory undertakers and others to keep their land clear, as far as practicable, from litter and refuse relate to the 'relevant land' of those organisations and individuals. Relevant land is defined in section 86 as:

> (4) Subject to subsection (8) below, land is 'relevant land' of a Principal Litter Authority if, not being relevant land falling within subsection (7) below, it is open to the air and is land (but not a highway or in Scotland a public road) which is under the direct control of such an authority to which the public are entitled or permitted to have access with or without payment *(see note at (8) below).*
>
> (5) Land is Crown Land if it is land:
>
> (a) occupied by the Crown Estate Commissioners as part of the Crown Estate;
>
> (b) occupied by or for the purposes of a government department or for naval, military or air force purposes; or
>
> (c) occupied or managed by anybody acting on behalf of the Crown; and is 'relevant Crown land' if it is Crown land which is open to the air and is land (but not a highway or in Scotland a public road) to which the public are entitled or permitted to have access with or without payment; and 'the appropriate Crown authority' for any Crown land is the Crown Estate Commissioners, the minister in charge of the government department or the body which occupies or manages the land on the Crown's behalf, as the case may be *(see note at (8) below).*
>
> (6) Subject to subsection (8) below, land is 'relevant land' of a designated statutory undertaker if it is the land which is under the direct control of any statutory undertaker of any description which may be designated by the Secretary of State, by order, for the purposes of this Part, being land to which the public are entitled or permitted to have access

with or without payment or, in such cases as may be prescribed in the designation order, land in relation to which the public have no such right or permission. *(See the Litter (Statutory Undertakers) (Designation and Relevant Land) Order 1991 (as amended 1992).)*

(7) Subject to subsection (8) below, land is 'relevant land' of a designated educational institution if it is open to the air and is land which is under the direct control of the governing body of or, in Scotland, of such body or of the educational authority responsible for the management of, any educational institution or educational institution of any description which may be designated by the Secretary of State, by order, for the purposes of this Part. (*The Litter (Designated Educational Institutions) Order 1991.)*

(8) The Secretary of State may, by order, designate descriptions of land which are not to be treated as relevant Crown land or as relevant land of Principal Litter Authorities, of designated statutory undertakers or of designated educational institutions or of any description of any of them. *NB. The Litter (Relevant Land of Principal Litter Authorities and Relevant Crown Land) Order 1991 excludes from the interpretation of relevant land as respects both PLAs and the Crown, land below the high water spring tide level.*

(9) Every highway maintainable at the public expense other than a trunk road which is a special road is a 'relevant highway' and the local authority which is, for the purposes of this Part, 'responsible' for so much of it as lies within its area is, subject to any order under subsection (11) below *(The Highway Litter Clearance and Cleaning (Transfer of Duties) Order 1991)*:

(a) in Greater London, the Council of the London borough or the Common Council of the City of London;
(b) in England, outside Greater London, the Council of the district;
(c) in Wales, the Council of the County or County borough; and
(d) the Council of the Isles of Scilly.

(10) In Scotland, every public road other than a trunk road which is a special road is a 'relevant road' and the local authority which is, for the purposes of this Part, 'responsible' for so much of it as lies within its area is, subject to any order under subsection (11) below, the council constituted under section 2 of the Local Government etc. (Scotland) Act 1994.

(11) The Secretary of State may, by order, as respects relevant highways or relevant roads, relevant highways or relevant roads of any class or any part of a relevant highway or relevant road specified in the order, transfer the responsibility for the discharge of the duties imposed by section 89 below from the local authority to the highway or roads authority; but he shall not make an order under this subsection unless:

(a) (except where he is the highway or roads authority) he is requested to do so by the highway or roads authority;
(b) he consults the local authority; and

(c) it appears to him to be necessary or expedient to do so in order to prevent or minimise interference with the passage or with the safety of traffic along the highway or, in Scotland, road in question; and where, by an order under this subsection, responsibility for the discharge of those duties is transferred, the authority to which the transfer is made is, for the purposes of this Part, 'responsible' for highway, road or part specified in the order. *(See various Highway Litter Clearance and Cleaning (Transfer of Duties) Orders.)*

(12) A place on land shall be treated as 'open to the air' notwithstanding that it is covered if it is open to the air on at least one side.

(13) The Secretary of State may, by order, apply the provisions of this Part which apply to refuse to any description of animal droppings in all or any prescribed circumstances subject to such modifications as appear to him to be necessary. *(See the Litter (Animal Droppings) Order 1991.)*

(14) Any power under this section may be exercised differently as respect different areas, different descriptions of land or for different circumstances.

(section 86)

Statutory undertaker

Statutory undertaker means:

(a) any person authorised by any enactment to carry on any railway, light railway, tramway or road transport undertaking;
(b) any person authorised by any enactment to carry on any canal, inland navigation, dock, harbour or pier undertaking; or
(c) any relevant airport operator (within the meaning of Part V of the Airports Act 1986) (section 98(6)).

LITTER ABATEMENT ORDERS

References

Environmental Protection Act 1990, sections 89 and 91.
Code of Practice on Litter and Refuse DETR under section 89 of Environmental Protection Act, April 2006.
DEFRA Guidance on the Clean Neighbourhoods and Environment Act 2005 – Litter and Refuse 2007.

Extent

This provision applies in England, Wales and Scotland but not Northern Ireland.

Duty to keep land clear of litter

The following are placed under a duty to keep their relevant land (definition on page 485) clear of litter and refuse, as far as is practicable:

(a) local authorities as respects highways;
(b) the Secretary of State as respects trunk roads or other highways for which he is responsible;
(c) principal litter authorities (PLAs) as respects their relevant land;
(d) the Crown as respects its relevant land;
(e) designated statutory undertakers as respects their relevant land;
(f) governing bodies as designated educational institutions as respects their relevant land (section 89(1)).

In respect of (a) and (b), each local authority and the Secretary of State is also required to keep highways and roads clean, as far as is practicable (section 89(2)).

Litter standards

Regard must be had to the character and use of the land in question as well as to the measures which are practicable in the circumstances (section 89(3)).

The Secretary of State has prescribed standards to which the responsible bodies must have regard in discharging their duties in a Code of Practice made under section 89(7). These establish what are reasonable and generally acceptable levels of cleanliness to be met and are admissible in court in any proceedings for a litter abatement order (section 91(11)). See Code of Practice on Litter and Refuse DETR under section 89 of Environ mental Protection Act, April 2006 and DEFRA Guidance on the CNEA 2005 – Litter and Refuse 2007.

Litter abatement order

Any person aggrieved (other than a PLA) by the defacement by litter or refuse of any of the land set out under 'Duty to keep land clear of litter' (above) or, where appropriate, aggrieved by a highway not being clear may complain to a magistrates' court (section 91(1) and (2)).

The person must give the responsible person or body five days' written notice of his intention to make the complaint (section 91(5)).

If the court is satisfied that the land or highway is defaced by litter and refuse or the highway is not clear and that the person has not complied with his statutory duty, taking due account of the Code of Practice, it may make a litter abatement order (section 91(6)).

Effect of a litter abatement order

The litter abatement order requires the litter or refuse to be cleared away or cleaning to be undertaken within a specified period, failing which the person or body to whom the order is addressed is guilty of an offence (section 91(9)).

FC122 Litter abatement orders (LAOs) – Sections 89 and 91 Environmental Protection Act 1990

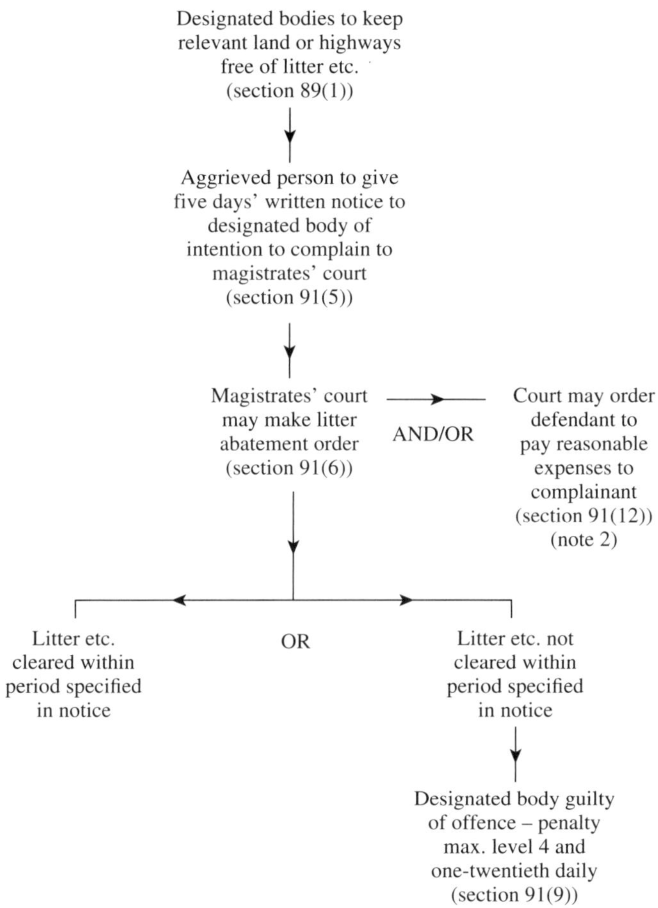

Notes

1. This procedure allows the public to bring to the attention of a court the fact that a responsible body is not complying with its responsibilities under section 89. For powers of a principal litter authority in relation to other bodies, see FC123.
2. Even if a litter abatement order is not made, the court may require the complainant to be compensated if, although the litter etc. had been cleared away, it had been present at the time the complaint was made and there were reasonable grounds for bringing the complaint (section 91(12)).

Payment of expenses

Whether or not the court makes a litter abatement order, it may order the defendant authority to pay expenses to the complainant where it feels that there were reasonable grounds for bringing the complaint and that the defacement or uncleanliness existed at the time the complaint was made (section 91(12)).

LITTER ABATEMENT NOTICES

References

Environmental Protection Act 1990, sections 89 and 92, as amended by the Anti-Social Behaviour Act 2003 and the Clean Neighbourhoods and Environment Act 2005 (CNEA 2005).
Code of Practice on Litter and Refuse issued April 2006 by the DETR under section 89 of the Environmental Protection Act.
DEFRA Guidance on the CNEA 2005 – Litter and Refuse 2007.

Extent

This provision applies in England, Wales and Scotland but not Northern Ireland.

Duty to keep land clear of litter

Designated bodies are placed under a duty to keep their land or highways free of litter or certain highways clean, as far as is reasonably practicable and subject to guidance in the Code of Practice (section 89).

These duties are outlined more fully under 'Litter abatement orders' on page 488.

Relationship of litter abatement orders and litter abatement notices

Litter abatement orders are made by a magistrates' court against the designated body (including PLAs) upon complaint by an aggrieved person other than a PLA. Litter abatement notices are served by PLAs (other than an English county or joint boards) against any other designated bodies.

Designated bodies other than PLAs are therefore subject to action for non-compliance with their duty to keep land etc. free of litter and clean, both from certain PLAs and from the public.

Scope

Where a PLA (other than an English county or joint board) is satisfied that any of the following land is defaced by litter or refuse (definition on page 484), or that such defacement is likely to recur to the extent that the statutory duty under section 89(1) is breached, it may serve a litter abatement notice.

FC123 Litter abatement notices – Section 89 Environmental Protection Act 1990

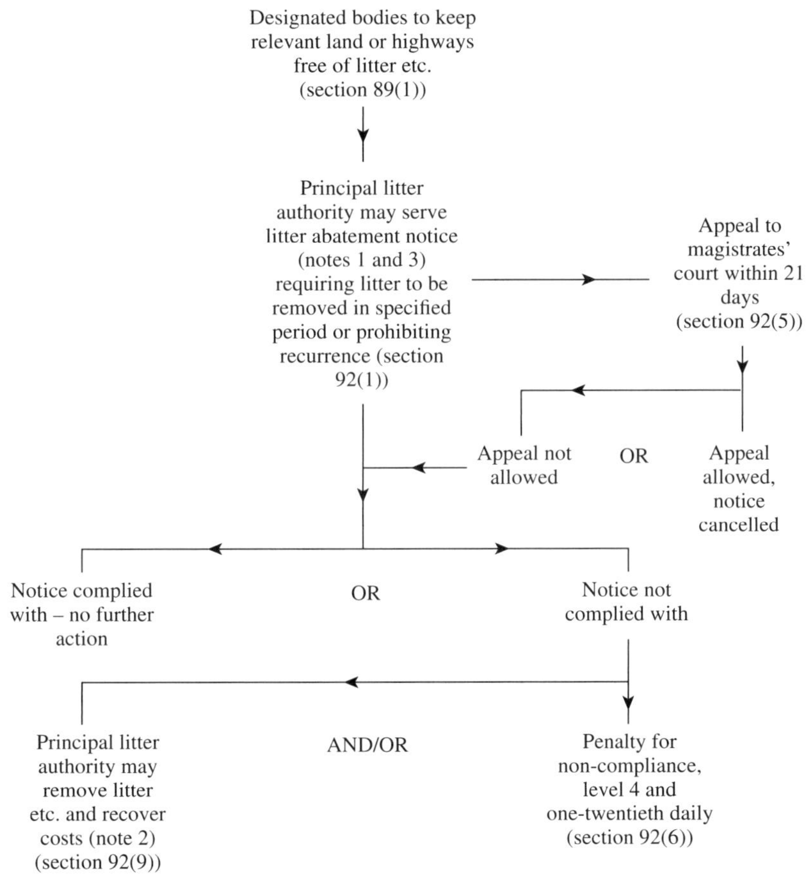

Notes

1. The power to serve litter abatement notices is given to principal litter authorities other than English county councils and joint boards (section 92(1)).
2. The principal litter authorities' powers of default may not be exercised in respect of certain Crown land or specified land of statutory undertakers (section 92(10)).
3. For the service of notices under the Environmental Protection Act, see pages 24–26, 56, 140.

The land is any relevant land (definition on page 485) of:

(a) the Crown;
(b) a designated statutory undertaker;
(c) a designated educational institution (section 92(1)).

Litter abatement notices

The notice may either require the litter etc. to be cleared within a specified period or prohibit the land from becoming defaced again (section 92(2)). Failure to comply brings a maximum penalty of level 4 and one-twentieth daily, and the PLA has powers of default, except in the case of Crown land and that of statutory undertakers, and cost recovery (section 92(6) and (9)).

Notices are served on:

(a) the Crown authority;
(b) the designated statutory undertaker;
(c) the education authority (section 92(3)).

Appeals

Appeal is to the magistrates' court within 21 days of the date of service, and compliance with the duty under section 89(1) constitutes a statutory defence against the litter abatement notice (section 92(4) and (5)).

LITTER CLEARING NOTICES (LCNs)

References

Environmental Protection Act 1990, section 92A.
The Environmental Offences (Fixed Penalties) (Miscellaneous Provisions) Regulations 2007.
The Environmental Offences (Use of Fixed Penalty Receipts) Regulations 2006.
DEFRA Guidance on the Clean Neighbourhoods and Environment Act 2005 (CNEA 2005) – Litter and Refuse 2006 and Fixed Penalty Notices 2006.

Extent

This provision applies to England, Scotland and Wales.

Scope

This procedure allows a PLA (see pages 484, 499) to serve a litter clearing notice (LCN) where there is a need to clear litter and refuse from any land in the open air and they are of the view that it is causing defacement so as to be detrimental to the amenity of the area (section 92A(1)–(2)).

FC124 Litter clearing notices (LCNs) – Section 92A Environmental Protection Act 1990

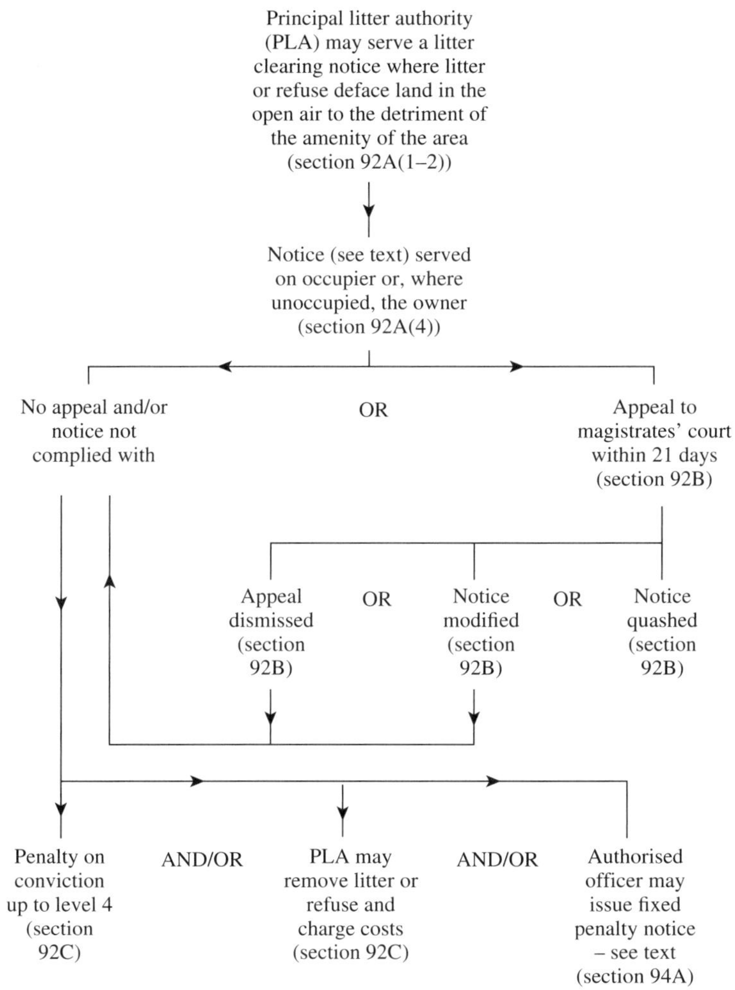

Land of the following descriptions is exempted from this provision:

(a) a highway maintainable at public expense;
(b) land under the direct control of a principal litter authority;
(c) Crown land;
(d) relevant land of a statutory undertaker;
(e) relevant land of an educational institution;
(f) land which is covered but open on at least one side and to which the public are not entitled or permitted to have access, with or without payment (section 92A(11)).

Litter clearing notices

The notice is to be served on the occupier of the land or, where there is no occupier, on the owner (section 92A(4)). It will require the person:

(i) to clear the litter or refuse; and
(ii) if the LA is satisfied that the land is likely to become defaced again, to take reasonable steps to prevent that happening (section 92A(3)).

Under (i) above, the notice may specify the period in which it must be complied with, not less than 28 days from service, and indicate the standard of compliance required (section 92A(5)–(6)). In forming the requirements, the PLA must have regard to the DEFRA Guidance on the CNEA 2006 (section 92A(7)).

Where the local authority is unable to establish the occupancy or ownership, the notice may be served by posting it on the land (section 92A(9)).

Appeals

There is a right of appeal against the notice to the magistrates' court within 21 days of service. The grounds are:

(a) there is a material defect or error in the notice or in connection with it;
(b) the notice should have been served on another person;
(c) the land is not defaced so as to be detrimental to the amenity of the locality;
(d) the action required is unfair or unduly onerous.

The court may quash or modify the notice or dismiss the appeal (section 92B).

Enforcement

If the notice is not complied with, the person, on conviction, is subject to a fine not exceeding level 4 and the local authority is empowered to enter the land and clear the litter and refuse and recover a reasonable charge (section 92C). Alternatively, an authorised officer of the PLA may serve a fixed-penalty notice that, if paid within 14 days, will settle the matter. The level of penalty is at the discretion of the PLA between £75 and £110, with discounts for early payment (section 94A).

STREET LITTER CONTROL NOTICES

References

Environmental Protection Act 1990, sections 93, 94 and 95, as amended by the Clean Neighbourhoods and Environment Act 2005 (CNEΛ 2005).
The Street Litter Control Notices Order 1991 (amended 1997).
The Environmental Offences (Fixed Penalties) (Miscellaneous Provisions) Regulations 2007.
The Environmental Offences (Use of Fixed Penalty Receipts) Regulations 2006.
DEFRA Guidance on the CNEA 2005 – Litter and Refuse 2006 and Fixed Penalty Notices 2006.

Extent

This procedure is available in England, Wales and Scotland only.

Scope

Principal litter authorities (other than an English county or joint board) may, in order to prevent accumulations of litter or refuse in and around any street or land in the open air adjacent to any street or around vehicles, stalls or other moveable structures, serve street litter control notices (SLCNs).

The circumstances which must occur before a notice can be issued are:

(a) the premises are one of a description which has been prescribed by the minister (see below); and
(b) the premises must have a street frontage; and either
(c) there is recurrent defacement by litter or refuse of the street or open land in the vicinity; or
(d) the condition of open land in the vicinity of the premises is detrimental to the amenities of the neighbourhood; or
(e) the conditions at the premises produce quantities of litter or refuse which cause defacement of the street or open land in the vicinity of the premises (section 93(1) and (2)).

Through the Street Litter Control Notices Order 1991 made under section 94(1) of the Environmental Protection Act 1990, the minister has designated the following descriptions of land and premises which may be the subject of SLCNs.

Prescribed commercial and retail premises

A SLCN may be issued in respect of commercial and retail premises of the following descriptions:

(a) premises used wholly or partly for the sale of food or drink for consumption off the premises;

FC125 Street litter control notices – Sections 93 to 95 Environmental Protection Act 1990

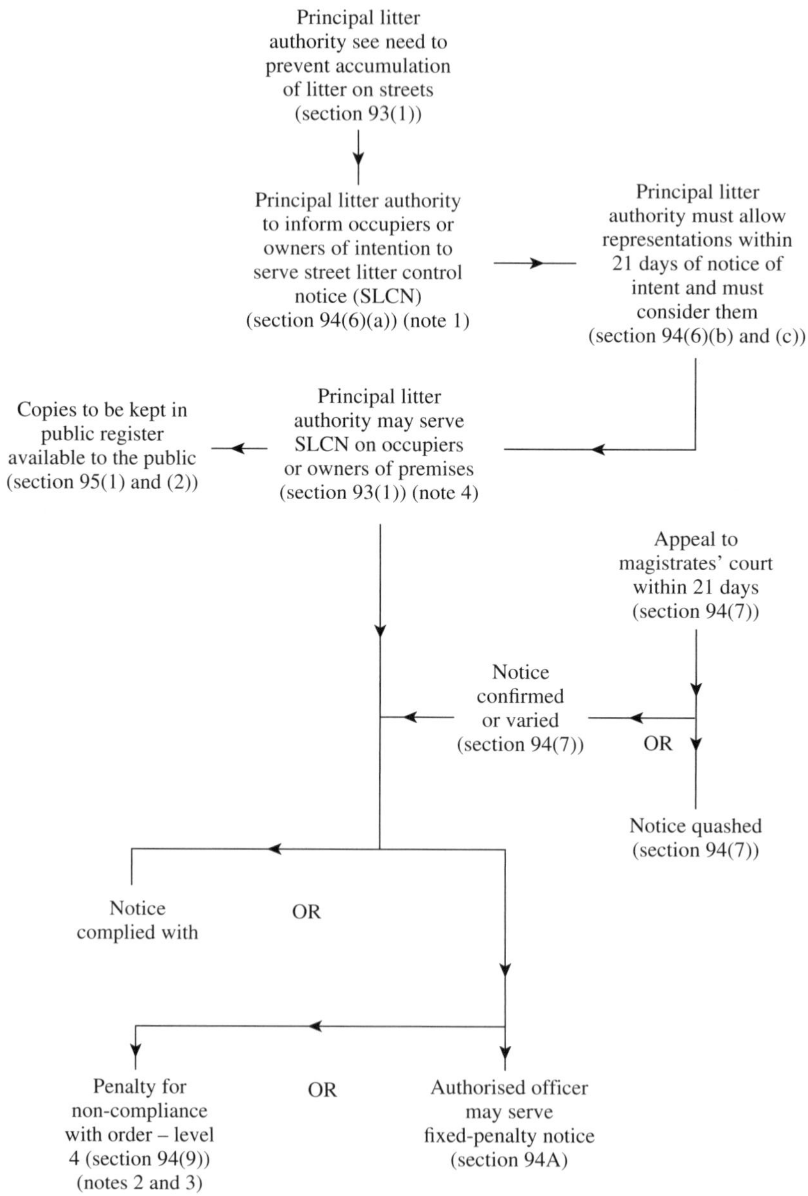

Notes

1. This notification does not need to be in writing but it is suggested that this should be the case.
2. No daily penalty is specified.
3. There is no provision for action in default and recharging.
4. For the service of notices under the Environmental Protection Act 1990, see pages 24–26, 56, 140.

(b) premises used wholly or partly for the sale of food or drink for consumption on premises;
(c) service stations and other premises on which fuel for motor vehicles is sold to the public;
(d) premises used wholly or partly as a cinema, theatre, concert hall, bingo hall, casino, dance hall, swimming bath, skating rink, gymnasium or area for other indoor or outdoor sports or recreations, or as an amusement arcade or centre;
(e) banks, building society offices or other premises with automated teller machines located on an outside wall of the premises;
(f) premises in respect of which there is for the time being in force a betting office licence granted under schedule 1 to the Betting, Gaming and Lotteries Act 1963;
(g) premises used wholly or partly for the sale of tickets or changes in any lottery; or
(h) premises used wholly or partly for the sale of goods of any description which are displayed on open land adjacent to the street, or on the street (article 2 of Street Litter Control Notices Order 1991).

Prescribed descriptions of land

Land which may be included in an area of open land specified in a SLCN is land which is part of the premises in respect of which the notice is issued and, subject to below, land of the following descriptions:

(a) land which is part of a street, other than a carriageway when it is open to vehicular traffic;
(b) relevant land of a principal litter authority; and
(c) land under the direct control of any other local authority.

The land described above may be specified:

(a) in a street litter control notice issued in respect of premises described in category (e) above if the land is within ten metres of those premises;
(b) in a street litter control notice issued in respect of any other premises, if the land is within 100 metres of the premises (article 3 of Street Litter Control Notices Order 1991).

Consultation

Before serving the SLCN, the PLA must notify those on whom notices are to be served and consider any representation made by them within 21 days of that notification (section 94(6)).

Responsible person

Notices are to be served on the occupier of the premises or, where the premises are unoccupied, on the owner (section 93(2)).

Content of notices

The notice *must*:

(a) specify the premises;
(b) state the grounds upon which the notice is issued;
(c) specify all areas of open land adjoining or in the vicinity of the frontage;
(d) specify in relation to that area the PLA's reasonable requirements (section 93(3)).

The requirements in (d) must relate to the clearing of litter and refuse from the specified area of open land and may include:

(a) the provision and emptying of litter receptacles;
(b) the carrying out of actions within a specified period;
(c) regular actions at such times and intervals as specified (section 94(4)).

In forming its requirements, the PLA must have regard to its own duties under Part 4 of the Environmental Protection Act 1990, including those under section 85 in respect of its own land and highways (section 94(5)).

Appeals

Recipients of a SLCN may appeal to the magistrates' court who may confirm, vary or quash the notice (section 94(7)).

Enforcement

Failure to comply with the order is an offence punishable by a maximum penalty of level 4 (section 94(9)). Alternatively, the PLA may issue a fixed-penalty notice of between £75 and £110.

Public register

The PLA must keep a register including copies of all SLCNs (and associated court orders). The register must be available to the public free of charge at all reasonable times and the PLA must provide copies of entries on payment of a reasonable charge (section 95).

Definition

Street means a relevant highway, a relevant road or any other highway or road over which there is a right of way on foot (section 93(4)).

LITTER OFFENCES

References

Environmental Protection Act 1990, sections 87 and 88, as amended by the Clean Neighbourhoods and Environment Act 2005 (CNEA 2005).

The Environmental Offences (Fixed Penalties) (Miscellaneous Provisions) Regulations 2007.
The Environmental Offences (Use of Fixed Penalty Receipts) Regulations 2006.
DEFRA Guidance on the CNEA 2005 – Litter and Refuse 2006 and Fixed Penalty Notices 2006.

Litter offence

Any person who throws down, drops or otherwise deposits litter (definition on page 484) and leaves anything which causes the defacement of a place by litter is guilty of an offence (section 87(1)).

This applies to land anywhere in the open air regardless of ownership, except in a covered place open to the air on at least one side and to which the public does not have access. It also applies to dropping litter into water such as rivers or lakes and on beaches above the low water mark. The offence is not committed if the littering was authorised by law or was done by or with the consent of the owner, occupier or any other authority or person having control of the area. Such consent may only be given in relation to a watercourse, lake or pond if the same person owns all the surrounding land (section 87).

Fixed-penalty schemes

As an alternative to proceeding summarily for the littering offence, an authorised officer of a litter authority may serve a fixed-penalty notice on the offender (section 88(1)), which requires the payment of the fine to the local authority within 14 days (section 88(2)). The level of the fixed penalty may be set by the local authority at between £50 and £80, with discounts for early payment and, if paid within 14 days, no further action can be taken. Where an authorised officer proposes to issue a fixed-penalty notice, he may require the person to give his name and address and it is an offence not to do so or to give false information (section 88).

Definitions

Authorised officer means an officer of a litter authority or someone not an employee of the local authority who is authorised in writing by the authority for the purpose of issuing notices under this section (section 88(10)).
Litter authority for the purpose of this procedure is:

(a) any principal litter authority, other than an English county council or a joint board;
(b) any English county council or joint board designated by the Secretary of State, by order (no such orders have been made), in relation to such area as is specified in the order (not being an area in a National Park);
(c) the Broads Authority (section 88(9)).

FC126 Litter offences – Sections 87 and 88 Environmental Protection Act 1990

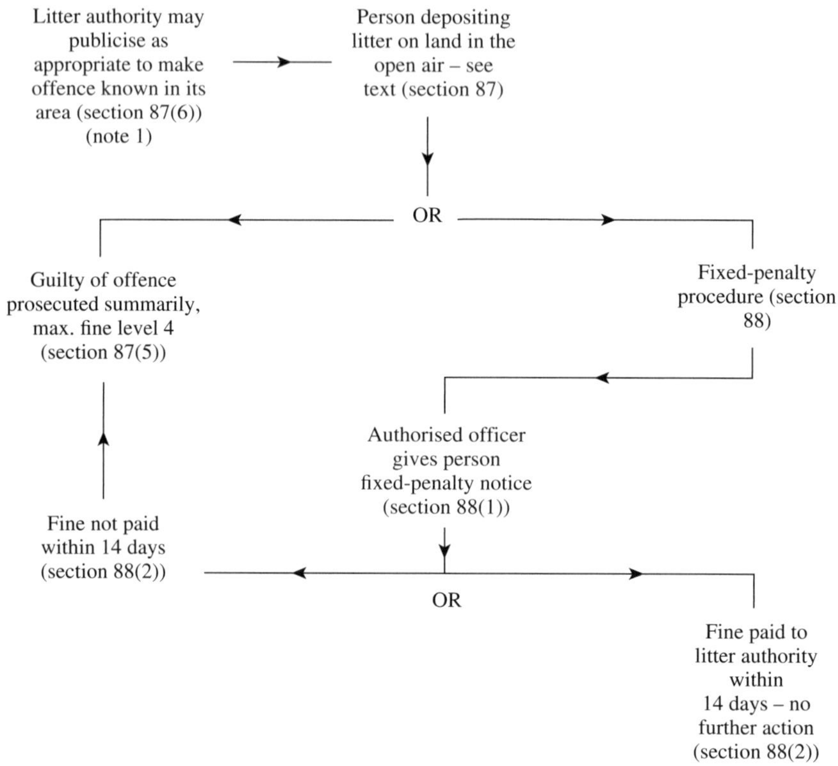

Notes

1. Publicity is not a prerequisite to action under sections 87 or 88.
2. For controls over the fouling of land by dogs, see FC101.

REMOVAL OF NOXIOUS MATTER

Reference

Public Health Act 1936, section 79.
Note: Although provision was made in the Control of Pollution Act 1974, schedule 4 for section 79 of the Public Health Act 1936 to be repealed, this has not been implemented and this procedure remains available.

Extent

This provision applies to England and Wales.

Scope

The procedure may be used where the proper officer of the local authority considers that any accumulation of noxious matter ought to be removed. The word 'noxious' is not defined, but according to the Concise Oxford Dictionary means 'harmful, unwholesome' (section 79(1)).

Notice

The notice is served by the proper officer on either the owner or the occupier of the premises concerned, requiring removal of the accumulation within 24 hours (section 79(1)). There is no provision for appeal or for penalties.

Default

If the accumulation is not removed within 24 hours from the service of the notice, the proper officer may have it removed (section 79(1)).

Recovery of costs

The local authority can recover from either the owner or the occupier the expenses of any reasonable actions taken by its proper officer in acting in default (section 79(2)).

Definition

Proper officer means the specified officer in the Act (Local Government Act 1972, schedule 29, part 1).

FC127 Removal of noxious matter – Section 79 Public Health Act 1936

Noxious matter on any premises (Public Health Act 1936, section 79(1))

↓

Notice from proper officer to owner or occupier requiring removal within 24 hours (section 79(1))

↓

If notice not complied with

↓

Proper officer may remove accumulation (section 79(1))

↓

Local authority may recover expenses from owner or occupier (section 79(2))

Notes

1. For accumulations constituting a statutory nuisance, see FC120.
2. For removal of rubbish resulting from demolition, see FC62.
3. For removal of abandoned articles, see FC129.
4. For removal of abandoned cars, see FC132A and 132B.

ACCUMULATIONS OF RUBBISH

Reference

Public Health Act 1961, section 34.

Scope

The procedure deals with any rubbish which is in the open air and which is seriously detrimental to the amenities of the neighbourhood (section 34(1)).

Notices

Notices must be in writing (Public Health Act 1936, section 283), specify the steps which the local authority proposes to take and indicate the provisions for counter-notice and appeals.

The section makes provision for the service of a counter-notice from the person initially served with the notice. This should state that the person will do the work required to comply with the notice. The local authority can take no action until 28 days from the service of the notice to allow the potential service of a counter-notice or an appeal to the magistrates' court (section 34(2)).

Appeals

Appeal is to the magistrates' court within 28 days of service on the grounds that the local authority was not justified in taking action under this section, or that the steps which it proposes to take are unreasonable (section 34(2)(b)).

There is no power for the local authority to recover costs.

Definition

Rubbish means rubble, waste paper, crockery and metal, and any other kind of refuse (including organic matter), but does not include any material accumulated for, or in the course of, any business.

REMOVAL AND DISPOSAL OF ABANDONED REFUSE

References

Refuse Disposal (Amenity) Act 1978, section 6.
Removal of Refuse Regulations 1967.

Extent

This procedure applies in England, Wales and Scotland but not in Northern Ireland (section 13(5)).

FC128 Accumulations of rubbish – Section 34 Public Health Act 1961

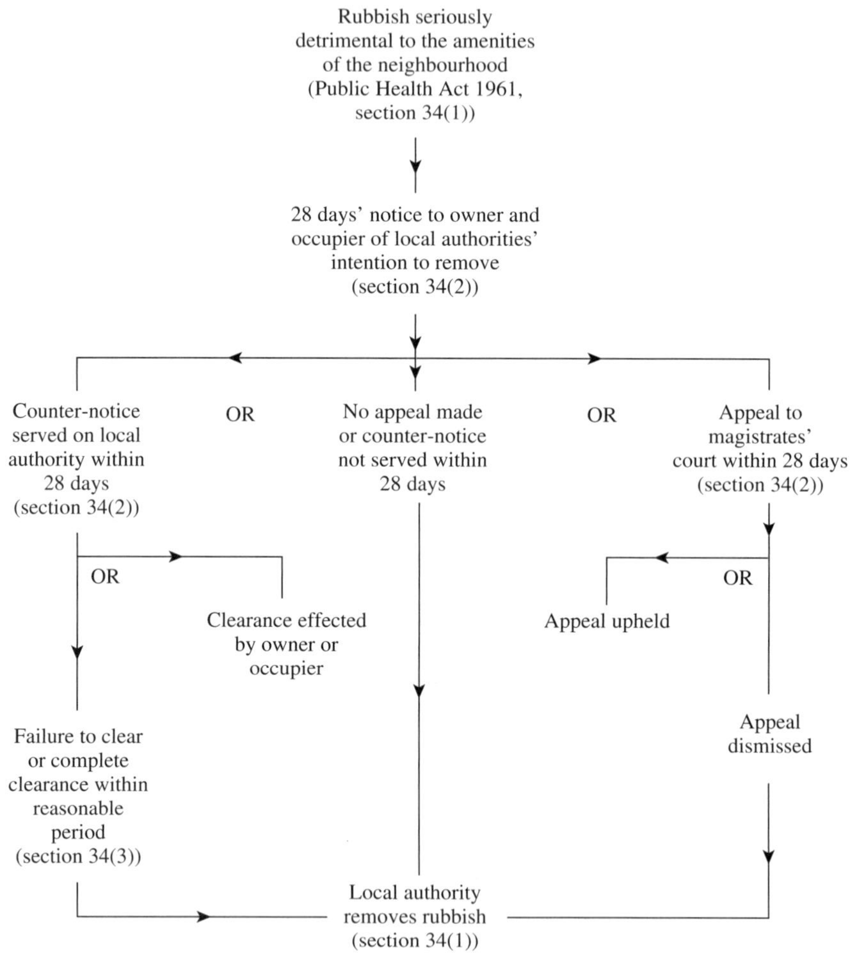

Notes

1. Where rubbish constitutes a statutory nuisance, see FC120.
2. For proper officer's power relating to noxious matter, see Public Health Act 1936, section 79, FC127.
3. For local authority power relating to removal of manure, see Public Health Act 1936, section 80.
4. For removal of rubbish resulting from demolition, see FC62.
5. There is no power for the local authority to recover its costs.
6. The Town and Country Planning Act 1990, section 215, allows a local authority to deal with a site which adversely affects the neighbourhood.

Scope

These provisions deal with anything (other than motor vehicles) abandoned without lawful authority on any land in the open air or on any other land forming part of a highway (section 6(1)).

Unlike the provisions relating to abandoned cars etc., there is no duty placed here on the local authority to act; the powers are discretionary.

Unoccupied land

Where the land in question appears to be unoccupied, the local authority may immediately remove the articles without notice (section 6(1)).

Occupied land

In this situation, the local authority is first required to serve notice on the occupier stating its intention to remove the abandoned articles. If the occupier objects in writing to the proposal within 15 days of service of the notice, the local authority cannot proceed further unless other legislative procedures are available for the particular circumstances (notes to FC129). If no objection within the 15-day period (see Removal of Refuse Regulations 1967, regulation 4) is received, the local authority may remove the articles (section 6(2)).

Notice

The form of the notice is specified in the Removal of Refuse Regulations 1967, which, although made under the Civic Amenities Act 1967 by a provision now repealed, has been continued in force by section 12(3) of the Refuse Disposal (Amenity) Act 1978. With appropriate rewording, this form should be used.

Recovery of costs

The local authority is entitled to recover its costs of removal from either:

(a) the person who left the articles on the land; or
(b) any person convicted under section 2(1) of the 1978 Act for illegally abandoning the articles (section 6(4)).

Sums are recoverable as a simple contract debt (section 5(2)) and courts making a conviction under section 2(1) may, on application by the local authority, order the person convicted to reimburse to the local authority their costs of removal (section 6(6)).

Penalties

This procedure itself carries no penalties but prosecution for abandoning the articles may be taken under section 2(1)(b) of the 1978 Act, with penalties not exceeding level 4 on the standard scale or imprisonment for up to three months, or both.

FC129 Removal and disposal of abandoned refuse – Section 6 Refuse Disposal (Amenity) Act 1978

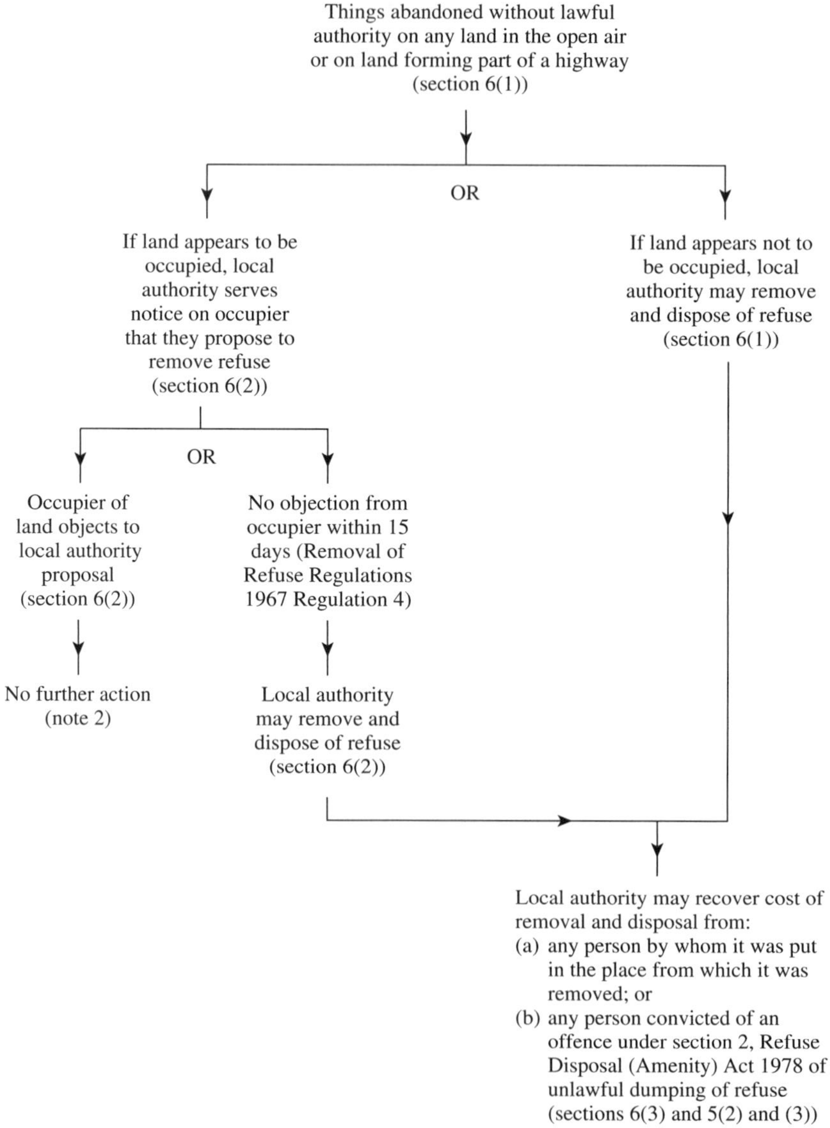

Notes

1. This procedure does not apply to abandoned cars – see FC132.
2. The provisions do not provide for any action beyond this point but see following notes for alternative procedures.
3. For accumulations constituting a statutory nuisance, see FC120.
4. For removal of rubbish resulting from demolition, see FC62.
5. For removal of rubbish seriously detrimental to the amenities of a neighbourhood, see FC128.
6. This procedure applies in Scotland but not in Northern Ireland.

REMOVAL OF CONTROLLED WASTE ON LAND

References

Environmental Protection Act 1990, section 59, as amended by the Anti-Social Behaviour Act 2003 and the Clean Neighbourhoods and Environment Act 2005(CNEA 2005).
DEFRA Guidance on the CNEA 2005 – Waste 2006.

Extent

This provision applies to England, Wales and Scotland.

Scope

These provisions apply whenever controlled waste (see 'The definition of waste' on pages 482–483) has been deposited on land without the authority of a waste management licence under section 35, and may be operated by either a waste collection authority (WCA) or the Environment Agency as the waste regulation authority (WRA) (section 59(1)). Section 59A enables the Secretary of State to issue directions as to which categories of waste should be given priority under this procedure.

Removal by notice

The WCA or WRA may serve written notice on the occupier of the land requiring:

(a) removal of the waste within a specified period of not less than 21 days; and/or
(b) the taking of specified steps to eliminate or reduce the consequences of the deposit (section 59(1)).

Where:

(a) there is no occupier; or
(b) the occupier cannot be found; or
(c) the occupier has successfully appealed against a notice served on him,

the notice may be served on the owner (section 59ZA).

Where there is non-compliance, the WCA or WRA may remove the waste and/or prosecute. Costs of acting in default may be recovered (section 59(5), (6) and (7)).

Removal etc. by a WCA or WRA

This may be effected, without recourse to the notice procedure, where:

(a) there is no occupier of the land affected; or

FC130 Removal of controlled waste on land – Section 59 Environmental Protection Act 1990

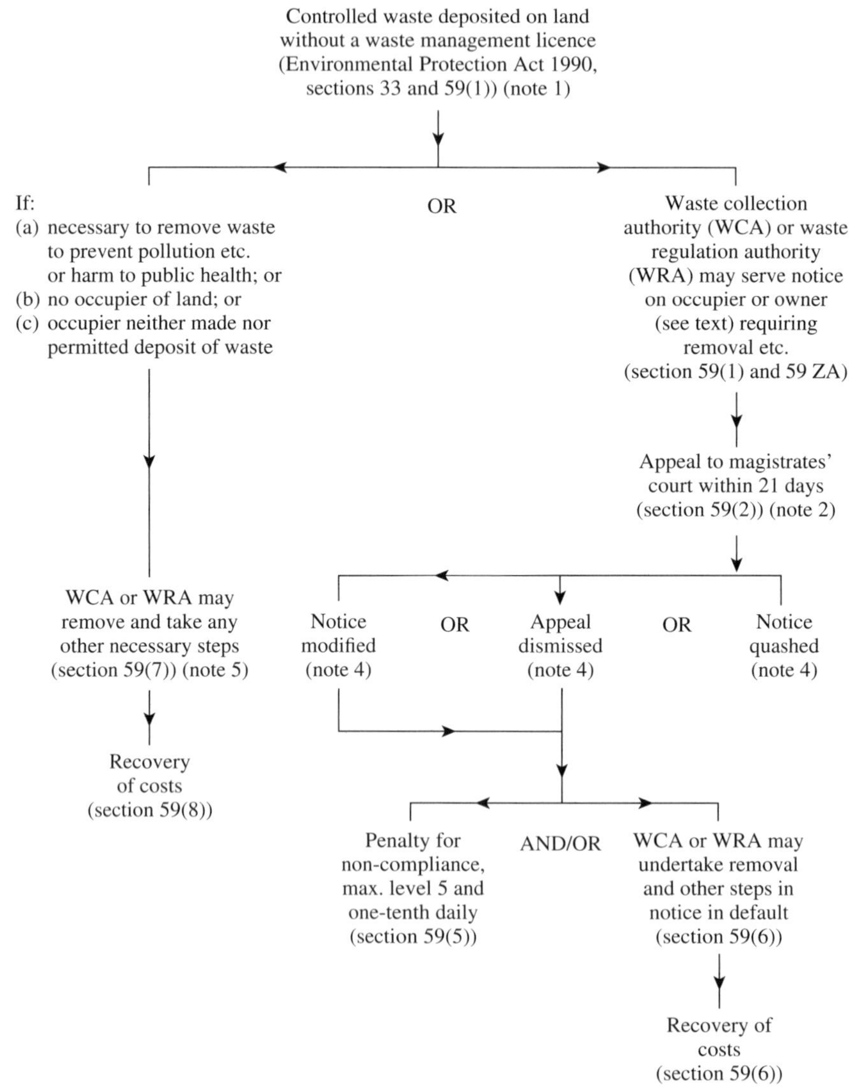

Notes

1. For definition of controlled waste, see page 482.
2. The notice is suspended pending determination of the appeal (section 59(4)).
3. For procedure for removing abandoned vehicles, see FC132A and FC132B.
4. Either party may appeal to the Crown Court against the decision of the magistrates' court (section 73(1)).
5. Once removed, the waste belongs to the WCA or WRA and may be dealt with accordingly (section 59(9)).
6. For service of notices, see page 24–26, 56, 140.

(b) the occupier neither made nor knowingly permitted the deposit; or
(c) removal, and/or steps to eliminate or reduce the consequences of the deposit is necessary forthwith to remove or prevent pollution of land, water or air or harm to human health – see definitions, pages 484, 512, 517 (section 59(7)).

Recovery of costs both of removing the waste and of taking steps to eliminate or reduce the consequences of the deposit is possible from any person who caused, or knowingly permitted, the deposit (section 59(8)).

Appeals

Appeal is to the magistrates' court within 21 days of service and no particular grounds of appeal are specified. The court is required to quash the notice if either:

(a) the appellant neither deposited nor knowingly caused or permitted the deposit; or
(b) there is a material defect in the notice (section 59(3)).

The notice is suspended pending determination of the appeal (section 59(4)).

ABANDONED TROLLEYS

References

Environmental Protection Act 1990, section 99 and schedule 4, as amended by the Clean Neighbourhoods and Environment Act 2005.
DEFRA Guidance on the Clean Neighbourhoods and Environment Act 2005 – Abandoned Trolleys 2006.

Extent

This section applies to England, Scotland and Wales but not Northern Ireland.

Scope

This procedure is discretionary and the powers may only be used after the scheme has been adopted by resolution of the local authority (section 99(1)).

Consultation

Before passing a resolution adopting the scheme, the local authority must consult with persons likely to be affected or their representatives, e.g. supermarket operators (section 99(3)).

There is also a duty on the local authority once a scheme has been adopted to consult the same people from time to time on the operation of the scheme (section 99(4)).

FC131 Abandoned trolleys – Section 99 Environmental Protection Act 1990

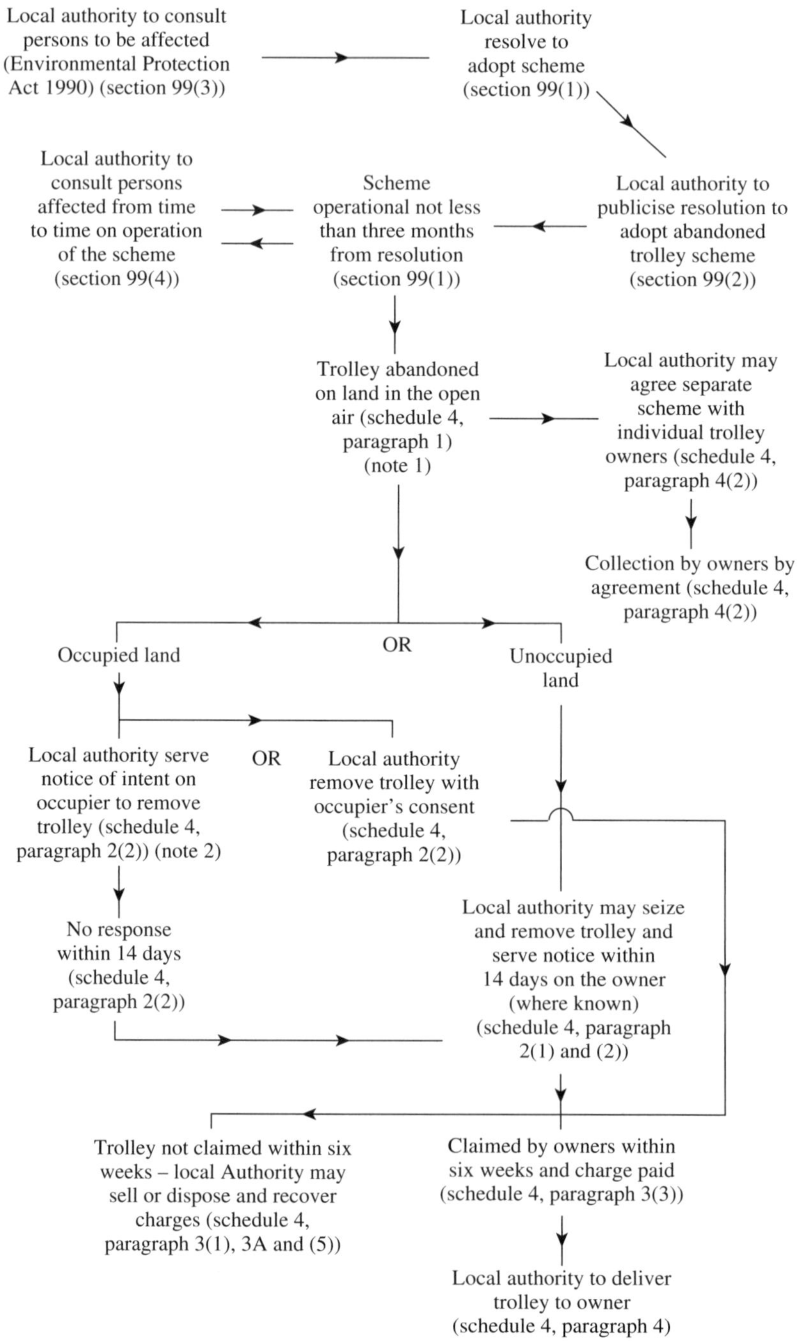

Notes

1. For exceptions, see text.
2. For service of notices under Environmental Protection Act 1990, see pages 24–26, 56, 140.

Publicity

Following the resolution of adoption, the scheme must be publicised in at least one local newspaper (section 99(2)).

Voluntary schemes

The local authority is able to reach separate agreements with each owner of trolleys whereby the operators agree to arrange collection themselves outside of the statutory scheme. In such cases, if abandoned trolleys are collected by the local authority, for any trolleys removed by them within the times specified in the statutory scheme, no charge is payable (schedule 4, paragraph 4(2)).

Abandoned trolleys

Trolleys which may be dealt with by the local authority under a scheme are those found to be abandoned in the open air except:

(a) on land in which the trolley owner has a legal estate;
(b) in a trolley-park in an off-street parking place;
(c) in any other trolley park designated by the local authority; or
(d) on land used for the purposes of the undertaking by the owner of the luggage trolley (schedule 4, paragraph 1).

Removal

On unoccupied land, the local authority may remove the trolley immediately. However, on occupied land a notice of intention must be served on the occupier, who has 14 days to respond. In the absence of a response within that period, the local authority may remove the trolley (schedule 4, paragraph 2(2)).

Retention, return and disposal

Once seized and removed, the local authority must serve notice within 14 days on the owner notifying him of the removal of the trolley, the place where it is being kept and notification that, if not claimed within six weeks, it will dispose of it (schedule 4, paragraph 3(2)).

If claimed within the six weeks, the trolley is delivered back to its owner by the local authority and a charge made (schedule 4, paragraph 3(3) and (4)).

If not claimed within the six-week period and the local authority has made reasonable enquiries to ascertain ownership, the local authority may sell or otherwise dispose of the trolley (schedule 4, paragraph 3(1)(b) and (5)).

Charges

Charges to be levied by the local authority should be such as to recover the costs of removing, storing and disposing of trolleys under the scheme (schedule 4, paragraph 3A and 4(2)).

Definitions

The scheme applies to both shopping and luggage trolleys defined as:

Luggage trolley means a trolley provided by a person carrying on an undertaking of any railway, light railway, tramway, road transport undertaker or airport operator to travellers for use by them for carrying their luggage to, from or within the premises used for the purpose of the undertaking, not being a trolley which is power-assisted.
Shopping trolley means a trolley provided by the owner of a shop to customers for use by them for carrying goods purchased at the shop, not being a trolley which is power-assisted (schedule 4, paragraph 5).

ABANDONED VEHICLES

References

Refuse Disposal (Amenity) Act 1978, sections 3 to 5, as amended by the Clean Neighbourhoods and Environment Act 2005.
Removal and Disposal of Vehicles Regulations 1986 (as amended).
Road Traffic Regulations 1984.
Removal, Storage and Disposal of Vehicles (Prescribed Sums and Charges) Regulations 2008.
Guidance to LAs on the Clean Neighbourhoods and Environment Act 2005 – Abandoned Vehicles, DEFRA, 2006.

Extent

This procedure applies in England and Wales (slightly different provisions for Scotland) but not in Northern Ireland (section 13(5)).

Scope

Local authorities are placed under a statutory duty to remove motor vehicles unlawfully abandoned on any land in the open air or on any other land forming part of a highway except that, where the cost of removal to the nearest convenient carriageway would be unreasonably high, the procedure is discretionary (section 3(1) and (3)).

There is no legal definition of 'abandoned' but it is suggested that this would include:

(a) untaxed, with any or some of the following:
(b) no current vehicle keeper;
(c) stationary for a significant time;
(d) significantly damaged, run down or unroadworthy;
(e) burned out;
(f) lacking one or more number plates;
(g) containing waste.

FC132A Part 1 Abandoned vehicles: removal – Sections 3 and 4 Refuse Disposal (Amenity) Act 1978

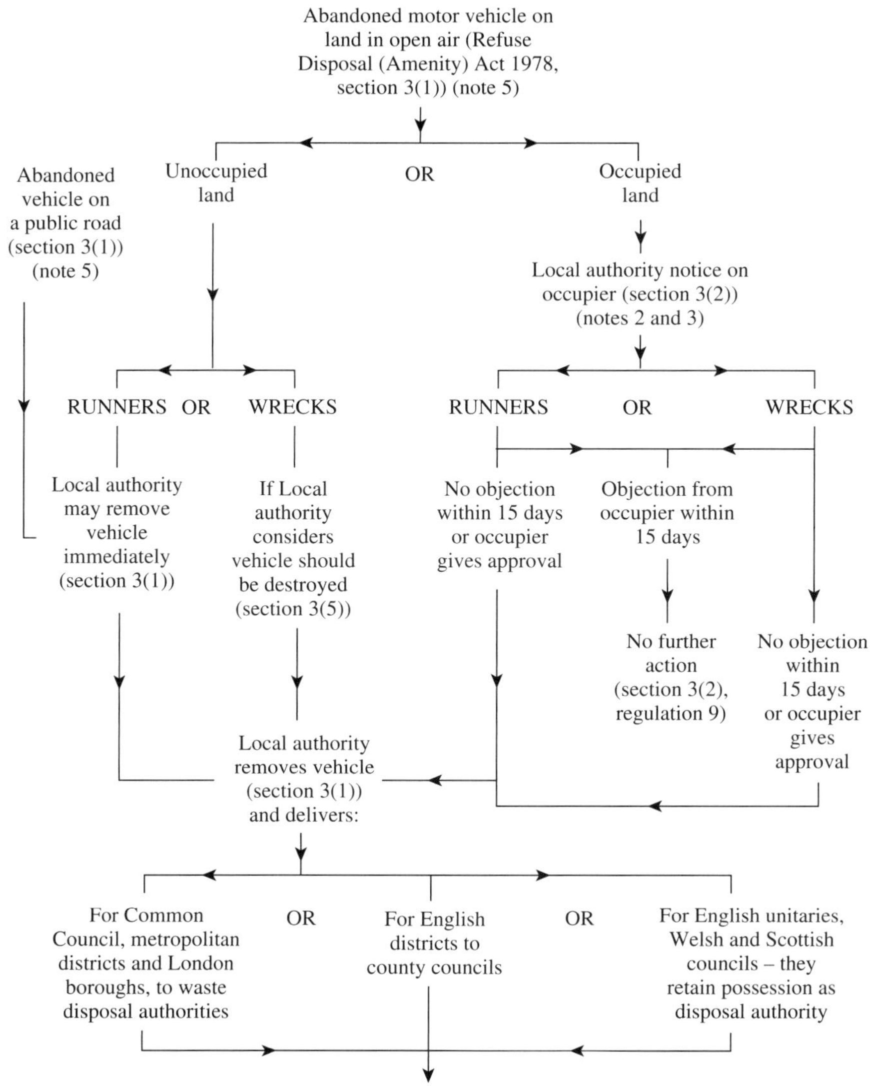

(Continued on next page)

Notes

1. All regulation numbers refer to the Removal of Vehicles Regulations 1986.
2. If the land is occupied by the local authority itself, no 'occupier' notice is required.
3. For powers to remove vehicles illegally, obstructively or dangerously parked, see Road Traffic Regulations Act 1984 and Removal of Vehicles Regulations 1986 (as amended 1993).
4. For power of entry, see section 8.
5. There is a penalty for this offence and an alternative fixed-penalty procedure – see text.

FC132B Part 2 Abandoned vehicles: disposal

(Continued from previous page)

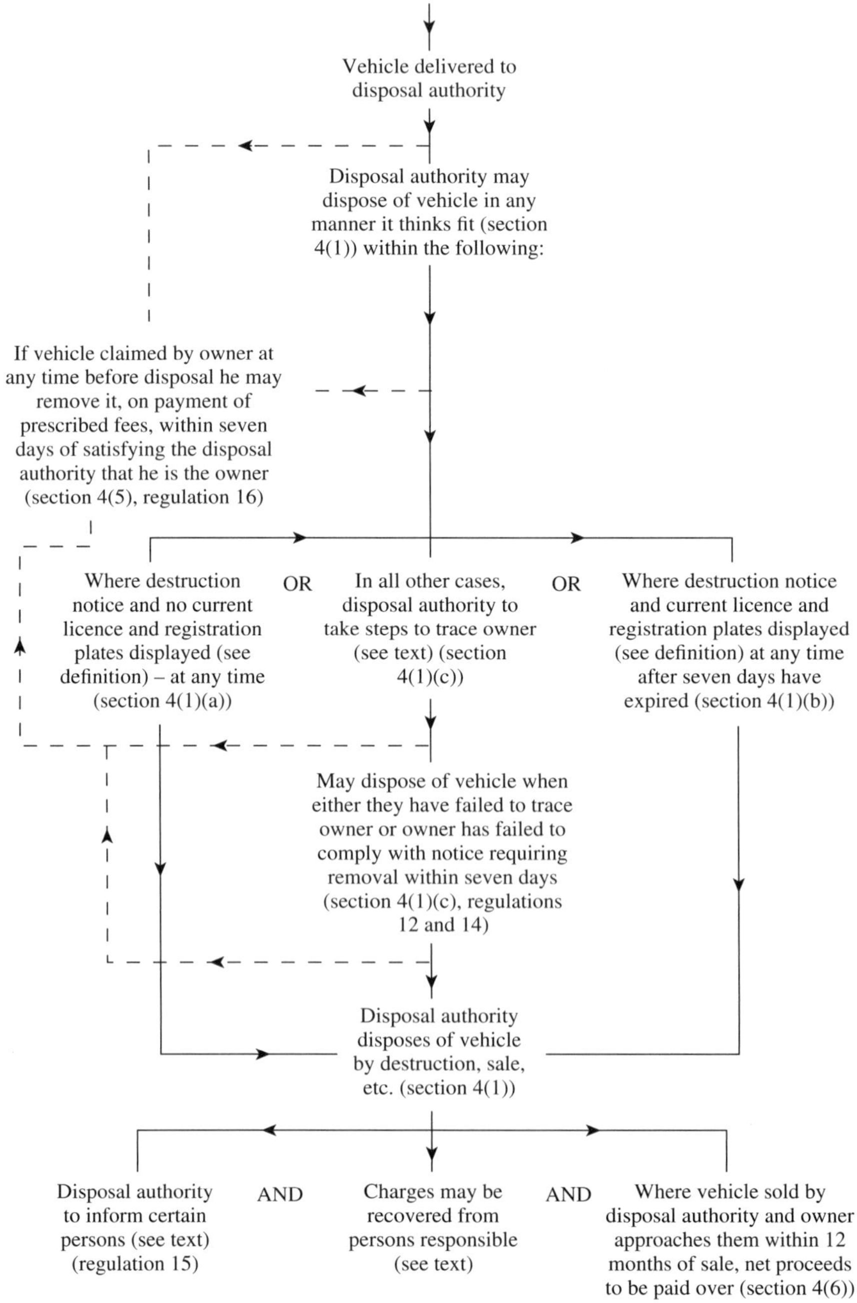

The removal of vehicles broken down or illegally, obstructively or dangerously parked are matters for the police and traffic wardens under the Removal and Disposal of Vehicles Regulations 1986 (as amended). Traffic officers have powers to remove cars in certain circumstances under the Removal and Disposal of Vehicles (Traffic Offices) (England) Regulations 2008.

Unoccupied land

If the vehicle is on unoccupied land, it may be removed immediately without notice (section 3(1)).

Occupied land

If the vehicle is situated on occupied land, a further notice procedure is required. In this case, the local authority must give at least 15 days' written notice to the occupier (section 3(2)).

Vehicles abandoned on roads

No notice of intended removal is required for a vehicle abandoned on a road within the meaning of the Road Traffic Regulation Act 1984. In essence, this is any length of highway or any road to which the public has access and will therefore include all highways to which the public has a right to pass. Access roads on housing estates and car parking areas are therefore likely to be covered by this definition (section 3(2A)).

Delivery of vehicle to disposal authority

Having removed the vehicle, the local authority must deliver it by arrangement:

(a) in London and English metropolitan areas, to the waste disposal authorities; or
(b) in the rest of England, to the county council.

In England, Wales and Scotland, the unitary authority retains possession and continues to operate the procedure from that point as refuse disposal authority (section 3(7)).

Arrangements

Where two authorities are involved, delivery to the disposal authority is required to be in accordance with arrangements agreed between the local authority and the disposal authority, and these arrangements may include sharing of expenses incurred or income received. In default of agreement, there is provision for arbitration or, in the case of London, reference to the Secretary of State (sections 3(7)).

Penalties

No penalties are specified for not recovering vehicles after notices have been served but it is an offence to abandon a vehicle in the open air or on land forming part of a highway, and penalties are not exceeding level 4 on the standard scale and/or imprisonment for up to three months (section 2(1)).

Fixed-penalty notices

Fixed-penalty procedures in sections 2A, 2B and 2C of the Refuse Disposal (Amenity) Act 1978 are available to local authorities for the offence of abandoning a vehicle. The penalty is set at £200 but may be amended by order, and the local authority may treat the amount as having been paid if a lesser amount is paid before the end of a shorter period as it may specify.

Safe custody

For vehicles other than those intended for destruction with the appropriate notice attached, the local authority or disposal authority having custody of the vehicle is responsible for taking reasonable steps to ensure its safe custody (section 3(8)). In the event of this being in question, approaches may be made by the person concerned to the Secretary of State, who must hold a public inquiry (sections 1(5) and 3(7)).

Disposal of vehicles

The disposal authority may dispose of vehicles delivered to them in any manner it thinks fit in the timescales shown in FC132, part 2 (section 4(1)). The sale of a vehicle by the local authority, e.g. at an auction, may be an acceptable form of disposal for a suitable vehicle. This may include the sale of a vehicle at auction.

The steps to be taken to trace the owner (i.e. other than where a destruction notice is given) are specified in regulation 12 of the Removal and Disposal of Vehicles Regulations 1986.

Notices to owners requiring removal of the vehicle are to be served by either:

(a) delivering it to the owner;
(b) leaving it at the usual or last known place of abode;
(c) sending it by a pre-paid registered letter or recorded delivery addressed to him at his usual or last known abode;
(d) where the owner is an incorporated company or body, delivering it to the secretary or clerk or sending it addressed to that person by either pre-paid registered letter or recorded delivery at their registered or principal office (regulation 13).

The persons to whom information must be given following the disposal of the vehicle are detailed in regulation 15 of the Removal and Disposal of Vehicles Regulations 1986.

Charges for storage and removal

Statutory charges are prescribed from time to time by the Removal, Storage and Disposal of Vehicles (Prescribed Charges) Regulations 2008. Charges may be recovered from the person responsible:

(a) the owner at the time the vehicle was abandoned unless he shows that he was not aware of or concerned in the abandonment; or
(b) any person who put it in the place from which it was removed; or
(c) persons convicted of abandoning the vehicle under section 2(1) (section 5(4) of the Refuse Disposal (Amenity) Act 1978).

Definitions

Motor vehicle means a mechanically propelled vehicle intended or adapted for use on roads, whether or not it is in a fit state for such use, and includes any trailer intended or adapted for such use as an attachment to such a vehicle, any chassis or body, with or without wheels, appearing to have formed part of such a vehicle or trailer and anything attached to such a vehicle or trailer (section 11(1)). This definition includes caravans.

Note – The Removal and Disposal of Vehicle (Amendment) (England) Regulations 2012 extend the power of constables in England to require the removal of vehicles from land other than a public road. The amendments also make further provision for the case where a vehicle may be removed from land other than such a road. On the date of publication of this book, they were not yet in force.

RECEPTACLES FOR COMMERCIAL AND INDUSTRIAL WASTE

References

Environmental Protection Act 1990, section 47, as amended by the Clean Neighbourhoods and Environment Act 2005 (CNEA 2005).
The Environmental Offences (Fixed Penalties) (Miscellaneous Provisions) Regulations 2007.
DEFRA Guidance on the CNEA 2005 – Waste 2006 and Fixed Penalty Notices 2006.

Extent

This provision applies to England, Wales and Scotland.

Scope

This procedure may be applied where:

(a) commercial or industrial waste is stored at a premises; and
(b) if not stored in receptacles of a particular kind, the waste is likely to cause nuisance or be detrimental to the amenities of the neighbourhood (section 47(2)).

FC133 Receptacles for commercial and industrial waste – Section 47 Environmental Protection Act 1990

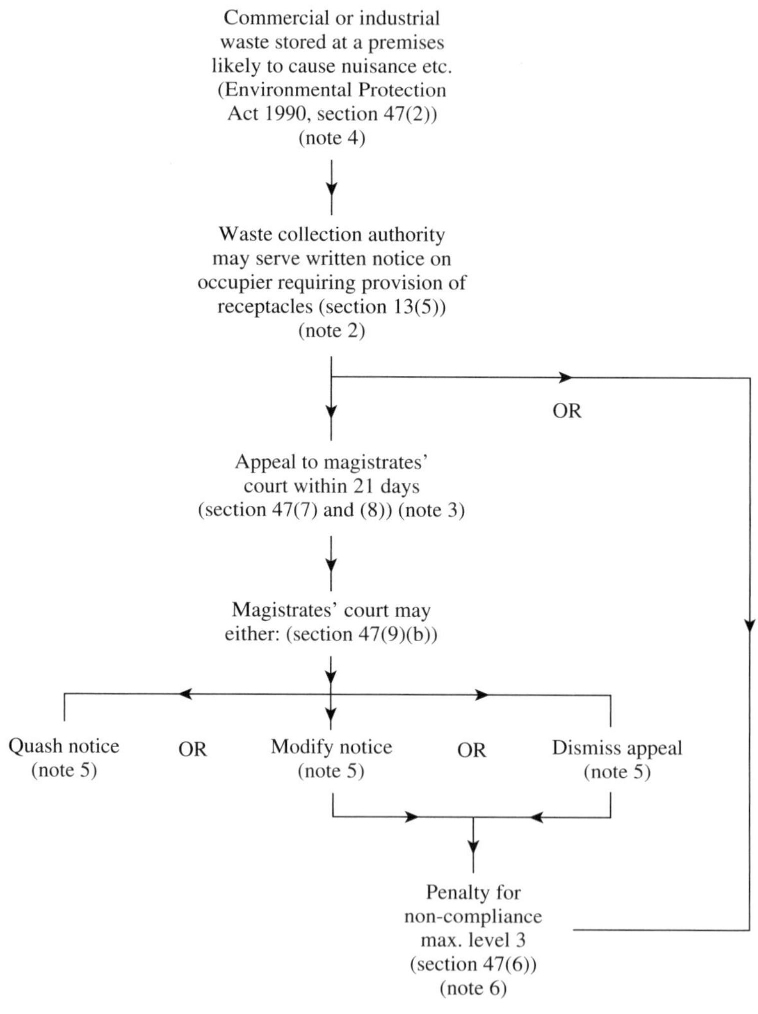

Notes

1. For provision of receptacles for household waste, see FC134.
2. For service of notices, see pages 24–26, 56, 140.
3. The notice is suspended pending determination of the appeal (section 47(9)(1)).
4. For definition of industrial and commercial waste, see pages 482–483.
5. Either party may appeal to the Crown Court against the decision of the magistrates' court (section 73(1)).
6. There is a fixed-penalty option – see text.

Power of waste collection authority (WCA) to provide receptacles

A WCA has power to supply receptacles for commercial and industrial waste (see 'The definition of waste' on pages 482–483) which the WCA has been requested to collect. A charge for a receptacle for industrial waste is mandatory, but the charge for a receptacle for commercial waste is discretionary (section 47(1)). The charge may be on a purchase or rental basis.

WCAs are under a duty to collect commercial waste when asked to do so by an occupier but have discretion on the collection of industrial waste. Where collections are made, the WCA must make a charge (section 45).

Contents of notice

The notice may include requirements covering:

(a) size, construction and maintenance of the receptacle (the kind and number must be reasonable);
(b) the siting of the receptacle for emptying and access for that purpose;
(c) placing of the receptacle on the highway, with the consent of the highway authority and in accordance with arrangements for damage liability;
(d) the substances which may or may not be put into the receptacles and the precautions to be taken;
(e) the steps to be taken by occupiers to facilitate collection (section 47(3) and (4)).

The time allowed for compliance (where appropriate) must be reasonable, and should not be less than the period within which an appeal may be made, i.e. 21 days.

Appeals

Appeal may be made to the magistrates' court within 21 days of receipt of the notice on the following grounds:

(a) any requirement is unreasonable;
(b) the waste is not likely to cause nuisance or be detrimental to the amenities of the neighbourhood (section 47(7) and (8)).

Upon appeal, the notice is suspended until the appeal is determined and, in any subsequent proceedings, no question concerning the reasonability of any requirement can be entertained (section 47(9)(c)).

Penalties

A person not complying with the requirements of the notice without reasonable excuse is liable, on conviction, to a fine not exceeding level 3 (section 47(6)).

Fixed-penalty notices are available to an authorised officer where a person has not complied with the notice. This offers the offender the opportunity to discharge his liability to conviction by payment of the fixed amount (section 47ZA).

The amount is at the discretion of the local authority between £75 and £110, with discounts for early payment (Environmental Offences (Fixed Penalties) (Miscellaneous Provisions) Regulations 2007 as amended).

Definition

Receptacle includes a holder for receptacles (section 47(10)).

RECEPTACLES FOR HOUSEHOLD WASTE

References

Environmental Protection Act 1990, section 46, as amended by the Clean Neighbourhoods and Environment Act 2005 (CNEA 2005).
The Environmental Offences (Fixed Penalties) (Miscellaneous Provisions) Regulations 2007.
DEFRA Guidance on the CNEA 2005 – Waste 2006 and Fixed Penalty Notices 2006.

Extent

This provision applies to England, Wales and Scotland.

Scope

This procedure may be used by a waste collection authority (WCA) to require the occupier of any premises from which it has the duty to collect household waste ('The definition of waste' on pages 482–483) to place the waste for collection in receptacles of the kind and number specified in the notice (section 46(1)).

WCAs have a duty to collect household waste. No charge may be made except in prescribed cases detailed in the Controlled Waste Regulations 1992 (section 45).

Content of notice

The kind and number of receptacles to be used must be reasonable, but separate receptacles can be required for waste which is to be recycled (section 46(2)).

The notice may include requirements dealing with:

(a) the size, construction and maintenance of receptacles;
(b) the placing of them to facilitate emptying and access for that purpose;
(c) the placing of receptacles on the highway (with the consent of the highway authority and with an arrangement for damage liability);
(d) limitation of the waste which may or may not be put into the receptacles and the precautions to be taken with particular substances or articles; and
(e) the steps to be taken by occupiers to facilitate the collection of the waste (section 46(4) and (5)).

FC134 Receptacles for household waste – Section 46 Environmental Protection Act 1990

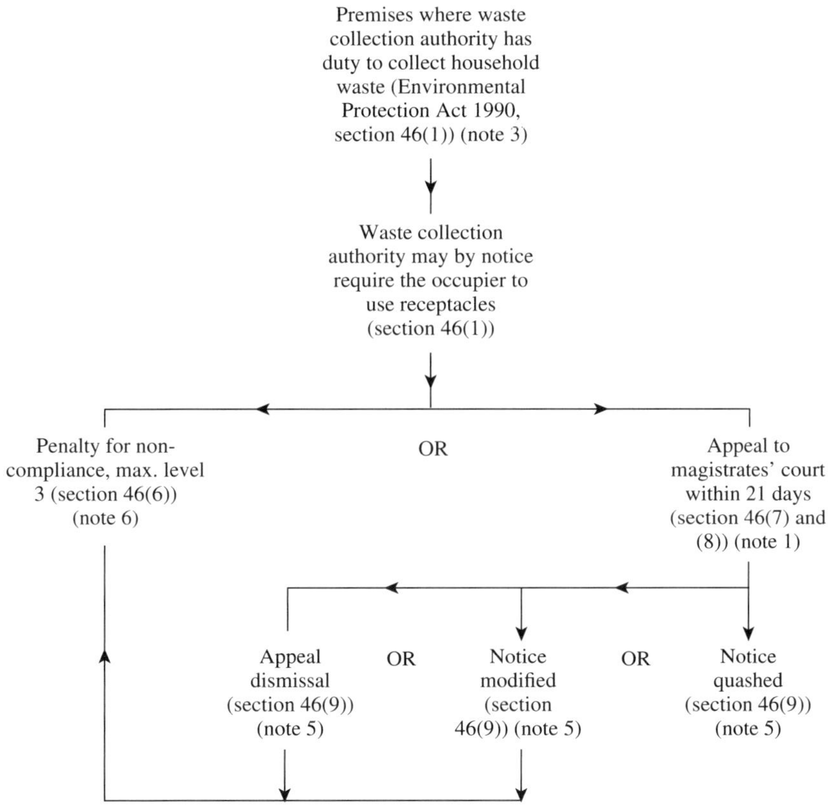

Notes

1. The notice is suspended pending determination of the appeal (section 46(9)(a)).
2. For the provision of receptacles for industrial and commercial waste, see FC133.
3. For definition of waste, see pages 482–483.
4. For service of notices, see pages 24–26, 56, 140.
5. Either party may appeal to the Crown Court against the decision of the magistrates' court (section 73(1)).
6. There is a fixed-penalty option – see text.

Provision of receptacles

The WCA has a choice of the way in which the necessary receptacles are to be provided:

(a) by the WCA free of charge;
(b) with the occupier's agreement, by the WCA on payment by the occupier (single purchase or periodic payments);
(c) by the occupier, including where the occupier refuses an agreement under (b) (section 46(3)).

Appeal

The occupier may appeal to the magistrates' court within 21 days of the notice being served or, where required to enter into an agreement under (b) above, within 21 days of the expiry of the period allowed by the WCA to enter into the agreement.

The grounds of appeal are:

(a) that any requirement is unreasonable; and
(b) that the receptacles already used are adequate (section 46(7) and (8)).

Upon appeal, the notice is suspended until the appeal is determined and in any subsequent proceedings for non-compliance with the notice, no question regarding the reasonability of any requirement can be entertained (section 46(9)).

Penalties

A person not complying with the requirements of the notice without reasonable excuse is liable, on conviction, to a fine not exceeding level 3 (section 46(6)). Fixed-penalty notices are available to an authorised officer where a person has not complied with the notice. This offers the offender the opportunity to discharge his liability to conviction by payment of the fixed amount (section 47ZA). The amount is at the discretion of the local authority between £75 and £110, with discounts for early payment (Environmental Offences (Fixed Penalties) (Miscellaneous Provisions) Regulations 2007).

Definition

Receptacle includes a holder for receptacles (section 46(10)).

SCRAP METAL DEALERS

Reference

Scrap Metal Dealers Act 1964.

Extent

The provisions apply in England and Wales but not in Northern Ireland or Scotland (section 11(3)).

FC135 Scrap metal dealers – Scrap Metal Dealers Act 1964

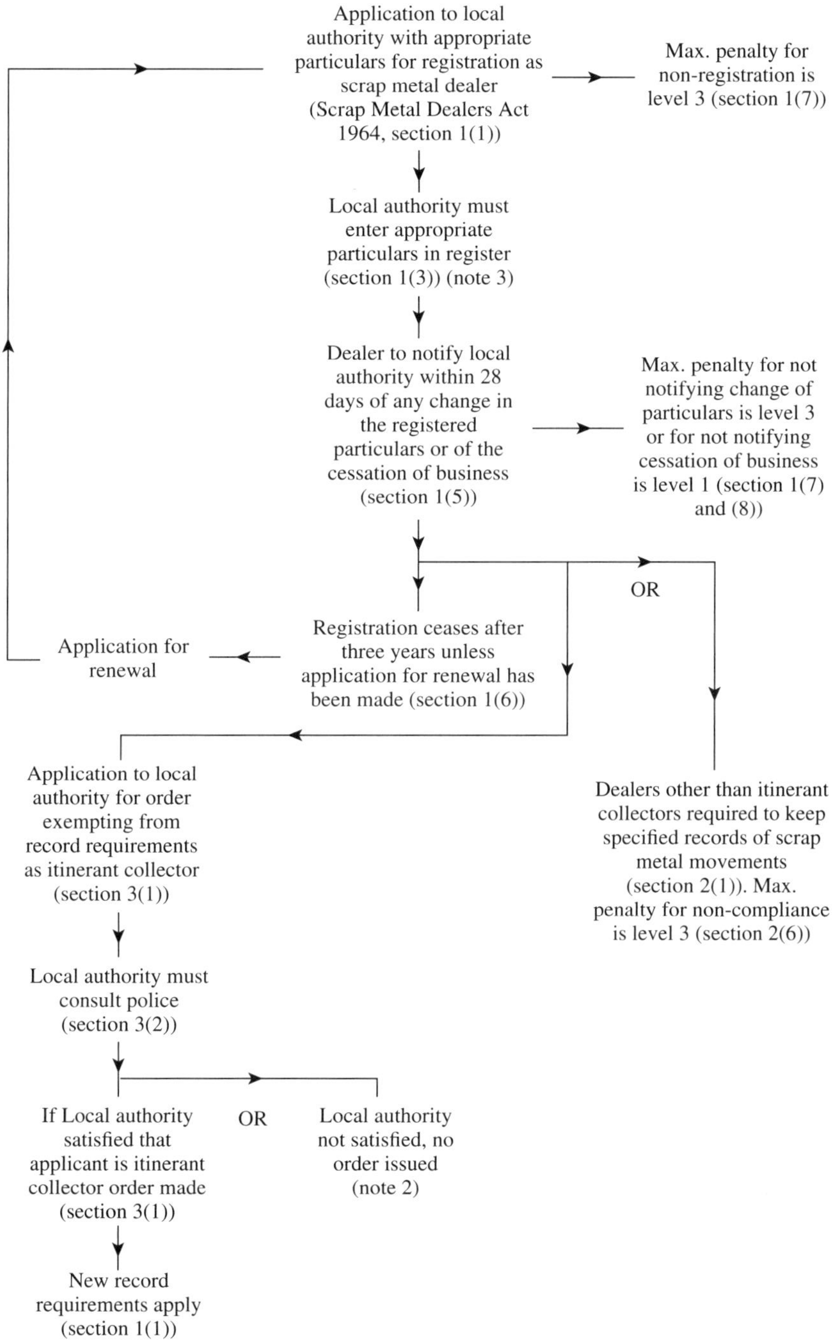

Notes

1. If the local authority is itself acting as a scrap metal dealer, although this procedure as a whole does not apply, particulars must be kept in the register (section 1(10)).
2. There is no appeal against this decision.
3. There is no discretion here, provided the appropriate particulars have been submitted.

Scope

Scrap metal dealers (definition below) are required to register with the local authority (see definitions below) in whose area they are carrying on the business (section 1(1)). The business is said to be carried on in the area of a particular local authority if either:

(a) he occupies a place as a scrap metal store in that area; or
(b) he resides in that area *but* does not occupy a scrap metal store in that area or elsewhere; or
(c) a place is occupied by him wholly or partly for his scrap metal business in that area but he does not occupy a scrap metal store in that area or elsewhere (section 1(2)).

Appropriate particulars

These are the details which must be included in the application for registration and which form the information contained within the register. The details are:

(a) full name of the dealer;
(b) the address:
    (i) for an individual, the usual place of residence;
    (ii) for a body corporate, its registered or principal office;
(c) the address of each place occupied as a scrap metal store;
(d) if no scrap metal stores are occupied, a statement to that effect;
(e) if a place other than a scrap metal store is occupied by him, the address of that place (section 1(4)).

No registration fee is payable.

Records

Each dealer must keep a book containing records of all scrap metal received, processed and dispatched and these are modified in relation to itinerant collectors certified as such by order of the local authority (sections 2 and 3).

A police constable has the power to enter a place at all reasonable times, examine these records and take copies or extracts (section 6(1)). These powers are not available to local authority officers.

Restrictions by courts

If a person is convicted of not being registered, of not keeping the required records or of an offence of dishonesty, the court may by order, effective for up to two years, impose restrictions on him:

(a) no scrap metal to be received between 6 pm and 8 am; and/or
(b) scrap metal received, unless processed, to be kept for at least 72 hours.

The maximum penalty for non-compliance with an order is level 3 (section 4).

Power of entry

In addition to the powers given to a police constable ('Records' above), an authorised officer of a local authority may enter any place at any reasonable time if he has reasonable grounds for believing that the place is being used as a scrap metal store and that particulars are not entered in the register. On application, a magistrates' court may issue a warrant to enter, by force if necessary. The maximum penalty for obstruction is level 1 (section 6).

Local authority officers have no right to enter registered places.

Definitions

Article includes any part of an article.
Itinerant collector means a person regularly engaged in collecting waste materials, and old, broken, worn out or defaced articles, by means of visits from house to house.
Place includes any land, whether consisting of enclosed premises or not.
Scrap metal includes any old metal and any broken, worn out, defaced or partly manufactured articles made wholly or partly of metal, and any metallic wastes, and also includes old, broken, worn out or defaced tooltips or dies made of any of the materials commonly known as hard metal or of cemented or sintered metallic carbides.
Scrap metal dealer – a person carries on business as a scrap metal dealer if he carries on a business which consists wholly or partly of buying and selling scrap metal, whether the scrap metal sold is in the form in which it was bought or otherwise, other than a business in the course of which scrap metal is not bought except as materials for the manufacture of other articles and is not sold except as a by-product of such manufacture or as surplus materials bought but not required for such manufacture.
Scrap metal store means a place where scrap metal is received or kept in the course of the business of a scrap metal dealer.

Any reference in the Act to *metal*, except in the phrases 'hard metal' and 'metallic carbides', shall be taken as a reference to any of the following metals: aluminium, copper, iron, lead, magnesium, nickel, tin and zinc, or, subject to the next following subsection, to brass, bronze, gunmetal, steel, white metal or any other alloy of any of the said metals.

For the purpose of this Act, a substance being an alloy referred to in the last preceding subsection shall not be treated as being such an alloy if, of its weight, 2 per cent or more is attributable to gold or silver or any one or more of the following metals: platinum, iridium, osmium, palladium, rhodium and ruthenium (section 9).

WATER INDUSTRY ACT 1991
GENERAL PROCEDURAL PROVISIONS

The following provisions are applicable to the use by local authorities of all procedures dealt with in this chapter.

Information

A local authority may by notice require any person to provide such information as it may reasonably require to fulfil its powers and duties with respect to both public and private water supplies. The maximum penalty for non-compliance with such notices is level 5 (section 85).

Power of entry etc.

A person designated in writing by the local authority may:

(a) enter premises supplied with private or public water to see if any action is required of the local authority;
(b) carry out such inspections, measurements and tests, including the taking of samples of water, land and articles, as is considered appropriate.

Unless the situation is one of emergency, power of entry cannot be required unless it is requested at a reasonable time and after having given at least 24 hours' notice to the occupier in relation to non-business premises. Persons convicted of obstructing an local authority officer are subject to penalties not exceeding level 5 (section 84(3) and schedule 6).

Where the local authority shows that entry has been refused, or is anticipated, or the premises are unoccupied or the case is one of emergency, a Justice of the Peace may issue a warrant to authorise entry, if need be by force (schedule 6, paragraph 2).

Definitions

Owner in relation to any premises, means the person who:

(a) is for the time being receiving the rackrent of the premises whether on his own account or as the agent or trustee for another person; or
(b) would receive the rackrent if the premises were let at a rackrent (section 219(1)).

Private supply means:

1. subject to subsection (2) below, a supply of water provided otherwise than by a water undertaker (including a supply provided for the purposes of the bottling of water), and cognate expressions shall be construed accordingly;
2. for the purposes of any reference … to a private supply, or to supplying water by means of a private supply, water shall be treated as supplied to

any premises not only where it is supplied from outside those premises, but also where it is abstracted, for the purpose of being used or consumed on those premises, from a source which is situated on the premises themselves; and for the purposes of this subsection water shall be treated as used on any premises where it is bottled on those premises for use or consumption elsewhere.

(section 93(1) and (2))

Water undertaker – a company may be appointed:

(a) by the Secretary of State; or
(b) with the consent of or in accordance with a general authorisation given by the Secretary of State, by the Director General of Water Services, to be the water undertaker ... for any area of England and Wales (section 6(1)).

CONTROL OVER PUBLIC WATER SUPPLIES

References

Water Industry Act 1991, sections 77 and 78.
Water Supply (Water Quality) Regulations 2010.
Water Supply (Water Quality) (Amendment) Regulations 2011.

Extent

These provisions apply to England and Wales.

Scope

Public supplies are those provided by an appointed water undertaker (WU), and each water undertaker is placed under a statutory duty to supply water for domestic or food production purposes which is wholesome and to ensure that, in general, there is no deterioration in the quality of water from any of its sources (section 68(1)).

In this connection, domestic purposes include drinking, washing, cooking, central heating and sanitary purposes (section 218(1)). 'Food production purposes' means the manufacturing, processing, preserving or marketing purposes with respect to food or drink for which water supplied to food production premises may be used; and 'food production premises' means premises used for the purposes of a business of preparing food or drink for consumption otherwise than on premises (section 93).

Enforcement

Those duties placed on water undertakers are enforceable by the Secretary of State and the Director General of Water Services through enforcement orders (section 18).

Wholesomeness

Water is to be judged as being wholesome if it satisfies the requirements of regulation 4 of the Water Supply (Water Quality) Regulations 2010 or regulation 4 of the Private Water Supplies Regulations 2009 as appropriate (section 67(1)).

Local authority role

Local authorities are placed under a statutory duty to take all such steps as they consider appropriate for keeping themselves informed about the wholesomeness and sufficiency of all public water supplies (section 77(1)).

In this connection, local authorities:

(a) must make arrangements with the water undertaker to be notified of events giving rise to significant risks to health; and
(b) may take and have analysed by designated officers such public water samples as they may reasonably require (regulations 37 and 38, Water Supply (Water Quality) Regulations 2000).

Powers of the local authority

There are three situations which the local authority deals with in relation to public supplies:

1. Where a local authority believes that:
   (a) the water is, or is likely to become, unwholesome;
   (b) the water is insufficient for domestic purposes;
   (c) the water is, or is likely to cause, a danger to life or health because of unwholesomeness or insufficiency; or
   (d) the quality of sources of water are deteriorating,

   the local authority is required to inform the water undertaker. Where the local authority is not satisfied that all appropriate remedial action is being taken by the water undertaker, it must notify the Secretary of State, who may then use his powers of enforcement (section 78).
2. The local authority is required to notify the water undertaker wherever it (the water undertaker) is satisfied that it is not reasonably practicable at reasonable cost to provide, or to maintain, a piped supply of wholesome water that is sufficient for domestic purposes, and to request the water undertaker to provide a supply of water to particular premises by means other than pipes where:
   (a) this is practicable at reasonable cost; and
   (b) the unwholesomeness or insufficiency is such as to cause danger to life and health.

   In these circumstances, the local authority is required to pay to the water undertaker the appropriate water charges, but may recover these from the owners or occupiers of the premises benefiting from the supply. The water undertaker is under a duty to comply with a local authority request which fulfils the criteria outlined above (section 79).

3. Where the local authority is of the opinion that a supply is so polluted as to be prejudicial to health, it may apply to a magistrates' court for an order closing or restricting that supply. This procedure is detailed in FC139 (Public Health Act 1936, section 140).

CONTROL OVER PRIVATE WATER SUPPLIES

References

Water Industry Act 1991, sections 80–83.
Water Supply (Water Quality) Regulations 2000.
The Private Water Supplies Regulations 2009.
Circular 24/91 DoE, Private Water Supplies.

Extent

These provisions apply to England and Wales.

Scope

A private water supply is a supply of water provided otherwise than by a water undertaker and includes a supply provided for the purposes of the bottling of water. This definition is extended to cover water sources abstracted both outside and inside of premises, and water bottling in premises for use or consumption elsewhere (sections 93(1) and (2)). The Food Safety Act 1990 extended the definition to cover water supplied for food production (also see page 177).

Local authorities are under the same general duty to monitor the wholesomeness and sufficiency of private supplies as they are in relation to public supplies (section 77(1)), but in this case they have direct and sole enforcement powers. The Private Water Supplies Regulations 2009 specify the scope of monitoring distributed by a person other than a water undertaker or licensed water supplies and for large supplies and supplies to commercial or public premises, the keeping of records and their notification to the Secretary of State (regulations 7 to 13).

Water Industry Act 1991: information

The local authority may serve a notice on any person requiring him to provide any information it may reasonably require in order to fulfil its responsibilities for private supplies (section 85(1)).

Improvement notices

Where a local authority is satisfied that a private water supply is not wholesome or sufficient for domestic or food production purposes, it may serve an improvement notice on the relevant person (section 80(1)).

By reason of a direction issued by the Secretary of State on 25 August 1989, before serving any notice, local authorities must ascertain the terms of any agreement, contract, licence or other document relating to the terms upon which the water is supplied, and have regard to this in determining the content of any notice.

Standards

Water from private supplies is to be judged as being wholesome for domestic purposes if it satisfies the requirements of the tables in schedule 1 of the Private Water Supplies Regulations 2009 (section 67 Water Industry Act 1991 and regulation 4).

Relevant person

The persons upon whom the notice should be served are the owners *and* occupiers of premises served by the supply, the owner *and* occupiers of premises where the source is situated *and* other persons who exercise powers of management or control of that source (section 80(7)).

Notice requirements

A notice must:

(a) identify the reasons why the supply is considered to be unwholesome and/or insufficient;
(b) specify the steps required to provide a wholesome and sufficient supply;
(c) allow a period of not less than 28 days for representation or objections to be received;
(d) indicate the need for referral to the Secretary of State if representations or objections are made within the specified period (section 80(2)).

The notice may specify one or more of the following:

(a) the steps to be taken by the local authority themselves;
(b) the steps to be taken by the person specified in the notice within a specified period;
(c) the payments to be made to another relevant person or the local authority for expenses in meeting the local authority requirements;
(d) the payments to be made by the local authority to a person for the costs of meeting the notice requirements (section 80(3));

and may require:

(a) a supply of water to be provided by the water undertaker or another person; and
(b) steps to be taken to ensure that this is done (section 80(6)).

Operation of notices

Where no representation or objections are made to the local authority within the period specified in the notice, the notice becomes effective at the end of the period allowed for such representation to be made (section 81(1)).

FC136 Private water supply: notice procedure – Regulation 18 Private Water Supplies Regulations 2009

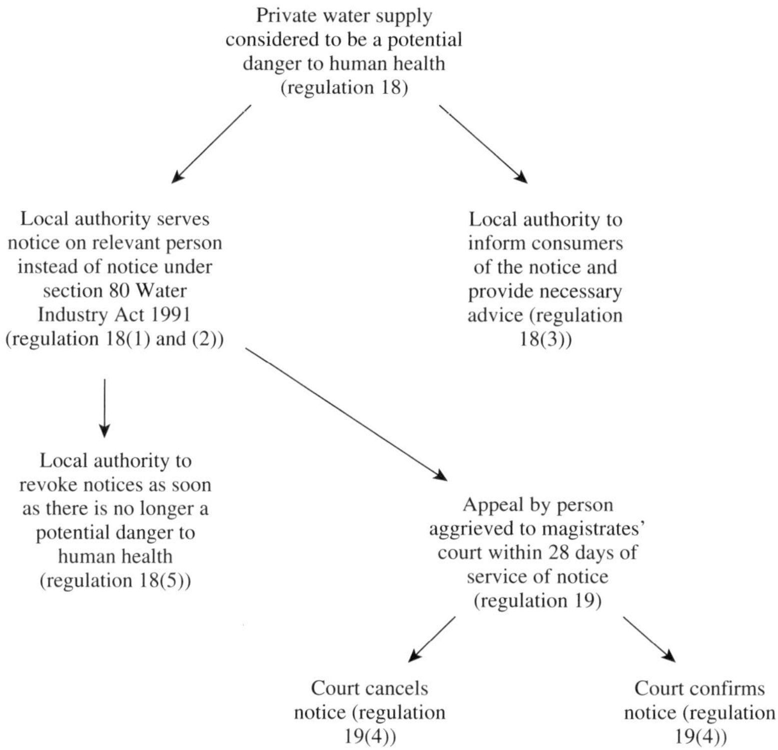

Notes

1. Offence to breach notice or fail to comply with it (regulation 18(6)).
2. Penalties – fine and/or imprisonment (regulation 20(1) and 20(2) for body corporate).
3. Local authority can charge fees for risk assessment, sampling, investigation, granting authorisation and analysing samples (schedule 5).

Where representations etc. are made and not withdrawn, then the local authority must refer the matter to the Secretary of State, who may then either:

(a) confirm the notice with or without modification; or
(b) not confirm the notice.

Where the notice is confirmed, it becomes effective on a date as specified by the Secretary of State in his confirmation (sections 81(2), (3) and (4)).

Enforcement

The local authority's powers of enforcement where an improvement notice is not complied with are restricted to the undertaking of required works and the recovery of related costs (section 82).

Private Water Supplies Regulations 2009

These regulations replace 1991 regulations and implement European Council Directive 98/83/EC.

Regulations 7 to 11 of the Private Water Supplies Regulations 2009 also make it a duty of local authorities to monitor private water supplies, and take and analyse samples of certain private supplies. The regulations also lay down a sampling regime for private supplies (schedule 3).

Risk assessment

Every five years, the local authority must carry out a risk assessment of each private supply that supplies water to any premises (other than a supply to a single dwelling not used for any commercial activity) to establish whether there is a significant risk of supplying water that would constitute a potential danger to human health. It must also carry out a risk assessment of a private supply to a single dwelling not used for any commercial activity if requested to do so by the owner or occupier of that dwelling (regulation 6).

Investigation

If a local authority considers a private water supply to be a potential danger to health, it must carry out an investigation as to why it is unwholesome or fails one of the parameters in the regulations (regulation 15). It must also inform people likely to consume the water that it is a potential danger to human health (regulation 14).

If it is in the pipework of a single dwelling, the local authority must inform the people concerned and give necessary advice. If the local authority cannot solve the problem, it must grant an authorisation for different standards (in effect, a relaxation in standards) or serve a notice under regulation 18 or section 80 of the Water Industry Act 1991.

If an authorisation is granted, the person granted it must take action in accordance with the timetable in the authorisation. If the supply exceeds 1,000 m^3 a day

FC137 Private water supply: action in the event of failure – Part 3 Private Water Supplies Regulations 2009

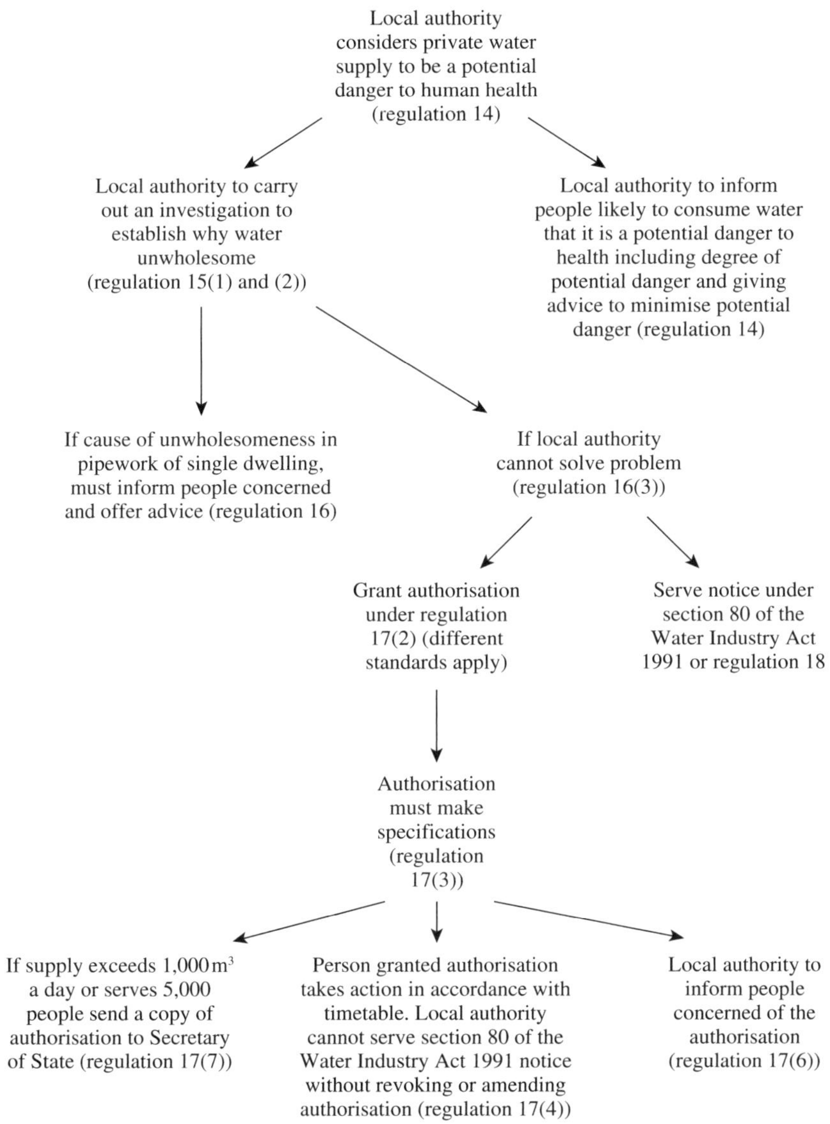

Notes

1. Local authority must keep progress of remedial action under review (regulation 17(8)).
2. Local authority can grant authorisation for up to three years with prior consent from Secretary of State and send copy to Secretary of State (regulation 17(9)).
3. Local authority may revoke or amend the authorisation at any time and in particular if the timetable for remedial action is not followed.

or serves more than 5,000 people, it must send a copy of the authorisation to the Secretary of State (regulation 17(7)).

The local authority must also inform people concerned of the authorisation (regulation 17(6)).

Notice

If any private supply of water intended for human consumption constitutes a potential danger to human health, a local authority must serve a notice on the relevant person.

The notice must:

(a) identify the private supply;
(b) state the grounds for serving the notice;
(c) prohibit or restrict the use of that supply;
(d) specify what other action is necessary to protect human health.

The local authority must promptly inform consumers of the issue of the notice and provide any necessary advice. The notice may be subject to conditions and may be amended by further notice at any time. The local authority must revoke the notice as soon as there is no longer a potential danger to human health. It is an offence to breach a notice served by the local authority (regulation 18). Any person aggrieved by such notice can appeal (regulation 19). Penalties for breach of the notice are included in regulation 20.

The local authority must make and maintain records (regulation 12 and schedule 4), and must send a copy of the records to the Secretary of State (regulation 13).

WATER SUPPLY FOR NEW HOUSES

Reference

Building Act 1984, section 25.

Extent

These provisions apply to England and Wales.

Satisfactory water supply

Before approving plans submitted for approval under the building regulations, the local authority must be satisfied that the water supply proposed will be sufficient and wholesome for domestic purposes and provided:

(a) either by connecting to a piped supply provided by the water undertakers, or
(b) if (a) is not reasonable, by otherwise taking water into the house, by pipe, e.g. from a well supply; or
(c) if neither (a) nor (b) can be reasonably required, by providing a supply within a reasonable distance of the house (section 25(1)).

FC138 Water supply for new houses – Section 25 Building Act 1984

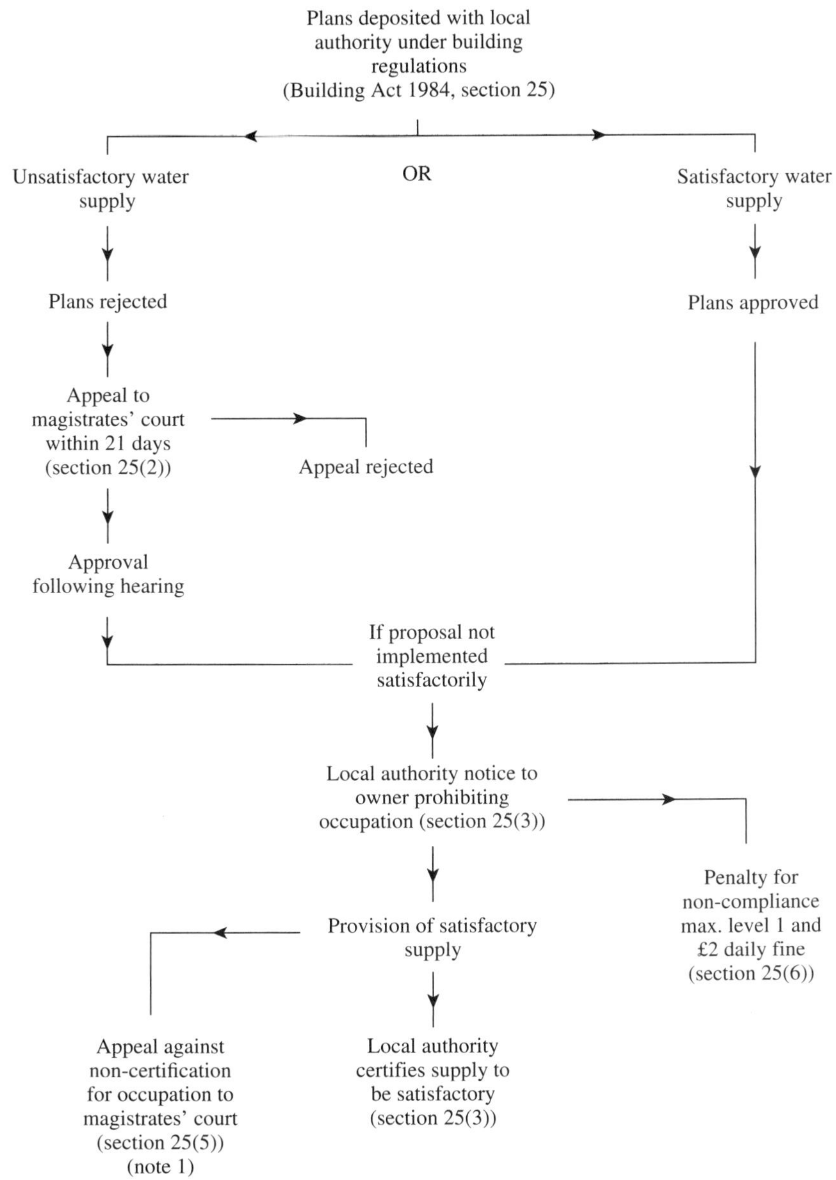

Notes

1. The court may either dismiss the appeal or authorise the occupation of the house, the latter having the same effect as certification by the local authority.
2. For closure/restriction of polluted supplies, see FC139.

In deciding whether or not water is 'wholesome', the standards of section 67 of the Water Industry Act 1991 and of any regulations made under that section (i.e. the Water Supply (Water Quality) Regulations 2000 (as amended) and the Private Water Supply Regulations 2009) apply here (section 25(7)).

Notices

Notices requiring prohibition of occupation in the event of the supply not being sufficient or wholesome, despite the passing of plans, must be in writing and served on the owner (section 92).

AUTHORISATIONS FOR QUALITY STANDARDS FOR PUBLIC WATER SUPPLIES

The Secretary of State deals with authorisation of the temporary supply of public water that is not wholesome. Such authorisations are time related and cannot extend beyond three years.

In making an application to relax the standards of the Water Supply (Water Quality) Regulations 2000, the applicant must send a copy to the local authority, which may make representations to the Secretary of State within 30 days. If the Secretary of State is minded to revoke or modify the authorisation, he must give the local authority at least six months' notice unless immediate action is necessary in the interests of public health (2000, regulations 20–24).

CLOSURE OR RESTRICTION OF POLLUTED WATER SUPPLY

Reference

Public Health Act 1936, section 140.

Extent

These provisions apply to England and Wales.

Scope

The section applies to wells, tanks or any other source of water supply not vested in the local authority which are used, or likely to be used, for domestic purposes or the preparation of food or drink for human consumption and which are so polluted as to be prejudicial to health (section 140(1)).

It would seem that this procedure is available to a local authority and some port health authorities[1] to deal with water supplied by statutory water undertakings operating under the Water Industry Act 1991 and therefore provides the only direct, statutory remedy available to local authorities in respect of such public water supplies (see above page 536).

FC139 Closure or restriction of polluted water supply – Section 140 Public Health Act 1936

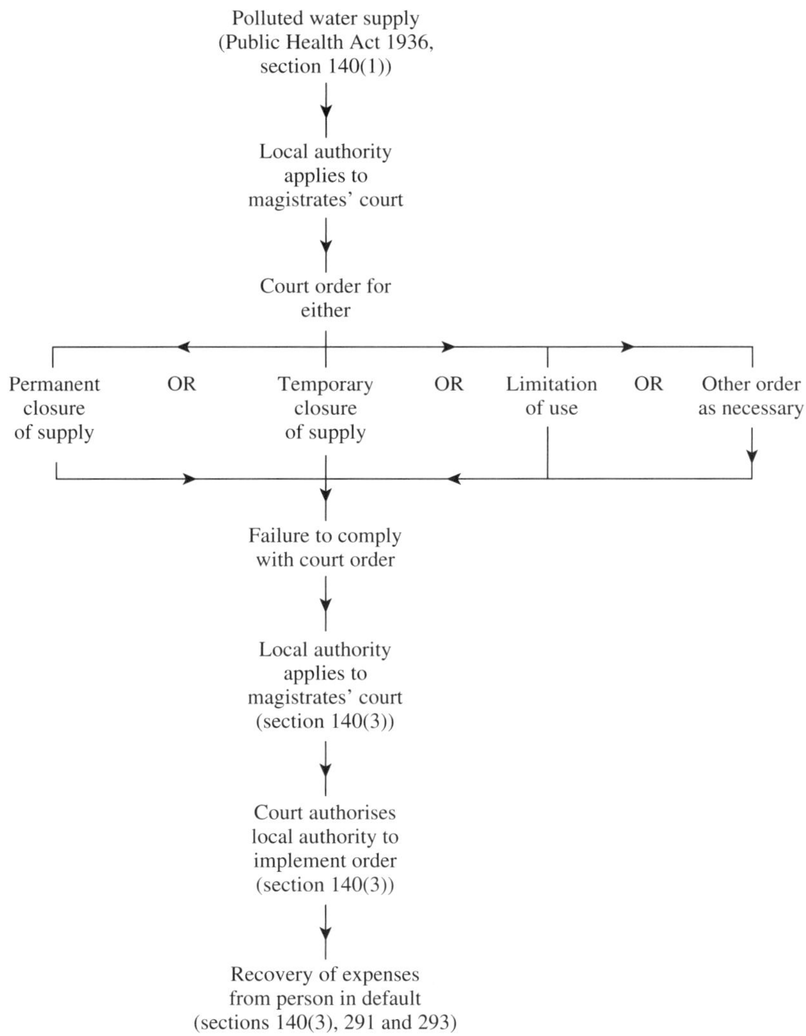

Note

1. Bristol Port Health Authority Order 2010; Cornwall Port Health Authority Order 2010; Cowes Port Health Authority Order 2010; Heysham Port Health Authority Order 2011; Portsmouth Port Health Authority Order 2010; Southampton Port Health Authority Order 2010.

Persons responsible

The court order following applications by the local authority may be addressed to owners or occupiers of the premises to which the source of supply belongs or to any other person having control over it (section 140(1) and (2)).

Court orders

These may order any of the following:

(a) permanent closure or cutting off of the source; or
(b) temporary closure or cutting off of the source; or
(c) restriction of use for certain purposes only; or
(d) any other measure to prevent injury or damage to health of persons using water or consuming food or drink prepared from it (section 140(2)).

Works in default

If the court order is not complied with, the local authority can apply to the court to authorise them to do what is necessary to remedy the problem and it may claim expenses reasonably incurred in default of the order.

Chapter 7

PUBLIC SAFETY

ALCOHOL AND ENTERTAINMENT LICENCES
LICENSING ACT 2003

References

The Licensing Act 2003.
The Anti-Social Behaviour Act 2003, sections 40–41.
The Licensing Act 2003 (Permitted Temporary Activities (Notices)) Regulations 2005 as amended.
The Licensing Act 2003 (Personal Licences) Regulations 2005 as amended.
The Licensing Act 2003 (Premises Licences and Club Premises Certificates) Regulations 2005 as amended.
The Licensing Act 2003 (Fees) Regulations 2005 as amended.
The Licensing Act 2003 (Hearings) Regulations 2005 as amended.
The Licensing Act 2003 (Licensing Authority's Register) (Other Information) Regulations 2005.
The Licensing Act 2003 (Summary Review of Premises Licences) Regulations 2007.
Guidance to local authorities and police officers was issued by the Secretary of State under section 182 of the Licensing Act 2003 and approved by Parliament in October 2012 (The Guidance).

Background

The 2003 Act amalgamated six previous licensing regimes (alcohol, public entertainment, cinemas, theatres, late-night refreshment houses and night cafes) into a single integrated scheme and the separate licensing procedures for each have been repealed. The general framework of the Act is shown in FC140.

Extent

The provisions of the Licensing Act 2003 apply in England only.
The Anti-Social Behaviour Act 2003 applies in England, Wales and Northern Ireland (section 181).

FC140 Licensing Act 2003 – Framework

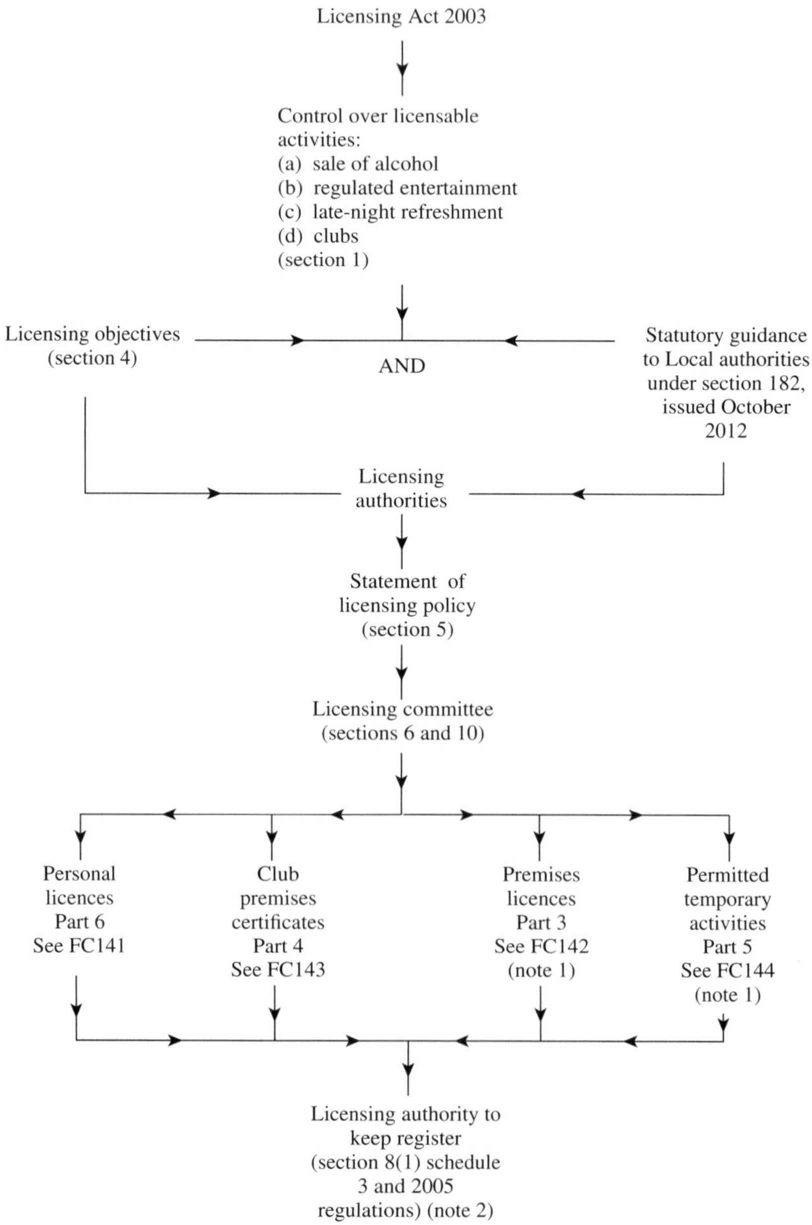

Notes

1. Magistrates' court and police have powers of closure – see text (sections 160 and 161).
2. The Licensing Act 2003 (Licensing Authority's Register) (Other Information) Regulations 2005.

Scope and definitions

The Licensing Act 2003 applies to 'licensable activities', which are defined as:

(a) the sale by retail of alcohol;
(b) the supply of alcohol by clubs;
(c) the provision of regulated entertainment; and
(d) the provision of late-night refreshment (section 1).

'Regulated entertainment' is fully defined in schedule 1. The term includes 'entertainment' and to come within this definition the first test is that either must be provided for:

- the public or sections of the public;
- exclusively for members of a qualifying club and their guests; or
- if neither of these applies, for consideration and a view to profit.

The second test is that the premises on which the entertainment is provided is made available for the purpose (schedule 1, paragraphs 1–3).

'Entertainment' means the following:

1. performance of a play;
2. exhibition of a film;
3. indoor sporting events;
4. boxing or wrestling entertainment;
5. performance of live music;
6. playing of recorded music;
7. performance of dance; and
8. events which have the purpose of entertaining an audience or spectators (schedule 1, paragraph 2).

'Plays', 'film exhibitions', 'indoor sporting events' and 'boxing or wrestling entertainment' are defined in part 3 of schedule 1.

Exempted regulated entertainment

The following are exempted from controls on regulated entertainment:

(a) film exhibitions for advertisement, information and exhibition;
(b) film exhibitions at museums and art galleries;
(c) use of TV or radio receivers;
(d) religious services and places of worship;
(e) garden fetes;
(f) Morris dancing;
(g) vehicles in motion (schedule 1, part 2).

Such licensable activity may only be carried on under and in accordance with a premises licence (PL), a temporary event notice (TEN) or, where appropriate, a club premises certificate (CPC), and to operate in a non-authorised premises constitutes an offence (section 2).

In addition for those selling or supplying alcohol in licensed premises, except qualifying clubs, a personal licence is required.

Late-night refreshment

This is defined as taking place where:

(i) a person at any time between 11 pm and 5 am supplies hot food or hot drink to the public, on or from any premises and whether for consumption on or off the premises; or
(ii) between those hours when members of the public are admitted to any premises the person supplies or is willing to supply hot food or hot drink for consumption on or off the premises.

There are exemptions from this definition in relation to supplying certain clubs, hotels and employees and to premises licensed under the Greater London Council (General Powers) Act 1966 and the London Local Authorities Act 1995 (schedule 2).

Licensing objectives

The licensing objectives that must be promoted by the licensing authority and licence holders and to which they must have regard to in their policy framework document are:

- the prevention of crime and disorder;
- public safety;
- the prevention of public nuisance;
- the protection of children from harm (section 4).

Statements of licensing policy

Each local authority must prepare and publish a statement of licensing policy every five years and in doing so must consult widely, including the police, fire and rescue authority, primary care trust or local health board, representatives of holders of licences issued under the Act, representatives of local businesses and the public (section 5). The statement must have regard to the licensing objectives of the Act (section 4) – see above – and to the guidance issued under section 182.

Licensing committee

Each local authority is required to establish a licensing committee of between 10 and 15 members of the authority to discharge the council's licensing functions except that of the preparation and publication of the statement of policy which must be approved by the whole council. The committee may delegate functions to sub-committees and/or officers, except in the latter case where objections or representations have been made (sections 6, 9 and 10).

TYPES OF LICENCES

Personal licences (PeLs) (FC141)

These are licences permitting an individual to sell alcohol by retail or to supply alcohol by or on behalf of a club. They do not authorise the sale of alcohol

FC141 Personal licences – Sections 111 to 135 Licensing Act 2003

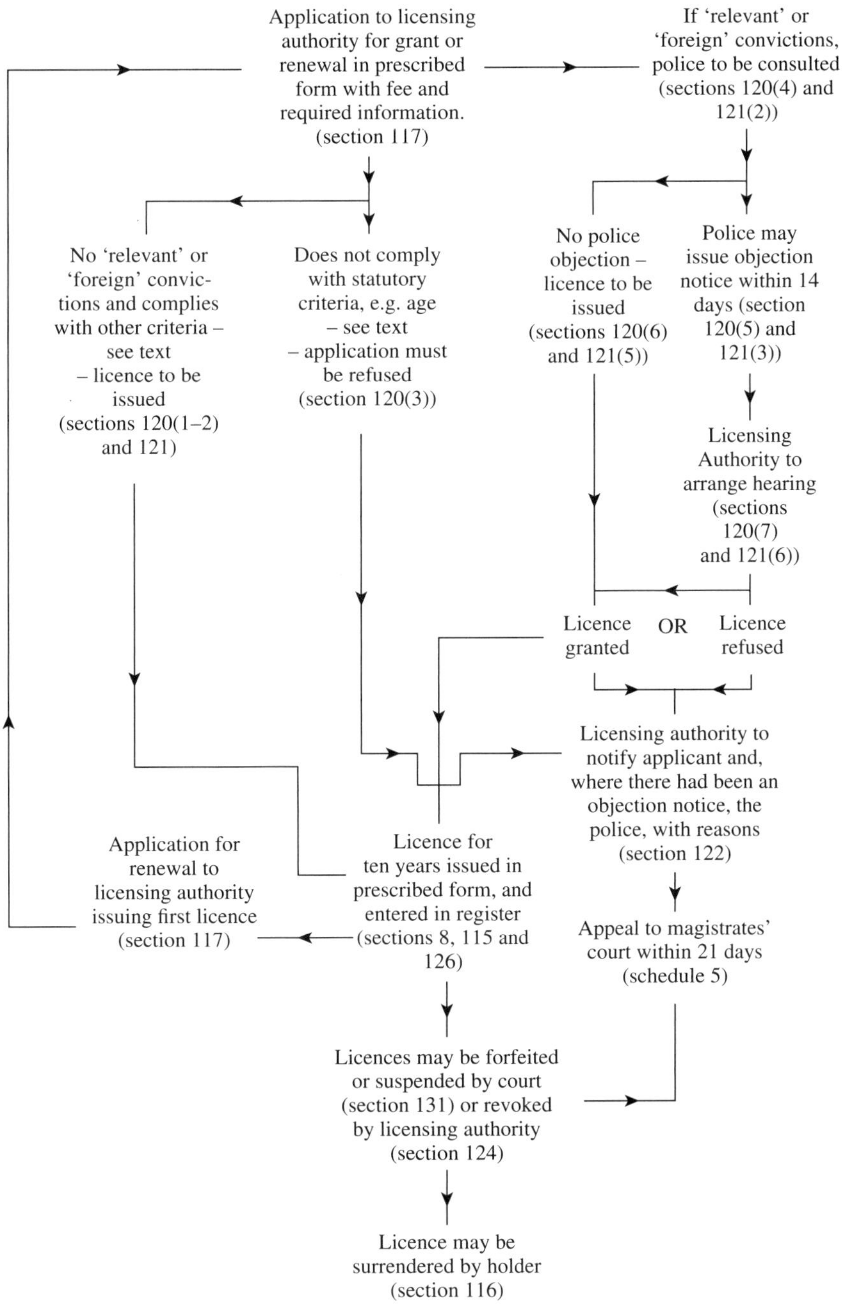

anywhere but only from premises with a premises licence (section 111). Personal licences last for ten years, unless surrendered or revoked, and are renewable. They attach to the person and are transferable to other licensed premises. In each licensed premises, the 'designated premises supervisor' must have a personal licence. Others employed there may also hold a personal licence and, although there is no requirement for all serving alcohol to do so, all sales or supplies of alcohol must be authorised by a personal licence holder.

Applications for grant and renewal of personal licences must be in the form prescribed in the Licensing Act 2003 (Personal Licences) Regulations 2005 and be accompanied by the prescribed fee. Information about any criminal record, details of licensing qualifications and two photographs taken in accordance with detailed instructions are required. Applications are to be made to the licensing authority for the area in which the applicant resides and applications for renewal are to be made to the same licensing authority. Applications for renewal can only be made within the first two months of a three-month period before the date of expiration (section 117).

Consideration of applications is biased towards the applicant and applications must be approved so long as:

(i) the applicant is aged 18 or over;
(ii) the applicant possesses a qualification or is of a description prescribed by the Secretary of State;
(iii) the applicant has not forfeited a personal licence within five years of the application; and
(iv) he has not been convicted of any relevant offence or any foreign offence (section 120).

Where the applicant has been convicted of 'relevant' or 'foreign' offences, the Licensing Act must consult the police before any licence is granted (sections 120 and 121). 'Relevant offences' are those detailed in schedule 4 and include:

- those involving serious crime;
- those involving serious dishonesty;
- those involving the supply of drugs;
- certain sexual offences;
- offences created by the Licensing Act 2003.

'Foreign offences' are those, other than 'relevant offences', under the law of any place outside England and Wales (section 113).

The police may issue a notice of objection if it is considered that the grant of the licence would undermine the prevention of crime and disorder. In the absence of such objection, the licence must be granted. If the licensing authority receives a police objection notice, it must arrange a hearing and must then give reasons for its decision on the application (section 120).

Where the application is one for renewal, the application must be approved unless there have been 'relevant' offences since the original grant of the licence or its renewal. Where there are such convictions, the police must be consulted and may issue an objection notice where the crime prevention objective is threatened. If objection is made, it is to be followed by a hearing (section 121).

Where the licensing authority becomes aware of 'relevant' or 'foreign' convictions during the life of a personal licence, it must notify the police, who may issue an objection notice. The licensing authority must then arrange a hearing and the licence may be revoked. Decisions must be notified to the police and the holder with reasons (section 124).

The licence holder must notify the licensing authority of any 'relevant' or 'foreign' conviction in the life of the licence unless the court has forfeited or suspended the licence (section 132).

Where the holder of a personal licence is convicted of a 'relevant' offence, the court may forfeit the licence or suspend it for up to six months. Such orders may be suspended to allow time for appeal (section 129). The court must notify the licensing authority where the holder of a personal licence is convicted of a 'relevant' offence (section 131) and the holder must notify the licensing authority of convictions. Notification of decisions of the licensing authority is to be given to both the applicant and the police (section 122).

Appeals against the refusal of the grant, or renewal of a personal licence or its revocation, may be made to the magistrates' court within 21 days. The police may also appeal against the grant or renewal of a personal licence despite an objection notice from them (schedule 5, part 3, paragraph 17).

Premises licences (PLs) (FC142)

This is a licence which authorises the use of specified premises for the carrying on of one or more 'licensable activities' (section 11). A premises licence is effective until it is revoked or surrendered and is not time related unless, at the request of the applicant, it is for a limited period (section 26).

The applicant will normally be the person carrying on or proposing to carry on the business involving 'licensable activities' (section 16). The application must be in the prescribed form and accompanied by:

(a) an operating schedule, including:
    (i) opening hours;
    (ii) duration of licence if time-related licence required by applicant;
    (iii) details of proposed 'designated premises supervisor';
    (iv) whether alcohol is to be supplied;
    (v) how the 'licensing objectives' will be achieved;
(b) a plan of the premises;
(c) in the case of premises supplying alcohol, the consent of the 'designated premises supervisor' named in the application (section 17);
(d) the prescribed fee.

Applications must be advertised and arrangements made for representations to be made in accordance with the Licensing Act 2003 (Premises Licences) Regulations 2005 (section 17(5)).

If representations are not received and the application has been properly made, the licence must be granted (section 18(1)). Where 'relevant' representations (i.e. made by an 'interested party' or a 'responsible authority' relating to the likely effect of the proposal on the 'licensing objectives') are made, the

FC142 Premises licences – Sections 11 to 59 Licensing Act 2003

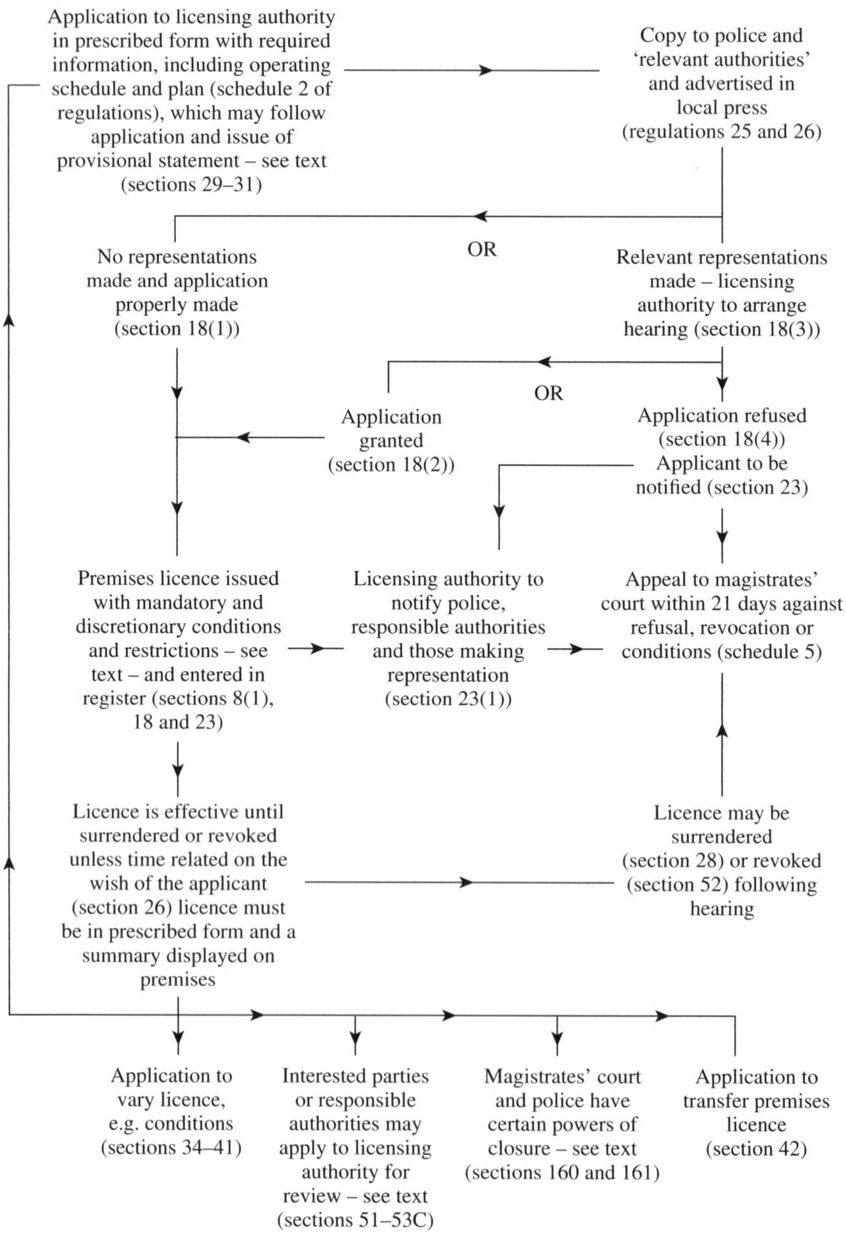

Note

1. Regulation numbers are of the Licensing Act 2003 (Premises Licences and Club Premises Certificates) Regulations 2005 as amended.

licensing authority must arrange a hearing unless all parties agree this to be not required (section 18).

In granting the licence, the licensing authority may attach conditions that relate to the promotion of the 'licensing objectives' and these may include:

- ruling out certain licensable activities;
- rejecting a specified premises supervisor,

but they must be consistent with those conditions listed by the applicant in the operating schedule. The licensing authority may impose different conditions on different parts of the premises and on different licensable activities (section 18(10)) – also see 'Conditions' on pages 553, 569.

Alternatively, and following the hearing, the licensing authority may wish to refuse to grant a licence (section 18).

Mandatory conditions

The licensing authority is required to apply prescribed conditions where applications relate to matters including:

- the supply of alcohol;
- exhibition of films;
- door supervision; and
- plays (sections 19–22).

Discretionary conditions

Conditions relevant to those set out in the application may be attached – see 'Conditions' on pages 553, 569.

The form of licence, and a summary of it, must be displayed prominently at the premises, as specified in the regulations.

Provisional statements (PS)

Where a premises is to be constructed, extended or altered for use for licensable activities, application may be made to the licensing authority for a provisional statement. This should be in the prescribed form and accompanied by the fee. The application must include details of the work to be carried out (section 29). This allows the person to establish if the works will be acceptable to the licensing authority when completed and lead to the issue of a premises licence or variation.

Applications for provisional statements must be advertised in the prescribed manner to give responsible authorities, interested parties and local residents the chance to make representations (section 30).

Where no representations are made, the licensing authority is to issue a provisional statement indicating that no representations have been made (section 31(1–2)). Where there are 'relevant representations' (see above), the licensing authority must arrange a hearing unless this is agreed as being unnecessary by all parties (section 31(3), (6) and (7)). Following the hearing, the licensing authority must notify the decision to the applicant, police and anyone who made representations (section 31).

Variations

There is provision for premises licences to be varied on application made in the prescribed way. This may include variation in any way, e.g. alteration of conditions or of the licensable activities, but not:

(i) to change the designated premises supervisor, for which there is a separate procedure in section 37;
(ii) to extend the date of expiration of the premises licence where there is a time limit; or
(iii) to vary substantially the premises.

Applications are subject to fee and advertising requirements (section 34).

If there are no representations, the licensing authority must make the variation subject to the mandatory conditions. Where relevant representations are made, a hearing must be held unless dispensed with by agreement of all parties. Notification of the decision is to be made to the applicant, police and all those making representations (section 36).

Transfers

Application may be made to the licensing authority for the transfer of a premises licence using the prescribed form and accompanied by the fee. The application must be copied to the police who may object within 14 days if they believe that the transfer would undermine the crime protection objective (section 42).

The licensing authority must not consider the application unless:

(a) the transfer is to have immediate effect;
(b) the existing holder has consented; or
(c) the applicant has taken all reasonable steps to gain that consent and is in a position to use the premises immediately (section 44).

Where the police have objected, a hearing must be held (section 44(5)) and the applicant and police notified of the decision with the reasons for it (section 45).

Reviews

Interested parties or a responsible authority (which includes a local authority as environmental health authority) may apply to the licensing authority at any time for a review, e.g. residents concerned about public nuisance or police who consider that measures to prevent crime or disorder are not adequate. Hearings must then be held and the licensing authority must, if it considers necessary for the promotion of licensing objectives:

- modify the conditions;
- exclude any licensable activity;
- remove the designated supervisor;
- suspend the licence for no more than three months;
- revoke the licence.

The licensing authority decision following reviews does not have effect until after the period for appeal has expired or the appeal has been heard (sections 51–53C).

Noise at night

For the procedure to deal with excessive noise at night from licensed premises, see FC28.

Club premises certificates (CPCs) (FC144)

These give authorisation for the 'qualifying clubs' (defined in sections 61–62) to use club premises for certain 'qualifying club activities', which are:

(a) the supply of alcohol to a member of the club and to their guests for consumption on the premises (supply off the premises is only allowed in certain circumstances – section 73); and
(b) the provision of regulated entertainment for the members and their guests.

Such premises will include party political clubs, the British Legion, working men's clubs and sports clubs.

The holding of a club premises certificate exempts the club from the requirements for any member or employee to be the holder of a personal licence and have a designated premises supervisor in order to supply or sell alcohol. However, there are special provisions concerning conditions to be attached regarding the sale of alcohol, including its management by a committee (section 64).

Clubs with a club premises certificate are not subject to immediate closing orders by the police and are not affected by magistrates' orders for an area – see below in relation to premises licences and temporary event notices.

Applications for a club premises certificate are to be made in the form prescribed in schedule 9 of the Licensing Act 2003 (Premises Licences and Club Premises Certificates) Regulations 2005 and must be accompanied by the necessary supporting information, including an operating schedule, a plan of the premises and a copy of the club rules. The operating schedule will set out the activities to be undertaken, the hours of opening and the arrangements proposed to achieve the licensing objectives. The application has to be advertised (see applications below) and is subject to representations and hearings as for premises licences above (section 71). In the absence of relevant representations, the application must be granted (section 72). Conditions relevant to those set out in the application may be attached – see 'Conditions' on pages 553, 569 – and the club premises certificate will also be subject to mandatory conditions dealing with the exhibition of films (sections 73–75). Club premises certificates may be subject to variation as for premises licences.

The procedure set out in FC28 for dealing with excessive noise at night applies to premises holding a club premises certificate.

Permitted temporary activities (PTAs) (FC144)

A permitted temporary activity is one that goes ahead in accordance with a temporary event notice submitted by the 'premises user' to the licensing authority and where:

(i) the temporary event notice has been acknowledged by the licensing authority;

FC143 Club premises certificates (CPC) – Sections 60 to 97 Licensing Act 2003

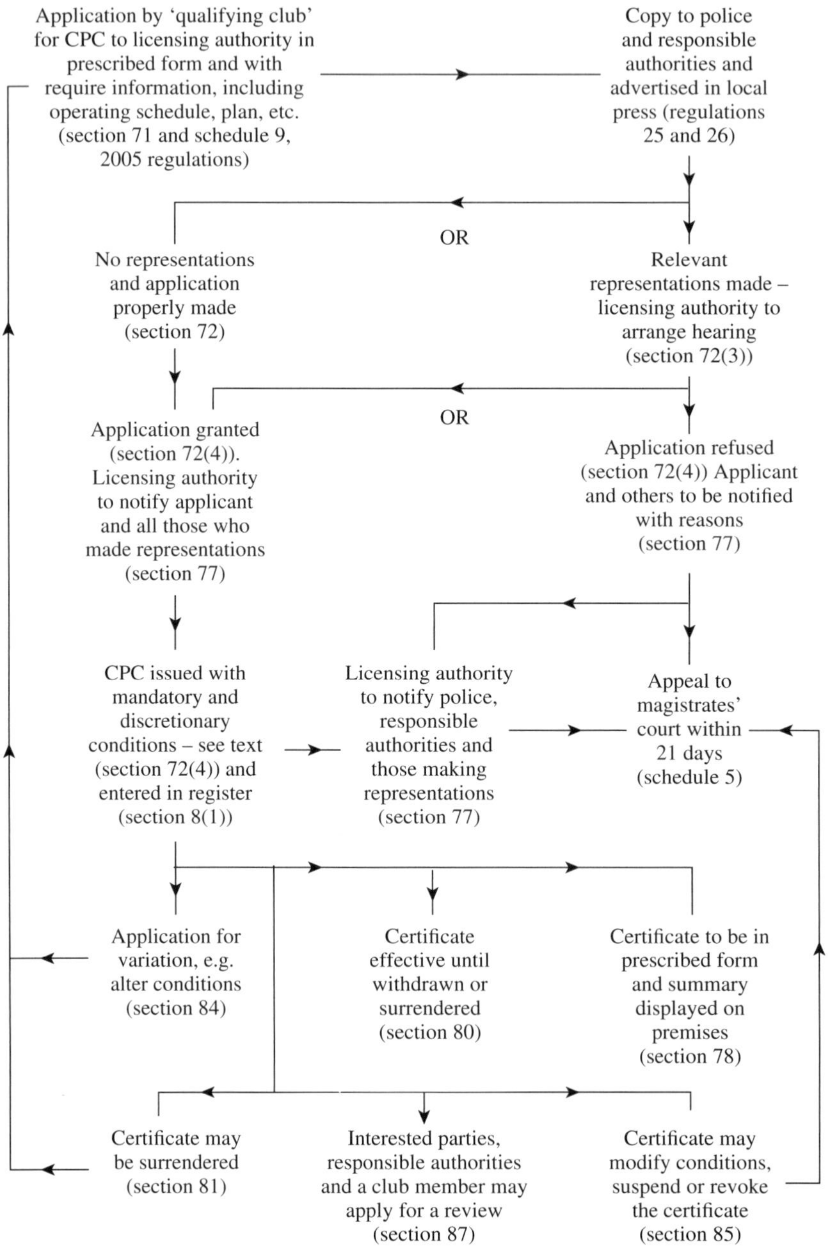

Note

1. The regulations in this procedure refer to the Licensing Act 2003 (Premises Licences and Club Premises Certificates) Regulations 2005.

(ii) the temporary event notice has not been subsequently withdrawn by the premises user; and
(iii) the licensing authority has not issued a counter-notice (see below) (section 98).

A temporary event notice allows premises users over 18 to carry out licensable activities on a temporary basis for no more than 96 hours subject to conditions and restrictions specified in the Act, including a maximum attendance of 500 – see below. Examples of situations where these might be used are:

(i) for a personal licence holder to carry out activities at a premises which does not have a premises licence;
(ii) for a person who does not hold a personal licence to carry out licensable activities at a premises which may or may not be licensed (section 100).

Temporary event notice (TEN)

This must be submitted by the premises user to the licensing authority in the prescribed form, and containing specified information, and copied to the chief of police at least ten days before the event in a form prescribed by the Secretary of State in the Licensing Act 2003 (Permitted Temporary Activities) (Notices) Regulations 2005. This will include details of:

(a) the licensable activities to be carried out;
(b) the total length of the event, which must not exceed 96 hours;
(c) the times;
(d) the maximum number of people to be admitted, which must be less than 500;
(e) whether alcohol is to be served (section 100).

The licensing authority must acknowledge the temporary event notice before the time of the event by returning the duplicate copy of it (section 102).

The temporary event notice may be withdrawn up to 24 hours before the event and, in this case, the intended duration does not count towards maximum periods above (section 103).

A permitted temporary activity event does not require the specific authorisation of the licensing authority. The temporary event notice is a notification process and not an application for approval. It is for the licensing authority to check that the various limitations as imposed by the Act on frequency of use etc. are not to be breached and to deal with any police objections on the grounds of a threat to crime prevention objective. The licensing authority may only intervene if a counter-notice is necessary – see below.

Counter-notice (CN)

The licensing authority must issue a counter-notice to the applicant where:

(i) the premises user is a personal licence holder who has already given 50 temporary event notices or ten late temporary event notices within the same calendar year; or

FC144 Permitted temporary activities (PTA) – Sections 98 to 110 Licensing Act 2003

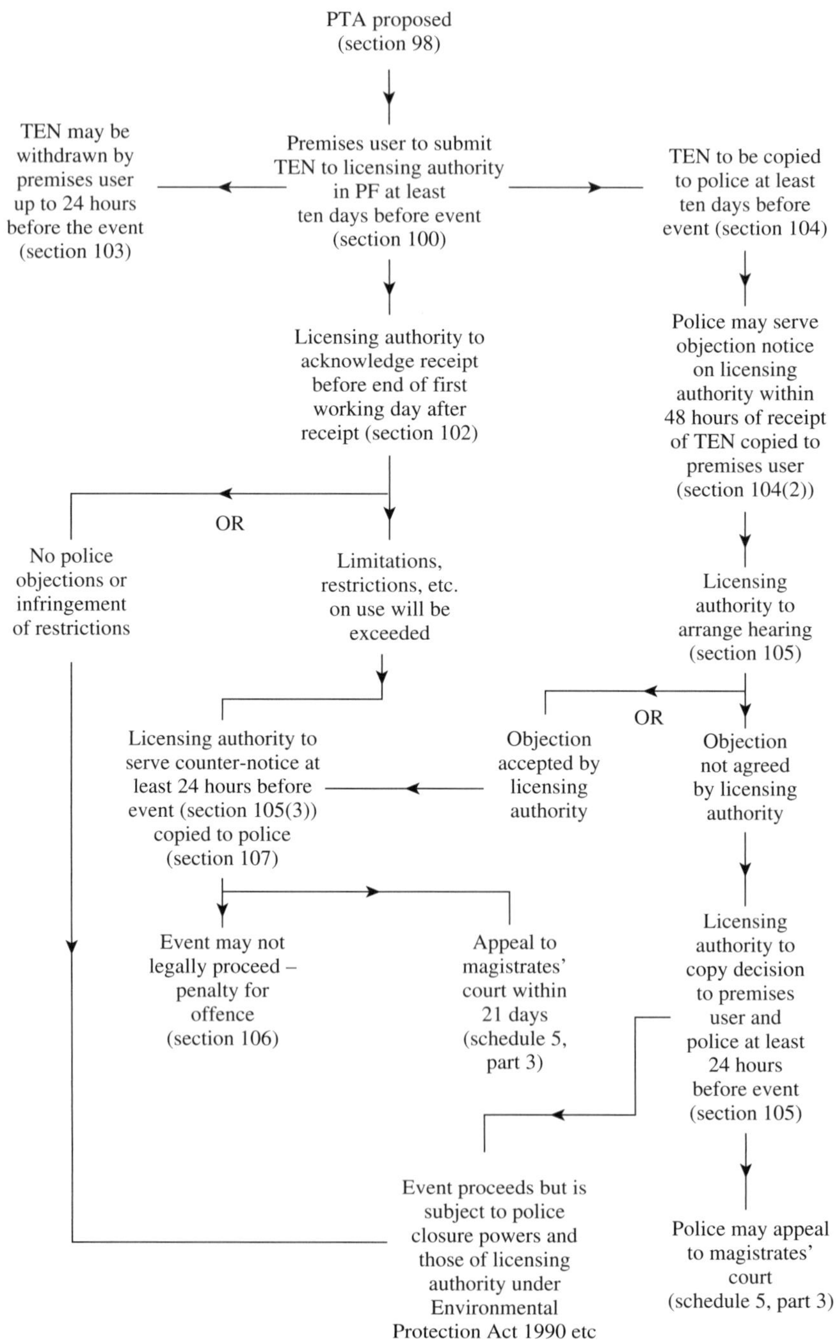

Notes

1. TEN = temporary event notice.
2. PF = prescribed form.

(ii) the premises user does not hold a personal licence but has already given five temporary event notices or two late temporary event notices in that year; or
(iii) 12 temporary event notices have been given for the same premises in that year; or
(iv) temporary event notices are given for events at the same premises falling on more than 21 days in a year (section 107); and
(v) there is a police objection to a temporary event notice (see below).

The licensing authority must serve a counter-notice at least 24 hours before the date of the event (section 104). The effect of the counter-notice is that the event may not legally proceed (section 106).

The licensing authority has no powers to attach additional terms, conditions or limitations to those that are applied by the Act. Problems including noise would have to be dealt with either under other powers, including the Environmental Protection Act 1990, or by exercising closure powers – see FC145.

Police objections to temporary event notices

If, after receiving the copy of the temporary event notice, the police believe that the event would undermine the crime prevention objective, they must notify the licensing authority and the premises user by objection notice within 48 hours of receiving it, giving reasons for the objection (section 104).

The licensing authority must then arrange a hearing unless all parties agree this to be unnecessary. If the licensing authority accepts the police objection, it must issue a counter-notice (see above) to prevent the event proceeding. If the licensing authority does not accept the objection, it must notify the police. Any decision or counter-notice must be given to the premises user at least 24 hours before the event, otherwise it will be able to go ahead (section 105).

In the period between the service of a counter-notice and a hearing, the police and the applicant are able to agree such amendments to the temporary event notice as would allow the objection to be withdrawn (section 106).

Conditions attached to premises licences and club premises certificates

In addition to the mandatory conditions – see page 547 – the licensing authority may only attach conditions, restrictions or limitations so far as they are necessary for the promotion of the licensing objectives. Such conditions etc. should initially emerge from the applicant's own risk assessment of the premises and then be incorporated into the operating schedule to form part of the application. Early consultation with responsible authorities and residents is desirable and the applicant's proposed conditions etc. will also form part of the statutory consultation that forms part of the application process and in any subsequent hearings. The licensing authority will therefore be unable to include any discretionary conditions unless the need for them has arisen through the making of relevant representations and then only in relation to the licensing objective under which the representation was made (sections 18, 71 and 72).

Chapter 10 of the guidance provides help on conditions.

There is no provision that provides for the attachment of conditions etc. to permitted temporary activities.

Closure of licensed premises (FC145)

1 By the environmental health authority

Under the Anti-Social Behaviour Act 2003, the chief executive officer of the licensing authority responsible for environmental health functions may make a closure order for a premises subject to a premises licence or temporary event notice where:

(a) a public nuisance is being caused; and
(b) closure is necessary to prevent that nuisance (section 40(1)).

This power may be delegated to an environmental health officer (section 41(2)).

The period of closure cannot exceed 24 hours and begins at the time the manager receives written notice of the order (section 40(3)).

After service of the notice, the chief executive or environmental health officer, if authorised:

(i) may cancel the order in writing;
(ii) must cancel it if it is no longer necessary; and
(iii) must give notice as soon as is reasonably practicable to the licensing authority (section 41(1)).

Persons who, without reasonable excuse, open the premises in contravention of the notice are guilty of an offence and liable to a penalty of imprisonment for up to three months or a fine not exceeding £20,000, or both (section 40(5)).

2 By a magistrates' court

Upon application by the police and where it is considered necessary to prevent disorder, a magistrates' court is able to close all premises with a premises licence or temporary event notice (but not a club premises certificate) within a particular geographical area for a period not exceeding 24 hours (section 160) and it is an offence to open the premises or event during that period.

3 By the police

A senior police officer has the power to make a closure order in respect of a specified premise subject to a premises licence or temporary event notice (but not a club premises certificate) for up to 24 hours where there is actual or likely disorder such that public safety is threatened or where it is necessary to prevent a nuisance due to noise from the premises (section 161). The closure order may

FC145 Closure of licensed premises – Sections 40 and 41 Anti-Social Behaviour Act 2003

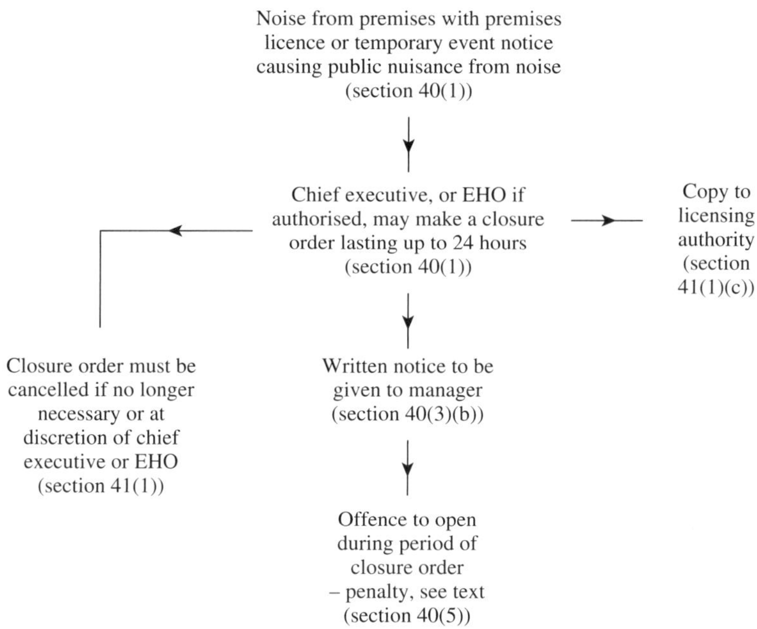

Note

1. There is no appeal against this procedure.

be extended in certain circumstances by further periods of 24 hours (section 162).

The police must apply to the magistrates' court as soon as possible after the closure order has been made to allow the court to arrange a hearing. Notice of this must be given to the licensing authority. The court may either:

(a) revoke the closure order; or
(b) order its extension until the licensing authority has conducted a review (see above) (sections 164–165).

Where the licensing authority receives notice from the court that a closure order is being considered, it must review the premises licence. Having done so, the options available to the licensing authority to pursue the licensing objectives are:

(i) revocation of the licence;
(ii) modification of licence conditions;
(iii) exclusion of certain licensable activities;
(iv) removal of the designated premises supervisor (section 167).

Applications

The forms of application, details to be included and other detailed matters for procedures dealt with here are contained in the following regulations:

- Licensing Act 2003 (Permitted Temporary Activities) (Notices) Regulations 2005;
- Licensing Act 2003 (Personal Licences) Regulations 2005;
- Licensing Act 2003 (Premises Licenses and Club Premises Certificates) Regulations 2005.

The latter set of regulations includes provisions that require the advertising of applications for premises licences, provisional statements and club premises certificates. Regulations 25 and 26 require advertisement of applications:

(a) by posting a notice outside of the premises concerned for at least 28 days after submission of the application; and
(b) inserting a notice in a local newspaper at least once in the ten days following the submission.

Applications must also be copied to each responsible authority. These are the:

- police;
- fire and rescue authority;
- health and safety enforcing authority;
- planning authority;
- environmental health authority;
- bodies representing the interests of children;
- other licensing authorities where part of the premises is in another area;
- and for vessels:
  (a) navigation authority;
  (b) Environment Agency;

(c) British Waterways Board (section 13(4)) (Licensing Act 2003 (Premises Licences and Club Premises Certificates) Regulations 2005).

Fees

Fees are payable to the licensing authority in the following cases:

1. Premises licences and club premises certificates:
   (a) grant and variation of licence;
   (b) an annual fee.
   These fees are calculated on the basis of bands that relate to rateable value and capacity of the premises.
2. Permitted temporary activities, personal licences and miscellaneous notices:
   (a) applications;
   (b) notices, e.g. change of name or address;
   (c) applications for provisional statements etc.
   These are fixed amounts.

Exemptions from the application and annual fee requirements are made but only in relation to regulated entertainment at:

- schools and colleges where the entertainment is on behalf of the school or college;
- in church halls, village halls and the like (Licensing Act 2003 (Fees) Regulations 2005).

Hearings

Throughout the procedures involved in the Licensing Act 2003, provision is made for the licensing authority to arrange for a hearing to deal with representations that have been made about the various applications to them. The Licensing Act 2003 (Hearings) Regulations 2005 as amended makes detailed requirements as to how these hearings are to be arranged and conducted, including:

- notice of hearing;
- rights of attendance and the making of representations;
- procedures at the hearing;
- determination and the time at which these must be made;
- notices of determination;
- records of proceedings.

The schedules indicate to whom notices of hearing and determinations should be sent, the timing of these and the documents that the licensing authority must send out. Proceedings must be in public unless there is a greater public interest served in not doing so.

Register

Each licensing authority is required to keep a register which records details of the licences and other authorisations issued, of the temporary event notices received and of notices and applications received. The form of the register is set out in schedule 3 (section 8(1)).

In addition, the register must also contain the details laid down in the Licensing Act (Licensing Authority's Register) (Other Information) Regulations 2005, e.g. operating schedules and lists of plans and works to be undertaken.

The register must be available to the public free of charge during office hours and a copy of information must be supplied on request (for a fee) (section 8(3–5)).

Appeals

The provisions for appeals against the various decisions made by licensing authorities in dealing with all types of licences are contained in schedule 5. Depending upon the particular circumstances, appeals may be instigated by applicants, holders of licences, the police and persons making 'relevant representations'. Generally appeals are to be made within 21 days from the day on which the decision was notified and, in relation to the making of temporary event notices, not later than five working days before the event is to begin. All appeals are to be made to the magistrates' court (schedule 5).

The magistrates' court may dismiss the appeal, alter the decision or remit the matter back to the licensing authority (section 181).

Offences

An offence is committed if a person carries on a licensable activity without a premises licence, a club premises certificate or a temporary event notice and the penalty upon summary conviction is imprisonment for up to six months and/or a fine not exceeding £20,000 (section 136).

GAMBLING ACT 2005

Responsible authorities must be consulted on Gambling Act and Licensing Act applications. The responsible authority for the prevention of risk of pollution of the environment or harm to human health is a statutory consultee for these purposes and in local authorities this is placed in the environmental health service.

Licensing committee

Each licensing authority must establish a licensing committee consisting of between 10 and 15 members of the authority (section 6, Licensing Act 2003). This committee will deal with all licences and permits, including those required by the Gambling Act 2005.

The licensing objectives for gambling are:

(a) preventing gambling from being a source of crime or disorder, being associated with crime or disorder or being used to support crime;
(b) ensuring that gambling is conducted in a fair and open way; and

(c) protecting children and other vulnerable persons from being harmed or exploited by gambling.

The Gambling Commission issues guidance to local authorities on operating the Act and, in particular, the principles to be applied by local authorities in exercising their functions under the Act – 'Guidance to licensing authorities (4th edition)' (with sections 20 and 27 updated and published in February 2013).

The functions of licensing authorities with respect to gambling are:

- to licence premises for gambling activities (casinos, adult gaming centres, betting shops, tracks, bingo premises, family entertainment centres);
- to consider notices given for the temporary use of premises for gambling;
- to grant permits for gaming and gambling machines in clubs and miners' welfare institutes;
- to regulate gaming and gaming machines in alcohol-licensed premises;
- to grant permits to family entertainment centres (FECs) for the use of certain lower-stake gaming machines;
- to grant permits for prize gaming;
- to consider occasional use notices for betting at tracks;
- to register small societies' lotteries.

Definitions

Betting means making or accepting a bet on:

(a) the outcome of a race, competition or other event or process;
(b) the likelihood of anything occurring or not occurring; or
(c) whether anything is or is not true (section 9).

Betting intermediary means a person who provides a service designed to facilitate the making or acceptance of bets between others. Acting as a betting intermediary is providing facilities for betting (section 13).
Casino is an arrangement whereby people are given an opportunity to participate in one or more casino games.
Casino game means a game of chance which is not equal chance gaming (section 7).
Complex lottery is where persons are required to pay in order to participate in the arrangement and, in the course of the arrangement, one or more prizes are allocated to one or more members of a class and the prizes are allocated by a series of processes, and the first of those processes relies wholly on chance.
Equal chance gaming is when gaming does not involve playing or staking against a bank, and the chances are equally favourable to all participants (section 8).
Facilities for gambling means:

(a) inviting others to gamble in accordance with arrangements made by him;
(b) providing, operating or administering arrangements for gambling by others; or

(c) participating in the operation or administration of gambling by others.

But it does not include providing an article other than a gaming machine to a person who intends to use it, or may use it, in the course of any of the above activities or making facilities for remote communication available for use by persons carrying on any of those activities, or persons gambling in response to those activities with certain conditions (section 5).
Gambling means gaming, betting and participating in a lottery (section 3).
Game of chance includes a game that involves both an element of chance and an element of skill, a game that involves an element of chance that can be eliminated by superlative skill and a game that is presented as involving an element of chance, but does not include a sport.
Gaming means playing a game of chance for a prize.
Interested party – a person is an interested party in relation to a premises licence or in relation to an application for or in respect of a premises licence if, in the opinion of the licensing authority which issues the licence or to which the application is made, the person:

(a) lives sufficiently close to the premises to be likely to be affected by the authorised activities;
(b) has business interests that might be affected by the authorised activities; or
(c) represents persons who satisfy paragraph (a) or (b) (section 158).

Premises licence – a licence which states that it authorises premises to be used for:

(a) the operation of a casino (a 'casino premises licence');
(b) the provision of facilities for the playing of bingo (a 'bingo premises licence');
(c) making Category B gaming machines available for use (an 'adult gaming centre premises licence');
(d) making Category C gaming machines available for use (a 'family entertainment centre premises licence'); or
(e) the provision of facilities for betting, whether by making or accepting bets, by acting as a betting intermediary or by providing other facilities for the making or accepting of bets (a 'betting premises licence') (section 150).

Prize in relation to gaming (except in the context of a gaming machine) means money or money's worth, and includes both a prize provided by a person organising gaming and winnings of money staked (section 6).
Prize in relation to lotteries includes any money, articles or services – whether or not described as a prize, and whether or not consisting wholly or partly of money paid, or articles or services provided, by the members of the class among whom the prize is allocated (section 14).
Provisional statement provides a level of assurance about the outcome of a subsequent premises licence application and provides the holder with some protection against representations when they make an application for a premises licence.

Responsible authorities

(a) a licensing authority in England and Wales in whose area the premises are wholly or partly situated;
(b) the Commission;
(c) the chief officer of police or chief constable for a police area in which the premises are wholly or partly situated;
(d) the fire and rescue authority for an area in which the premises are wholly or partly situated;
(e) the local planning authority, for an area in which the premises are wholly or partly situated;
(f) an authority which has functions by virtue of an enactment in respect of minimising or preventing the risk of pollution of the environment or of harm to human health in an area in which the premises are wholly or partly situated;
(g) a body which is designated in writing, by the licensing authority for an area in which the premises are wholly or partly situated, as competent to advise the authority about the protection of children from harm;
(h) Her Majesty's Commissioners of Customs and Excise; and
(i) any other person prescribed for the purposes of this section by regulations made by the Secretary of State (section 157).

Simple lottery is where persons are required to pay in order to participate in an arrangement and, in the course of the arrangement, one or more prizes are allocated to one or more members of a class, and the prizes are allocated by a process which relies wholly on chance (section 14).

All sections referred to above are from the Gambling Act 2005.

PREMISES LICENCES (GAMBLING)

References

Gambling Act 2005.
Gambling Act 2005 (Premises Licences and Provisional Statements) Regulations 2007.
Gambling Act 2005 (Mandatory and Default Conditions) (England and Wales) Regulations 2007.
Gambling Act 2005 (Operating Licence Conditions) Regulations 2007.
Gambling Act 2005 (Premises Licences) (Review) Regulations 2007.
Gambling (Premises Licence Fees) (England and Wales) Regulations 2007.
Gambling (Operating Licence and Single-Machine Permit Fees) Regulations 2006.

Extent

These provisions apply to England, Wales and Scotland.

FC146 Premises licences – Gambling Act 2005 and Gambling Act 2005 (Premises Licences and Provisional Statements) Regulations 2007

Licensing authority

Applicant

Application for premises licence (sections 151 and 159)

Notice within seven days to responsible authorities (regulations 12-16)

Applicant publishes notice of application in local newspaper within 28 days

Notice on premises or where premises is to be built

Representations from interested parties and responsible authorities within 28 days (regulation 15)

Licensing authority intends to attach or remove conditions to licence

Licensing authority need not hold hearing with permission of applicant, interested parties and responsible authorities if licensing authority thinks representations are vexatious etc. (section 162)

Notify all parties (section 162 and regulation 12)

Notice of hearing to interested parties (regulation 6)

Grant licence (section 163)

Issue licence (sections 164, 165 and 169)

Reject licence (section 163)

Reasons for objection (sections 164, 165 and 169)

Give notice to applicant, commission, chief police officer and HM Customs and Excise (sections 164, 165 and 169)

Licensing authority on notification or application can amend, vary or transfer licence (sections 186, 187 and 188)

Applicant surrenders licence or licence lapses (sections 192 and 194)

Licensing authority to notify commission, chief police officer and HM Customs and Excise (section 194)

Note

1. Sections refer to the Gambling Act 2005 and regulations to Gambling Act 2005 (Premises Licences and Provisional Statements) Regulations 2007.

Scope

Most functions under the Act are usually delegated to a licensing committee in England and Wales. Licensing authorities can determine an application for a premises licence, for the variation or transfer of a premises licence, for a provisional statement and for a review of a premises licence (section 154). Wider delegations are available in Scotland (section 155).

Principles to be applied

A licensing authority permits the use of premises for gambling in accordance with any relevant codes of practice, reasonably consistent with the licensing objectives and in accordance with the licensing statement published by the authority. In determining whether to grant a premises licence, a licensing authority cannot consider the expected demand for the facilities which it is proposed to provide (section 153).

Register

A licensing authority must maintain a register of premises licences and other gambling-related licences issued by the authority. The register and information must be available for inspection by members of the public at all reasonable times, and at the request of a member of the public make arrangements for a copy of an entry in the register, or of information, to be made available on the payment of a fee (section 156).

Making an application

A person may apply to a licensing authority for a premises licence to carry on an activity to:

- operate a casino;
- provide facilities for the playing of bingo;
- make a gaming machine available for use;
- provide other facilities for gaming; or
- provide facilities for betting (section 37(1)).

An application may be made only by a person who holds an operating licence which authorises him to carry on the activity on the premises where the licence is sought and has a right to occupy the premises to which the application relates.

An application must be made in the prescribed form and manner, and contain or be accompanied by the prescribed information or documents with the prescribed fee (section 159).

Form of licence

A premises licence must specify:

(a) the name of the person to whom it is issued;

(b) a home or business address of that person;
(c) the premises to which it relates;
(d) the activities for which it authorises the premises to be used;
(e) any condition attached by the licensing authority;
(f) any exclusion of a default condition effected by the licensing authority.

It must also include a plan of the premises and the date the licence will expire (section 151).

For the detail of the form and content of a premises licence, see the Gambling Act 2005 (Premises Licences and Provisional Statements) Regulations 2007 as amended.

Note – a premises licence cannot refer to more than one activity except in the case of a track, but special arrangements must be made for this (section 152).

Notice of application

Applications for a premises licence, variation, transfer, reinstatement of a premises licence and for a provisional statement must give notice of the application to each of the responsible authorities. The notice must be given within seven days of the date on which the application is made.

The applicant must publish notice of his application in a local newspaper and display the notice on the exterior of the premises to which the application relates for 28 consecutive days starting on the day on which the application is made to the licensing authority.

If the premises has not yet been built, the applicant must display the notice at a place which is as near as reasonably practicable to the premises or proposed premises where it can conveniently be read by members of the public.

Where a person fails to give proper notice of his application or to publish the notice of application within the relevant period in the correct form or manner, the responsible authority concerned must make any representations to the licensing authority within 28 days of receiving the notice.

The licensing authority cannot grant the application until the correct notice has been given by the applicant.

An application or a notice must be made or given in writing, which includes facsimile transmission or electronic mail with the prescribed fee (regulations 12 to 16 of the Gambling Act 2005 (Premises Licences and Provisional Statements) Regulations 2007).

Representations

An interested party or responsible authority may make representations in writing to the licensing authority concerning an application (section 161), within 28 days beginning on the date on which the application was made to the licensing authority (regulation 15 of the Gambling Act 2005 (Premises Licences and Provisional Statements) Regulations 2007).

HEARINGS ABOUT GAMBLING PREMISES LICENCES

In this section, regulation references are to the Gambling Act 2005 (Proceedings of Licensing Committees and Sub-Committees) (Premises Licences and Provisional Statements) (England and Wales) Regulations 2007.

Requirement for hearing

A licensing authority must hold a hearing to consider an application for a premises licence if:

(a) an interested party or responsible authority has made representations about the application;
(b) the authority proposes to attach a condition to the licence; or
(c) the authority proposes to exclude a condition that would otherwise be attached to the licence.

However, a licensing authority may determine an application for a premises licence without a hearing:

(a) with the consent of the applicant and any interested party or responsible authority who has made representations about the application;
(b) if the authority thinks that the representations are vexatious, frivolous or will certainly not influence the authority's determination of the application.

If a licensing authority proposes to determine an application without a hearing, they shall as soon as is reasonably practicable notify any person who made representations (section 162).

The hearing must be held subject to the Gambling Act 2005 (Proceedings of Licensing Committees and Sub-Committees) (Premises Licences and Provisional Statements) (England and Wales) Regulations 2007.

The hearing is to be held as soon as is reasonably practicable after the expiry of any period for representations (regulation 4).

Notice of hearing

The committee must give notice of any hearing, specifying the date on which, the place at which and the time when the hearing is to take place, stating that the relevant committee will make available the relevant documents to any person who has made representations and the licensee (for transfers) so that it is received no later than ten working days before the hearing is to be held (regulation 5).

Information and documents to accompany the notice of hearing

The notice of hearing must be accompanied by information in writing explaining the following:

(a) the consequences where a party informs the relevant committee that he does not wish to attend or be represented at the hearing, or fails to inform the relevant committee whether he wishes to attend or be represented at the hearing;

(b) the requirements imposed on the relevant committee in conducting a hearing;
(c) the consequences where a party has indicated that he wishes to attend or be represented at the hearing, but fails to attend or be represented at the hearing;
(d) the procedure to be followed at the hearing;
(e) the time limit and method, if any, by which a party should inform the relevant committee that he wishes to attend or address the hearing;
(f) the time limit and method, if any, by which a party should inform the relevant committee that he wishes to be assisted or represented by another person;
(g) the time limit and method, if any, by which a party should inform the licensing authority that he will want to call a witness to give evidence at the hearing, and the matters in relation to which he wishes that witness to give evidence;
(h) the time limit and method, if any, by which a party should inform the relevant committee that he wishes to withdraw any representations;
(i) the time limit and method, if any, by which a party should inform the relevant committee that he is willing to consent to the application being determined without a hearing;
(j) the matters, if any, on which the relevant committee considers at the time that it will want clarification at the hearing from a party.

The documents must be sent to the applicant, and the licensee (for an application or a review), and a person who has made representations (where requested), the licensee (regulation 6).

Power to postpone

The committee may postpone a hearing to a specified date, or arrange for a hearing to be held on a date specified by the committee if it considers it necessary to enable it to consider any information or documents provided by any party in response to a notice, or at the hearing, or having regard to the ability of any party, person representing a party or witness to attend the hearing.

Where the committee has adjourned a hearing to a specified date or specified an additional date, it must notify the parties of the new date, time and place for the hearing (regulation 7).

Proceedings of a relevant committee in conducting a hearing

The hearing must take place in public; however, the committee may direct that all or part of a hearing must be in private if it is satisfied that it is necessary in all the circumstances of the case, having regard to any unfairness to a party that is likely to result from a hearing in public and the need to protect, as far as possible, the commercial or other legitimate interests of a party (regulation 8).

The committee must allow a party to attend a hearing and be assisted or represented by any person, whether or not that person is legally qualified.

At the beginning of the hearing, the committee must explain the procedure that it proposes to follow in conducting the hearing.

In conducting a hearing, the committee must ensure that each party is given the opportunity to address the relevant committee or call witnesses to give evidence on any matter relevant to the application or review, or any representations made on the application or review.

The committee must also allow any party to question any other party or person representing a party and take into consideration documentary or other information in support of the application or representations produced by a party before the hearing; or at the hearing, with the consent of all the other parties attending the hearing (regulation 4).

Failure of parties to attend the hearing

The committee can proceed with a hearing in the absence of a party or a party's representative if the party has:

(a) informed the committee that he does not intend to attend or be represented at the hearing;
(b) failed to inform the committee whether he intends to attend or be represented at the hearing; or
(c) left the hearing in circumstances enabling the committee reasonably to conclude that he does not intend to participate further.

If a party has indicated that he does intend to attend or be represented at the hearing, but fails to attend or be represented, the relevant committee may adjourn the hearing to a specified date if it considers it to be in the public interest, or it may proceed with the hearing in the party's absence.

Where the hearing proceeds in the absence of a party, the committee must consider the application or representations made by that party.

Where the committee adjourns the hearing to a specified date, it must notify the parties of the date, time and place to which the hearing has been adjourned (regulation 10).

Exclusion of disruptive persons

The committee can require any person attending the hearing who, in their opinion, is behaving in a disruptive manner to leave the hearing and can refuse to permit him to return, or permit him to remain or return only on such conditions as the committee specifies.

Where a person is required to leave the hearing, the committee must permit him to submit in writing, before the end of the hearing, any information which he would have been entitled to give orally had he not been required to leave and it must take into account that information in reaching a determination of the application or review (regulation 11).

Procedure where a hearing is not to take place

Where the parties have notified the committee that they consent to the application or review being determined without a hearing, the committee must notify all the parties that the hearing has been dispensed with, and determine the application or review (regulation 12).

Determination of an application or a review

Following a hearing, the committee must determine the application or review within five working days after the hearing (regulation 13).

Power to extend time

The committee may extend the five-day time limit for a specified period where it considers an extension to be in the public interest. Where the committee has extended the time limit, it must give a notice of the extension to the parties, stating the period of the extension and the reasons for it (regulation 14).

Record of proceedings

The committee must ensure that a record of the hearing is taken in a permanent and intelligible form and that the record is kept for six years from the date that the application or review is finally determined (regulation 15).

Irregularities and clerical mistakes

The committee may disregard any irregularity resulting from a failure to comply with a provision of the regulations where that irregularity comes to its attention prior to it making a determination of the application or review. If the committee considers that any person may have been prejudiced by any such irregularity, it must take such steps that it considers necessary to remedy the consequences of the irregularity before reaching its determination (regulation 16).

The committee may correct clerical mistakes in any document recording a determination of the committee or errors arising in such a document from an accidental slip or omission (regulation 17).

Determination of application

On considering an application for a premises licence, a licensing authority must grant it or reject it (section 163).

Where a licence is granted or rejected, they must give notice of the grant to the applicant, the Commission, any person who made representations about the application, the chief officer of police and Her Majesty's Commissioners of Customs and Excise.

The licensing authority must issue a premises licence to the applicant with a summary of the terms and conditions of the licence in the prescribed form and include reasons for conditions and, if representations were made about the application, the authority's response to the representations. Where the application is rejected, the reasons for objection must be given to the applicant etc. (sections 164, 165 and 169).

OTHER PROVISIONS REGARDING GAMBLING PREMISES LICENCES

Conditions

Conditions cannot be imposed requiring membership in clubs and other bodies, nor can conditions be included imposing limits on stakes, fees, winnings or prizes (section 170 and 171). Conditions on licences are included in the Gambling Act 2005 (Mandatory and Default Conditions) (England and Wales) Regulations 2007.

Door supervision

Where a condition for door supervision is attached to a premises licence, if the person carrying out the guarding is required to hold a licence under the Private Security Industry Act 2001 (c. 12) authorising the guarding, it becomes a condition of the premises licence (section 178).

Annual fee

Premises licences are subject to an annual fee (section 184).

Availability of licence

The licence must be kept on the premises it refers to and made available on request to a constable, an enforcement officer or an authorised local authority officer. It is an offence if he fails, without reasonable excuse for failure, to comply with these requirements and persons guilty of an offence are liable, on summary conviction, to a fine not exceeding level 2 on the standard scale (section 185).

Change of circumstance

Where the holder of a premises licence ceases to reside or attend at the address specified in the licence, he must notify the licensing authority and inform it of a home or business address at which he resides or attends, accompanied by a fee.

Where notification is accompanied by the licence, the licensing authority must make the alteration to the information contained in the licence and return the licence to the licensee.

It is an offence to fail to comply with these requirements and persons guilty of an offence are liable on summary conviction to a fine not exceeding level 2 on the standard scale (section 186).

Variation

The holder of a premises licence may apply to the licensing authority to vary the licence by:

(a) adding, amending or removing an authorised activity;
(b) amending another detail of the licence;
(c) adding, amending, excluding or removing a condition attached to the licence.

The application for variation must be accompanied by a statement of the variation sought, the licence to be varied, or both a statement explaining why it is not reasonably practicable to produce the licence and an application for the issue of a copy of the licence.

In granting an application for variation, a licensing authority must specify a time when the variation shall begin to have effect and may make transitional provision (section 187).

Transfer

A person may apply to a licensing authority for a premises licence to be transferred to him. An application for transfer must specify the time when the transfer is to take effect and be accompanied by a written statement by the licensee consenting to the transfer.

A licensing authority shall grant an application for transfer unless they think it would be wrong to do so, having regard to any representations.

On the grant of an application for the transfer of a premises licence, the licensing authority must alter the licence so that the applicant for the transfer becomes the licensee, specify in the licence the time when the transfer takes effect and make the any other alteration of the licence as appears to them to be required (section 188).

An application for transfer must be accompanied by the licence, or both a statement explaining why it is not reasonably practicable to produce the licence and an application by the licensee for the issue of a copy of the licence.

If an application for transfer states that the applicant has failed to contact the licensee having taken all reasonable steps to do so, the licensing authority must disregard the requirement for a written statement by the licensee consenting to the transfer and take all reasonable steps to notify the licensee, or notify the applicant, of their determination and the reasons for it.

At the request of the applicant, the licence shall have effect as if the applicant for transfer were the licensee during the period beginning with the receipt of the application for transfer by the licensing authority, and ending with the determination of the application by the licensing authority (section 189).

Copy of licence

Where a premises licence issued, or a summary given, is lost, stolen or damaged, the licensee may apply to the licensing authority for a copy accompanied by the prescribed fee.

If satisfied that the licence or summary to which the application relates has been lost, stolen or damaged and, where the licence or summary has been lost or stolen, that the loss or theft has been reported to the police, after granting the application, the authority must issue a copy of the licence or summary to the applicant certified by the authority as a true copy, and the form in which the licence had effect before the loss, theft or damage (section 190).

Surrender

A premises licence shall cease to have effect if the licensee notifies the licensing authority of his intention to surrender the licence and gives the licensing authority either the licence or a written statement explaining why it is not reasonably practicable to produce the licence.

The licensing authority must notify the Commission, the chief officer of police and Her Majesty's Commissioners of Customs and Excise (section 192).

Revocation for failure to pay fee

Where the holder of a premises licence fails to pay the annual fee, the licensing authority must revoke the licence but the licensing authority may ignore this if they think that a failure to pay is attributable to administrative error (section 193).

Lapse

A premises licence will lapse if the licensee dies, the licensee becomes, in the opinion of the licensing authority, incapable of carrying on the licensed activities by reason of mental or physical incapacity, the licensee becomes bankrupt or the licensee's estate sequestrated and, in the case of a company or other legal body, it ceases to exist, or goes into liquidation.

Where the premises licence has lapsed, they must notify the Commission, the chief officer of police and Her Majesty's Commissioners of Customs and Excise (section 194).

Reinstatement

For six months after the lapse, a person may apply to the licensing authority for the licence to be reinstated with the applicant as the licensee.

The licensing authority must grant an application for reinstatement unless they think it would be wrong to do so, having regard to any representations.

The licensing authority must alter the licence so that the applicant for reinstatement becomes the licensee, specify in the licence that the reinstatement takes effect at the time when the application is granted and make any other alteration of the licence as appears to them to be required (section 195).

The application for reinstatement of a premises licence must be accompanied by the licence or both a statement explaining why it is not reasonably practicable to produce the licence and an application for the issue of a copy of the licence.

The licence remains valid while the application is being considered by the licensing authority (section 196).

REVIEW OF GAMBLING PREMISES LICENCES

In this section, reference to sections are to the Gambling Act 2005 and regulations to the Gambling Act 2005 (Premises Licences) (Review) Regulations 2007.

These regulations make provision about the form of, and procedure for, applications for a review of a premises licence.

Application for review

A responsible authority or interested party may apply to the licensing authority for a review by the authority of a premises licence.

The application must be made in the prescribed form and manner, specify the grounds on which the review is sought and contain or be accompanied by the prescribed information or documents (section 197).

Giving of notice of an application

A person making an application must give notice (seven days) of the application to the licence holder and the responsible authorities. The application must be in the specified form, including the period of 28 days when representations about the application may be made to the licensing authority by the person who holds the premises licence, a responsible authority or an interested party in relation to the premises (regulation 4). An application and associated notices must be made or given in writing, which includes facsimile transmission or electronic mail (regulation 12).

Publication of notice of an application and notice of review by the licensing authority

The authority must publish notice of the application

- (a) either:
  - (i) in a local newspaper or, if there is none, a local newsletter, circular or similar document, circulating within the licensing authority's area on at least one occasion within ten working days after the application is made to the authority; or
  - (ii) on the licensing authority's internet website for at least 28 consecutive days; and
- (b) by displaying the notice at a place, for 28 consecutive days:
  - (i) which is as near as reasonably practicable to the premises; and
  - (ii) where it can conveniently be read by members of the public (regulations 5 and 8).

Rejection of application

A licensing authority may reject an application for the review of a premises licence if they think that the grounds on which the review is sought:

(a) do not raise an issue relevant to the principles to be applied;
(b) are frivolous or vexatious;
(c) will certainly not cause the authority to wish to:
  - revoke the licence;
  - suspend the licence for a specified period not exceeding three months;
  - exclude a condition attached to the licence or remove or amend an exclusion;
  - add, remove or amend a condition;

(d) are substantially the same as the grounds specified in an earlier application; or
(e) are substantially the same as representations made in relation to the application for the premises licence.

A licensing authority shall consider the length of time that has elapsed since the making of the earlier application or since the making of the representations.

If a licensing authority considers that paragraphs (a) to (e) apply to some but not all of the grounds on which a review is sought, they may reject the application (section 198).

If the licensing authority does not reject the application, they must grant it (section 199).

Initiation of review by licensing authority

A licensing authority may review the use of a premises licences of a particular class and, in particular, arrangements made by licensees to ensure compliance with the conditions attached.

A licensing authority may review any matter connected with the use of premises in reliance on a premises licence if the authority has reason to suspect that the premises may have been used in purported reliance on a licence but not in accordance with a condition of the licence, or for any reason (which may relate to the receipt of a complaint about the use of the premises) it thinks that a review would be appropriate.

Before reviewing a premises licence, the licensing authority must give notice of their intention to hold the review to the licensee and publish notice of their intention to hold the review (section 200).

Review

As soon as the period for representations has expired, the licensing authority must review the premises licence.

The purpose of the review will be to consider whether to:

(a) revoke the licence;
(b) suspend the licence for a specified period not exceeding three months;

FC147 Review of a premises licence – Gambling Act 2005 and Gambling Act 2005 (Premises Licences) (Review) Regulations 2007

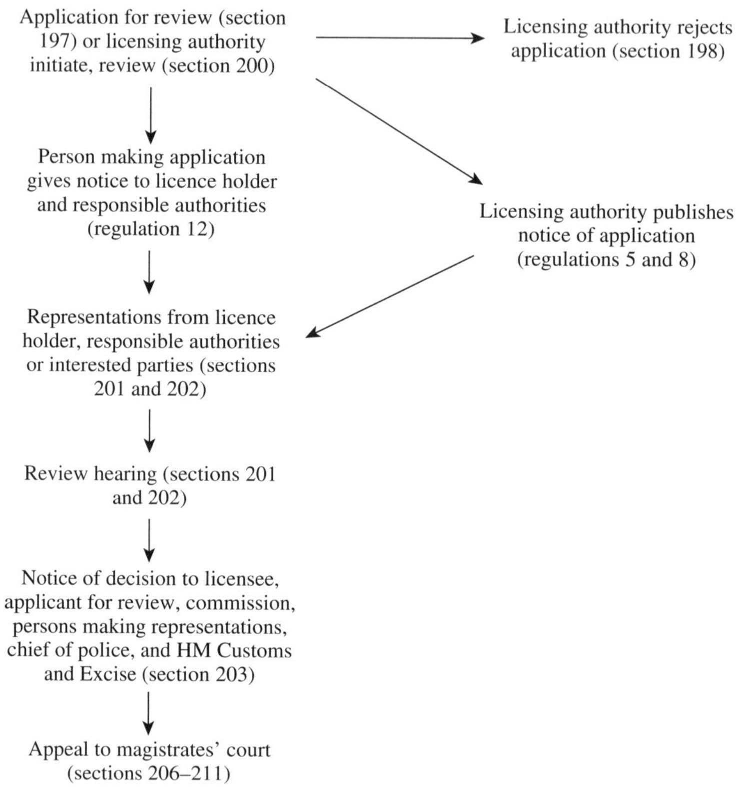

Note

1. References to sections refer to the Gambling Act 2005 and regulations refer to the Gambling Act 2005 (Premises Licences) (Review) Regulations 2007.

(c) exclude a condition attached to the licence or remove or amend an exclusion;
(d) add, remove or amend a condition,

specifying the time at which the action will take effect.

The licensing authority must hold a hearing about the review unless the applicant for the review and each person who has made representations about the review have consented to the conduct of the review without a hearing, or the licensing authority thinks that each representation made about the review is frivolous, vexatious or will certainly not influence the review.

The licensing authority must consider any representations and any grounds specified in the application for the review (sections 201 and 202).

As soon as possible after completion of a review, the licensing authority must give notice of their decision on the review to the licensee, the applicant for the review, the Commission, any person who made representations, the chief officer of police and Her Majesty's Commissioners of Customs and Excise.

The notice must be in the prescribed form and give the authority's reasons for their decision (section 203).

Failure to give proper notice of an application

Where a person fails to give proper notice or publish a proper notice of an application within the requirements of these regulations as to the form and manner in which it is to be given, the person holding the licence may make representations about the application within 28 days of receiving the notice. The licensing authority cannot grant the application until notice has been given by the applicant in accordance with the regulations and the 28-day period has elapsed (regulations 6 and 7).

There are similar provisions for failure to properly publish notice of intention to hold a review (regulation 10).

PROVISIONAL STATEMENTS REGARDING GAMBLING PREMISES LICENCES

Application

A person may make an application for a provisional statement about premises:

(a) that he expects to be constructed;
(b) that he expects to be altered; or
(c) that he expects to acquire a right to occupy.

An application for a provisional statement must include plans and other information about the construction, alteration or acquisition (section 204).

Where a licensing authority issues a provisional statement or an application is made for a premises licence in respect of the premises, the licensing authority must disregard any representations made in relation to the application for the premises licence unless they think that the representations address matters that could not have been addressed in representations in relation to the application for the provisional statement, or reflect a change in the applicant's circumstances.

The licensing authority may refuse the application or grant it on terms or conditions not included in the provisional statement, but only on consideration of the above matters (section 205).

Rights of appeal

Where a licensing authority rejects an application, the applicant may appeal.

Where a licensing authority grants an application, a person who made representations in relation to the application, and the applicant, may appeal.

Where a licensing authority takes action as a result of a review of a premises licence or determines to take no action as a result of a review, the licensee, a person who made representations in relation to the review, the person who applied for the review and the Commission may appeal.

Where a licensing authority takes action or makes a determination regarding transfer, the licensee or the applicant for transfer may appeal (section 206).

Appeals are heard in the magistrates' court. The magistrates' court may dismiss the appeal, substitute for the decision appealed against any decision that the licensing authority could have made, send the case to the licensing authority to decide in accordance with a direction of the court or make an order about costs (section 207).

Any decision of the licensing authority will have no effect while an appeal could be brought, or has been brought and has not yet been either finally determined or abandoned. However, the licensing authority making a determination may direct that the action can be taken, in which case the magistrates' court or sheriff determining an appeal may make any order that it or he thinks appropriate (section 208).

A party to an appeal to the magistrates' court or sheriff may appeal on a point of law to the High Court or to the Court of Session (section 209).

TYPES OF GAMBLING PREMISES LICENCES

Gaming machines

An adult gaming centre premises licence authorises the holder to make up to 20 per cent Category B machines available for use on the premises and to make any number of Category C or D gaming machines available for use on the premises.

Family entertainment centre

A family entertainment centre premises licence authorises the holder to make any number of Category C and D gaming machines available for use on the premises.

Bingo licences

A bingo premises licence authorises the holder to make available for use on the premises a number of Category B gaming machines not exceeding 20 per cent of the total number of gaming machines which are available for use on the premises and any number of Category C and D gaming machines available for use on the premises.

Betting licences

A betting premises licence authorises the holder to make up to four gaming machines, each of which must be of Category B, C or D, available for use (section 172).

Virtual gaming

A casino premises licence and a betting premises licence authorises the holder to make facilities available for betting on the outcome of a virtual game, race, competition or other event or process (section 173).

Casino licences

A casino premises licence for a *regional casino* using at least 40 gaming tables authorises the holder to make gaming machines available for use on the premises provided that each gaming machine is of Category A, B, C or D, and the number of gaming machines is not more than 25 times the number of gaming tables used in the casino, and no more than 1,250.

A casino premises licence for a *large casino* using at least one gaming table, or for a regional casino using fewer than 40 gaming tables, authorises the holder to make gaming machines available for use on the premises provided that each gaming machine is of Category B, C or D, and the number of gaming machines is not more than five times the number of gaming tables used in the casino, and no more than 150.

A casino premises licence for a *small casino* using at least one gaming table authorises the holder to make gaming machines available for use on the premises provided that each gaming machine is of Category B, C or D, and the number of gaming machines is not more than twice the number of gaming tables used in the casino, and no more than 80 (Gambling Act 2005 (Gaming Tables in Casinos) (Definitions) Regulations 2009).

A casino premises licence may be issued only in respect of a regional casino, a large casino or a small casino.

A casino premises licence authorises the holder to use the premises to make available any number of games of chance other than casino games and to use the premises for the provision of facilities for bingo, betting, or both (bingo is not authorised in a small casino) (section 174).

CASINO LICENCE APPLICATIONS

References

Gambling Act 2005 (Gaming Tables in Casinos) (Definitions) Regulations 2009.
Gambling (Inviting Competing Applications for Large and Small Casino Premises Licences) Regulations 2008.
Gambling (Geographical Distribution of Large and Small Casino Premises Licences) Order 2008.
Categories of Casino Regulations 2008.

Extent

These provisions apply to England, Wales and Scotland.

Applications for a casino licence (Gambling Act 2005, schedule 9)

The Gambling (Inviting Competing Applications for Large and Small Casino Premises Licences) Regulations 2008 and the Gambling (Geographical Distribution of Large and Small Casino Premises Licences) Order 2008 together provide the method of application and geographical spread and number of casinos throughout the country.

Where a licensing authority is able to grant one or more casino licences, but not all of the competing applications, the licensing authority must first consider whether they would grant a premises licence. They must disregard whether any of the other competing applications is more deserving of being granted. Each competing applicant is an interested party in relation to each of the other competing applications, and the application is subject to the requirements of a premises licence (paragraphs 3 and 4).

If a licensing authority determines that they would grant a number of competing applications greater than the number in the casino limits in the Gambling Act 2005, section 175, they can do so.

The licensing authority:

(a) must determine which of the competing applications would, in the authority's opinion, be likely if granted to result in the greatest benefit to the authority's area;
(b) may enter into a written agreement with an applicant, whether as to the provision of services in respect of the authority's area or otherwise;
(c) may determine to attach conditions to any licence issued so as to give effect to an agreement entered into under paragraph (b); and
(d) may have regard to the effect of an agreement entered into under paragraph (b) in making the determination specified in paragraph (a).

Having determined to grant one or more applications, the authority must grant that application and reject the other competing applications (paragraph 5).

The applicant can appeal against the decision of the licensing authority (paragraph 8).

This procedure also applies to provisional statements (paragraphs 9 and 10).

Resolution not to issue casino licences

A licensing authority may resolve not to issue casino premises licences.

A resolution must apply to the issue of casino premises licences generally, must specify the date on which it takes effect and lasts for three years. The authority may revoke the resolution at any time. The resolution must be published by being included in a statement (section 166).

TEMPORARY USE OF PREMISES FOR GAMBLING

References

Gambling Act 2005.
Gambling Act 2005 (Temporary Use Notices) Regulations 2007.

Extent

These provisions apply to England and Wales. Similar provisions apply in Scotland.

Scope

Premises can be used temporarily for gambling if a temporary use notice is in effect and the activity is carried on in accordance with the terms of the notice.

Nature and form of notice

A temporary use notice is a notice given by the holder of an operating licence and stating his intention to carry on one or more specified prescribed activities. A notice given by a person may specify an activity only if the person's operating licence authorises him to carry on the activity (section 215).

A temporary use notice must:

- be in the prescribed form;
- specify the activity to be carried on in reliance on the notice;
- specify the premises on which the activity is to be carried on;
- specify the period of time during which the notice is to have effect;
- specify times of day during that period at which the activity is to be carried on;
- specify periods during the previous 12 months during which a temporary use notice has had effect in respect of the premises or any part of the premises;
- specify the date on which the notice is given;
- contain any other prescribed information (section 216).

A temporary use notice will have effect during the period specified in the notice (section 217).

FC148 Temporary use of premises for gambling – Gambling Act 2005 and Gambling Act 2005 (Temporary Use Notices) Regulations 2007

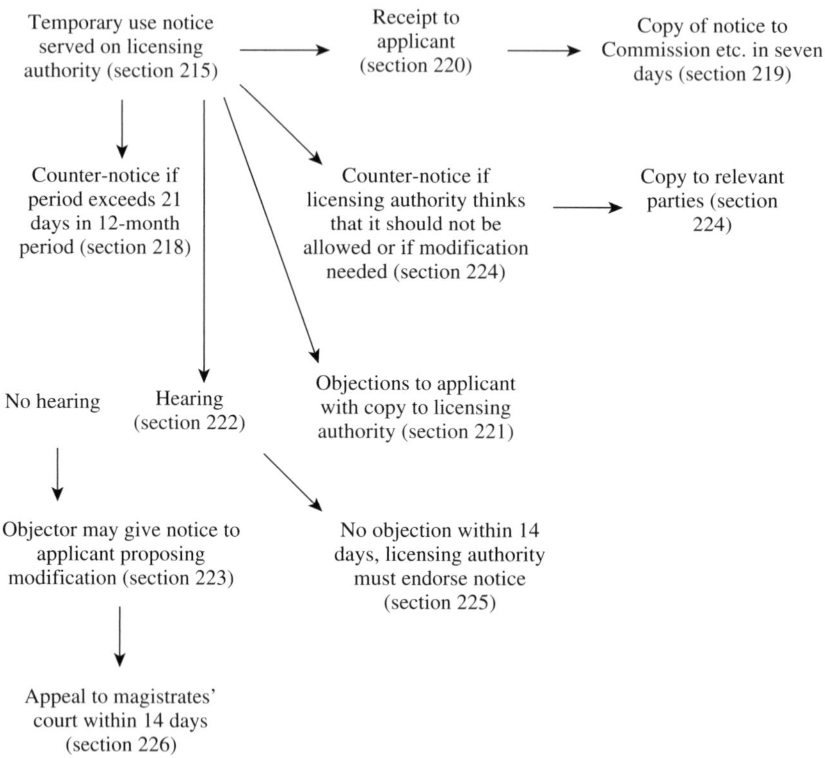

Note

1. Reference to sections are to the Gambling Act 2005 and to regulations are to the Gambling Act 2005 (Temporary Use Notices) Regulations 2007.

A set of premises may not be the subject of temporary use notification for more than 21 days in a period of 12 months. More than one temporary use notice can be in place if the aggregate of the periods for which the notices have effect does not exceed 21 days.

If a temporary use notice is given to a licensing authority and the period would be exceeded, the licensing authority must give a counter-notice stating the temporary use notice is not valid.

Provisions exist for use of the premises for a period which would not exceed the time limits. This would be included in the counter-notice.

Where there is a choice as to which part of the specified period to exclude, the licensing authority must consult the person who gave the temporary use notice before giving a counter-notice (section 218).

Giving notice

A temporary use notice must be given to the licensing authority. It must be given three months before the use is proposed.

The notice must be accompanied by a copy of the notice and the relevant fee.

A copy of the notice must be given to the Commission, the chief officer of police and Her Majesty's Commissioners of Customs and Excise within seven days of the date of the notice (section 219).

Acknowledgment of notice

The licensing authority must send a written acknowledgment of receipt of the notice to the person who gave it (section 220).

Objections

If any person thinks that the licensing objectives would not be met, he may give a notice of objection to the person who gave the temporary use notice within 14 days of the temporary use notice.

That person must give a copy of the notice to the licensing authority to which the temporary use notice was given. A notice of objection must state that the person giving the notice objects to the temporary use notice, and the person's reasons (section 221).

Hearing of objections

The licensing authority must hold a hearing at which the person who gave the notice, the person who objected and any other person who was entitled to receive a copy of the notice may make representations about the notice of the objection; however, if the parties agree in writing, the hearing need not be heard if a hearing is unnecessary (section 222).

Modification by agreement

If the hearing has not taken place, the objector may, by notice in writing to the person who gave the temporary use notice, propose a modification of the notice. If the person who gave the notice accepts the modification, he must give a new notice incorporating the modification, and the objection will be treated as withdrawn (section 223).

Counter-notice

If the licensing authority thinks that the temporary use notice should not be allowed or should have effect only with modification, the authority may give a counter-notice to the person who gave the notice.

A counter-notice may provide for the temporary use notice:

(a) not to have effect;
(b) to have effect only for a specified activity;
(c) to have effect only for an activity carried on during a specified period of time or at specified times of the day;
(d) to have effect with a specified condition.

A counter-notice must be in the prescribed form, contain the prescribed information, be given as soon as is reasonably practicable and state the licensing authority's reasons for giving it.

A copy must be given to the relevant parties (section 224).

Dismissal of objection

Where a counter-notice has not been served, the licensing authority must give notice of their determination to the person who gave the temporary use notice and to each person who received a copy of the notice (section 225).

Appeal

The person who gave the temporary use notice and a person who was entitled to receive a copy of the temporary use notice may appeal against the decision of the licensing authority in the magistrates' court within 14 days of the appellant receiving notice of the decision. The magistrates' court may dismiss the appeal, direct the licensing authority to take action of a specified kind, return the case to the licensing authority to act in accordance with the direction of the court or make an order about costs.

A party to an appeal may bring a further appeal to the High Court on a point of law (section 226).

Endorsement of notice

If no objection is made within a 14-day period, the licensing authority must endorse the copy submitted and return the endorsed copy to the person giving the notice. If a notice of objection is pending when the 14-day period expires, the licensing authority must either give a counter-notice or endorse the copy as above (section 227).

Consideration by licensing authority: timing

Temporary use notice proceedings must be dealt with within six weeks of receiving the notice (proceedings are: considering whether to give a notice of objection, holding a hearing or agreeing to dispense with a hearing and giving a counter-notice) (section 228).

Availability of notice

A person who gives a temporary use notice must arrange for a copy of the notice to be displayed prominently on the premises at any time when an activity is being carried on and arrange for the notice endorsed by the licensing authority to be produced on request to a constable, an officer of customs and excise, an enforcement officer or an authorised local authority officer.

It is an offence to fail to comply, liable on summary conviction to a fine not exceeding level 2 on the standard scale (section 229).

Withdrawal of notice

A temporary use notice can be withdrawn by the person who makes notification and the notice will have no effect and no further proceedings will take place (section 230).

Register

A licensing authority must maintain a register of temporary use notices given to them, make the register and information available for inspection by members of the public at all reasonable times and make arrangements for the provision of a copy of an entry in the register, or of information, to a member of the public on request on payment of a fee (section 234).

The Gambling Act 2005 (Temporary Use Notices) Regulations 2007 makes provision for the form and content of notices, fees to be charged and the way in which notices are given to licensing authorities.

GAMING AND GAMING MACHINES IN ALCOHOL-LICENSED PREMISES

References

Gambling Act 2005, schedules 12 and 13.

Gambling Act 2005 (Licensed Premises Gaming Machine Permits) (England and Wales) Regulations 2007 *(these regulations make provision for licensed premises gaming machine permits, including the form of a permit and various fees).*

Gambling Act 2005 (Club Gaming and Club Machine Permits) Regulations 2007 *(these regulations make provision for club gaming and club machine permits, including the application form, the form of each permit and various fees).*

Gambling Act 2005 (Exempt Gaming in Alcohol-Licensed Premises) Regulations 2007 *(these regulations prescribe the maximum amounts that may be staked and won in equal chance games).*

Gambling Act 2005 (Gaming in Clubs) Regulations 2007 *(these regulations prescribe the kinds of gaming for which a club may be established or conducted to provide facilities, in order to be a members' club or a commercial club).*

Gambling Act 2005 (Exempt Gaming in Clubs) Regulations 2007 *(these regulations prescribe the maximum amounts that may be staked and won and the maximum participation fee that may be charged in games of equal chance).*

Gambling Act 2005 (Club Gaming Permits) (Authorised Gaming) Regulations 2007 *(these regulations prescribe the games of chance that may be played under a club gaming permit and the maximum amount that may be charged by way of a participation fee in those games).*

Gambling Act 2005 (Gaming Machines) (Definitions) Regulations 2007 *(these regulations assign meanings to the terms 'dual-use computer' and 'domestic computer').*

Categories of Gaming Machines Regulations 2007 *(these regulations provide the necessary definitions for each category, A to D).*

Extent

These provisions apply to England, Wales and Scotland.

Club and gaming permits – schedule 12, Gambling Act 2005

A members' club or miners' welfare institute can apply to a licensing authority for a club gaming permit and a club machine permit. A commercial club can apply for a club machine permit (paragraph 1).

The application must:

(a) be made to a licensing authority in whose area the premises are wholly or partly situated;
(b) specify the premises for which the permit is sought;
(c) be made in the prescribed form and manner;
(d) contain or be accompanied by the prescribed information and documents; and
(e) be accompanied by the prescribed fee (£100 or £200) (paragraph 1).

The applicant must, within the prescribed time, send a copy of the application and of any accompanying documents to the Commission and the chief officer of police. Failure to comply with these requirements will mean any permit granted has no effect (paragraphs 2 and 3).

A person who receives a copy of an application for a permit may object to the application within the prescribed period of time and in the prescribed manner (paragraph 4).

The licensing authority must consider the application and grant or refuse it (paragraph 5).

The form, manner and fees for permit applications is included in the Gambling Act 2005 (Club Gaming and Club Machine Permits) Regulations 2007.

Refusal

The licensing authority may refuse an application for a permit only on one or more of the following grounds:

(a) that the applicant is not a members' club, a miners' welfare institute or a commercial club as the case may be;
(b) that the premises on which the applicant conducts its activities are used wholly or mainly by children, by young persons or by both;
(c) that an offence, or a breach of a condition of a permit, has been committed in the course of gaming activities carried on by the applicant;
(d) that a permit held by the applicant has been cancelled in the last ten years; or
(e) that an objection to the application has been made (paragraph 6).

Before refusing an application for a permit, a licensing authority must hold a hearing to consider the application and any objection made.

However, a licensing authority may dispense with the requirement for a hearing with the consent of the applicant and any person who has made an objection (paragraph 7).

Where a licensing authority grants an application for a permit, they must give notice of their decision and, where an objection was made, of the reasons for it to the applicant, to the Commission and the chief officer of police and issue the permit to the applicant (paragraph 8).

Where a licensing authority rejects an application for a permit, they must give notice of their decision and the reasons for it to the applicant, the Commission and the chief officer of police (paragraph 9).

If an applicant for a permit is the holder of a club premises certificate, he need not make an application for a club gaming permit and a club machine permit but the licensing authority must give notice of their decision as above (this does not apply to objectors).

The authority to whom an application is made shall grant it unless they think:

(a) that the applicant is established or conducted wholly or mainly for the purposes of the provision of facilities for gaming, other than gaming of a prescribed kind, and also provides facilities for gaming of another kind;
(b) that a club gaming permit or club machine permit issued to the applicant has been cancelled in the last ten years.

This paragraph does not apply to Scotland (paragraph 10).

A permit must be in the prescribed form and must specify:

(a) the name of the club or institute in respect of which it is issued;
(b) the premises to which it relates;
(c) whether it is a club gaming permit or a club machine permit;
(d) the date on which it takes effect; and
(e) other information as may be prescribed.

The permit is valid for ten years (paragraph 17).

If the application for the permit was made with reference to a club premises certificate, the permit must also identify the certificate (paragraph 11).

The permit holder must keep the permit on the premises (paragraph 12).

It is an offence for the occupier of premises to which a permit relates to fail to produce the permit on request for inspection by a constable or an enforcement officer and is liable on summary conviction to a fine not exceeding level 2 on the standard scale (paragraph 13).

An annual fee is payable to the licensing authority (paragraph 14).

The permit holder must apply for changes to the permit, together with the prescribed fee and either the permit or a statement explaining why it is not possible to produce the permit. The licensing authority must issue a varied copy of the permit. But if the authority thinks that it would refuse an application for the permit were it made anew, they can refuse the application for variation and cancel the permit (paragraph 15).

Where a permit is lost, stolen or damaged, the holder may apply to the issuing licensing authority for a copy, with the prescribed fee. The permit must be issued if the licensing authority is satisfied that the permit has been lost, stolen or damaged and, where lost or stolen, that the loss or theft has been reported to the police (paragraph 16).

The permit lapses if the holder ceases to be a member of the relevant club (paragraph 18).

A permit ceases when surrendered by notice given by the holder to the licensing authority (paragraph 19).

The licensing authority must notify the Commission and the chief officer of police where it believes that the permit has lapsed or it receives notice of surrender (paragraph 20).

Cancelling a permit

The licensing authority can cancel it if it thinks that the premises on which the holder of the permit conducts its activities are used wholly or mainly by children, by young persons or by both, or that an offence, or a breach of a condition of a permit, has been committed in the course of gaming activities carried on by the permit holder.

Before cancelling a permit, the licensing authority must:

(a) give the holder of the permit at least 21 days' notice of the authority's intention to consider cancelling the permit;
(b) consider any representations made by the holder;

(c) hold a hearing if the holder requests one; and
(d) comply with any prescribed requirements for the procedure to be followed in considering whether to cancel a permit.

If a licensing authority cancels a permit, they must give notice of the cancellation and the reasons for it to the holder, the Commission and the chief officer of police. The cancellation of a permit does not take effect for 21 days or until any appeal brought has been determined (paragraph 21).

The licensing authority must cancel the permit if the holder fails to pay the annual fee, but not if they think that a failure to pay is an administrative error (paragraph 22).

Forfeiture

The court may order forfeiture of the permit if a holder or an officer of the permit holder is convicted of an offence. The court must notify the licensing authority of the order (paragraph 23).

Renewal

An application for the renewal of a permit can only be made between three months and six weeks before its expiry (paragraph 24).

Appeals

Where a licensing authority:

- rejects an application for the issue or renewal of a permit, the applicant may appeal;
- grants an application for the issue or renewal of a permit in relation to which an objection was made, the person who made the objection may appeal;
- cancels a permit, the holder may appeal;
- determines not to cancel a permit, any person who made representations to the authority in connection with their consideration whether to cancel the permit may appeal.

Appeals are to the magistrates' court within 21 days of the appellant receiving notice of the decision against which the appeal is brought.

The magistrates' court may:

(a) dismiss the appeal;
(b) substitute for the decision appealed against any decision that the licensing authority could have made;
(c) restore a permit;
(d) return the case to the licensing authority to decide in accordance with a direction of the court;
(e) make an order about costs (paragraph 25).

Register

The licensing authority must maintain a register of permits issued by the authority, together with prescribed information, and make the register and information available for inspection by members of the public at all reasonable times, and make arrangements for the provision of a copy of an entry in the register, or of information, to a member of the public on request, who must pay the prescribed fee (paragraph 26).

Licensed premises gaming machines permits – schedule 13, Gambling Act 2005

A person who applies to a licensing authority for an on-premises alcohol licence or who holds an on-premises alcohol licence may apply for a licensed premises gaming machine permit but not if a premises licence is in effect (paragraph 1).

An application for a permit must:

(a) be made in the form and manner required by the licensing authority;
(b) specify the premises;
(c) specify the number and category of gaming machines in respect of which the permit is sought;
(d) contain or be accompanied by such other information or documents as the licensing authority may direct; and
(e) be accompanied by the prescribed fee (paragraph 2).

On considering the licensing objectives regarding the application for a permit, the licensing authority must:

(a) grant the application;
(b) refuse the application; or
(c) grant it in respect of:
 (i) a smaller number of machines than that specified in the application;
 (ii) a different category of machines from that specified in the application; or
 (iii) both (paragraph 4).

On granting an application, the licensing authority must issue a permit to the applicant.

If refused, the licensing authority must notify the applicant of the refusal and the reasons for it (paragraph 5).

The licensing authority may grant an application only if the applicant holds an on-premises alcohol licence.

The licensing authority cannot refuse an application or grant an application in respect of a different category or smaller number of gaming machines than that specified in the application unless they have:

(a) notified the applicant of their intention to refuse the application or grant the application in respect of:
 (i) a smaller number of machines than that specified in the application;

(ii) a different category of machines from that specified in the application; or
(iii) both; and
(b) given the applicant an opportunity to make oral or written representations, or both (paragraph 6).

A permit must be in the prescribed form and must specify:

(a) the person to whom it is issued;
(b) the premises;
(c) the number and category of gaming machines which it authorises; and
(d) the date on which it takes effect (paragraph 7).

The holder of a permit must keep it on the premises (paragraph 8).

The permit holder must pay an annual fee (paragraph 9).

It is an offence to fail to produce the permit on request for inspection by a constable, an enforcement officer or an authorised local authority officer and persons guilty of an offence are liable on summary conviction to a fine not exceeding level 2 on the standard scale (paragraph 10).

Lost, stolen or damaged permits can be copied on payment of a relevant fee.

The licensing authority must consider an application and grant it if satisfied that the permit has been lost, stolen or damaged and, where the permit has been lost or stolen, that the loss or theft has been reported to the police (paragraph 11).

A permit shall cease to have effect:

(a) if an on-premises alcohol licence ceases to have effect with respect to the premises to which it relates, or the permit holder ceases to be the holder of an on-premises alcohol licence (paragraph 13);
(b) if the permit holder gives to the licensing authority notice of surrender and either the permit or a statement explaining why it is not reasonably practicable to produce the permit (paragraph 14).

Cancellation, variation and transfer

The licensing authority may vary the number or category (or both) of gaming machines authorised by the permit on application from the permit holder (paragraph 15).

The licensing authority may cancel the permit, or may vary the number or category (or both) of gaming machines authorised by it, if they think that:

(a) it would not be reasonably consistent with pursuit of the licensing objectives for the permit to continue to have effect;
(b) gaming has taken place on the premises in purported reliance on the permit but otherwise than in accordance with the permit or a condition of the permit;
(c) the premises are mainly used or to be used for making gaming machines available; or
(d) an offence under this Act has been committed on the premises.

Before cancelling or varying a permit, a licensing authority must give the permit holder at least 21 days' notice of the authority's intention to consider cancelling or varying the permit, consider any representations made by the holder and hold a hearing if the permit holder requests one.

If the licensing authority cancels or varies a permit, they must give notice of the cancellation or variation and the reasons for it to the permit holder and the Commission.

The cancellation or variation of a permit shall not take effect for 21 days without an appeal being brought, or until any appeal brought has been determined (paragraph 16).

Failure to pay the annual fee will result in the licensing authority cancelling it, but not if they think that a failure to pay is attributable to administrative error (paragraph 17).

Where a permit holder, or the officer of a permit holder, is convicted of a relevant offence, the court may order forfeiture of the permit. The court must notify the licensing authority of the forfeiture (paragraph 18).

A person may apply for the transfer of a permit to him and must supply the permit, or statement explaining why it is not reasonably practicable to produce the permit, with the application (paragraph 19).

Appeals

The applicant for, or holder of, a permit may appeal if the licensing authority:

(a) rejects an application for a permit;
(b) grants an application for a permit in respect of a smaller number of machines than that specified in the application or a different category of machines from that specified in the application (or both); or
(c) gives a notice that the permit is to be cancelled.

The appeal must be made to the magistrates' court within 21 days of receiving notice of the decision against which the appeal is brought.

The magistrates' court may:

(a) dismiss the appeal;
(b) substitute for the decision appealed against any decision that the licensing authority could have made;
(c) restore a permit;
(d) return the case to the licensing authority to decide in accordance with a direction of the court;
(e) make an order about costs (paragraph 21).

The licensing authority must maintain a register of permits, make the register and information available for inspection by members of the public at all reasonable times and make arrangements for the provision of a copy of an entry in the register, or of information, to a member of the public on request, on payment of the relevant fee (paragraph 22).

Offences

Making machine available for use

It is an offence to make a gaming machine available for use by another without an operating licence, except:

- in family entertainment centres;
- where no prize or limited prize is made available.

The requirements of club gaming permits, club machine permits, automatic entitlement to gaming machine permits and licensed premises gaming permits still apply (section 242).

Linked machines

It is an offence to make a gaming machine ('the first gaming machine') available for use by another, and the amount or value of a prize available through use of the first gaming machine is or may be wholly or partly determined by reference to use made of another gaming machine ('the linked gaming machine'). This does not apply in a licensed casino premises (section 245).

Penalty

Persons guilty of an offence are liable on summary conviction to imprisonment for up to 51 weeks (six months in Scotland), a fine not exceeding level 5 on the standard scale, or both (section 246).

Family entertainment centre gaming machine permit

It is not an offence where a Category D gaming machine is available for use with a family entertainment centre gaming machine permit.

A family entertainment centre gaming machine permit is a permit issued by a licensing authority authorising a person to make Category D gaming machines available for use in a specified family entertainment centre (section 247 and schedule 10).

No prize or limited prize

It is not an offence for a gaming machine to be used by an individual and the individual does not, by using the machine, acquire an opportunity to win a prize or a prize of a value in excess of the amount that he pays for, or in connection with, his use of the machine (sections 248 and 249).

PERMITS FOR GAMING AND GAMBLING MACHINES IN CLUBS AND MINERS' WELFARE INSTITUTES

Exempt gaming

Gaming is exempt in these premises:

1. where the arrangements for the gaming satisfy the prescribed requirement amounts that may be staked, or the amount or value of a prize;
2. where no amount is deducted or levied from sums staked or won;
3. where any prescribed participation fee does not exceed such maximum or there is no fee;
4. where a game played on one set of premises is not linked with a game played on another set of premises;
5. where each person who participates is a member of the club or institute who applied for membership, was nominated for membership or became a member, at least 48 hours before he participates, or is a guest of a member of the club or institute who would be entitled to participate;
6. in alcohol-licensed premises, children and young persons are excluded from participation (section 269 and 279).

Club gaming permit

A club gaming permit is a permit issued by a licensing authority authorising the provision of facilities for gaming:

(a) on premises on which a members' club or a miners' welfare institute operates; and
(b) in the course of the activities of the club or institute.

A club gaming permit authorises:

(a) making up to three gaming machines available for use, each of which must be of Category B, C or D;
(b) the provision of facilities for gaming which are exempt but the stakes and prizes are within the prescribed amounts; and
(c) the provision of facilities for games of chance, of such class or description as are in accordance with the conditions specified below.

Those conditions are:

(a) that prescribed participation fee is charged;
(b) that amount is deducted or levied from sums staked or won is in accordance with regulations;
(c) that the public and children are excluded from any area of the club's or institute's premises where gaming is taking place (section 271 and Gambling Act 2005 (Club Gaming and Club Machine Permits) Regulations 2007).

Club machine permit

A club machine permit is a permit issued by a licensing authority authorising up to three gaming machines, each of which must be of Category B, C or D, to be made available for use on premises on which a members' club, a commercial club or a miners' welfare institute operates, and in the course of the activities of the club or institute. Only members and their guests may use the machines. Children cannot use Category B or C gaming machines (section 273 and Gambling Act 2005 (Club Gaming and Club Machine Permits) Regulations 2007).

The provisions of schedule 12 of the Gambling Act 2005 on page 584 apply to these clubs.

Gaming machines: automatic entitlement

Clubs and institutes can make one or two gaming machines, each of which is of Category C or D, available for use on premises provided that the person who holds the on-premises alcohol licence or the relevant Scottish licence sends the licensing authority written notice of his intention to make gaming machines available for use, with the prescribed fee.

The club or institute must comply with the relevant Code of Practice about the location and operation of a gaming machine (section 282).

Licensed premises gaming machine permits

It is an offence to make a gaming machine of Category C or D available without obtaining a licensed premises gaming machine permit.

A licensed premises gaming machine permit is a permit issued by a licensing authority authorising a person to make gaming machines of Category C or D (or both) available for use on premises. See Gambling Act 2005 (Exempt Gaming in Alcohol-Licensed Premises) Regulations 2007.

The provisions of schedule 13 of the Gambling Act 2005 on page 588 apply to these clubs but do not apply in Scotland (section 282).

Removal of exemption

A licensing authority may make an order removing exemptions from specified premises.

A licensing authority may make an exemption order only if they think that:

(a) the application of the exemption is not reasonably consistent with pursuit of the licensing objectives;
(b) gaming has taken place on the premises in purported reliance on the exemption but in breach of a condition of that exemption;
(c) the premises are mainly used or to be used for gaming; or
(d) an offence under this Act has been committed on the premises.

Before making an order, a licensing authority must:

(a) give the holder of the on-premises alcohol licence or of the relevant Scottish licence at least 21 days' notice of the authority's intention to consider making an order;
(b) consider any representations made by the licensee;

(c) hold a hearing if the licensee requests one; and
(d) comply with any prescribed requirements for the procedure to be followed in considering whether to make an order.

If a licensing authority makes an order, it must give the licensee a copy of the order and written reasons for the decision to make the order.

A licensee may appeal against the making of an order to the magistrates' court within 21 days of receiving a copy of the order against which the appeal is brought.

The magistrates' court may dismiss the appeal, allow the appeal and quash the order made by the licensing authority and make an order about costs (section 284).

Definitions

Adult gaming centre means premises which have an adult gaming centre premises licence (section 237).

Commercial club means a club:

(a) which is established and conducted wholly or mainly for purposes other than the provision of facilities for gaming;
(b) which is not established with the purpose of functioning only for a limited period of time; and
(c) which has at least 25 individual members (section 267).

Family entertainment centre means premises (other than an adult gaming centre) wholly or mainly used for making gaming machines available for use (section 238).

Gaming machine means a machine which is designed or adapted for use by individuals to gamble (section 235).

Members' club means a club:

(a) which is established and conducted wholly or mainly for purposes other than the provision of facilities for gaming;
(b) which is established and conducted for the benefit of its members;
(c) which is not established with the purpose of functioning only for a limited period of time; and
(d) which has at least 25 individual members (section 266).

Miners' welfare institute means an association:

(a) which is established and conducted for social and recreational purposes;
(b) whose affairs are managed by a group of individuals of whom at least two-thirds are miners' representatives;
(c) which operates on premises, the use of which is regulated in accordance with a charitable trust; and
(d) whose trust has received money from the Miners' Welfare Fund, the Coal Industry Social Welfare Organisation (section 268).

Prize in relation to a gaming machine includes any money, article, right or service won, whether or not described as a prize, but does not include an opportunity to play the machine again (section 239).

PERMITS TO FAMILY ENTERTAINMENT CENTRES (FECs) FOR THE USE OF CERTAIN LOWER-STAKE GAMING MACHINES

References

Gambling Act 2005, schedule 10.
Gambling Act 2005 (Family Entertainment Centre Gaming Machine) (Permits) Regulations 2007.

Extent

These provisions apply to England, Wales and Scotland.

Application

Only adults who occupy or propose to occupy the premises may make an application for the use of the premises as an unlicensed family entertainment centre (paragraph 2).

An application for a permit must:

(a) be made in the form and manner required by the licensing authority;
(b) specify the premises for which the permit is sought;
(c) contain or be accompanied by such other information or documents required by the licensing authority;
(d) include the prescribed fee (paragraph 5).

A licensing authority may prepare a statement of principles that they propose to apply in issuing permits. The statement may, in particular, specify matters that the licensing authority proposes to consider in determining the suitability of an applicant for a permit. The authority need not (but may) have regard to the licensing objectives and have regard to any relevant guidance (paragraph 7).

On considering an application for a permit, a licensing authority may grant or refuse it, but it cannot attach conditions.

The licensing authority must issue a permit to the applicant or, if refused, must notify the applicant of the refusal and the reasons for it (paragraph 8).

A licensing authority may grant an application for a permit only if they are satisfied that the applicant intends to use the premises as an unlicensed family entertainment centre and if they have consulted the relevant chief officer of police. The permit has effect for ten years and must be displayed on the premises (paragraphs 9, 12 and 19).

A licensing authority cannot refuse an application unless they have notified the applicant of their intention to refuse the application and of their reasons, and given the applicant an opportunity to make oral or written representations, or both (paragraph 10).

A permit must be in the prescribed form and must specify:

(a) the person to whom it is issued;
(b) the premises; and
(c) the date on which it takes effect.

If the person to whom a permit is issued changes his name or wishes to be known by another name, he can send the permit to the licensing authority with the prescribed fee and a request that a new name be substituted for the old name. The licensing authority must comply with the request and return the permit to the holder (paragraph 11).

Lapsing

Permits lapse:

- if the holder ceases to occupy the premises (paragraph 13);
- if the licensing authority notifies the holder that the premises are not being used as a family entertainment centre (paragraph 14);
- if the holder dies;
- if the holder becomes, in the opinion of the licensing authority as notified to him, incapable of carrying on the activities authorised by the permit by reason of mental or physical incapacity;
- if the holder becomes bankrupt.

The permit may still be valid for six months following the lapse and have effect to the personal representatives of the holder, the trustee of the bankrupt's estate or the liquidator of the company, where relevant (paragraph 15).

Surrender

A permit ceases to have effect if the holder gives a notice of surrender to the licensing authority together with either the permit or a statement explaining why it is not reasonably practicable to produce the permit (paragraph 16).

Forfeiture

Where the holder of a permit is convicted of a relevant offence, the court by or before which he is convicted may order forfeiture of the permit. The court must notify the licensing authority of the forfeiture.

The terms on which forfeiture is ordered includes a requirement that the permit holder delivers the permit to the licensing authority or a statement explaining why it is not reasonably practicable to produce the permit (paragraph 17).

The holder of a permit may apply to the licensing authority for renewal of the permit between six and two months of the expiry date.

A licensing authority may refuse an application for renewal of a permit only on the grounds that an authorised local authority officer has been refused access to the premises without reasonable excuse or that renewal would not be reasonably consistent with pursuit of the licensing objectives (paragraph 18).

It is an offence for an occupier of a premises to fail to produce the permit on request for inspection by a constable, enforcement officer or authorised local authority officer, liable on summary conviction to a fine not exceeding level 2 on the standard scale (paragraph 20).

Where a permit is lost, stolen or damaged, the holder may apply to the licensing authority for a copy, with a prescribed fee (paragraph 21).

Appeal

The applicant for, or holder of, a permit may appeal to the magistrates' court within 21 days of notification, if the licensing authority rejects an application for the issue or renewal of a permit or the permit lapses.

The magistrates' court may dismiss the appeal, substitute for any decision that the licensing authority could have made, restore a permit, return the case to the licensing authority to decide in accordance with a direction of the court and make an order about costs (paragraph 22).

Register

The licensing authority must maintain a register of permits issued by the authority, make the register and information available for inspection by members of the public at all reasonable times and make arrangements for the provision of a copy of an entry in the register, or of information, to a member of the public on request, on payment of the prescribed fee (paragraph 23).

PERMITS FOR PRIZE GAMING

References

Gambling Act 2005, schedule 14.

Gambling Act 2005 (Prize Gaming) (Permits) Regulations 2007 *(these regulations make provisions about the fees relating to a prize gaming permit and the form of the permit).*

Gambling Act 2005 (Limits on Prize Gaming) Regulations 2009 *(these regulations prescribe the limits on participation fees, and maximum prize amounts and values in respect of prize gaming).*

Extent

These provisions apply in England, Wales and Scotland.

Prize gaming permits

Premises or facilities for prize gaming or equal chance gaming in an adult gaming centre, travelling fair or licensed family entertainment centre must have a prize gaming permit and comply with the conditions in that permit (sections 289, 290 and schedule 14).

Such facilities are available in a bingo hall where it has a bingo premises licence (section 291).

Conditions for prize gaming

Participation fees must be within the prescribed limits.

All the chances to participate in a particular game must be acquired or allocated on one day and in the place where the game is played.

The game must be played entirely on that day, and the result of the game must be made public in the place where the game is played, and as soon as is reasonably practicable after the game ends, and in any event on the day on which it is played.

A prize for which a game is played, or the aggregate of the prizes where all the prizes are money, must not exceed the prescribed amount, and in any other case, must not exceed the prescribed value. Participation in the game by a person does not entitle him or another person to participate in any other gambling (section 293).

Application

A person who applies for an on-premises alcohol licence or who holds one may apply for a licensed premises gaming machine permit (schedule 14, paragraph 1).

An application for a permit must be made in the form and manner that the licensing authority directs, include the prescribed fee and specify the premises for which the permit is sought, the number and category of gaming machines for which the permit is sought and any other information or documents that the licensing authority directs (schedule 14, paragraph 2).

The licensing authority must consider the application having regard to the licensing objectives, any relevant guidance issued by the Commission and such other matters as they think relevant.

After considering the application, the licensing authority must grant or refuse the application, or grant the permit for a smaller number of machines than that specified in the application, a different category of machines from that specified in the application, or both (schedule 14, paragraph 4).

The licensing authority cannot attach conditions to a permit. It must issue a permit to the applicant or, where refused, notify the applicant of the refusal, and the reasons for it (schedule 14, paragraph 5).

The licensing authority must notify the applicant of its intention to refuse the application or that it intends to grant a permit for a smaller number of machines or a different category of machines from that specified in the application, or both, and the applicant must be given an opportunity to make oral or written representations (schedule 14, paragraph 6)

Permit

A permit must be in the prescribed form and must specify:

(a) the person to whom it is issued;
(b) the premises;
(c) the number and category of gaming machines which it authorises; and
(d) the date on which it takes effect.

If the person to whom a permit is issued changes his name or wishes to be known by another name, he may send the permit to the licensing authority with the prescribed fee and a request that a new name be substituted for the old name, and the licensing authority must issue the amended permit (schedule 14, paragraph 7).

The holder of a permit must keep it on the premises (schedule 14, paragraph 8).

Annual fees must be paid by the permit holder (schedule 14, paragraph 9).

It is an offence for the occupier of a premises to fail to produce the permit on request for inspection by an enforcement officer or an authorised local authority officer, liable on summary conviction to a fine not exceeding level 2 on the standard scale (schedule 14, paragraph 10).

Where a permit is lost, stolen or damaged, the holder may apply to the licensing authority for a copy, paying the prescribed fee (schedule 14, paragraph 11).

A permit ceases if:

- an on-premises alcohol licence ceases to have effect;
- the permit holder ceases to be the holder of an on-premises alcohol licence (schedule 14, paragraph 13);
- the permit holder gives a notice of surrender to the licensing authority, including either the permit or a statement explaining why it is not reasonably practicable to produce the permit (schedule 14, paragraph 14).

The permit holder may apply to the licensing authority to vary the number or category (or both) of gaming machines authorised by the permit (schedule 14, paragraph 15).

Cancelling and variation

The licensing authority may cancel a permit, or may vary the number or category (or both) of gaming machines authorised by it, if they think that:

(a) it would not be reasonably consistent with the licensing objectives for the permit to continue to have effect;
(b) gaming has taken place on the premises in purported reliance on the permit but otherwise than in accordance with the permit or a condition of the permit;
(c) the premises are mainly used or to be used for making gaming machines available; or
(d) an offence under this Act has been committed on the premises.

Before cancelling or varying a permit, the licensing authority must:

(a) give the permit holder at least 21 days' notice of the authority's intention to consider cancelling or varying the permit;
(b) consider any representations made by the holder;
(c) hold a hearing if the permit holder requests one; and
(d) comply with any prescribed requirements for the procedure to be followed in considering whether to cancel or vary a permit.

The licensing authority must give notice of the cancellation or variation and the reasons for it to the permit holder, and the Commission, subject to any appeal (schedule 14, paragraph 16).

If the permit holder fails to pay the annual fee, the licensing authority must cancel it unless failure to pay is attributable to administrative error (schedule 14, paragraph 17).

Forfeiture

Where a permit holder, or the officer of a permit holder, is convicted of a relevant offence, the court by or before which he is convicted may order forfeiture of the permit. The court can apply terms on the permit. The court must notify the licensing authority of the court's action (schedule 14, paragraph 18).

Transfer

A person may apply for the transfer of a permit to him if he is applying for the transfer of an on-premises alcohol licence. He must supply the permit, or a statement explaining why it is not reasonably practicable to produce the permit, with his application. The permit cannot be issued unless the alcohol licence is approved (schedule 14, paragraph 19).

Appeal

The applicant for, or holder of, a permit may appeal to the magistrates' court if the licensing authority:

- (a) rejects an application for a permit;
- (b) grants an application for a permit in respect of a smaller number of machines than that specified in the application or a different category of machines from that specified in the application (or both); or
- (c) gives a notice to cancel the permit,

within 21 days of the action by the licensing authority.

The magistrates' court may:

- (a) dismiss the appeal;
- (b) substitute for the decision appealed against any decision that the licensing authority could have made;
- (c) restore a permit;
- (d) return the permit to the licensing authority to decide in accordance with a direction of the court;
- (e) make an order about costs (schedule 14, paragraph 21).

Register

The licensing authority must maintain a register of permits issued by the authority, make it and information available for inspection by members of the public at all reasonable times and make arrangements for the provision of a copy of an entry in the register, or of information, to a member of the public on request, with payment of the relevant fee (schedule 14, paragraph 22).

Definition

Prize gaming is where neither the nature nor the size of a prize played for is determined by reference to the number of persons playing, or the amount paid for or raised by the gaming (section 288).

REGISTERING SMALL SOCIETY LOTTERIES

References

Gambling Act 2005, schedule 11, part 5.
Gambling Act 2005 (Incidental Non-Commercial Lotteries) Regulations 2007
Guidance to Licensing Authorities (4th edition) *(sets out the principles that licensing authorities should adopt in registering small society lotteries).*

Extent

These provisions apply to England, Wales and Scotland.

Scope

Society lotteries are lotteries promoted for the benefit of a non-commercial society. A society is non-commercial if it is established and conducted:

- for charitable purposes;
- for the purpose of enabling participation in, or of supporting, sport, athletics or a cultural activity;
- for any other non-commercial purpose other than that of private gain.

A small society lottery:

- does not have proceeds that exceed £20,000 for a single draw;
- does not have aggregated proceeds from lotteries in excess of £250,000 in any one year.

Small society lotteries do not require a licence but must be registered with the licensing authority in the area where the principal office of the society is located.

Licensing authorities should inform a society that they must apply for an operating licence if, in the course of running a small society lottery, proceeds exceed £20,000 for a single draw or aggregated proceeds from lotteries exceed £250,000 in any one year.

Society lotteries (schedule 11, part 4, Gambling Act 2005)

The arrangements for a small society lottery must ensure that at least 20 per cent of the proceeds of the lottery are applied to a purpose for which the society is conducted (paragraph 33).

The maximum prize in a small society lottery is £25,000 (paragraph 34).

FC149 Registration of small society lotteries – Schedule 11, Part 5 Gambling Act 2005

Statement to licensing authority (lottery arrangements, proceeds, amounts deducted, expenses) (paragraph 39) → Licensing authority notifies Commission if a large lottery (paragraph 39)

↓

Application for registration (paragraph 42)

↓

- Licensing authority enters into register (paragraphs 44 and 46)
- If licensing authority intends to refuse application, must give applicant opportunity to make representations (paragraph 49)
- Refused if licence revoked or refused in last five years (paragraph 47)

Licensing authority revokes registration (paragraph 50)

↓

Specifies when revocation takes place – immediately or within two months (paragraph 50)

↓

Notifies applicant of revocation or refusal (paragraph 51)

↓

Society appeals to magistrates' court (paragraph 51)

↓

Magistrates' court affirms it, reverses it or other order (paragraph 51)

Society applies for registration to be cancelled (paragraph 52)

↓

Licensing authority cancels registration

↓

Licensing authority notifies applicant and Commission of cancellation (paragraph 53)

The arrangements for a small society lottery may include a rollover (paragraph 35).

A lottery ticket in a small society lottery must identify the promoting society, state the price of the ticket, the name and an address of a member of the society who is designated as having responsibility within the society for the promotion of the lottery, or the external lottery manager, and either state the date of the draw or enable the date of the draw in the lottery to be determined (paragraph 36).

The price for each ticket must be the same and must be paid to the promoter of the lottery before any person is given the ticket or any right in respect of membership of the class among whom prizes are to be allocated (paragraph 37).

The promoting society of a small society lottery must be registered with a local authority (paragraph 38 and part 5).

The society must send a statement to the local authority at least three months prior to the date of the lottery, which includes:

(a) the arrangements for the lottery (including the dates on which tickets were available for sale or supply, the dates of any draw and the arrangements for prizes, including any rollover);
(b) the proceeds of the lottery;
(c) the amounts deducted by the promoters of the lottery in respect of the provision of prizes (including the provision of prizes in accordance with any rollover);
(d) the amounts deducted by the promoters of the lottery in respect of other costs incurred in organising the lottery;
(e) any amount applied to a purpose for which the promoting society is conducted; and
(f) whether any expenses in connection with the lottery were defrayed otherwise than by deduction from proceeds, and, if they were:
    (i) the amount of the expenses; and
    (ii) the sources from which they were defrayed.

The statement must be signed by two adult members of the society who are appointed for the purpose in writing by the society or, if it has one, its governing body, and accompanied by a copy of the appointment (paragraph 39).

If after receiving a statement the local authority thinks that the lottery to which the statement relates was a large lottery, they must notify the Commission in writing (including a copy of the statement) (paragraph 40).

A lottery is exempt from licensing if it is promoted wholly on behalf of a non-commercial society, it is a small lottery and conditions of a small society lottery are satisfied (paragraph 30).

A large lottery is where its proceeds may exceed £20,000 and within a calendar year the aggregate of the promoting society's proceeds from society lotteries exceeds £250,000 (paragraph 31).

Registering with the licensing authority (schedule 11, part 5, Gambling Act 2005)

An application for registration must:

(a) be in the prescribed form;
(b) specify the purposes for which the society is conducted;
(c) contain such other information, and be accompanied by such documents, that may be prescribed; and
(d) be accompanied by the prescribed fee (paragraph 42).

After receipt of an application, the local authority must enter the applicant in a register and notify the applicant and the Commission of the registration (paragraph 44 and 46).

The local authority must refuse an application for registration if, in the last five years, an operating licence held by the applicant for registration has been revoked or refused (paragraph 47).

A local authority may refuse an application for registration if they think that the applicant is not a non-commercial society, a person who will or may be connected with the promotion of the lottery has been convicted of a relevant offence or information provided in or with the application for registration is false or misleading (paragraph 48). However, the local authority cannot refuse an application for registration unless they have given the applicant an opportunity to make representations (paragraph 49).

A local authority may revoke a registration if they think that they would be obliged or permitted to refuse an application for the registration were it being made anew.

Where a local authority revokes a registration, they must specify whether the revocation takes effect immediately or within two months of the date of notification and must allow the society an opportunity to make representations (paragraph 50).

The local authority must notify the applicant society of the refusal or revocation and the society may appeal to a magistrates' court within 21 days of notification of the refusal or revocation of registration.

The magistrates' court may affirm the local authority's decision, reverse it or make any other order (paragraph 51).

A registered society may apply in writing to the registering authority for the registration to be cancelled (paragraph 52). The local authority must then cancel the registration, notify the formerly registered society of the cancellation and notify the Commission of the cancellation (paragraph 53).

An annual fee is payable to the registering local authority. If a registered society fails to pay the annual fee, the local authority may cancel the society's registration and must notify the formerly registered society and the Commission (paragraph 54).

Statements must be retained for at least 18 months and be made available for inspection by members of the public at all reasonable times. Arrangements must be made for the provision of a copy of it, or part of it, to any member of the public on request, on payment of a fee (paragraph 55).

Promotion of or facilitating a lottery

It is an offence to promote or facilitate a lottery which is not exempt without an operating licence. It is not an offence if a person acts on behalf of the licence holder so long as he is not an external lottery manager (sections 258 and 259).

Misusing profits of lottery

It is an offence if a person uses any part of the profits of a lottery for a purpose other than that stated on the lottery tickets, or in an advertisement for the lottery (section 260). This does not apply to:

(a) an incidental non-commercial lottery;
(b) a private society lottery; and
(c) a small society lottery.

It is an offence if a person uses any part of the profits of a lottery for a purpose other than one for which the lottery is permitted to be promoted (section 261 and schedule 11).

Breach of condition

A non-commercial society commits an offence if:

(a) a lottery, purporting to be an exempt lottery, is promoted on the society's behalf wholly or partly at a time when the society is not registered with a local authority;
(b) the society fails to comply with the registration records' requirements; or
(c) the society provides false or misleading information (section 262).

Penalty

Persons guilty of offences are liable on summary conviction to imprisonment for a term not exceeding 51 weeks (six months in Scotland), a fine not exceeding level 5 on the standard scale, or both (section 263).

Definitions

Draw includes any process by which a prize in the lottery is allocated (section 255).
External lottery manager is a person who makes arrangements for a lottery on behalf of a society or authority of which he is not a member, an officer or an employee under a contract of employment (section 257).
Lottery ticket means a document or article that confers, or can be used to prove, membership of a class for the purpose of the allocation of prizes in a lottery (section 253).
Proceeds and profits of a lottery means the aggregate of amounts paid for lottery tickets. The profits of a lottery are:

(a) the proceeds of the lottery, minus

(b) amounts deducted by the promoters of the lottery for prizes, sums to be made available for allocation in another lottery in accordance with a rollover, or other costs reasonably incurred in organising the lottery (section 254).

Promoting a lottery – a person promotes a lottery if he:

(a) makes arrangements for the printing of lottery tickets;
(b) makes arrangements for the printing of promotional material;
(c) arranges for the distribution or publication of promotional material;
(d) possesses promotional material with a view to its distribution or publication;
(e) makes other arrangements to advertise a lottery;
(f) invites a person to participate in a lottery;
(g) sells or supplies a lottery ticket;
(h) offers to sell or supply a lottery ticket;
(i) possesses a lottery ticket with a view to its sale or supply;
(j) does or offers to do anything by virtue of which a person becomes a member of a class among whom prizes in a lottery are to be allocated; or
(k) uses premises for the purpose of allocating prizes or for any other purpose connected with the administration of a lottery (section 252).

Rollover means an arrangement whereby the fact that a prize is not allocated or claimed in one lottery increases the value of the prizes available for allocation in another lottery (section 256).

CONTROLS OVER TRADING ON SUNDAYS AT LARGE SHOPS

Reference

Sunday Trading Act 1994 as amended by the Regulatory Reform (Sunday Trading) Order 2004.

Extent

These procedures apply only in England and Wales (section 9(4)).

Scope

By virtue of this Act, Sunday trading is deregulated except in relation to large shops where generally trading is prohibited except in accordance with the 'six-hour trading' procedures in FC150.

A local authority may designate its area to be a loading control area in order to control loading and unloading before 9 am on Sundays in accordance with procedures in FC151.

Large shops

A 'large shop' is one which has a relevant floor area exceeding 280 square metres.

FC150 Controls over trading on Sundays at large shops – Schedules 1 and 2 Sunday Trading Act 1994

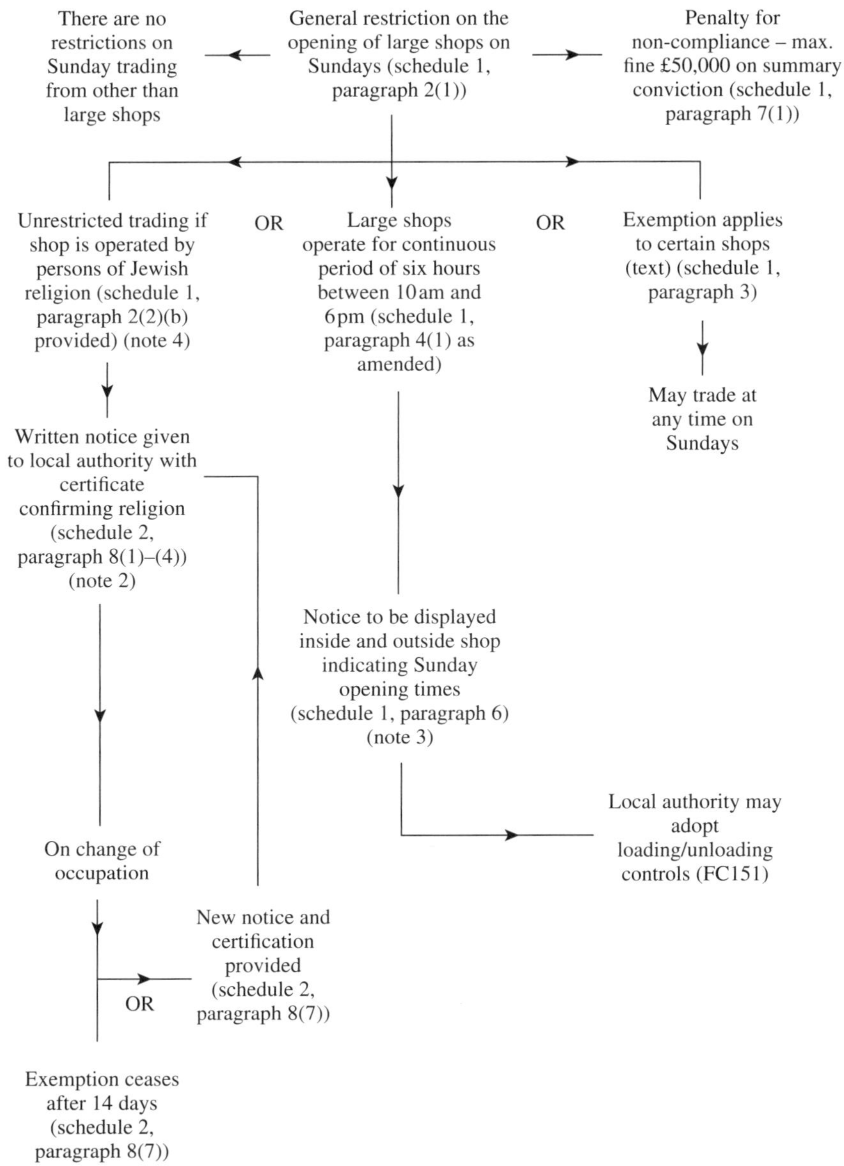

Notes

1. Penalty not exceeding level 5 for making false statements (schedule 2, paragraph 8(10)) and the local authority may cancel notice (schedule 2, paragraph 8(11)).
2. A fresh certificate is not required if previously provided for that occupier – this is at the local authority's discretion (schedule 2, paragraph 8(8)).
3. Penalty not exceeding level 2 for non-compliance (schedule 1, paragraph 7(2)).
4. These shops must be closed on the Jewish Sabbath – sundown on Friday to sundown on Saturday.

The 'relevant floor area' is defined as being the internal floor area of so much of the shop as consists of or comprises in a building excluding any part of the shop which throughout the week ending with the Sunday in question is used neither for the serving of customers in connection with the sale of goods nor for the display of goods.

Outlets which offer a service, e.g. hairdressers, shoe repair shops, restaurants, public houses, etc., are therefore excluded from Sunday trading restrictions (schedule 1, paragraph 1).

Exempted large shops

Some specified types of large shops are exempted from the six-hour restriction. These are:

(a) farm shops;
(b) motor and cycle supply shops;
(c) stands at exhibitions;
(d) shops at railway stations and airports;
(e) petrol filling stations;
(f) pharmacists for the sale of medicines and medicinal and surgical appliances;
(g) shops servicing ocean-going ships (schedule 1, paragraph 3).

Shops occupied by persons observing the Jewish Sabbath are also exempt from the six-hour limit on Sunday trading provided the procedure identified in FC150 has been followed and the shop is closed on the Jewish Sabbath.

The certificate required to be sent to the local authority with the notification is one to be signed by an authorised person (a minister or secretary of a synagogue or a person nominated for the purpose by the Board of Deputies of British Jews) and indicates that the person giving the notice is of the Jewish religion. Such exemption is also available to persons of any religious body who regularly observe the Jewish Sabbath but certification is required from the religious body concerned (schedule 2, paragraphs 8 and 9).

The six-hour provision

Unless exempt or occupied by Jews, large shops may trade for six continuous hours on Sundays between 10 am and 6 pm. The choice of the six-hour trading period is a matter for each trader (schedule 1, paragraph 4 as amended).

Shops using the six-hour trading provision must display conspicuous notices both inside and outside of the building specifying the permitted Sunday opening times (schedule 1, paragraph 6).

Controls over loading and unloading

FC151 shows the procedure through which local authorities may, at their discretion, operate a scheme of control over loading/unloading of goods from vehicles before 9 am on Sundays at large shops that are permitted to open. There are no such restrictions on small shops.

FC151 Controls over loading and unloading at large shops on Sundays – Section 2 Sunday Trading Act 1994

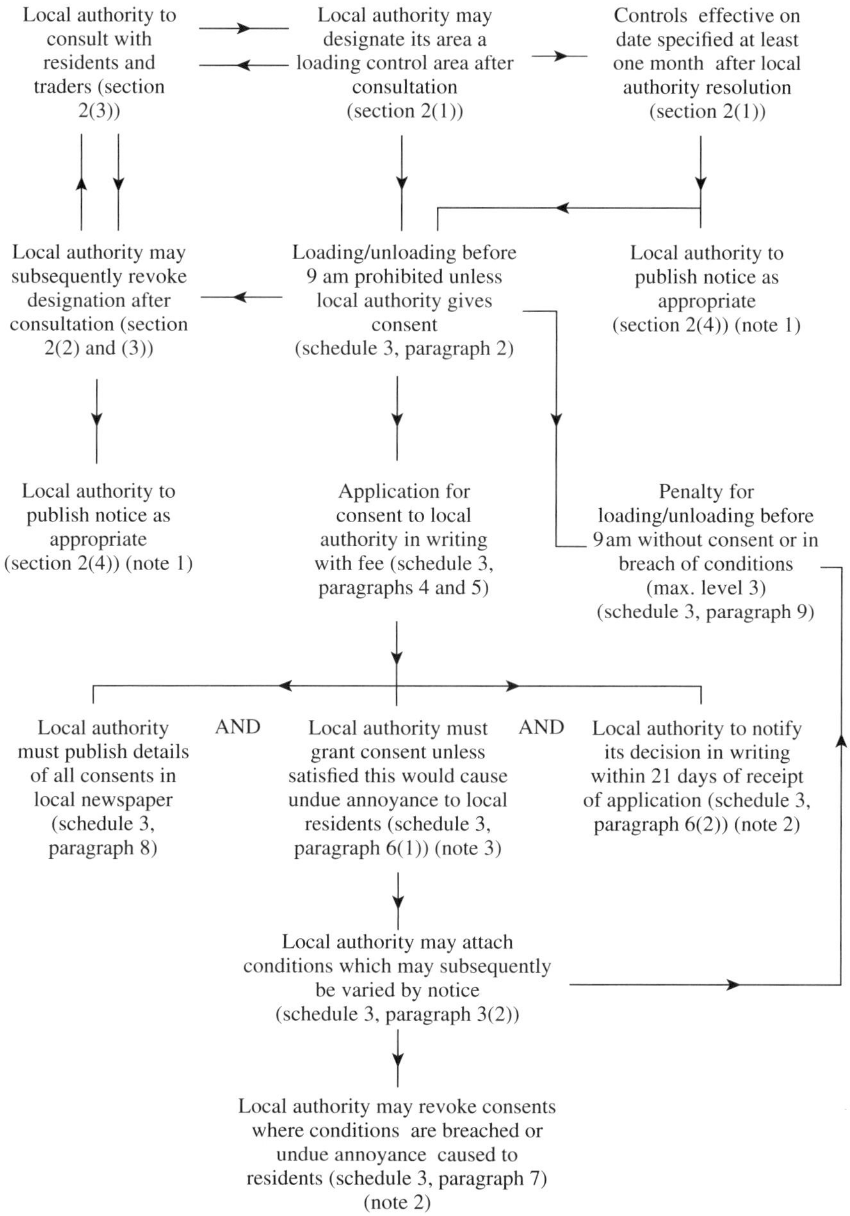

Notes

1. No particular means of publication is specified; this is for the local authority to decide.
2. There is no provision for appeal against a refusal of consent, conditions attached or revocation.
3. There is no requirement on a local authority to keep a register of consents.

The purpose of the procedure is to allow local authorities to prevent local residents from being caused undue annoyance and requires the local authority, once a loading control area has been declared, to sanction loading and unloading at each premises before 9 am and specifying appropriate conditions. A scheme can only apply to the whole of the local authority area.

Before adopting or subsequently revoking a scheme, the local authority must consult people likely to be affected, residents and traders, and must publicise its decision to adopt or revoke (section 2 and schedule 3).

Enforcement

Local authorities have a duty to enforce these provisions and must appoint inspectors who have powers of entry to large shops to check on possible contraventions, inspect and take copies of records (including computerised records) and to take measurements and photographs (schedule 2).

Definitions

Retail sale means any sale other than a sale for use or resale in the course of a trade or business, and references to retail purchase shall be construed accordingly.

Sale of goods does not include:

(a) the sale of meals, or refreshments or intoxicating liquor for consumption on the premises on which they are sold; or
(b) the sale of meals or refreshments prepared to order for immediate consumption off those premises.

Shop means any premises where there is carried on a trade or business consisting wholly or mainly of the sale of goods.
Stand in relation to an exhibition means any platform, structure, space or other area provided for exhibition purposes (schedule 1, paragraph 1).

STREET TRADING

Reference

Local Government (Miscellaneous Provisions) Act 1982, section 3 and schedule 4.

Adoption

The procedures only operate through a specific resolution of adoption by the local authority (section 3). There are no requirements about the advertisement etc. of any intention to adopt these provisions.

Extent

This procedure applies in England and Wales but does not apply in Greater London (for which see the London Local Authorities Act 1990) or in Scotland and Northern Ireland.

Designation

Following adoption, the local authority may designate streets or parts of streets to be either:

(a) prohibited streets in which street trading is not allowed;
(b) licence streets in which licences to trade are required; and
(c) consent streets where the consent requirements operate.

These designations may be rescinded or changed from one type to another at any time using the full procedure (schedule 4, paragraph 2).

Notice of intention to designate is required with copies to the chief police officer and to the highways authority. If the designation is to be for licence streets, the consent of the highways authority is required.

The notice must contain a draft of the resolution and say that written representations may be made during a specified period not less than 28 days from publication of the notice (schedule 4, paragraph 2(6)).

Scope

Street trading is defined as being the selling or exposing or offering for sale of any article (or living thing) in a street, but the following are not considered to be street trading:

(a) trading by a person acting as a pedlar under the authority of a pedlar's certificate granted under the Pedlars Act 1871;
(b) anything done in a market or fair, the right to hold which was acquired by virtue of a grant (including a presumed grant) or acquired or established by virtue of an enactment or order;
(c) trading in a trunk road picnic area provided by the Secretary of State under section 112 of the Highways Act 1980;
(d) trading as a news vendor;
(e) trading which:
   (i) is carried on at premises used as a petrol filling station; or
   (ii) is carried on at premises used as a shop or in a street adjoining premises so used and as part of the business of the shop;
(f) selling things, or offering, or exposing them for sale, as a roundsman;
(g) the use for trading under Part VIIA of the Highways Act 1980 of an object or structure placed on, in or over a highway;
(h) the operation of facilities for recreation or refreshment under Part VIIA of the Highways Act 1980;
(i) the doing of anything authorised by regulations made under section 5 of the Police, Factories, etc. (Miscellaneous Provisions) Act 1916.

The reference to trading as a news vendor in (d) above is a reference to trading where:

(a) the only articles sold or exposed or offered for sale are newspapers or periodicals; and
(b) they are sold or exposed or offered for sale without a stall or receptacle for them or with a stall or receptacle for them which does not:
  (i) exceed 1 m in length or width or 2 m in height;
  (ii) occupy a ground area exceeding 0.25 m^2; or
  (iii) stand on the carriageway of a street (schedule 4, paragraph 1(2) and (3)).

Licences

Applications must be in writing and must give:

(a) full name and address of applicant;
(b) street, day and times of proposed trading;
(c) description of articles, stalls or containers;
(d) any other particulars required by the council, which can include two photographs of the applicant (schedule 4, paragraph 3(2) and (3)).

Unless one or more of the following grounds of refusal are applicable, the local authority must grant the licence and may even grant it if the grounds of refusal are available:

(a) that there is not enough space in the street for the applicant to engage in the trading in which he desires to engage without causing undue interference or inconvenience to persons using the street;
(b) that there are already enough traders trading in the street from shops or otherwise in the goods in which the applicant desires to trade;
(c) that the applicant desires to trade on fewer days than the minimum number specified in a resolution under schedule 4, paragraph 2(11);
(d) that the applicant is unsuitable to hold the licence by reason of having been convicted of an offence or for any other reason;
(e) that the applicant has at any time been granted a street trading licence by the council and has persistently refused or neglected to pay fees due to them for it or charges due to them under schedule 4, paragraph 9(6), for services rendered by them to him in his capacity as licence holder;
(f) that the applicant has at any time been granted a street trading consent by the council and has persistently refused or neglected to pay fees due to them for it;
(g) that the applicant has without reasonable excuse failed to avail himself to a reasonable extent of a previous street trading licence (schedule 4, paragraph 3(6)).

Also licences must be refused if the applicant is under 17 years of age or the location is covered by a control order (road-side sales) under section 7 of the Local Government (Miscellaneous Provisions) Act 1976.

FC152 Street trading: designation of – Schedule 4 of Local Government (Miscellaneous Provisions) Act 1982 (LG(MP)A)

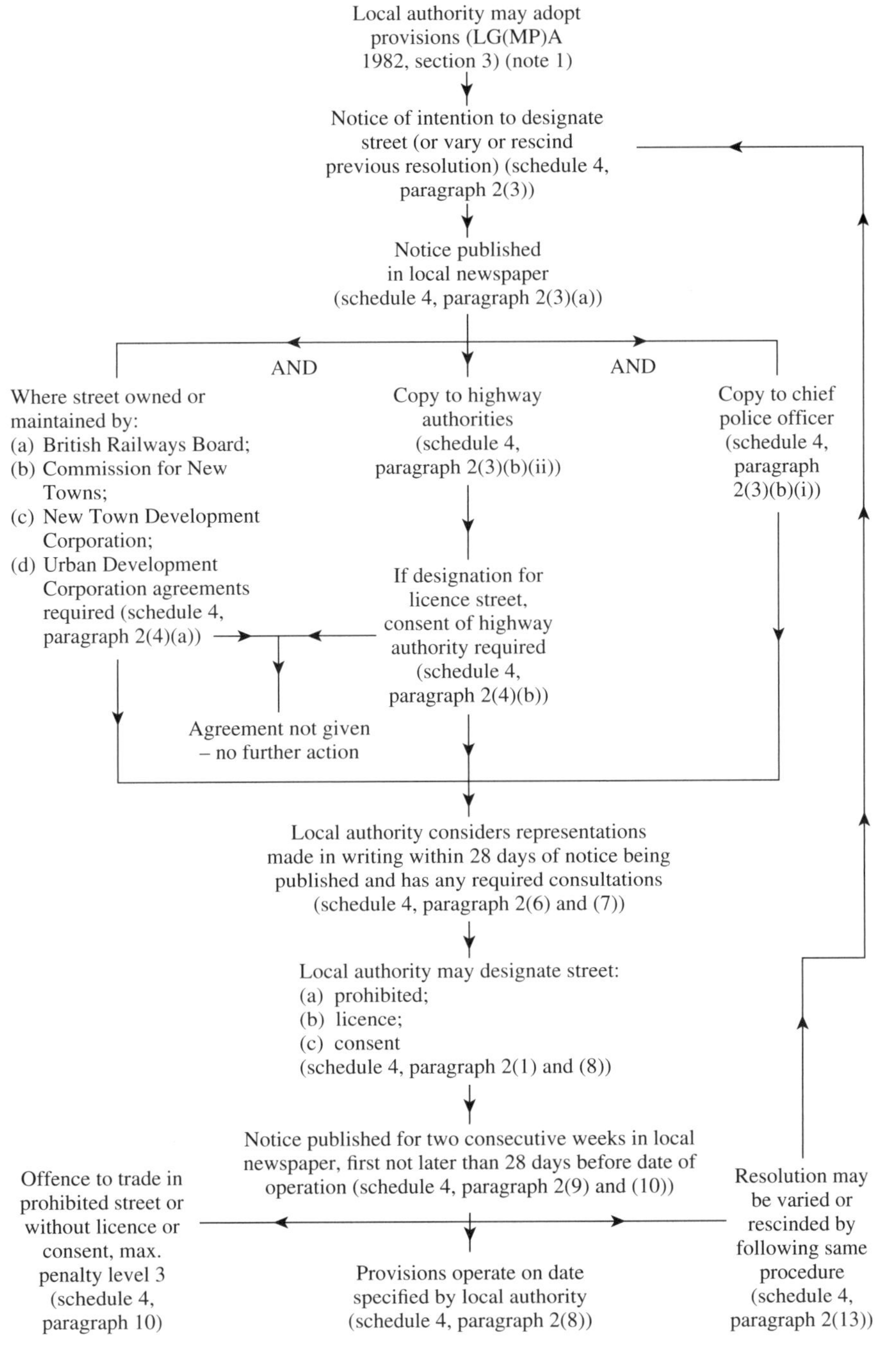

Note

1. No advertisement, publicity, etc. required at this stage.

FC153 Street trading: licences – Schedule 4 of Local Government (Miscellaneous Provisions) Act 1982 (LG(MP)A)

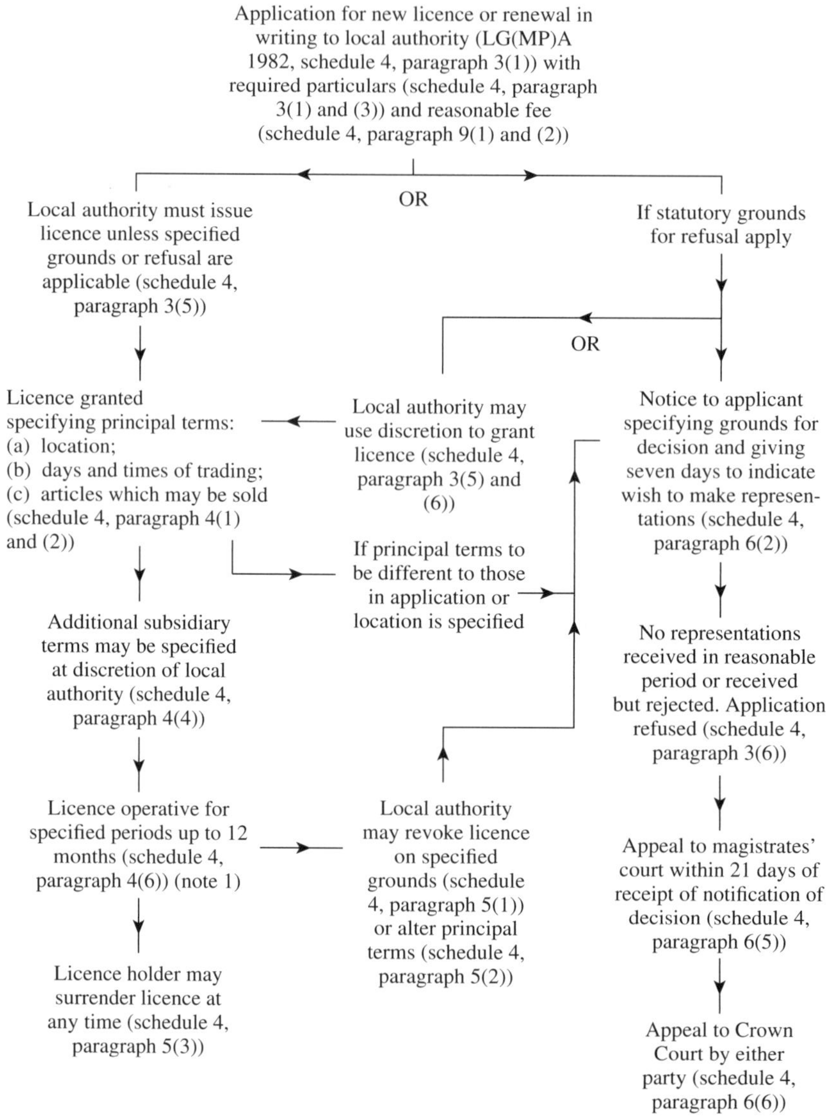

Note

1. The maximum penalty for breaching the principal terms of the licence is level 3. There is no offence committed in breaching the subsidiary terms but this could be taken into account in any consideration of renewal or revocation (schedule 4, paragraph 10(4)).

FC154 Street trading: consents – Schedule 4 of Local Government (Miscellaneous Provisions) Act 1982 (LG(MP)A)

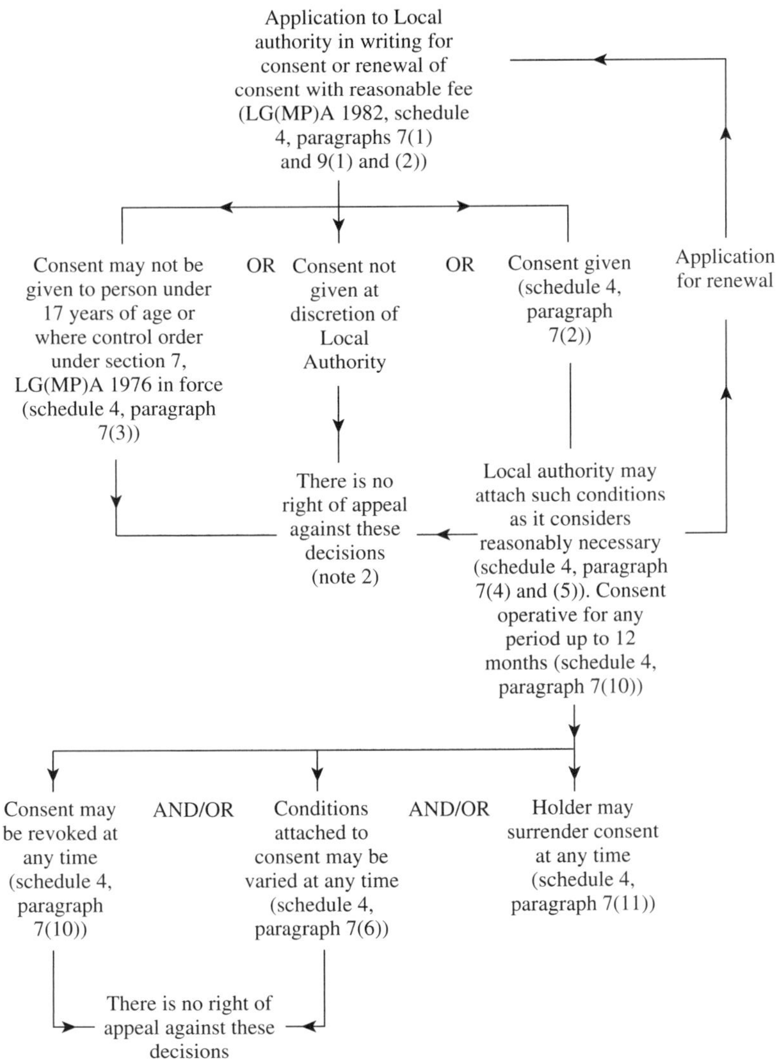

Notes

1. The maximum penalty for contravening a condition of consent is level 3 (schedule 4, paragraph 10(4)).
2. Although no appeal procedure exists, in the case of *R* v *Bristol City Council, ex parte Pearce and Another* (1984), the judge commented that the local authority should tell the applicant of the content of their objections and give him an opportunity to comment.

The licence issued must state:

(a) the street in which and days and times between which the holder is able to trade; and
(b) the description of articles in which he may trade and may also state a particular location for trading (schedule 4, paragraph 4(1)).

These are known as the principal terms of the licence and a breach may result in prosecution (schedule 4, paragraph 4(1)–(3)).

In addition, subsidiary terms may be applied at the discretion of the local authority and these may include:

(a) specifying the size and type of any stall or container which the licence holder may use for trading;
(b) requiring that any stall or container so used shall carry the name of the licence holder or the number of his licence, or both; and
(c) prohibiting the leaving of refuse by the licence holder or restricting the amount of refuse which he may leave or the places in which he may leave it (schedule 4, paragraph 4(5)).

Licences may be revoked if:

(a) owing to circumstances which have arisen since the grant or renewal of the licence, there is not enough space in the street for the licence holder to engage in the trading permitted by the licence without causing undue interference or inconvenience to persons using the street;
(b) the licence holder is unsuitable to hold the licence by reason of having been convicted of an offence or for any other reason;
(c) since the grant or renewal of the licence, the licence holder has persistently refused or neglected to pay fees due to the council for it or charges due to them under schedule 4, paragraph 9(6), for services rendered by them to him in his capacity as licence holder; or
(d) since the grant or renewal of the licence, the licence holder has without reasonable excuse failed to avail himself of the licence to a reasonable extent (paragraph 5(1)),

and the local authority may also vary the principal terms by altering the days or times of trading or restricting the type of goods sold, subject to notice and appeal (schedule 4, paragraph 5(2)).

The licence street provisions are most appropriate for the formalised street market situations and imply a positive will to promote trading on behalf of the local authority to the extent that refusal powers are limited.

Consents

There are no specified particulars for consent applications, although they must be in written form (schedule 4, paragraph 7(1)). Unless the applicant is under 17 years of age or the location is covered by a control order under section 7 of the Local Government (Miscellaneous Provisions) Act 1976 (roadside sales) in which situations refusal is mandatory, consents are entirely at the discretion of the local authority as it sees fit. If consent is given, the local authority may attach conditions at its discretion, including those to prevent obstruction and nuisance or annoyance.

Specific consent to trade from a vehicle or portable stall is required. Consent procedures are most applicable where trading is to be itinerant or infrequent and there is no appeal against refusals or conditions to be applied. There is no requirement in these provisions to hear applicants but, in the case of R v Bristol City Council, ex parte Pearce and Another (1984), it was indicated that the local authority should tell applicants of the contents of any objections and give them an opportunity to comment (schedule 4, paragraph 7).

Fees

The local authority may charge such fees as they consider reasonable for both licences and consents with different fee levels being possible for different types of licence/consents, periods, streets and articles. Fees may be paid in instalments and the initial application need only be accompanied by a deposit at the discretion of the local authority. Fees are returnable on surrender, revocation and refusal.

Separate and additional charges may be levied on licence holders for the collection of refuse, street sweeping and any other services rendered by the local authority (schedule 4, paragraph 9).

Definition

Street includes:

(a) any road, footway, beach or other area to which the public have access without payment; and
(b) a service area as defined in section 329 of the Highways Act 1980; and also includes any part of a street (schedule 4, paragraph 1(1)).

STREET COLLECTIONS

References

Local Government Act 1972, section 254.
Charitable Collections (Transitional Provisions) Order 1974 (articles below).

Extent

These provisions apply to England and Wales.

Scope

A licensing authority may, by a resolution, adopt the model street collection regulations. The resolution must send a copy to the Secretary of State (there is no need for confirmation) and insert an advertisement in two newspapers circulating within the area with a copy of the resolution and stating that a copy of the regulations will be provided free of charge to any person on application to the authority (section 5).

Model Street Collection Regulations

A promoter must apply in writing at least one month before the date for a proposed collection to obtain a permit to collect in a street and public place (other than a collection taken at a meeting in the open air) (articles 2 and 3). Collections must be made on the day and between the hours stated in the permit (article 4). The collection may be limited to streets or public places as the licensing authority thinks fit (article 5).

The promoter must provide written authority to a collector before collection is carried out (article 6).

The licensing authority may waive the requirement for a collector to remain stationary if in connection with a procession (article10).

Payment for services connected with the collection may be approved by the licensing authority.

Within one month after the date of any collection, the person to whom a permit has been granted must provide the licensing authority with:

(a) a statement showing the amount received and the expenses and payments in connection with the collection certified by that person and a qualified accountant;
(b) a list of the collectors;
(c) a list of the amounts contained in each collecting box; and
(d) a description of the proper application of the proceeds of the collection.

They must also publish in a newspaper or newspapers a statement showing the name of the person to whom the permit has been granted, the area to which the permit relates, the name of the charity or fund to benefit, the date of the collection, the amount collected and the amount of the expenses and payments in connection with the collection.

The licensing authority may extend the period of one month (article 16).

It is an offence to contravene any of the regulations and any person guilty of an offence will be liable on summary conviction to a fine not exceeding level 1 on the standard scale (article 18).

Definitions

Collecting box means a box or other receptacle for the reception of money from contributors (article 1).
Collection means a collection of money or a sale of articles for the benefit of charitable or other purposes.
Permit means a permit for a collection.
Promoter means a person who causes others to act as collectors.

SEX ESTABLISHMENT LICENCES

Reference

Local Government (Miscellaneous Provisions) Act 1982, section 2 and schedule 3.

Extent

These provisions apply to England and Wales.

Scope

A local authority may resolve that the provisions of the control of sex establishments in schedule 3 of the Act may apply to their area.

Once the resolution has been passed, the provisions come into force one month after the resolution is passed.

The local authority must publish notice that they have passed a resolution in two consecutive weeks in a local newspaper circulating in their area at least 28 days before the day of the provisions coming into force (section 2).

Provisions introduced in 2011 bring the licensing of lap dancing premises and similar venues in line with other sex establishments (sex shops and sex cinemas).

The provisions allow the licensing authority to prescribe a wider range of conditions on the licences of sexual entertainment venues than those available under the Licensing Act 2003 and allow local people to oppose an application for a sex establishment licence if they have legitimate concerns that a sexual entertainment venue would be inappropriate, given the character of an area.

Licences must be renewed at least annually and local people have the opportunity to raise objections (if any) at that time.

Licensing provisions

Sex establishments must only operate under a licence granted by a licensing authority (paragraph 6). An application to operate a sex establishment must be made to the appropriate authority. An application must be made in writing and must include:

(a) the full name of the applicant or body corporate;
(b) his permanent address or address of registered or principal office, the full names and private addresses of the directors or other persons responsible for its management;
(c) his age;
(d) the full address of the premises;
(e) where an application relates to a vehicle, vessel or stall, where it is to be used as a sex establishment (paragraph 10).

The authority may grant to any applicant, and from time to time renew, a licence (with terms, conditions and restrictions) to use any premises, vehicle, vessel or stall specified in it for a sex establishment (paragraph 8).

FC155 Sex establishment licences – Section 2 and Schedule 3 Local Government (Miscellaneous Provisions) Act 1982

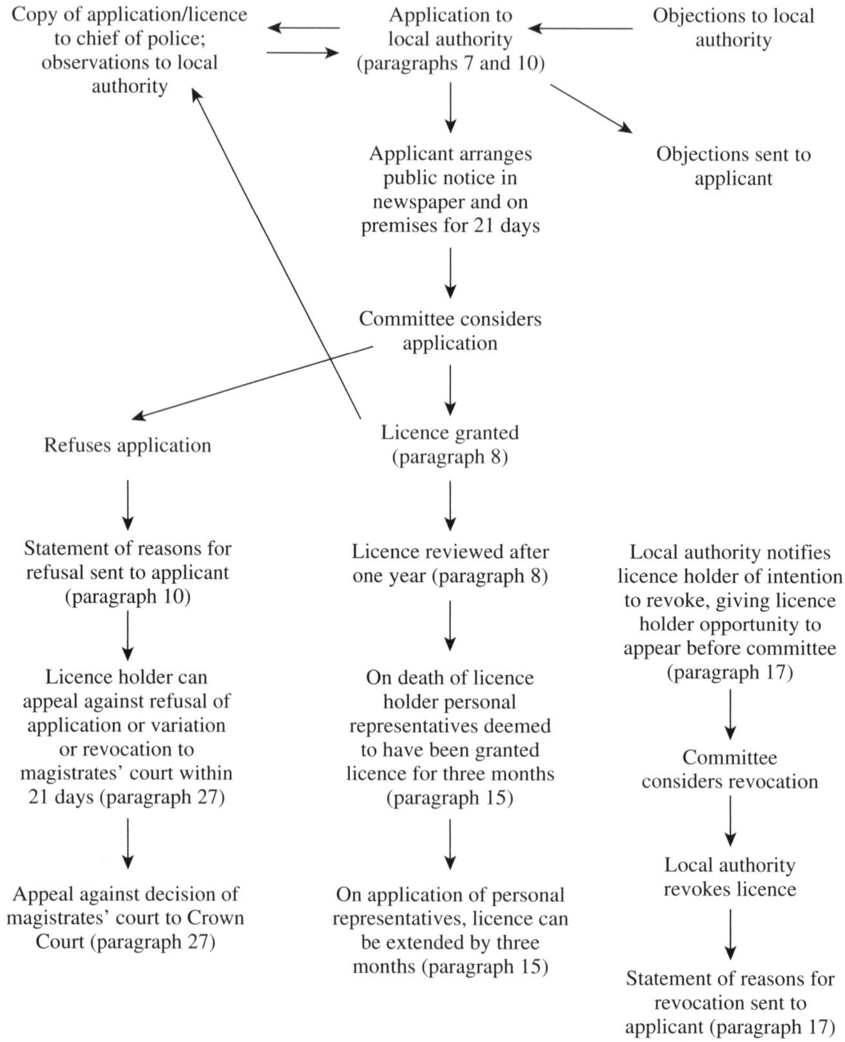

The licence remains in place for one year (paragraph 8) and the authority may, where they think fit, transfer the licence to any other person on application (paragraph 9).

The applicant must give public notice of the application by publishing an advertisement in a local newspaper circulating in the appropriate authority's area not later than seven days after the date of the application. For premises, notice must be displayed for 21 days beginning with the date of the application on or near the premises and in a place where the notice can conveniently be read by the public. The notice must be prescribed by the authority.

A copy of an application for the grant, renewal or transfer of a licence must be sent to the chief officer of police within seven days.

Any person objecting to an application for the grant, renewal or transfer of a licence must give notice in writing of his objection to the authority, stating the grounds of the objection within 28 days of the date of the application. The authority then must give notice in writing of the general terms of the objection to the applicant. The authority must not, without the consent of the person making the objection, reveal the objector's name or address to the applicant.

The authority must have regard to any observations made by the chief officer of police and any objections of which notice has been sent to them before determining the application.

The authority must give an opportunity of being heard by a committee of the authority:

(a) before refusing to grant a licence, to the applicant;
(b) before refusing to renew a licence, to the holder;
(c) before refusing to transfer a licence, to the holder and the person to be transferred to.

Where the appropriate authority refuses to grant, renew or transfer a licence, they must give the applicant a statement in writing of the reasons for their decision (paragraphs 10 and 17).

Licences remain in force where an application has been made for its renewal or transfer until the withdrawal of the application or its determination (paragraph 11).

Waiver

The authority may waive the requirement of a licence in any case where they consider that to require a licence would be unreasonable or inappropriate. Where the authority grants an application for a waiver, they must give the applicant for the waiver notice that they have granted his application. The authority may at any time terminate the waiver notice by giving 28 days' notice to the person in receipt of the waiver notice (paragraph 7).

Granting licences

Licences cannot be granted to:

(a) a person under the age of 18; or
(b) a person who is for the time being disqualified due to revocation of a licence; or
(c) a person, other than a body corporate, who is not resident in an European Economic Area state or was not so resident throughout the period of six months immediately preceding the date when the application was made; or
(d) a body corporate which is not incorporated in an European Economic Area state; or
(e) a person who has, within a period of 12 months immediately preceding the date when the application was made, been refused the grant or renewal of a licence for the premises, vehicle, vessel or stall in respect of which the application is made, unless the refusal has been reversed on appeal.

Refusal

The authority may refuse an application for the grant or renewal of a licence if:

(a) the applicant is unsuitable to hold the licence by reason of having been convicted of an offence or for any other reason;
(b) on granting, renewing or transferring the licence, the business to which it relates would be managed by or carried on for the benefit of a person, other than the applicant, who would be refused the grant, renewal or transfer of such a licence if he made the application himself;
(c) the number of sex establishments, or sex establishments of a particular kind, in the relevant locality at the time the application is determined is equal to or exceeds the number which the authority considers is appropriate for that locality (which may be nil);
(d) the grant or renewal of the licence would be inappropriate, having regard:
  (i) to the character of the relevant locality; or
  (ii) to the use to which any premises in the vicinity are put; or
  (iii) to the layout, character or condition of the premises, vehicle, vessel or stall in respect of which the application is made (paragraph 12).

Where the holder of a licence granted dies, the licence shall be deemed to have been granted to his personal representatives and will remain in force for three months and then expire. However, the authority may, on the application of the representatives, extend or further extend the period of three months if the authority is satisfied that the extension is necessary for the purpose of winding up the deceased's estate and that no other circumstances make it undesirable (paragraph 15).

The authority may, at the written request of the holder of a licence, cancel the licence (paragraph 16).

Revocation

The authority may, after giving the holder of a licence an opportunity of appearing before and being heard by them, revoke the licence. Where a licence is revoked, the authority must, if required to do so by the person who held it, give him a statement in writing of the reasons for their decision within seven days. Where a licence is revoked, its holder shall be disqualified from holding or obtaining a licence in the area of the appropriate authority for a period of 12 months beginning with the date of revocation (paragraph 17).

Variation

The licence holder may apply to the authority to vary the terms, conditions or restrictions on application. The authority may vary or refuse the application (paragraph 18).

Charging

The authority may charge a reasonable fee for determining the licence (paragraph 19).

Offences

It is an offence for:

(a) a person to knowingly use, or knowingly cause or permit the use of, any premises, vehicle, vessel or stall without a licence;
(b) a licence holder of a sex establishment to employ in the business of the establishment any person known to him to be disqualified from holding such a licence;
(c) a licence holder or servant or agent of the holder without reasonable excuse to knowingly contravene, or without reasonable excuse, knowingly permit the contravention of, a term, condition or restriction specified in the licence (paragraph 20);
(d) any person, in connection with an application for the grant, renewal or transfer of a licence, to make a false statement which he knows to be false in any material respect or which he does not believe to be true (paragraph 21);
(e) a licence holder without reasonable excuse to knowingly permit a person under 18 years of age to enter the establishment or to employ a person known to him to be under 18 years of age in the business of the establishment (paragraph 23).

A person guilty of an offence is liable on summary conviction to a fine not exceeding £20,000 or level 3 on the standard scale (paragraph 22).

Powers of entry

An authorised officer of a local authority may, at any reasonable time, enter and inspect any sex establishment in respect of which a licence is for the time being in force, with a view to seeing:

- (i) whether the terms, conditions or restrictions on or subject to which the licence is held are complied with;
- (ii) whether any person employed in the business of the establishment is disqualified from holding a licence;
- (iii) whether any person under 18 years of age is in the establishment; and
- (iv) whether any person under that age is employed in the business of the establishment.

An authorised officer may enter and inspect a sex establishment if he has reason to suspect that an offence has been, is being or is about to be committed, if he has a warrant granted by a Justice of the Peace. The authorised officer must produce his authority if required to do so by the occupier of the premises or the person in charge of the vehicle, vessel or stall in relation to which the power is exercised.

Any person who, without reasonable excuse, refuses to permit an authorised officer of a local authority to exercise any such power shall be guilty of an offence and shall for every such refusal be liable on summary conviction to a fine not exceeding level 5 on the standard scale (paragraph 25).

Seizure

A person acting under the authority of a warrant may seize and remove anything found on the premises concerned that the person reasonably believes should be forfeited. The person who, immediately before the seizure, had custody or control of anything seized may request any authorised officer who seized it to provide a record of what was seized. The authorised officer must provide the record within a reasonable time of the request being made (paragraph 25A).

Appeals

Where an application or request for variation is refused or a licence holder is aggrieved by any term, condition or restriction on the licence or the licence is revoked, the applicant or licence holder can appeal to a magistrates' court within 21 days of the relevant date.

An appeal against the decision of a magistrates' court may be brought to the Crown Court. The decision of the Crown Court is final and either court can make such order as it thinks fit. The authority must carry out the order of the court.

Where a licence is revoked or an application for the renewal of a licence is refused, the licence remains in force until the time for bringing an appeal has expired or until the determination or abandonment of the appeal. Where the licence holder makes an application to vary the licence or the authority imposes a term, condition or restriction other than one specified in the application, the licence is free of it until the appeal is determined (paragraph 27).

Definitions

Relevant entertainment means any live performance or any live display of nudity which is of such a nature that, ignoring financial gain, it must reasonably be assumed to be provided solely or principally for the purpose of sexually stimulating any member of the audience (whether by verbal or other means). An audience includes an audience of one.

Sex cinema is any premises, vehicle, vessel or stall used to a significant degree for the exhibition of moving pictures, by whatever means produced, which:

- are concerned primarily with the portrayal of, or primarily deal with, relate to or are intended to stimulate or encourage, sexual activity; or acts of force or restraint which are associated with sexual activity; or
- are concerned primarily with the portrayal of, or primarily deal with or relate to, genital organs or urinary or excretory functions.

A sex cinema does not include a dwelling-house to which the public is not admitted.

A premises shall not be treated as a sex cinema only if they are licensed under section 1 of the Cinemas Act 1985, of their use for a purpose for which a licence is required; or of their use for an exhibition to which section 6 of that Act (certain non-commercial exhibitions) applies given by an exempted organisation within the meaning of section 6(6) of that Act.

Sex establishment means a sex cinema, sex shop or sexual entertainment venue.

Sex shop means any premises, vehicle, vessel or stall used for a business which consists to a significant degree of selling, hiring, exchanging, lending, displaying or demonstrating:

- sex articles; or
- other things intended for use in connection with, or for the purpose of stimulating or encouraging, sexual activity; or acts of force or restraint which are associated with sexual activity.

No premises shall be treated as a sex shop by reason only of their use for the exhibition of moving pictures by whatever means produced.

Sexual entertainment venue means any premises at which relevant entertainment is provided before a live audience for financial gain of an organiser. For the provisions to apply, it is not necessary for the organiser to receive financial gain directly or indirectly from the performance or display of nudity.

Sexual entertainment venues: exempt premises – the following are not defined as sexual entertainment venues:

(a) sex cinemas and sex shops;
(b) premises at which the provision of relevant entertainment is such that:
  (i) there have not been more than 11 occasions on which relevant entertainment has been so provided which fall (wholly or partly) within the period of 12 months;

(ii) no occasion has lasted for more than 24 hours; and
(iii) no occasion has begun within the period of one month beginning with the end of any previous occasion on which relevant entertainment has been so provided.

PLEASURE BOAT LICENSING

Reference

Public Health Acts Amendment Act 1907, section 94.

Extent

These provisions apply in England, Scotland, Northern Ireland and Wales.

Scope

A local authority may grant licences for pleasure boats and pleasure vessels (with terms and conditions) to be let for hire or to be used for carrying passengers for hire. Licences also apply to persons in charge of or navigating such boats and vessels. The local authority may charge an annual fee for each type of licence.

The period of the licence is determined by the local authority and the licence can be suspended or revoked where they think it is necessary or desirable in the interests of the public.

It is an offence to let for hire pleasure boats not licensed under these provisions and any person guilty of an offence is liable to a penalty not exceeding level 3 on the standard scale. However it is a defence where the failure to comply is due to a reasonable excuse for the failure.

Any person aggrieved by the withholding, suspension or revocation of any licence may appeal to a court, held after the expiration of two clear days after such withholding, suspension or revocation, but the aggrieved person must give 24 hours' written notice of an appeal with details of the grounds of appeal. The court has the power to make an order as they see fit and to award costs. Costs are recoverable summarily as a civil debt. These provisions do not apply to pleasure boats etc. on any inland waterway owned or managed by the Canal & River Trust or licensed under other relevant regulations.

Definition

Let for hire means let for hire to the public (section 94).